Synthesis, Characterization and Application of Polymer-Based Materials

Synthesis, Characterization and Application of Polymer-Based Materials

Guest Editors

Nadya Vasileva Dencheva
Zlatan Zlatev Denchev

MDPI

Basel • Beijing • Wuhan • Barcelona • Belgrade • Novi Sad • Cluj • Manchester

Guest Editors

Nadya Vasileva Dencheva
Department of Polymer
Engineering
University of Minho
Guimaraes
Portugal

Zlatan Zlatev Denchev
Department of Polymer
Engineering
University of Minho
Guimaraes
Portugal

Editorial Office
MDPI AG
Grosspeteranlage 5
4052 Basel, Switzerland

This is a reprint of the Special Issue, published open access by the journal *Molecules* (ISSN 1420-3049), freely accessible at: https://www.mdpi.com/journal/molecules/special_issues/H9J5126872.

For citation purposes, cite each article independently as indicated on the article page online and as indicated below:

Lastname, A.A.; Lastname, B.B. Article Title. *Journal Name* **Year**, *Volume Number*, Page Range.

ISBN 978-3-7258-7228-2 (Hbk)
ISBN 978-3-7258-7229-9 (PDF)
https://doi.org/10.3390/books978-3-7258-7229-9

Contents

About the Editors

Nadya Vasileva Dencheva

Nadya Vasileva Dencheva is part of the Department of Polymer Engineering, Institute for Polymers and Composites, University of Minho, Portugal. She obtained her PhD in Polymer Science and Composites from the University of Minho. Her research focuses on polymer nanocomposites via in situ polymerization, reactive microencapsulation, the synthesis of electroconductive composites, advanced synchrotron WAXS/SAXS analysis of polymers, and molecular imprinting for selective adsorption and filtration of bioactive entities. Her work further encompasses the design of micro- and nanoparticles for protein and enzyme immobilization, biocatalysts for green chemistry applications, enzyme cascade reactions, and biosensors. She has authored more than 55 peer-reviewed papers and book chapters and holds six patents. She has presented approximately 70 communications at international conferences and has been actively involved in collaborative research projects in advanced functional polymer systems. Her scientific contributions have received more than 1100 citations, an h-index of 21, and an i10-index of 35, according to Google Scholar.

Zlatan Zlatev Denchev

Zlatan Zlatev Denchev is part of the Department of Polymer Engineering, Institute for Polymers and Composites, University of Minho, Portugal. His research focuses on the synthesis, advanced characterization, and processing of polymeric materials, with particular emphasis on microfibrillar and nanostructured composites, in situ polymerization strategies, reactive microencapsulation, electroconductive and EMI-shielding composites, biocatalyst immobilization for green chemistry applications, and advanced synchrotron WAXS/SAXS analysis of polymers. He has authored more than 130 peer-reviewed articles and book chapters, holds nine patents, and has delivered over 90 communications at international conferences, including invited and keynote lectures. He has supervised postdoctoral fellows, PhD and MSc students; coordinated several national and industrial research projects; and actively contributed to international scientific collaboration and editorial activities in the field of polymer science. His published works have received more than 3200 citations, an h-index of 30, and an i10-index of 77, according to Google Scholar.

Preface

Polymer-based materials continue to evolve rapidly, responding to increasing demands for sustainability, multifunctionality, and advanced manufacturing solutions. The contributions collected in this Special Issue reflect the diversity and the multifaceted development of contemporary polymer research, comprising molecular design, structural characterization, processing innovations and application-oriented original solutions.

The fourteen papers of this Special Issue demonstrate how polymer science addresses modern technological challenges through bio-based synthesis strategies, hybrid architectures, intelligent and stimuli-responsive systems, additive manufacturing, and machine learning-assisted optimization. The Issue also highlights the growing importance of non-destructive and in situ structural characterization methods, as well as the careful balance between processability, performance and environmental responsibility.

With more than 51,000 views and more than 130 citations so far, the first edition has attracted significant international attention. A second edition of this Special Issue is currently open, aiming to further expand the dialog on predictive modeling, real-time structural analysis, degradation mechanisms, and lifecycle considerations in advanced polymer systems.

This Special Issue is addressed to researchers, engineers, and industrial practitioners working in polymer chemistry, materials science, chemical engineering, and related interdisciplinary fields. It is our expectation that it will continue to foster interdisciplinary collaboration and support the rational design of next-generation polymer-based materials aligned with the global sustainability and innovation goals.

Nadya Vasileva Dencheva and Zlatan Zlatev Denchev
Guest Editors

molecules

MDPI

Editorial

Synthesis, Characterization, and Application of Polymer-Based Materials

Zlatan Zlatev Denchev * and Nadya Vasileva Dencheva *

IPC—Institute for Polymers and Composites, School of Engineering, Campus of Azurém, University of Minho, 4800 Guimarães, Portugal
* Correspondence: denchev@dep.uminho.pt (Z.Z.D.); nadiad@dep.uminho.pt (N.V.D.)

1. Introduction

The field of polymer-based materials is advancing rapidly, driven by the growing demand for sustainable technologies, enhanced functionalities, and cutting-edge manufacturing methods. This *Molecules* Special Issue, titled "Synthesis, Characterization, and Application of Polymer-Based Materials", presents a selection of contributions—research and review articles—that reflect both recent developments and persistent challenges in this dynamic area of materials science. The first edition of this Special Issue was closed on 30 September 2024.

Comprising three review articles and eleven original research papers, this first edition covers a broad range of topics spanning from eco-friendly synthesis and hybrid polymer systems to machine learning–assisted material design and performance optimization. A total of 67 authors from research centers in China, the USA, Japan, Portugal, and Poland contributed to this collaborative effort.

2. Coverage of Emerging Trends

A prominent theme throughout the Special Issue is the development of *sustainable and environmentally conscious polymer synthesis* strategies. For instance, Agbo et al. [Contribution 1] report the incorporation of a microalgae-derived bio-binder into epoxy composites, promoting the integration of renewable resources into high-performance polymer matrices. Similarly, Souza et al. [Contribution 2] investigate biomass mastication as a pre-treatment method to improve the downstream processing of polyhydroxyalkanoates (PHA) synthesized by mixed microbial cultures. Complementing these experimental studies, the review by Getya and Gitsov [Contribution 3] offers an in-depth overview of hybrid polymer networks constructed from natural macromolecules such as cellulose, chitin, alginic acid, gellan gum, lignin, and their derivatives. Collectively, these contributions align with global efforts to transition toward biodegradable, bio-based polymer systems, as recently reviewed by Chauhan et al. [1].

This Special Issue also highlights *advances in functional and stimuli-responsive polymers.* Motoyanagi et al. [Contribution 4] demonstrate the synthesis of cyclic poly(N-isopropylacrylamide)s via Ring Expansion Reversible Addition–Fragmentation Chain Transfer (RE-RAFT) polymerization, enabling precise control of thermo-responsive behavior. Wilczewski et al. [Contribution 5] present a curcuminoid-functionalized graphene system that enhances dispersion within PVC-based nanocomposites, improving stability and uniformity. Braz et al. [Contribution 6] introduce hybrid nanoflowers anchored on polyamide microparticles for efficient cascade biocatalysis. This work notably employs wide- and small-angle X-ray scattering, illustrating the power of advanced characterization techniques using synchrotron radiation and spectroscopy for investigating soft matter that includes polymer-based materials—an approach recently emphasized by Takahara et al. [2]. Contributions 4 through 6 collectively underscore a growing focus on "intelligent" polymers with adaptable and multifunctional properties, consistent with the recent review by Jingcheng et al. [3].

Several studies in this Special Issue focus on *application-driven innovations* across critical technological sectors. Wang et al. [Contribution 7] develop a solvent-free method for coating polyacrylonitrile onto $Li_{6\cdot4}La_3Zr_{1\cdot4}Ta_{0\cdot6}O_{12}$ (LLZTO), a garnet-type solid-state electrolyte known for its high ionic conductivity and thermal stability in lithium batteries. Ranaiefar et al. [Contribution 8] combine additive manufacturing and machine learning to optimize the mechanical performance of carbon-reinforced ABS honeycomb composites, demonstrating the relevance of polymer-based materials in lightweight structural engineering. Kamo and Matsumoto [Contribution 9] investigate the role of pore size in epoxy monoliths designed for use as sheet-type adhesives—an area where polymer architecture critically influences bonding strength and durability. Contributions 7–9 reflect the broad utility of advanced polymer materials in energy, structural, and adhesive applications.

Additional contributions explore the design of *polymer composites with enhanced durability and tailored properties*. Smejda-Krzewicka et al. [Contribution 10] formulate butyl rubber nanocomposites filled with phyllosilicates, improving chemical resistance and barrier performance. Cao et al. [Contribution 11] describe a fluorinated flow modifier for boron nitrite/polyphenylene sulfide (BN/PPS) composites, yielding materials with high thermal conductivity, favorable dielectric behavior, and excellent processability via melt blending. These two studies reaffirm the importance of polymer systems with engineered molecular architectures and processing routes, enabling customized performance for a wide range of applications.

An unconventional yet notable contribution is the work by Wu et al. [Contribution 12], who apply silicone oil emulsion polymerization to the conservation of waterlogged wooden artifacts. This study demonstrates the expanding role of polymer science in *cultural heritage preservation*.

The two concluding review articles in this Special Issue extend its thematic reach into the *innovative polymer processing* domain. Mahmood et al. [Contribution 13] provide a comprehensive overview of polymer composites in 3D/4D printing, including discussions of materials, printing strategies, and structural design. Zhan et al. [Contribution 14] review advances in polymer-reinforced SiO_2 aerogels, emphasizing improvements in both mechanical performance and fabrication processes.

Taken together, all contributions to this Special Issue reflect the decisive and multifaceted role of polymer-based materials in addressing current scientific and industrial challenges. This Special Issue also touches on several cross-cutting themes, including the application of machine learning in material design [Contribution 8], trade-offs between processability and performance (as seen in LLZTO-based systems and some adhesives [Contributions 7 and 9]), interface engineering in hybrid composites [Contributions 8–10], and the increasing use of advanced, non-destructive structural characterization [Contribution 6].

3. Editorial Perspective and Future Outlook

The fourteen papers contributed to this first edition of this Special Issue provide a representative snapshot of ongoing progress in polymer science and engineering. They effectively address critical challenges in polymer synthesis, material characterization, and the deployment of polymer-based systems in real-world applications. Strengths of this

Special Issue include a clear emphasis on sustainability [Contributions 1–3], functional versatility [Contributions 4–6], and advanced processing techniques [Contributions 7–11], including also the 4D printing [Contribution 13], in which the 3D printed object changes its shape, properties, or functionality over time in response to external stimuli.

Nonetheless, a closer examination reveals opportunities for further exploration in future editions or new dedicated Special Issues. In particular, the integration of predictive modeling frameworks, as recently reviewed by Li et al. [4], along with the broader use of in situ and real-time characterization techniques (such as synchrotron radiation, neutron scattering, and reflectometry), and more comprehensive studies on degradation pathways [5] and lifecycle assessments [6], are essential for enabling the rational design, scalable production, and environmental responsibility of advanced polymer systems.

As of approximately one year after its closure, the first edition of this Special Issue has received over 50,000 views and 89 citations, reflecting substantial interest and engagement from the scientific community. A second edition of this Special Issue is currently open, with a submission deadline in April 2026, and is available on the following page: https://www.mdpi.com/journal/molecules/special_issues/UKD1605F0N. (accessed on 1 August 2025)

We hope that both editions of this Special Issue will continue to stimulate interdisciplinary research, strengthen collaboration between academia and industry, and support the advancement of next-generation polymer materials aligned with global sustainability and innovation goals.

Author Contributions: Z.Z.D. and N.V.D.: methodology, validation, formal analysis, investigation, data curation, writing—original draft preparation, and writing—review and editing. All authors have read and agreed to the published version of the manuscript.

Funding: This research received no external funding.

Conflicts of Interest: The authors declare no conflicts of interest.

List of Contributions:

1. Agbo, P.; Mali, A.; Kelkar, A.D.; Wang, L.; Zhang, L. Injecting Sustainability into Epoxy-Based Composite Materials by Using Bio-Binder from Hydrothermal Liquefaction Processing of Microalgae. *Molecules* **2024**, *29*, 3656. https://doi.org/10.3390/molecules29153656.
2. Souza, H.K.S.; Matos, M.; Reis, M.A.M.; Covas, J.A.; Hilliou, L. Can Biomass Mastication Assist the Downstreaming of Polyhydroxyalkanoates Produced from Mixed Microbial Cultures? *Molecules* **2023**, *28*, 767. https://doi.org/10.3390/molecules28020767.
3. Getya, D.; Gitsov, I. Synthesis and Applications of Hybrid Polymer Networks Based on Renewable Natural Macromolecules. *Molecules* **2023**, *28*, 6030. https://doi.org/10.3390/molecules28166030.
4. Motoyanagi, J.; Bessho, K.; Minoda, M. Smart and Efficient Synthesis of Cyclic Poly(N-isopropylacrylamide)s by Ring Expansion RAFT (RE-RAFT) Polymerization and Analysis of Their Unique Temperature-Responsive Properties. *Molecules* **2024**, *29*, 5392. https://doi.org/10.3390/molecules29225392.
5. Wilczewski, S.; Skórczewska, K.; Tomaszewska, J.; Osial, M.; Dąbrowska, A.; Nikiforow, K.; Jenczyk, P.; Grzywacz, H. Graphene Modification by Curcuminoids as an Effective Method to Improve the Dispersion and Stability of PVC/Graphene Nanocomposites. *Molecules* **2023**, *28*, 3383. https://doi.org/10.3390/molecules28083383.
6. Braz, J.F.; Dencheva, N.V.; Malfois, M.; Denchev, Z.Z. Synthesis of Novel Polymer-Assisted Organic-Inorganic Hybrid Nanoflowers and Their Application in Cascade Biocatalysis. *Molecules* **2023**, *28*, 839. https://doi.org/10.3390/molecules28020839.
7. Wang, X.; Zhang, K.; Shen, H.; Zhang, H.; Yao, H.; Chen, Z.; Jiang, Z. Solvent-Free Method of Polyacrylonitrile-Coated LLZTO Solid-State Electrolytes for Lithium Batteries. *Molecules* **2024**, *29*, 4452. https://doi.org/10.3390/molecules29184452.

8. Ranaiefar, M.; Singh, M.; Halbig, M.C. Additively Manufactured Carbon-Reinforced ABS Honeycomb Composite Structures and Property Prediction by Machine Learning. *Molecules* **2024**, *29*, 2736. https://doi.org/10.3390/molecules29122736.
9. Kamo, Y.; Matsumoto, A. Control of Pore Sizes in Epoxy Monoliths and Applications as Sheet-Type Adhesives in Combination with Conventional Epoxy and Acrylic Adhesives. *Molecules* **2024**, *29*, 2059. https://doi.org/10.3390/molecules29092059.
10. Smejda-Krzewicka, A.; Irzmańska, E.; Mrozowski, K.; Adamus-Włodarczyk, A.; Litwicka, N.; Strzelec, K.; Szynkowska-Jóźwik, M.I. The New Elastomeric Compounds Made of Butyl Rubber Filled with Phyllosilicates, Characterized by Increased Barrier Properties and Hydrophobicity and Reduced Chemical Degradation. *Molecules* **2024**, *29*, 1306. https://doi.org/10.3390/molecules29061306.
11. Cao, B.; Huang, X.; Zhang, W.; Wu, P. Fluorine-Containing Flow Modifier for BN/PPS Composites Enabled by Low Surface Energy. *Molecules* **2022**, *27*, 8066. https://doi.org/10.3390/molecules27228066.
12. Wu, M.; Han, X.; Zhang, Z.; Wang, J. Effectiveness Evaluation of Silicone Oil Emulsion In Situ Polymerization for Dehydration of Waterlogged Wooden Artifacts. *Molecules* **2024**, *29*, 4971. https://doi.org/10.3390/molecules29204971.
13. Mahmood, A.; Perveen, F.; Chen, S.; Akram, T.; Irfan, A. Polymer Composites in 3D/4D Printing: Materials, Advances, and Prospects. *Molecules* **2024**, *29*, 319. https://doi.org/10.3390/molecules29020319.
14. Zhan, W.; Chen, L.; Kong, Q.; Li, L.; Chen, M.; Jiang, J.; Li, W.; Shi, F.; Xu, Z. The Synthesis and Polymer-Reinforced Mechanical Properties of SiO_2 Aerogels: A Review. *Molecules* **2023**, *28*, 5534. https://doi.org/10.3390/molecules28145534.

References

1. Chauhan, K.; Kaur, R.; Chauhan, I. Sustainable bioplastic: A comprehensive review on sources, methods, advantages, and applications of bioplastics. *Polym.-Plast. Technol. Mater.* **2024**, *63*, 913–938. [CrossRef]
2. Takahara, A.; Higaki, Y.; Hirai, T.; Ishige, R. Application of Synchrotron Radiation X-ray Scattering and Spectroscopy to Soft Matter. *Polymers* **2020**, *12*, 1624. [CrossRef] [PubMed]
3. Jingcheng, L.; Reddy, V.S.; Jayathilaka, W.A.D.M.; Chinnappan, A.; Ramakrishna, S.; Ghosh, R. Intelligent Polymers, Fibers and Applications. *Polymers* **2021**, *13*, 1427. [CrossRef] [PubMed]
4. Li, Q.; Zhan, S.; Liu, Z.; Dong, C.; Zhao, H.; Yue, T.; Zhao, Q.; Zhang, L.; Li, Y.; Liu, J. Developing Hybrid Machine Learning Frameworks for Polymer Property Prediction Based on Composition and Sequence Features. *J. Chem. Inf. Model.* **2025**, *65*, 1549–9596. [CrossRef] [PubMed]
5. Kari, F.G.; Delgado, A.G.; Nghiem, L.D. A critical review of the application of polymeric materials in environmental protection. *Environ. Sci. Process. Impacts* **2015**, *17*, 1259–1273. [CrossRef]
6. Ramesh, P.; Vinodh, S. State of art review on Life Cycle Assessment of polymers. *Int. J. Sustain. Eng.* **2020**, *13*, 411–422. [CrossRef]

molecules

MDPI

Article

Injecting Sustainability into Epoxy-Based Composite Materials by Using Bio-Binder from Hydrothermal Liquefaction Processing of Microalgae

Philip Agbo [1,†], Abhijeet Mali [1,†], Ajit D. Kelkar [2], Lijun Wang [3,*] and Lifeng Zhang [1,*]

1 Department of Nanoengineering, Joint School of Nanoscience and Nanoengineering, North Carolina A&T State University, 2907 E Gate City Blvd., Greensboro, NC 27401, USA
2 Department of Mechanical Engineering, College of Engineering, North Carolina A&T State University, 1601 E Market St., Greensboro, NC 27411, USA; kelkar@ncat.edu
3 Department of Natural Resources and Environmental Design, College of Agriculture and Environmental Sciences, North Carolina A&T State University, 1601 E Market St., Greensboro, NC 27411, USA
* Correspondence: lwang@ncat.edu (L.W.); lzhang@ncat.edu (L.Z.); Tel.: +1-336-285-3833 (L.W.); +1-336-285-2875 (L.Z.)
† These authors contributed equally to this work.

Abstract: We report a transformative epoxy system with a microalgae-derived bio-binder from hydrothermal liquefaction processing (HTL). The obtained bio-binder not only served as a curing agent for conventional epoxy resin (e.g., EPON 862), but also acted as a modifying agent to enhance the thermal and mechanical properties of the conventional epoxy resin. This game-changing epoxy/bio-binder system outperformed the conventional epoxy/hardener system in thermal stability and mechanical properties. Compared to the commercial EPON 862/EPIKURE W epoxy product, our epoxy/bio-binder system (35 wt.% bio-binder addition with respect to the epoxy) increased the temperature of 60% weight loss from 394 °C to 428 °C and the temperature of maximum decomposition rate from 382 °C to 413 °C, while the tensile, flexural, and impact performance of the cured epoxy improved in all cases by up to 64%. Our research could significantly impact the USD 38.2 billion global market of the epoxy-related industry by not only providing better thermal and mechanical performance of epoxy-based composite materials, but also simultaneously reducing the carbon footprint from the epoxy industry and relieving waste epoxy pollution.

Keywords: microalgae; epoxy resin; hydrothermal liquefaction; sustainability; composite

1. Introduction

Epoxy resin is a class of thermosetting pre-polymers that possesses excellent physical properties, such as good processability, excellent mechanical strength, great chemical resistance, high adhesion strength, superior thermal stability, and outstanding insulation capability after curing [1]. Epoxy resin is extensively used in a wide range of commercial products, including coatings, adhesives, and particularly polymer composite materials, in applications such as transportation, construction, aerospace, electronics, etc. Epoxy resin is generally derived from fossil fuel and produced through a series of chemical reactions. The most popular epoxy resin is diglycidyl ether of bisphenol A (DGEBA), which constitutes more than 75% of the epoxy market [2]. DGEBA is produced by a condensation reaction between bisphenol A and epichlorohydrin with a catalyst such as sodium hydroxide [3]. Epoxy resin must go through a curing process to solidify and achieve its final mechanical properties. In the curing process, a chemical compound termed as a curing agent or hardener with two or more functional groups (e.g., -NH_2) that can react with the epoxide functional groups () in epoxy resin is added and reacts with epoxy resin at an elevated temperature, forming a 3D crosslinked molecular network. It is noteworthy that the production of conventional epoxy resin and its curing agent requires a large amount of fossil fuel that can significantly contribute to greenhouse gas emissions and climate change. Furthermore, the epoxy products are not biodegradable, and they can stay in our environment for hundreds of years without significant degradation, posing a risk of contamination to air, soil, and water.

Sustainable or green epoxy resin has attracted significant attention in recent years [4,5], and has emerged not only in scientific publications [6,7], but also in real market such as Eco Green Resin from A. C. Composite Resins (Sacramento, CA, USA), Entropy Resin from Entropy Resins Inc. (Bay City, MI, USA), and Ecopoxy Resin from EcoPoxy Inc. (Morris, MB, Canada). Although the sources of these epoxy products are from biomass and sustainable, such as sugar [8], vegetable oil [9], and fatty acid [10], there are a few common limitations, including a requirement of a paired conventional curing agent, high cost of the epoxy resin, high viscosity of the epoxy resin in processing [11], and particularly low mechanical and thermal performance of the cured products compared to their conventional epoxy counterparts [12,13], which can be ascribed to the relatively flexible molecular backbone in those epoxy molecules.

Our research targeted the development of a sustainable epoxy system from renewable resources with high mechanical performance. It is known that algae are a renewable bioresource because of their high photosynthetic efficiency and high growth rates. It is reported that algae can be converted to bio-oil using a hydrothermal liquefaction (HTL) process in which algal biomass directly decomposes in water at a moderate temperature and pressure level [14,15]. The algal bio-oil has been described as similar to conventional petroleum and can be upgraded and used as transportation fuel [16]. However, the bio-oil contains a high nitrogen content, which derives from the protein component of algae. The nitrogen content imparts several undesirable properties for fuel purposes and needs to be reduced [17]. It is noteworthy that the nitrogen content in the bio-oil, such as amine ($-NH_2$) functional groups, could react with conventional epoxy resins just like their hardeners. In this research, we acquired a bio-oil from *Chlorella vulgaris*, a single-celled freshwater green microalga that grows naturally in lakes and ponds worldwide, through HTL processing and placed the bio-oil into a brand-new field, i.e., epoxy resin and composite materials. To differentiate its use from biofuel uses, we use the term bio-binder herein to refer to the bio-oil from HTL processing of microalgae. In this research, we incorporated the bio-binder into a representative conventional epoxy resin (EPON 862), investigated the thermal stability and mechanical properties of the epoxy/bio-binder system, and compared the results with those of EPON 862 with a paired hardener (EPICURE W). Our results indicated that the integration of the bio-binder from HTL processing of microalgae with conventional epoxy resin is a game-changing way to inject sustainability into current epoxy-based commercial products, aiming towards future environmentally friendly and high-performance epoxy resin-based composite materials.

2. Results and Discussion

2.1. Composition and Curing

To learn the composition of the bio-binder obtained from HTL processing of the green microalgae, we performed GC-MS and FTIR analysis (Figure 1). GC-MS analysis indicated that the bio-binder was a complicated mixture of alkanes, amides, esters, alcohols, amines, etc., as primary constituents. The characteristic FTIR spectrum of the bio-binder indicated certain functional groups that were consistent with the GC-MS analysis results. Specifically, the broad IR peak in the range of 3200 cm^{-1} to 3700 cm^{-1} could be attributed to mixed stretching vibrations involving both O-H and N-H bonds. The IR peaks in the region of 3000 cm^{-1} and 3200 cm^{-1} could be assigned to C-H stretching from the aromatic ring and alkene, while the peaks in the region of 2800 cm^{-1} to 3000 cm^{-1} represented C-H stretching vibrations in alkane. The IR peaks centered at ~1707 cm^{-1}, ~1650 cm^{-1}, and ~1553 cm^{-1} indicated C=O stretching in carboxylic acid, C=O stretching in amide (Amide I)/N-H bending, and C-N stretching/N-H bending (Amide II) along with aromatic ring stretching (C=C-C), respectively. The IR peaks around 1454 cm^{-1} and 1362 cm^{-1} could be assigned to C-H bending in alkane. The IR peak at 1219 cm^{-1} could be ascribed to aromatic and aliphatic C-O/C-N stretching. The IR peaks centered at 968 cm^{-1} and 700 cm^{-1} could represent C=C bending and C-H out-of-plane bending in the benzene ring. All these FTIR peaks indicated the presence of alkane, alkene, and aromatic compounds with hydroxyl, amide, amine, and carboxylic acid functional groups, which can react with the epoxide functional groups in epoxy resin [18].

Chemical Constituent in Bio-binder from GC-MS	
Alkane 42.81%	Esters 8.24%
Alkenes 0.63%	Carbonyls 1.27%
Alcohols 3.63%	Aromatics 2.28%
Amines 3.36%	Fatty Acids 1.58%
Amides 21.7%	

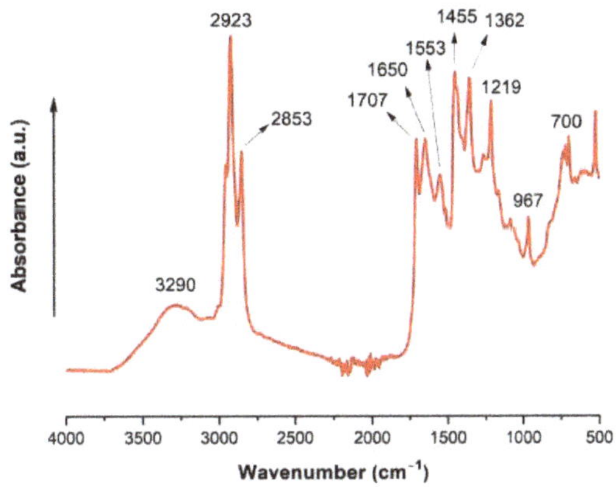

Figure 1. Chemical constituents from GC-MS (area percentage) and FTIR spectra of the bio-binder from HTL processing of the green microalgae.

Accordingly, we studied the composition of the epoxy/bio-binder systems with EPON 862, a popular conventional epoxy for composite use. It is well known that the curing cycle of epoxy resin plays a key role in achieving its desired mechanical property and long-term durability. Following the recommended curing cycle of EPON 862 from the manufacturer, we realized a practicable epoxy system with bio-binder compositions at 30 wt.%, 35 wt.%, and 40 wt.% of EPON 862. With extremely low percentages of the bio-binder, like 10 wt.% and 20 wt.% of the epoxy, the product is not usable (sticky). With high percentages of the bio-binder, like 45 wt.% and 50 wt.% of epoxy, the product is also not usable (too soft and cannot maintain integrity in mechanical tests).

Then, we used DSC to determine the optimal curing cycle for the epoxy/bio-binder system (Figure 2). Based on the recommended curing cycle of EPON 862, we studied five curing temperatures for our epoxy/bio-binder system, i.e., 200 °F, 250 °F, 300 °F, 350 °F, and 400 °F (93.3 °C, 121.1 °C, 148.9 °C, 176.7 °C, and 204.4 °C, correspondingly) and four different curing times, i.e., 1 h, 2 h, 3 h, and 4 h. The degree of cure of each sample was determined by using Equation (1) that took into account the total curing heat and the individual sample's curing heat in the curing process through DSC [19].

$$\%\ \text{Cure} = \left(1 - \frac{\text{DHR}}{\text{DHT}}\right) \times 100 \tag{1}$$

where DHR represents the heat of cure calculated from the area under the curing exothermic peak of the DSC curve of each individual cured epoxy/bio-binder sample, while DHT represents the total heat of cure calculated from the total area under the curing exothermic peak of the DSC curve of the correspondingly uncured epoxy/bio-binder sample. Across our epoxy/bio-binder system, the degree of cure increased with both the curing time and the temperature. The epoxy/bio-binder samples cured at 204.4 °C for 4 h exhibited the highest degree of cure, nearly reaching 100%. In the meantime, we acquired the glass transition temperatures (T_g) of all the cured epoxy/bio-binder samples under the studied curing temperatures and times (Table 1). Compared to the cured conventional counterpart, i.e., EPON 862 with the regularly paired EPIKURE W hardener (EPON 862 + EPIKURE

W), all the cured epoxy/bio-binder samples came with much lower T_gs, which could be attributed to the small molecules in the bio-binder that led to the plasticizer effect and, thus, reduced the T_gs of the correspondingly cured epoxy [20]. Nevertheless, the maximum T_gs of the cured epoxy/bio-binder samples were observed under the curing condition of 204.4 °C for 4 h. Therefore, we adopted this curing condition to prepare all the samples for the following tests.

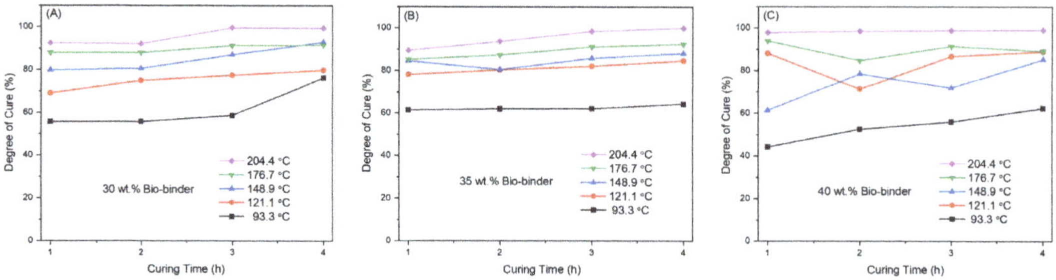

Figure 2. Degree of cure of the epoxy/bio-binder samples at bio-binder loadings of 30 wt.% (**A**), 35 wt.% (**B**), and 40 wt.% (**C**) of epoxy under different curing conditions.

Table 1. Glass transition temperatures (T_g) of conventional epoxy counterpart (EPON 862 + EPIKURE W) and the epoxy/bio-binder samples at bio-binder loadings of 30 wt.%, 35 wt.%, and 40 wt.% of epoxy under different curing conditions.

Bio-Binder Loading (wt.%)	Curing Time (h)	Curing Temperature				
		93.3 °C	121.1 °C	148.9 °C	176.7 °C	204.4 °C
0	1	24.8	47.5	72.4	84.8	129.2
	2	31.3	49.3	88.0	126.0	134.1
	3	36.8	56.3	97.3	135.8	141.0
	4	39.6	96.6	112.9	138.5	142.0
30	1	26.1	26.5	28.5	299.0	30.5
	2	26.4	28.4	28.6	32.4	33.2
	3	28.0	25.9	30.1	39.2	42.4
	4	28.4	28.8	33.3	39.9	45.4
35	1	27.1	29.3	29.4	30.4	31.2
	2	29.2	30.4	31.4	38.5	38.6
	3	30.8	31.3	38.1	39.4	42.4
	4	35.5	37.3	41.3	48.2	55.0
40	1	29.0	30.5	34.8	45.4	47.7
	2	30.1	32.7	39.5	42.9	51.2
	3	31.3	33.5	41.3	43.2	53.7
	4	38.0	47.6	48.8	49.5	56.3

2.2. Thermal Stability

We used thermogravimetric analysis (TGA) next to study the thermal stability of the epoxy/bio-binder system and compared it with the conventional counterpart (EPON 862 + EPIKURE W) (Figure 3A). The three epoxy/bio-binder samples with bio-binder loadings at 30 wt.%, 35 wt.%, and 40 wt.% of epoxy exhibited almost identical thermal behavior. Compared to the counterpart, the epoxy/bio-binder system showed a similar weight loss rate from room temperature to 250 °C, a slightly faster weight loss from 250 °C to 375 °C, and slower weight loss thereafter. The 60% and 80% weight loss temperatures increased from 394 °C and 450 °C for the conventional counterpart to 428 °C and 545 °C, respectively, for the epoxy/bio-binder sample with 35 wt.% bio-binder incorporation. We further performed derivative thermogravimetric analysis (Figure 3B). For the conventional epoxy counterpart, a peak representing the temperature of the maximum decomposition rate was observed around 382 °C. In the case of the epoxy/bio-binder sample with the bio-binder at loading of 35 wt.% of epoxy, the temperature of the maximum decomposition rate increased to around 413 °C. The faster weight loss of the epoxy/bio-binder samples from 250 °C to 375 °C could be attributed to small molecules in the bio-binder, while the crosslinking reaction between the bio-binder and epoxy resin enabled higher thermal stability afterwards.

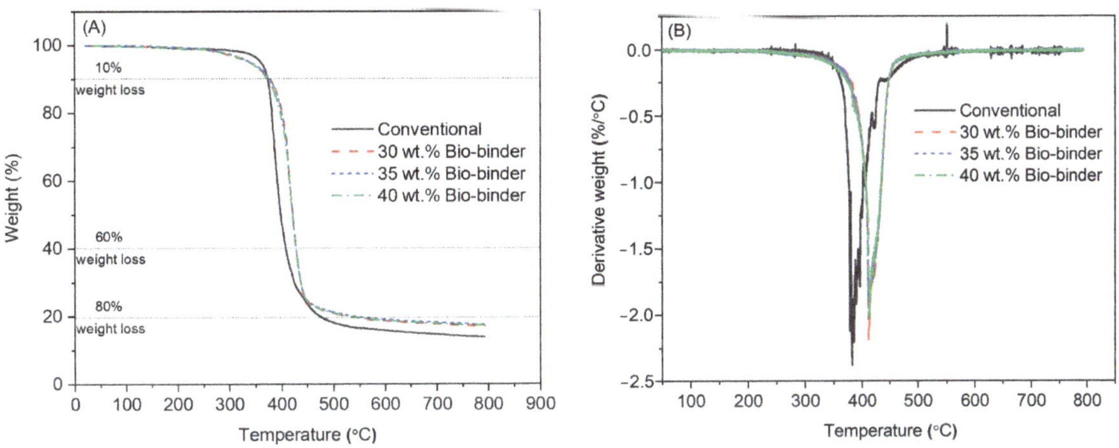

Figure 3. Weight loss (**A**) and derivative weight loss (**B**) against temperature of the conventional epoxy counterpart (EPON 862 + EPIKURE W) and the epoxy/bio-binder samples with bio-binder at loadings of 30 wt.%, 35 wt.%, and 40 wt.% of epoxy after curing from TGA analysis.

2.3. Mechanical Performance

To evaluate the mechanical performance of the epoxy/bio-binder system, we performed tensile and flexural tests on the cured epoxy/bio-binder composite materials to characterize their strength, stiffness, ductility, and toughness under conditions of tensile/stretching force, bending load, as well as sudden impact. With respect to the conventional counterpart, our epoxy/bio-binder samples, in most cases, showed improvements in all the mechanical properties (Figure 4). The epoxy/bio-binder sample with bio-binder at 35 wt.% of epoxy possessed the best mechanical properties. This epoxy/bio-binder sample exhibited tensile strength of 67.2 MPa, a Young's modulus of 3.44 GPa, elongation at break of 10.8%, and work of fracture of 1596 kJ/m^3 from the tensile test, which were improvements of ~16%, ~15%, ~39%, and ~63%, respectively, with respect to those of the conventional counterpart. This epoxy/bio-binder sample also demonstrated flexural strength of 118 MPa, a flexural modulus of 3.13 GPa, flexural strain at failure of 5.49%, and flexural work of fracture of 3718 kJ/m^3, which were improvements of ~14%, ~19%, ~22%, and ~64%, respectively, with respect to those of the conventional counterpart, as assessed through a 3-point bending test. Furthermore, this epoxy/bio-binder sample also showed an Izod impact strength of 21.9 J/m, an improvement of ~32% with respect to that of the conventional counterpart.

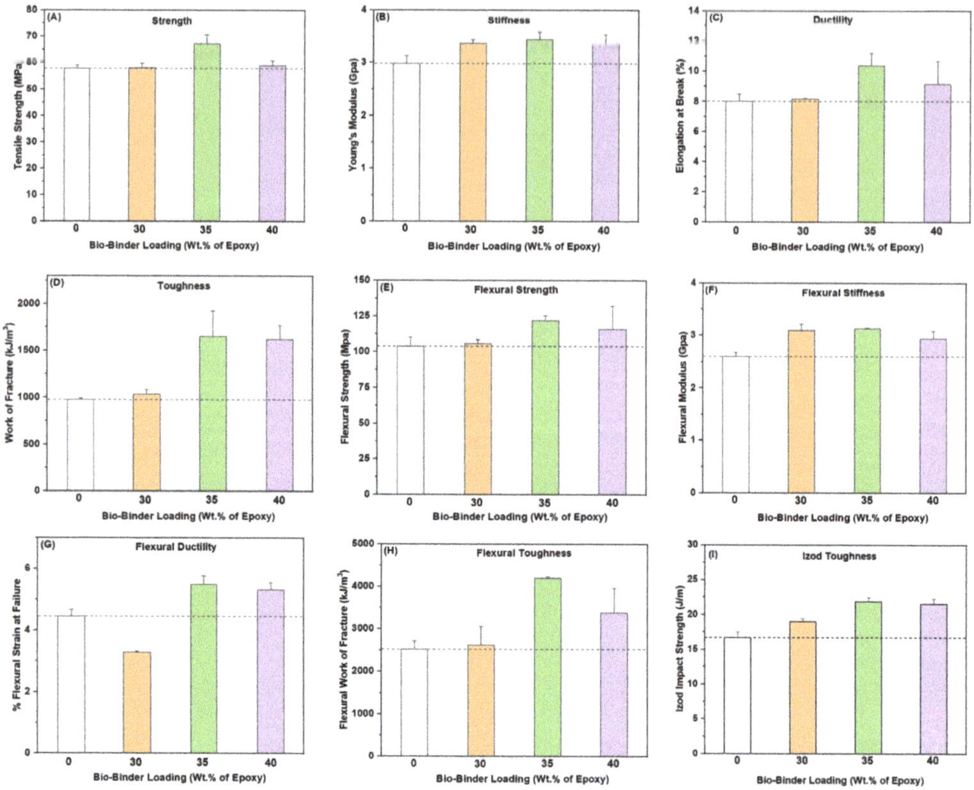

Figure 4. Tensile strength (**A**), Young's modulus (**B**), elongation at break (**C**), work of fracture (**D**), flexural strength (**E**), flexural modulus (**F**), flexural strain (**G**), flexural work of fracture (**H**), and Izod impact strength (**I**) of the conventional epoxy counterpart (EPON 862 + Epikure W) and the epoxy/bio-binder samples at bio-binder loadings of 30 wt.%, 35 wt.%, and 40 wt.% of epoxy after curing from tensile, three-point bending, and Izod impact tests, respectively.

To further understand the mechanical performance of the epoxy/bio-binder system, we investigated the fracture surface of the epoxy/bio-binder samples from the tensile test, three-point bending test, and Izod impact test using SEM (Figure 5). For the conventional epoxy counterpart, the fracture surface was relatively smooth with few fracture lines, indicating less energy consumption in the process of break and correspondingly low toughness. In comparison, the epoxy/bio-binder sample with the bio-binder at 35 wt.% of epoxy exhibited a much rougher surface, indicating more energy consumption during the fracture process and improved toughness from the bonding between the molecules of epoxy and the bio-binder.

As indicated earlier, the bio-binder is a mixture of various molecules, including both small-molecular-weight molecules and large-molecular-weight molecules, with some containing functional groups that can react with the epoxide functional groups in epoxy resin, participating in the crosslinking reaction in the process of curing. In the epoxy/bio-binder system, the final mechanical performance of the cured product is determined by the number of functional groups for robust epoxy curing (crosslinking) and the number of small molecules that can degrade mechanical properties of the cured epoxy/bio-binder. As a result, the best mechanical performance in the epoxy/bio-binder system was observed with the sample containing the bio-binder at 35 wt.% of epoxy.

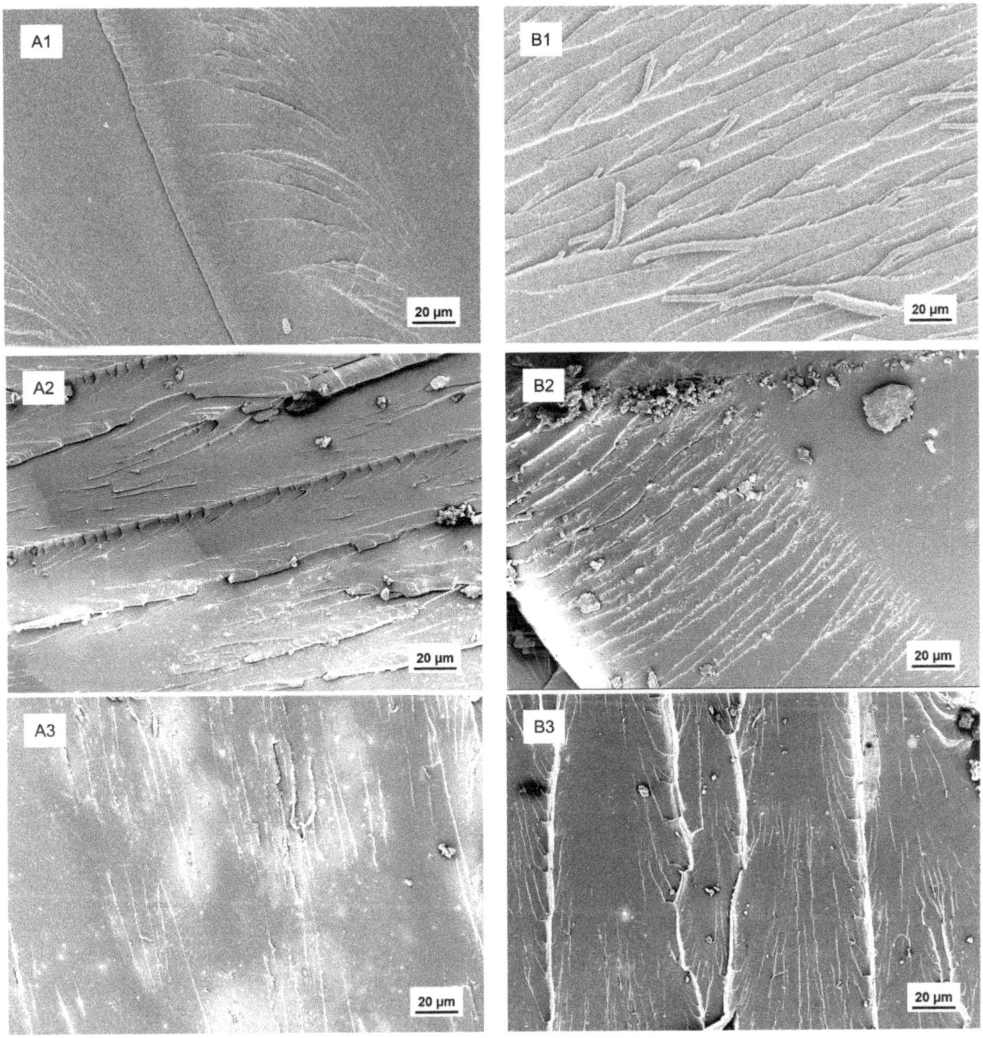

Figure 5. SEM of fracture surfaces of cured epoxy and epoxy/bio-binder after mechanical tests: conventional epoxy counterpart (EPON 862 + Epikure W) (**A**) and the epoxy/bio-binder sample at bio-binder loadings of 35 wt.% of epoxy (**B**). Labels "**1**", "**2**", and "**3**" indicate the samples from the tensile test, three-point bending test, and Izod impact test, respectively.

3. Material and Methods

3.1. Materials

A conventional epoxy resin (EPON 862, a phenol formaldehyde polymer glycidyl ether) and its curing agent, EPIKURE W (an aromatic amine), were purchased from Miller Stephenson (Danbury, CT, USA). A green microalga (*Chlorella vulgaris*) was acquired from nuts.com (Cranford, NJ, USA). Anhydrous acetone was purchased from Fisher Scientific (Waltham, MA, USA).

3.2. Bio-Binder Production

A 1 L Parr reactor (Parr Instrument Company, Moline, IL, USA) was loaded with the microalgae and DI water with a mass ratio of 1:9.6. The reactor was heated to 300

°C with a heating rate of 5 °C/min. The reaction was carried out for 30 min at 300 °C under a pressure of 1500 psi. The bio-binder was extracted from the oil phase, followed by centrifugation and filtration to remove the aqueous phase and solid particles (Figure 6A).

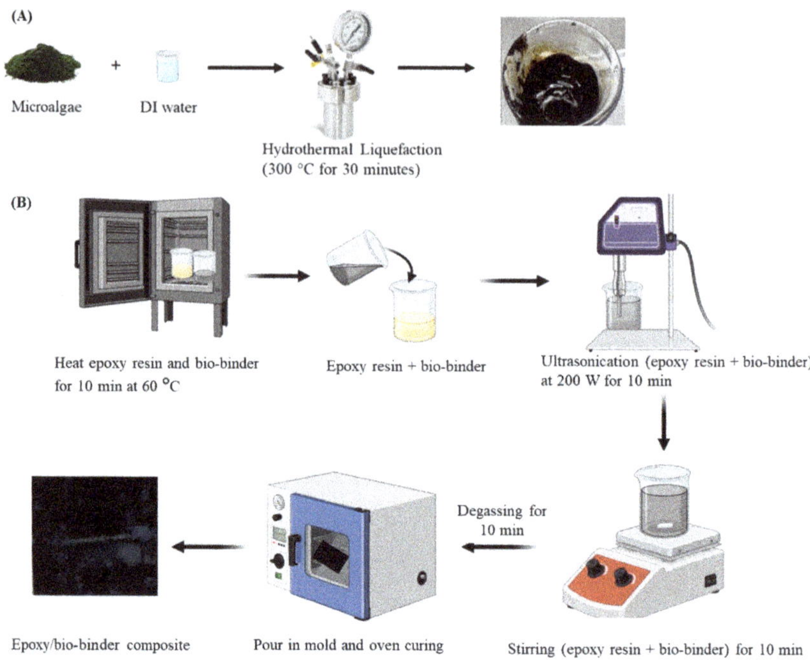

Figure 6. Schematic demonstrations for preparation of bio-binder from microalgae through hydrothermal liquefaction (**A**) and fabrication of epoxy/bio-binder composite (**B**).

3.3. Fabrication of Epoxy/Bio-Binder Composite

The bio-binder was added to EPON 862 at 30 wt.%, 35 wt.%, and 40 wt.% with respect to the epoxy, then subjected to vigorous ultrasonication using a 500 W QSONICA (Newtown, CT, USA) ultrasonic probe sonicator at 40% power for 10 min. A homogeneous mixture was achieved, and the mixed resin was degassed at 60 °C under a vacuum for 10–15 min to remove air bubbles. The degassed epoxy/bio-binder system was poured into a 6 in. × 6 in. mold and cured in an oven for the set curing cycle (Figure 6B). For comparison, a conventional counterpart was prepared from EPON 862 with the regularly paired EPIKURE W hardener using the same approach and cured under the recommended condition by the manufacturer.

3.4. Characterization

The functional groups of the bio-binder were characterized by an Agilent (Santa Clara, CA, USA) Varian 670 Fourier-transform infrared (FTIR) spectrometer. TA Q200 differential scanning calorimetry (DSC) and a TA instruments (New Castle, DE, USA) Q500 thermogravimetric analyzer (TGA) were used to study the curing cycle and the thermal stability of the epoxy/bio-binder systems, respectively. Tensile, flexural (three-point bending), and Izod impact tests were carried out using a universal material testing machine (UTM Instron 3384, Norwood, MA, USA) and an Izod impact tester (Tinius Olsen IT 503, Horsham, PA, USA) according to ASTM (West Conshohocken, PA, USA) standards D638-Type IV, D790, and D256, respectively.

4. Conclusions

In this research, we successfully improved the sustainability of epoxy resin by integrating the bio-binder from hydrothermal liquefaction processing of green microalgae. It is clear from our fundamental research that the epoxy/bio-binder system, particularly the epoxy/bio-binder sample with bio-binder loading at 35 wt.% of epoxy, outperformed the conventional epoxy resin with a paired curing agent in terms of thermal stability and mechanical property after curing. Compared to the neat epoxy resin (EPON 862), incorporation of the bio-binder at 35 wt.% of epoxy significantly enhanced the thermal stability of the resultant epoxy/bio-binder composite, as demonstrated by the rise in temperatures at 60% and 80% weight loss from 394 °C and 450 °C to 428 °C and 545 °C, respectively, as well as the increase in temperature at the maximum decomposition rate from 382 °C to 413 °C. In the meantime, the tensile strength, stiffness, ductility, and toughness of the epoxy/bio-binder composite increased by 16%, 15%, 39%, and 63%, respectively, and the flexural strength, stiffness, ductility, and toughness improved by 14%, 19%, 22%, and 64%, respectively. Furthermore, the Izod impact strength of this epoxy/bio-binder sample was boosted by 32%.

Undoubtedly, our epoxy/bio-binder system overcame the brittleness of conventional epoxy resins and has great potential to replace them in applications that are associated with mechanical performance and thermal stability, such as structural composite uses, on top of potentially lower cost and sustainable component. Unlike the existing green epoxy products, the highlight of our epoxy/bio-binder system is that this system eliminates the use of regular hardeners of epoxy resin, while it has an even higher mechanical performance than that of conventional epoxy resin, which will be competitive in the market.

With employment of the nanoarchitectonics method in our future research to tune the physical properties of our epoxy/bio-binder system, we anticipate that the application of our microalgae-derived bio-binder with epoxy resin will contribute to broader goals of sustainable epoxy-based composite development and greenhouse gas mitigation. Our epoxy/bio-binder system holds tremendous promise across various industries, such as automotive, aerospace, and construction, where sustainable materials with comprehensive mechanical performance are increasingly valued.

Author Contributions: P.A.: investigation, validation, writing—original draft; A.M.: investigation, validation, formal analysis, writing—original draft; A.D.K.: conceptualization, methodology, resources; L.W.: conceptualization, methodology, resources, supervision; L.Z.: conceptualization, methodology, resources, supervision, project administration, writing—review and editing, funding acquisition. All authors have read and agreed to the published version of the manuscript.

Funding: This research was funded by the National Science Foundation (CMMI 2000318).

Institutional Review Board Statement: Not applicable.

Informed Consent Statement: Not applicable.

Data Availability Statement: Data are contained within the article. The original contributions presented in the study are included in the article; further inquiries can be directed to the corresponding author/s.

Acknowledgments: This work was conducted at the Joint School of Nanoscience and Nanoengineering of North Carolina A&T State University, a member of Southeastern Nanotechnology Infrastructure Corridor (SENIC) and National Nanotechnology Coordinated Infrastructure (NNCI), which is supported by the National Science Foundation (ECCS-2025462).

Conflicts of Interest: The authors declare no conflicts of interest.

References

1. Pappa, C.; Feghali, E.; Vanbroekhoven, K.; Triantafyllidis, K.S. Recent advances in epoxy resins and composites derived from lignin and related bio-oils. *Curr. Opin. Green Sustain. Chem.* **2022**, *38*, 100687. [CrossRef]
2. Celikbag, Y.; Meadows, S.; Barde, M.; Adhikari, S.; Buschle-Diller, G.; Auad, M.L.; Via, B.K. Synthesis and characterization of bio-oil-based self-curing epoxy resin. *Ind. Eng. Chem. Res.* **2017**, *56*, 9389–9400. [CrossRef]
3. Petrie, E.M. *Epoxy Adhesive Formulations*; McGraw-Hill: New York, NY, USA, 2006; pp. 1–26.

4. Agbo, P.; Mali, A.; Deng, D.; Zhang, L. Bio-oil-based epoxy resins from thermochemical processing of sustainable resources: A short review. *J. Compos. Sci.* **2023**, *7*, 374. [CrossRef]
5. Goncalves, F.A.M.M.; Santos, M.; Cernadas, T.; Ferreira, P.; Alves, P. Advances in the development of biobased epoxy resins: Insight into more sustainable materials and future applications. *Int. Mater. Rev.* **2022**, *67*, 119–149. [CrossRef]
6. Dinu, R.; Lafont, U.; Damiano, O.; Mija, A. High glass transition materials from sustainable epoxy resins with potential applications in the aerospace and space sector. *ACS Appl. Polym. Mater.* **2022**, *4*, 3636–3646. [CrossRef]
7. Chong, K.L.; Lai, J.C.; Rahman, R.A.; Adrus, N.; Al-Saffar, Z.H.; Hassan, A.; Lim, T.H.; Wahit, M.U. A review on recent approaches to sustainable bio-based epoxy vitrimer from epoxidized vegetable oils. *Ind. Crops Prod.* **2022**, *189*, 115857. [CrossRef]
8. Rapi, Z.; Szolnoki, B.; Bako, P.; Niedermann, P.; Toldy, A.; Bodzay, B.; Keglevich, G.; Morosi, G. Synthesis and Characterization of biobased epoxy monomers derived from D-glucose. *Eur. Polym. J.* **2015**, *67*, 375–382. [CrossRef]
9. Marriam, F.; Irshad, A.; Umer, I.; Asghar, M.A.; Atif, M. Vegetable oils as bio-based precursors for epoxies. *Sustain. Chem. Pharm.* **2023**, *31*, 100935. [CrossRef]
10. Ortiz, P.; Vendamme, R.; Eevers, W. Fully biobased epoxy resins from fatty acids and lignin. *Molecules* **2020**, *25*, 1158. [CrossRef] [PubMed]
11. Sorrentino, L.; Turchetta, S.; Parodo, G.; Papa, R.; Toto, E.; Santonicola, M.G.; Laurenzi, S. RIFT process analysis for the production of green composites in flax fibers and bio-based epoxy resin. *Materials* **2022**, *15*, 8173. [CrossRef] [PubMed]
12. Thomas, J.; Bouscher, R.F.; Nwosu, J.; Soucek, M.D. Sustainable thermosets and composites based on the epoxides of norbornylized seed oils and biomass fillers. *ACS Sustain. Chem. Eng.* **2022**, *10*, 12342–12354. [CrossRef]
13. Hidalgo, P.; Salgado, L.; Ibacache, N.; Hunter, R. Influence of biochar and bio-oil loading on the properties of epoxy resin composites. *Polymers* **2023**, *15*, 1895. [CrossRef] [PubMed]
14. Rahman, Q.M.; Zhang, B.; Wang, L.; Joseph, G.; Shahbazi, A. A combined fermentation and ethanol-assisted liquefaction process to produce biofuel from *Nannochloropsis* sp. *Fuel* **2019**, *238*, 159–165. [CrossRef]
15. Chen, W.-T.; Qian, W.; Zhang, Y.; Mazur, Z.; Kuo, C.-T.; Scheppe, K.; Schideman, L.C.; Sharma, B.K. Effect of ash on hydrothermal liquefaction of high-ash content algal biomass. *Algal Res.* **2017**, *25*, 297–306. [CrossRef]
16. Muppaneni, T.; Reddy, H.K.; Selvaratnam, T.; Dandamudi, K.P.R.; Dungan, B.; Nirmalakhandan, N.; Schaub, T.; Holguin, F.O.; Voorhies, W.; Lammers, P. Hydrothermal liquefaction of Cyanidioschyzon merolae and the influence of catalysts on products. *Bioresour. Technol.* **2017**, *223*, 91–97. [CrossRef] [PubMed]
17. Jazrawi, C.; Biller, P.; He, Y.; Montoya, A.; Ross, A.B.; Maschmeyer, T.; Haynes, B.S. Two-stage hydrothermal liquefaction of a high-protein microalga. *Algal Res.* **2015**, *8*, 15–22. [CrossRef]
18. Liu, L.; Gao, S.; Jiang, Z.; Zhang, Y.; Gui, D.; Zhang, S. Amide-funcitonalized ionic liquid as curing agents for epoxy resin: Preparation, characterization, and curing behaviors with TDE-85. *Ind. Eng. Chem. Res.* **2019**, *58*, 14088–14097. [CrossRef]
19. Galehdari, N.A.; Kelkar, A.D. Effect of neutron radiation on the mechanical and thermophysical properties of nanoengineered polymer composites. *J. Mater. Res.* **2017**, *32*, 426–434. [CrossRef]
20. Wei, N.; Via, B.; Wang, Y.; McDonald, T.; Auad, M. Liquefaction and substitution of switchgrass (Panicum virgatum) based bio-oil into epoxy resins. *Ind. Crops Prod.* **2014**, *57*, 116–123. [CrossRef]

molecules

MDPI

Article

Can Biomass Mastication Assist the Downstreaming of Polyhydroxyalkanoates Produced from Mixed Microbial Cultures?

Hiléia K.S. Souza [1,*], Mariana Matos [2,3], Maria A.M. Reis [2,3], José A. Covas [1] and Loïc Hilliou [1,*]

[1] Institute for Polymers and Composites, Campus de Azurém, University of Minho, 4800-058 Guimarães, Portugal
[2] Associate Laboratory i4HB–Institute for Health and Bioeconomy, NOVA School of Science and Technology, NOVA University Lisbon, 2829-516 Caparica, Portugal
[3] UCIBIO—REQUIMTE, Applied Molecular Biosciences Unit, Department of Chemistry, NOVA School of Science and Technology, NOVA University Lisbon, 2829-516 Caparica, Portugal
* Correspondence: hileia.souza@dep.uminho.pt (H.K.S.S.); loic@dep.uminho.pt (L.H.)

Abstract: Polyhydroxyalkanoates (PHAs) are natural polyesters which biodegrade in soils and oceans but have more than double the cost of comparable oil-based polymers. PHA downstreaming from its biomass represents 50% of its overall cost. Here, in an attempt to assist downstreaming, mastication of wet biomasses is tested as a new mechanical continuous biomass pretreatment with potential for industrial upscaling. Downstreaming conditions where both product recovery and purity are low due to the large amount of treated wet biomass (50% water) were targeted with the following process: extraction of 20 g in 100 mL solvent at 30 °C for 2 h, followed by 4.8 h digestion of 20 g in 0.3 M NaOH. Under the studied conditions, NaOH digestion was more effective than solvent extraction in recovering larger PHA amounts, but with less purity. A nearly 50% loss of PHA was seen during digestion after mastication. PHAs downstreamed by digestion with large amounts of impurities started to degrade at lower temperatures, but their melt elasticity was thermally stable at 170 °C. As such, these materials are attractive as fully PHA-compatible processing aids, reinforcing fillers or viscosity modifiers. On the other hand, wet biomass mastication before solvent extraction improves PHA purity and thermal stability as well as the melt rheology, which recovers the viscoelasticity measured with a PHA extracted from a dried biomass.

Keywords: polyhydroxyalkanoates; extraction; digestion; bioplastic; mechanical pretreatment; rheology

1. Introduction

Polyhydroxyalkanoates (PHAs) are a family of natural polymers which are biosynthesized as energy and carbon storage granules with size of 0.2 to 0.5 micron [1] in the cells of many naturally occurring microorganisms. In contrast to other natural and abundant polymers, such as cellulose or starch, PHA are thermoplastic polyesters just like poly(lactic acids) (PLAs). In contrast to PLA, which requires industrial composting for biodegradation as defined by established standards [2], PHA is a more attractive biosourced polyester, as it naturally biodegrades in land, fields or the ocean [3]. However, by 2021, PHA production is a tenth of PLA production, and it remains a very marginal player in applications such as plastic packaging [4]. Plastic packaging is indeed a relevant application example for biosourced and biodegradable polymers: in the OECD countries, 31% of the plastics consumption is for packaging, which generates 40% of the total plastic waste [5], whereas packaging represents 50% of the biodegradable plastics market [4].

PHAs exhibit properties that hamper their acceptance by the plastic industry. First, PHAs are rather brittle, as they break at strains as low as 1 to 20% depending on their chemical structures [6,7], which prevents them from being used in many flexible packaging applications. Second, industrial polymer processing technologies require specific melt properties, such as viscosity and melt strength, in order to yield products with sufficient performance [8]. PHA melt viscosity is low, which suits injection molding of, for example, rigid packages such as jars, but its melt strength is not large enough to withstand more rheologically demanding processes such as extrusion film blowing [9]. More importantly, the PHA market price is two to five times the price of polyethylene [10,11], which makes nearly 50% of plastic packaging waste [5].

The recently established pilot scale production of PHA based on mixed microbial cultures (MMC) fed with carbon-rich industrial wastes (see [12] and references therein) gives good prospects for production cost reductions when compared to pure cultures grown on specific clean substrates. However, the extraction of MMC PHA from the fermented biomass takes a big share of the overall production cost. For instance, a technoeconomic analysis of the downstreaming of MMC polyhydroxybutyrate (PHB) produced from wastewater calculated that the extraction step can be two to four times more expensive than the fermentation step, contributing to a total production cost ranging from 1.5 to 2 €/kg PHB [13]. Therefore, the downstreaming of MMC PHA has received growing interest, as illustrated by the number of reviews published in the literature over the last couple of years (see e.g., [14–17]).

Among the various MMC PHA downstreaming processes that have been investigated, two are highlighted in Table 1: solvent extraction and NaOH digestion of the biomass. In solvent extraction, chemicals are used to weaken the cells' membrane and then solvate the PHA granules. The obtained PHA-rich liquid phase is then separated from the solid cell residues, and PHA is eventually recovered either by solvent evaporation or precipitation of the PHA solution into an antisolvent. Extraction with $CHCl_3$ has become a benchmark for evaluating new PHA downstreaming processes in laboratories because it guarantees the recovery of nearly pure MMC PHA (from 82.5 to 100% purity; see Table 1) with large molecular mass (no PHA degradation) from nearly all kinds of microbial strains [18]. However, because concentrated PHA solutions in $CHCl_3$ show large viscosities [19], a small amount of biomass must be treated in large volumes of solvents for achieving large recovery yields, as shown in Table 1. The water content in fresh biomasses also significantly lowers the solvent quality of PHA, which may explain a biopolyester recovery yield as low as 18% reported for wet biomasses containing 25 wt% PHA and treated with CH_2Cl_2 [20]. Indeed, very few studies targeted the enhancement of $CHCl_3$ extraction of MMC PHA using, for example, mechanical pretreatment of biomasses [21,22]. This is perhaps because extraction with chlorinated solvents is usually tagged as not suitable for industrial PHA production [14–18]. NaOH digestion of *Ralstonia eutropha* biomasses with rich PHA content has long been recognized as a PHA downstreaming route with industrial relevance [23]. However, this approach has been poorly tested for the recovery of MMC PHA, as seen in Table 1. A possible explanation for this is that cell breakage or hydrolysis in MMC is thought to be more difficult to achieve than in pure cultures [11,20] where cells can be selected for their fragility and high PHA content. Nevertheless, one may build on previous work performed on pure cultures treated with NaOH before continuous mechanical disruption [24] in an attempt to increase both the PHA purity and the recovery yield reported so far for MMC biomasses containing lower amounts of PHA [25–27].

Table 1. Product recovery and PHA purity reported for the MMC PHA biomass downstreaming with chlorinated solvents or NaOH digestion. The PHA content (PHA) in the initial biomass is given in weight percent (wt%), and the proportion of 3-hydroxyvalerate (HV) in the recovered poly(3-hydroxybutyrate-co-3-hydroxyvalerates) (PHBV) is given in %. The concentration c of dry biomass in chloroform is given in weight per volume percent (% *w*/*v*) in the corresponding solution (with $CHCl_3$, CH_2Cl_2 or NaOH with the corresponding molar concentration) and was computed from the reported type of biomass: dried (*d*), lyophylized (*l*) or wet (*w*), as computed from the referenced water content. The parameters of the downstreaming process are temperature (*T* in °C) and time (*t* in hours). PHA recovery and purity are given in %. "n.a." indicates data are not available.

PHA (wt%)	HV	Extraction			Recovery	Purity	Reference
		c (% *w*/*v*)	*T*	*t*			
Chlorinated solvent extraction							
70	31	4 (*l*) $CHCl_3$	80	48	82 [a]	95 [d]	[28]
66	18	2 (*l*) $CHCl_3$	80 (Soxhlet)	24	66 [b]	91 [d]	[29]
40	24	2.5 (*l*) CH_2Cl_2	50	4	52 [a]	94	[11]
25	5.6	n.a. (CH_2CL_2 + water)	61	0.5	18–30 [a]	n.a.	[20]
72	0	1 (*l*) CH_2Cl_2	25	12	56 [a]	98 [d]	[25]
72 [e]	66	2 [e] (*w*) $CHCl_3$ Sonication Beads mill Mechanical homogeneisation	61	2	31 [e] 7 [e] 24 [e] 44 [e]	94.5 [d] 90.6 [d] 93.3 [d] 88.3 [d]	[21]
n.a.	17.5	5 (*l*) $CHCl_3$ Microwave	100	2	36 [c] 1.5–38 [c]	<96 [d] >96 [d]	[22]
50	12	2.5 (*l*) $CHCl_3$	38	72	63.5 [a]	n.a.	[30]
38.7–56.1	7.7–20.2	2.9–7.1 (*d*) $CHCl_3$	61 (Soxhlet)	24	75.4–97.9 [f]	83–100	[31]
n.a.	4–5	5 (*d*) $CHCl_3$	60	2	32 [a]	>99 [d]	[32]
54	0	6.7 (*d*) $CHCl_3$	60	2	69 [a]	100 [d]	[33]
32	50	1 (*l*) $CHCl_3$	60 (reflux)	1	37.5 [b]	82.5	[34]
56 62	17 11	n.a. (*d*) $CHCl_3$ n.a. (*w*) $CHCl_3$	61 (Soxhlet)	8	96 [a]	98	[35]
50	25	20 (*d*) $CHCl_3$ 10 (*w*) $CHCl_3$ 10 (*w*) $CHCl_3$ and mastication	30	2	46.7 [a] 17.4 [a] 7.6–19.4 [a]	95.6 96.6 88.9–97.1	This study
NaOH digestion							
70	31	2 (*d*) in 0.3 M	30	4.8	100 [a]	100	[28]
65–73	0	2 (*w*) in 0.02 M to 1 M 2 (*l*) in 0.2 M	30	0.3–3 1	86–98.5 [a] 95.5 [a]	77–92 96	[25]
46	13	n.a. in 1 M	n.a.	3–24	80–87 [a]	75	[26]
64	13	2.5 (*l*) in 0.3 M	30	1	90 [a]	76	[27]
50	25	20 (*d*) in 0.3 M 10 (*w*) in 0.3 M and mastication	30	4.8	95 [a] 35.8 [a]	79 73.1	This study

[a]: pure PHA recovered over the PHA content in the biomass. [b]: pure PHA recovered over the biomass treated. [c]: extracted product over the biomass treated. [d]: after precipitation in nonsolvent. [e]: fresh (wet) biomass weight; PHA content in biomass and PHA recovery were computed from the ATR-FTIR analysis of the biomass after extraction. [f]: computation not detailed.

Here, the mastication of fresh MMC before biomass digestion by NaOH was tested. This continuous and upscalable process has never been used for the mechanical disruption of wet PHA biomasses before PHA recovery [14–18]. Furthermore, a recent evaluation of the environmental performances of different PHA downstreaming processes showed that mechanical disruption and the use of NaOH are more promising [36]. The aim of this study was to explore mastication as a new route to enhance the fragility of the cells, thereby facilitating the digestion of the non-PHA material by NaOH. To do this, a biomass containing 50% (weight basis) of MMC PHA produced from fruit waste at pilot scale was selected. The masticating of biomasses with different water contents was first tested for evaluating the feasibility of a continuous mastication process. Then, wet biomasses were masticated with NaOH before digestion. Next, the downstreaming performance of this new approach was compared with the performance of a NaOH digestion of the dried biomass. Additional extractions with $CHCl_3$ were performed for two purposes: (1) to understand the impact of using wet biomass, mastication or both on downstreaming performance and (2) to benchmark the MMC PHA's properties. Polymer characterization was focused on the thermal and melt properties in view to assess the melt processability of the produced biopolyesters. Results from this explorative study suggest that, although mastication does not assist the digestion of the specific biomass tested here, it permits the solvent extraction of PHAs with improved thermal and rheological properties.

2. Results and Discussion

2.1. Rheological Assessment of Biomass Mastication Ability

The purpose of the masticating unit used in the present study was to process solid-like materials. The screw and squeezing chamber dimensions, as well as the screw profile of the masticating unit, were thus designed to process solids. Therefore, the PHA enriched biomass should exhibit specific rheological properties to be efficiently masticated. Preliminary trials were performed with dried biomass (overnight at 60 °C in air ventilated oven) mixed with different amounts of distilled water in order to screen for the processability of the feeding material. Indeed, dried biomass cannot be processed in the masticating unit, as the powder could not be conveyed by the screw in the feeding zone. The best results were achieved with a feed consisting of 50 g dried biomass mixed with 50 mL water (50 wt% feed). The proper transport of material by the screw was limited either by larger biomass content because the material would stick to the barrel wall, or by smaller solid content, as the liquid would not flow through the squeezing zone at the tip of the screw. Figure 1 displays the rheological properties of the feed that showed the best processability, together with a feed formulated with 30 wt% dried biomass that could not be processed.

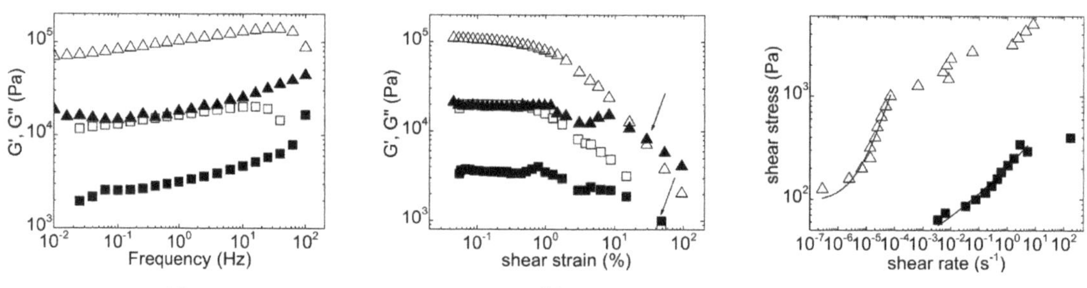

Figure 1. Mechanical spectra (**a**), dynamic stress sweeps (**b**) and flow curves (**c**) performed on biomass premixes with distilled water: 50 g with 50 mL (triangles) and 30 g with 70 mL (squares). In (**a**,**b**): G′ is displayed as open symbols; G″ is displayed as filled symbols. Arrows in (**b**) indicate the strains where slurries are fluidized, whereas lines in (**c**) are fits of Equation (1) to the data.

Both feeds displayed a viscoelastic and predominantly solid character. Both mechanical spectra showed a storage modulus G′ larger than the loss modulus G″ over the whole range of tested frequencies and with a pronounced frequency dependence for G′. The feed with 50 g/50 mL biomass was nearly 10 times more elastic (Figure 1a). The shear-induced fluidization of these solids was measured during the dynamic stress sweep, and it is represented by the crossover point between G′ and G″. The results show that less strain is needed to break up the 50 wt% feed into a fluid when compared with the more diluted feed (see arrows in Figure 1b), but the latter is still significantly less viscoelastic in the shear-induced liquid phase. Overall, the flow curves displayed in Figure 1c confirm this rheological behavior. For both feeds, the stress showed a plateau at lower shear rates, which is indicative of a solid with yield stress, whereas a non-Newtonian flow showed up at larger shear rates. This behavior is modeled by the Herschel-Bulkley equation, which reads as follows [37]:

$$\sigma = \sigma_y + K\dot{\gamma}^n \tag{1}$$

where σ is the shear stress, and σ_y is the stress where the solid yields and then flows with the shear rate $\dot{\gamma}$ with a flow consistency coefficient K and a shear thinning index n. The fits of Equation (1) to the data plotted in Figure 1c (before the shear rate regime where instabilities showed up, see below) returned yield stresses σ_y = 95 ± 33 Pa and σ_y = 39 ± 25 Pa for the feeds formulated with 50 wt% and 30 wt%, respectively. More importantly, the fits returned a $K \sim 3.5 \times 10^6$ Pa·s^{-n} for the 50 wt% feed, which is four orders of magnitude larger than that for the 30 wt% feed, and a thinning index n = 0.9 ± 0.1 against n = 0.3 ± 0.1 for the 30 wt%, indicating a more viscous material but with more flow ability for the 50 wt% feed. Overall, the rheological characterization is in harmony with the expected behavior for dense suspensions with various volume fractions of solids. Figure 1 indicates that 50 mL water added to the 50 g dried biomass is sufficient to act as a binder between biomass particles for the efficient conveying of the viscoelastic feed via the screw through the squeezing chamber. Adding larger amounts of water is detrimental to K and n, which become too low to allow continuous mastication.

Figure 1 also offers all of the rheological information needed to estimate the mechanical work transmitted to the 50 wt% biomass during mastication. The mechanical work per unit volume of biomass, W, is given by the product of stress and strain at play in the masticating unit, that is, $W = \sigma\gamma = \sigma\dot{\gamma}t_r$, where σ is the shear stress and $\dot{\gamma}$ is the average shear rate in the squeezing chamber, whereas t_r is the residence time in the masticating unit. With the 50 wt% biomass and a feeding rate of 10 mL per minute, the residence time of the material in the masticating unit was between 3 and 4 min, whereas the average shear rate in the squeezing unit is given by the following estimate [38]: $\dot{\gamma} = \frac{\pi DN}{60h}$, where N is the screw speed (in rpm) and D the external diameter of the screw in the masticating unit. Due to the conical shape of the screw in the squeezing chamber, a maximum and a minimum shear rate can be computed as 71 s^{-1} and 46 s^{-1}, respectively. Inserting these two shear rates in Equation (1), the corresponding shear stresses can be computed, and considering the residence times given above, a mechanical work W of the order of 1×10^9 J·kg^{-1} can be estimated for 1 kg of masticated biomass. This mechanical work seems, however, overestimated when compared to those delivered by mechanical homogenizers, bead mills or ultrasonicators, which have been used to assist PHA extraction [21,22] or digestion [24]. W is on the order of 10^6 J·kg^{-1} for the bead milling of NaOH digested biomasses [24], whereas mechanical works of 10^7 and 10^9 J·kg^{-1} were reported for mechanical homogenization and ultrasonication, respectively [39]. A possible explanation for the overestimation of W is given in Figure 1c. At shear rates corresponding to those taking place in the squeezing chamber of the masticating unit, the slurry with 50 wt% dried biomass actually did not flow as predicted by Equation (1). Instead, strain localization and nonhomogeneous flow, which are ubiquitous in the rheology of dense suspensions, took place as suggested by the stress plateau at larger shear rates in Figure 1c. Thus, a stress value on the order of 1 to 10 kPa should be considered to compute the mechanical work, which would then reach values between 7.5 and 126 kJ·kg^{-1} and compare to mechanical works computed for the melt mixing of, for example, polymer composites [40].

2.2. *Effect of Biomass Mastication on PHA Recovery*

Figure 2 presents a scheme summarizing the workflows used for the screening of wet biomass mastication to assist digestion with NaOH and for benchmarking with solvent extraction using $CHCl_3$. Mastication was tested with the productions of samples B, C1, C5, D1 and D5, whereas samples A, E and F were produced for benchmarking.

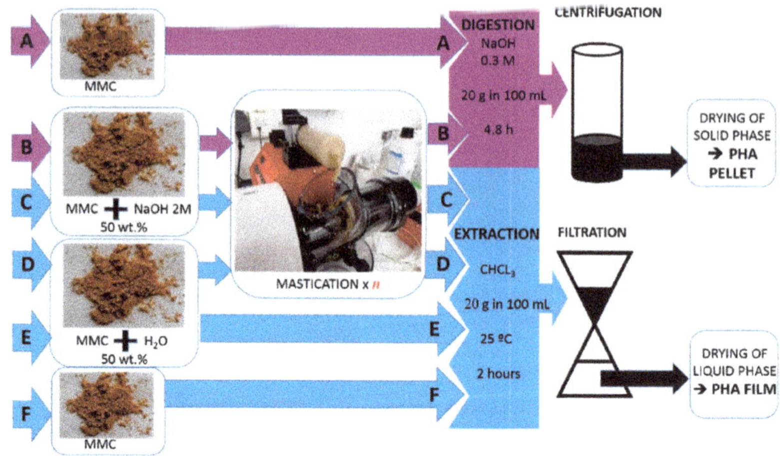

Figure 2. Flow chart of the downstreaming protocols used for digestion with 0.3 M NaOH or extraction with $CHCl_3$. The dried biomass (MMC) containing 50 wt% PHA was processed using routes A to F, yielding eight samples in the form of pellets (digestion, samples A and B) or films (extraction, samples C to F). Samples C1 and D1 were masticated with $n = 1$, whereas samples B, C5 and D5 were masticated with $n = 5$.

Figure 3 compares the product recovery achieved for all processed biomasses. We can conclude that mastication had a negative impact on the recovery of products obtained by digestion because the recovery for sample B was nearly half of that for dry biomass A. However, because the recovery for sample A was larger than 1, one may suspect that the corresponding pellet was made of a significant amount of non-PHA material. Indeed, pellets containing as much as 20% impurities were recovered recently from biomasses containing 52% of PHA using a digestion protocol where much less biomass was treated in 100 mL of NaOH [28]. The authors further reported that the digestion of fresh biomass resulted in a drop in product recovery, which is in harmony with the recovery achieved using sample B, i.e., a "reconstituted fresh biomass" containing 50 wt% water.

In order to assess whether mastication or handling of wet biomasses was responsible for the drop in product recovery, samples C1 to F were produced using the solvent extraction route. Nearly 20% recovery was achieved for sample C1, which matches the recovery obtained for the nonmasticated wet biomass E and for the masticated biomasses D1 and D5. This suggests that mastication does not impact on solvent extraction from wet biomasses. However, recovery drops below 10% for sample C5, indicating that longer wet biomass mastication in the presence of NaOH is responsible for the loss of downstreamed PHA. The mastication-assisted digestion of the wet biomass allowed downstreaming as much product as the solvent extraction from the dried biomass, as seen by comparing samples B and F in Figure 3. This suggests that for the present MMC biomass containing 50% PHA and for the large amount of biomass processed (between 10 and 20 g in 100 mL $CHCl_3$ or NaOH), a better product recovery rate was achieved via digestion of wet biomasses rather than extraction of a dried biomass. This result is solid, as experiments performed with dried biomasses show that product recovery was doubled when digestion was preferred to solvent extraction with $CHCl_3$ (see samples A and F). This is an important result considering the lower economic and environmental costs of NaOH digestion [36].

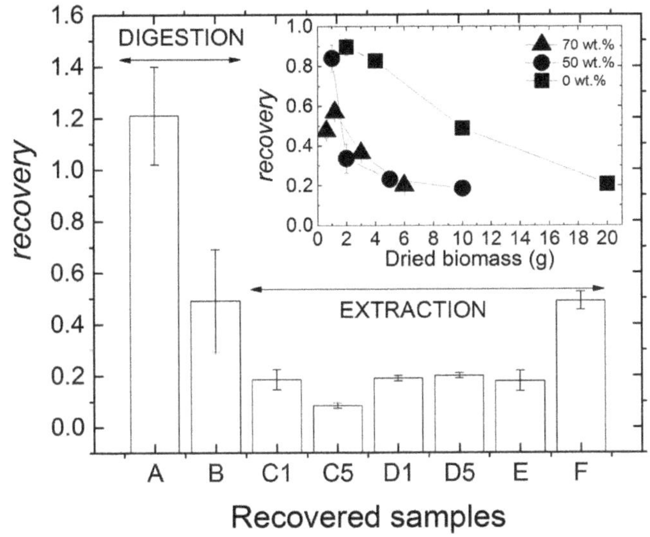

Figure 3. Product recovery achieved for all processed biomasses. Samples produced for benchmarking are: A for digestion with NaOH, and E and F for extraction in $CHCl_3$. Inset: product recovery as a function of the dried biomass weight used for the extraction with $CHCl_3$ of MMC PHA.

The levels of recovery reported in Figure 3 for extraction with $CHCl_3$ are significantly lower than those reported in Table 1. This was expected given the conditions of the benchmark extraction selected here, which target higher biomass concentration in chloroform, lower temperature and shorter extraction times compared to those listed in Table 1. The inset in Figure 3 clarifies this point because it shows that adding more biomass in chloroform is detrimental to the amount of PHA recovered. This limiting factor is usually explained by the increase in the viscosity of the PHA-rich solution in $CHCl_3$ [1]. Using biomass premixed with water had a negative impact on the amount of polymer recovered, which is possibly because water significantly lowers the quality of the solvent mixture (water plus chloroform) with respect to PHA. In addition, the ability of the water–chloroform mixture to make the cells' membranes more fragile to give access to the PHA granules may have been weakened as more water was added to the biomass. However, an important result from the inset in Figure 3 is the saturation in the recovery at nearly 20 % when water was added to the biomass. Thus, when more than 5 g of dry biomass were extracted with 100 mL $CHCl_3$, the water content in the biomass premix was no longer a limiting factor due to the plateauing of the recovery.

As for the mastication itself, though it did not assist the extraction in $CHCl_3$, no negative impacts were shown on the recovery rates reported for biomass sonication, bead milling [21] or microwave assisted extraction [22]. This can be explained by the mechanical work associated with mastication, which is orders of magnitude smaller than the mechanical works involved during ultrasonication or bead milling (see Section 2.1). In spite of mastication not being effective in enhancing product recovery, one may question how the characteristics of recovered films and pellets compare with those of the three benchmarked products A, E and F.

2.3. PHA Films and Pellets Characteristics

The chemical composition of films and pellets recovered after the downstream of the biomasses was assessed with FTIR spectroscopy. The spectra are displayed in Figure 4. All samples exhibited qualitatively similar DRIFT spectra (see Figure 4a) or FTIR spectra (see Figure 4b), which show the symptomatic bands for PHA (see e.g., [41,42] and references therein), namely the functional carbonyl bonding (C=O) at 1720 cm^{-1}, the ethyl (CH_2) and the methyl (CH_3) functional groups between 1458 cm^{-1} and 1380 cm^{-1} (which also show contributions from non-PHA residues such as collagen [20]) and the C-O-C stretching band in the region between 1279 cm^{-1} and 1182 cm^{-1} [43].

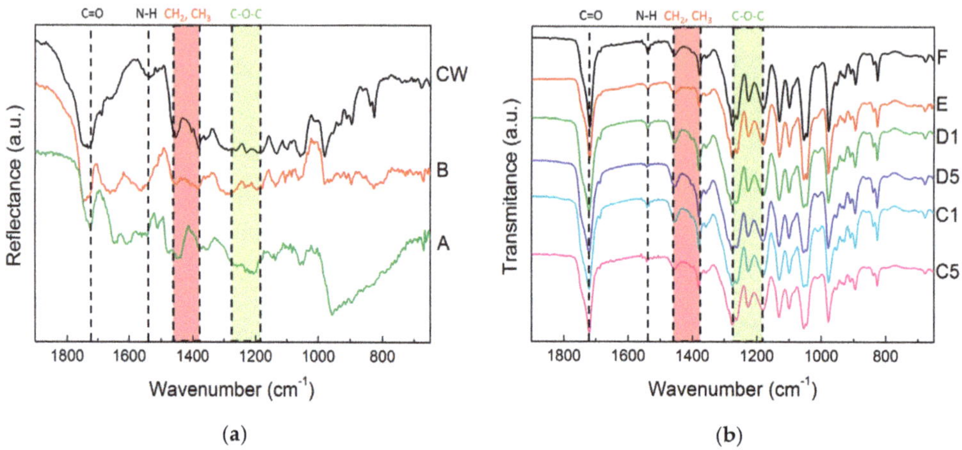

Figure 4. DRIFT spectra of pellets recovered from the digestion with NaOH compared to an MMW PHA produced from cheese whey and digested with NaClO (CW) (**a**) and FTIR spectra of films recovered from extraction in $CHCl_3$ (**b**).

In addition, a band at 1540 cm^{-1} assigned to amide II and found in several MMC PHA biomasses [20,44] showed up in all spectra presented in Figure 4. This band is a clear indication of the presence of non-PHA residues in the recovered films and pellets. This band and the C=O band were indeed used to estimate the PHA content (A_{PHA}) in biomasses, which was then validated by the PHA content measured from the TGA of the same biomasses [44]. Taking on board the approach reported by Matthew Chan et al. [44], the PHA content in the films and pellets can be computed as:

$$A_{PHA} = \frac{2 - logT_{1720}}{4 - logT_{1720} - logT_{1540}} \quad (2)$$

where T_{1720} is the intensity of the C=O band and T_{1540} is the intensity of the amide II band. The A_{PHA} of all films and pellets are reported in Table 2. As expected from the recovery data (see Figure 3), the lowest A_{PHA} values were computed from the DRIFT spectra of samples A and B. Thus, digestion with NaOH produced PHA with significant amounts of residues, which was expected because similar trends were noted for wet biomasses [35] or other biomasses with similar PHA content [26]. The FTIR analysis shows that mastication was effective in reducing the amount of non-PHA residues in the extracted films. This was not simply due to the presence of water in the biomass because sample E does not show a better A_{PHA} than sample F. Pretreating the biomass with NaOH prior to mastication led to a significant drop in the purity of the recovered PHA films. Note here that the low A_{PHA} values are not related to the use of DRIFT to acquire the spectra. Figure 4a shows a DRIFT spectrum of an MMC PHA produced from cheese whey using the same protocol as the one used in the present study [45]. This spectrum (labelled as CW in Figure 4a) is qualitatively similar to the other two DRIFT spectra, and the ratio A_{PHA} computed from the bands' intensities is 0.82. The latter is comparable to the A_{PHA} values computed from the FTIR spectra of samples E and F and is in line with the 89 % purity estimated from the TGA of this biopolyester [45].

Table 2. Samples produced from the downstreaming of corresponding biomasses with parameter n, with their purities quantified by FTIR (A_{PHA}) and TGA (wt%) as well as their melt properties (temperatures T_{deg}, T_g, T_c and T_m in °C, and dynamic shear viscosity $|\eta^*|$ in Pa·s).

Biomass	n	Sample	Purity		Melt Properties				
			A_{PHA}	wt%	T_{deg}	T_g	T_c	T_m	$\mid\eta^*\mid$ [d]
Digestion with NaOH									
A	0	A	0.48 [a]	79.2 ± 1.7	253.8 ± 1.4	-	-	110 [c]/159 [c]	71,560
B	5	B	0.53 [a]	73.1 ± 1.9	241 ± 1	-	-	106 [c]/151 [c]	23,350
Extraction in $CHCl_3$									
C	1	C1	0.94	96.5 ± 0.1	258.0 ± 1.3	−8.5	80.4	110/133	0.9
C	5	C5	0.94	88.9 ± 0.5	238.5 ± 1.6	−14 [b]	74 [b]	100 [b]/132	204
D	1	D1	0.95	97.1 ± 0.3	299.1 ± 0.6	−7.6	50.2	129/156	1241
D	5	D5	0.96	97.1 ± 0.3	301.7 ± 1.1	−7.2	49.2	130/157	1954
D	1	D1	0.95	97.1 ± 0.3	299.1 ± 0.6	−7.6	50.2	129/156	1241
E	0	E	0.87	96.6 ± 1.0	258.4 ± 0.9	−9.6	45.2	124/156	239
F	0	F	0.88	95.6 ± 0.7	306.0 ± 2.5	−9.9	43.5	128/156	2320

[a]: Computed from DRIFT spectra. [b]: Errors are 1, 2 and 5 °C for Tg, Tc and first Tm, respectively, whereas errors are 0.5 °C for all other samples. [c]: Measured from the first heating of the pellets, with an error of 1 °C. [d]: Measured at 170 °C and a frequency of 1 Hz.

The purity of all films and pellets was assessed using TGA. The curves of the selected samples are shown in Figure 5. For all of the film samples produced for benchmarking, the TGA curves show a single degradation process that started at 200 °C, whereas a second process initiated above 250 °C for the pellets (see Figure 5b). Based on the DRIFT information, this second process can be assigned to the rather large amount of non-PHA residues recovered after centrifugation and the washing of the pellets.

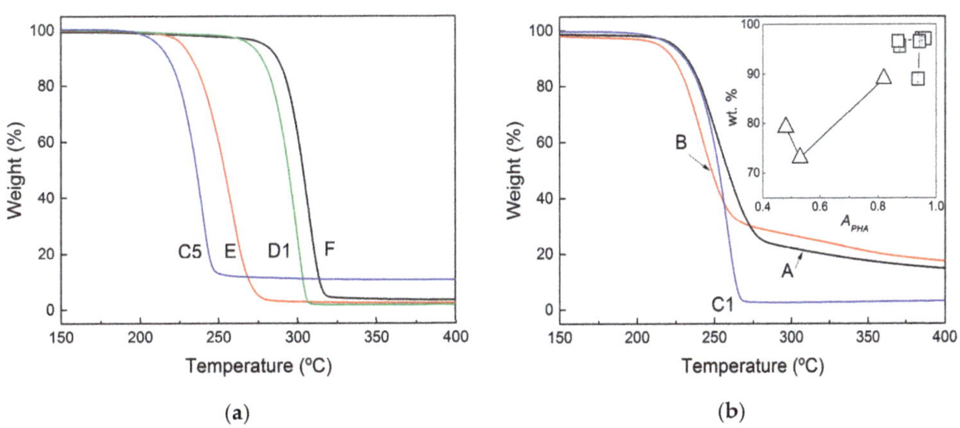

Figure 5. TGA curves of recovered films extracted with chloroform (**a**) and of pellets recovered after biomass digestion with NaOH (**b**). The inset in (**b**) relates the PHA purity inferred from both FTIR (squares) and DRIFT (triangles) analyses to the purity measured in weight percent (wt%) by using TGA.

The purity of pellets and films is indeed inferred from the amount of weighted material (wt%) at the end of the first drop in the TGA curves, which corresponds to PHA thermal degradation. These values are reported in Table 2, whereas the inset in Figure 5b presents a graphical comparison between such purity and the ratios of A_{PHA} computed from both FTIR and DRIFT analyses. The lower purity of the pellets correlates with lower values of

A_{PHA}, but for highly pure PHA samples (films with more than 90 wt% PHA), no correlation was found between A_{PHA} and TGA data. This is in contrast with the linear correlations found for PHA biomasses [44], which suggests that the FTIR methodology is not valid for assessing the purity of PHA materials showing less than 10 wt% purity (films recovered in the present study). This is understandable, as the two methods rely on sensitivities towards different types of impurities: infrared spectrometry detects proteins as residues, in contrast to the residual weight made of other nonvolatile (nonorganic) impurities measured with TGA. Nonetheless, the TGA analysis confirms the information conveyed via both FTIR and DRIFT analyses. The mastication of wet MMC PHA biomasses prior to $CHCl_3$ extraction enhanced the purity of the recovered polymer, whereas the use of NaOH during mastication or for biomass digestion was accompanied by a significant drop in product purity.

The TGA curves can also be used to estimate the thermal stability of the films and pellets by comparing the temperatures T_{deg} at which the rate of weight loss is at its maximum. Temperatures T_{deg} are listed in Table 2. Clearly, biomass digestion with NaOH gave products with bad thermal stability when compared to the benchmark, as shown by how samples A and B exhibited temperatures that were 50 to 60 °C below the T_{deg} measured for sample F, which shows the highest T_{deg}. Masticating with NaOH prior to extraction in $CHCl_3$ did not promote the thermal stability of the recovered products because the T_{deg} for samples C1 and C5 were in the same range as those exhibited by samples A and B. The larger amounts of non-PHA residues in these samples can explain the earlier onset for thermal degradation as reported elsewhere for MMC PHA produced from cheese whey and extracted with different amounts of impurities [45]. Smaller PHA molecular masses resulting from the chain's hydrolysis during NaOH digestion have also been pointed out to explain weaker thermal stability [25] because lower temperatures are needed to generate volatile products from the bond-breaking of smaller PHA chains [46]. In addition, differences in the HV content of downstreamed MMC PHA could be another reason for thermal degradation at lower temperatures due to the lower melting point of HV crystals, which triggers an earlier degradation of HV-rich PHA [47]. As for the solvent extraction, the use of a wet biomass (sample E) impacts negatively on the thermal stability of the extracted PHA film. However, biomass mastication reverts this trend, as samples D1 and D5 show a T_{deg} closer to the temperature measured for sample F.

The thermal properties measured with DSC are illustrated in Figure 6, which displays representative second heating curves. These curves present all thermal transitions reported for PHBV copolymers (see [45,48] and references therein), and the corresponding temperatures are reported in Table 2 for all samples. Starting from the lower temperatures, the glass transition ranged between $T_g = -14$ °C and $T_g = -7$ °C, the crystallization from the amorphous phase formed during the quenching of the melt was observed for all samples except the pellets at temperatures between $T_c = 43$ °C and $T_c = 80$ °C and the melting of the HV and HB crystals (2 distinct peaks) was seen at temperatures ranging from $T_m = 100$ °C to $T_m = 157$ °C.

As expected from an earlier study focused on the impact of impurities on MMC PHBV's thermal properties [45], the presence of non-PHA residues in its films and pellets shifted the glass transition and melting to lower temperatures, whereas crystallization occurred at higher temperatures. Note that the magnitudes of the shifts were consistent with the amount of impurity found in the samples to the point that the DSC curves for the pellets are featureless. In fact, pellet aging is needed for crystallization, and only the curves measured during the first heating of the pellets show the two melting processes (see the inset to Figure 6a). Overall, the temperatures listed in Table 2 suggest that the presence of impurities slows down crystallization and facilitates melting, which is probably because impurities act as defects that weaken the crystalline structure and decrease overall crystallinity [45]. However, it is known that larger HV content in the copolyester leads to a similar effect [49] and that PHBV with shorter chains equally drifts T_m to lower temperatures [50]. As such, one cannot exclude the possibility that digestion and mastication with

NaOH led to the selective recovery of PHBV with smaller molecular masses and containing more HV.

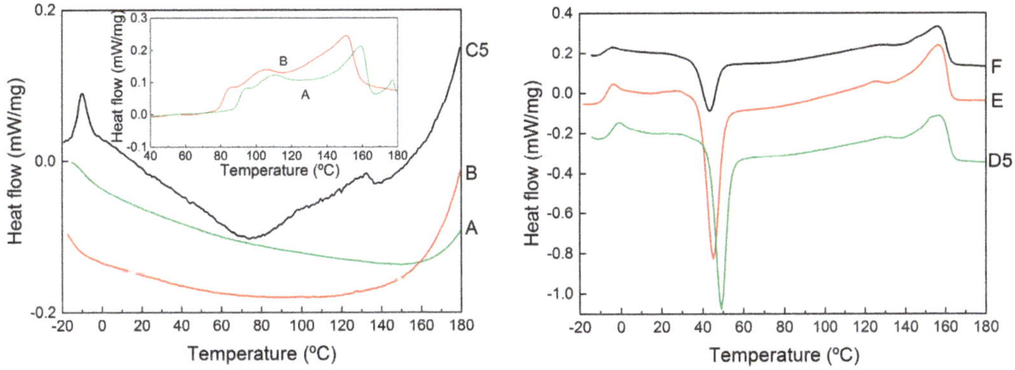

(**a**) (**b**)

Figure 6. DSC curves (measured during the second heating and vertically shifted for better visualization of the results) of samples recovered from the biomass digestion in NaOH (**a**) or extracted with $CHCl_3$ (**b**). The inset in (**a**) shows the DSC curves acquired during the first heating of the pellets.

Based on the data reported in Figure 6, the rheological response of the molten films and pellets was measured at 170 °C. Starting with samples produced for benchmarking, Figure 7 illustrates the thermal stability of the samples extracted in chloroform with no alkali pretreatment. The time dependence of the complex viscosity $|\eta^*|$ measured at 1 Hz is displayed in a semilogarithmic plot to highlight the Arrhenius type of the drop in viscosity at a later time after the initial temperature stabilization in the rheometer oven. The Arrhenius behavior of the viscosity relates to the random chain scission degradation mechanism of PHA [51], which describes a linear decrease over time of the PHA molecular mass. Because $|\eta^*|$ is a power law function of the polymer macromolecular mass, an exponential decrease of $|\eta^*|$ with time was recovered.

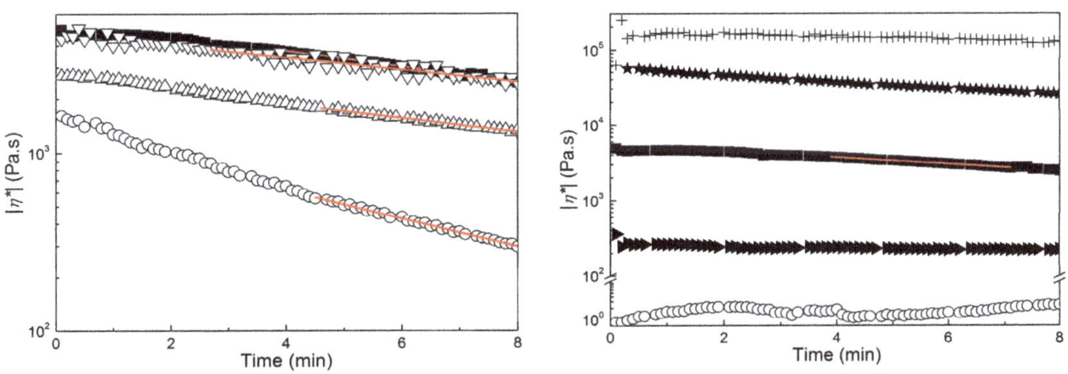

(**a**) (**b**)

Figure 7. Time dependence of the dynamic shear viscosity $|\eta^*|$ measured at 1 Hz and 170 °C. (**a**): Sample F (squares), sample E (circles), sample D1 (up triangles) and sample D5 (down triangles); (**b**): sample F (squares), sample C1 (circles), sample C5 (right triangles), sample B (stars) and sample A (crosses). Lines are Arrhenius fits to the data.

An Arrhenius fit to the data reveals that samples D1 and D5 were slightly more stable than sample F. This was expected because TGA results showed that sample F was recovered with a few more impurities, and these are known to be detrimental to the thermal stability of PHA melts [51,52]. Sample E degraded twice as fast, which is in harmony with the TGA indicating that it started to degrade at a much lower temperature. Thus, degradation at 170 °C was sped up for sample F as compared to other films. Biomass mastication with NaOH before extraction in chloroform allowed for the recovery of a virtually thermally stable product because $|\eta^*|$ was nearly constant for samples C1 and C5, whereas the viscosity of molten pellets A and B decayed at a much slower pace than sample F (see Figure 7b). This confirms that the nature of the impurities contained in the downstreamed products was more important than their amount when considering melt thermal stability [45], as all samples in Figure 7b were produced with much more impurity than sample F. In this respect, it is interesting to correlate the presence of fewer proteins in the impurities of samples D1 and D5 when compared to sample E, which otherwise shows the same level of impurity, with the improved thermal melt stability of D1 and D5.

However, such thermal enhancement comes at a high cost as shown in Figure 8, which presents the mechanical spectra of all products at 170 °C, and in Table 2, which reports the corresponding values of $|\eta^*|$ measured at 1 Hz. All of the mechanical spectra in Figure 8 were corrected for sample degradation using the Arrhenius equation and parameters computed from the fits performed in Figure 7. The mastication of the wet biomass before extraction in $CHCl_3$ had a positive effect on PHA's melt properties. The dynamic shear viscosity at 1 Hz was eight times larger after five passes in the masticating unit (compare $|\eta^*|$ for samples B and C5 in Table 2), and the apparent shear thinning with frequency of the PHA extracted from the dried biomass was recovered.

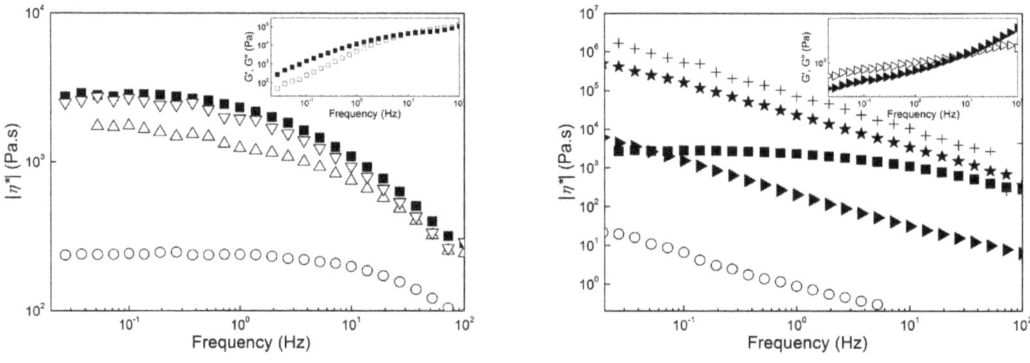

Figure 8. Mechanical spectra (dynamic shear viscosity $|\eta^*|$ as a function of frequency) measured at 170 °C. (**a**): Sample F (squares), sample E (circles), sample D1 (up triangles) and sample D5 (down triangles); (**b**): sample F (squares), sample C1 (circles), sample C5 (left triangles), sample B (stars) and sample A (crosses). Inset in (**a**): mechanical spectrum (storage modulus G′ (solid symbols) and loss modulus G″(empty symbols) as a function of the frequency) of sample F. Inset in (**b**): mechanical spectrum of sample C5.

The good matches between the mechanical spectra of samples F and D5 in Figure 8a suggests that the PHA in both samples showed nearly similar molecular mass distributions. Thus, the rheological data, together with the temperatures measured with DSC and TGA, indicate that samples D1, D5 and F contained essentially the same PHA. The use of NaOH during the downstreaming of PHA had a dramatic effect on their melt rheology because a significant drop in the dynamic shear viscosity accompanied by a change from a Rouse-like to a gel-like dynamic was seen (compare the spectra of samples F and C1 in Figure 8b).

A PHB produced from a biomass digested with NaOH was reported as not suitable for thermoplastic application because its molecular mass was smaller than that of a PHB extracted with chloroform from the same biomass [25]. Results for sample C1 confirm the negative effect of NaOH because the lower melt viscosity can be related to a significant drop in the PHA molecular mass. Sample C5 and the pellets extracted from the biomass digestion in NaOH exhibited a gel-like behavior. This is highlighted in the inset to Figure 8b for sample C5 with a storage modulus G′ larger than the loss modulus G″ at lower frequencies. This behavior contrasts with the terminal fluid-like behavior displayed by sample F in the inset to Figure 8a. To assign the gel-like behavior to the large amount of impurities in samples C1, B and A is straightforward because they could act as reinforcing solids that fill up the melt and, at the same time, impart thermal stability to the product's viscosity. Nonetheless, the rheology cannot discriminate between the reinforcing effect of impurities and a possible additional crosslinking between PHA chains triggered by NaOH, which further imparts an elastic behavior at lower frequencies.

Putting the rheological characteristics presented in Figures 7 and 8 into the perspective of an application such as melt extrusion, samples F, D1 and D5 showed viscoelastic behaviors which resembled those found in commercial PHB and PHBV [7,40,53] but with slightly improved thermal stability for materials produced using mastication. In contrast to this, the use of NaOH during mastication, biomass digestion, or both led to very thermally stable materials that seemed too elastic or not viscous enough to be extruded. Additional rheological testing under large deformation rates or melt extrusion trials should be performed before definitively disqualifying samples C5, B and A for such applications. However, they could be used as additives for PHA, in particular as fillers for melt reinforcement (samples A, B and C5) or as processing aids and plasticizers (samples D1 and C1) that are chemically similar to and thus compatible with other extrusion-grade PHAs.

3. Materials and Methods

3.1. MMC PHA Biomass Production

A biomass containing 50.0 ± 2.4% MMC PHA was produced from fruit waste using mixed microbial cultures in a three-stage pilot plant installed at NOVA.ID facilities. The fruit waste used for PHA production was supplied by a Portuguese juice fruit company, Sumol+Compal S.A., and it was collected from fruit pulp barrels that did not comply with the criteria to be processed by the company. This residue was characterized by having a high total COD content (174.9 ± 13 gCOD·L^{-1}) and essentially being made up of readily biodegradable sugars (84.8% of soluble COD), which is an important factor to increase acidogenic potential and, thus, polymer production. The feedstock was first fermented into PHA bioprecursors in a 60 L up flow anaerobic sludge blanket (UASB) reactor as described by Matos et al. [12]. This reactor was continuously operated using an organic loading rate (OLR) of 28 ± 2 gCOD·L^{-1}·d^{-1}, a pH of 5.4 ± 0.3, a hydraulic retention time (HRT) of 1 d and a temperature of 30.1 ± 0.2 °C. The fermented end-stream produced was subsequently fed to two reactors, and it was mainly comprised of lactate (1.0 ± 0.3 gCOD·L^{-1}), acetate (5.4 ± 0.7 gCOD·L^{-1}), propionate (2.1 ± 0.2 gCOD·L^{-1}), ethanol (0.8 ± 0.3 gCOD·L^{-1}), butyrate (9 ± 1 gCOD·L^{-1}) and valerate (2.7 ± 0.3 gCOD·L^{-1}).

An aerobic 100-L sequential batch reactor (SBR) was assembled to select an MMC with a high PHA storage capacity. The culture was fed with the fermented fruit waste under 12 h feast and famine cycles, as detailed elsewhere [12]. The HRT and sludge retention time (SRT) were set at 1 and 4 days, respectively, and the OLR was maintained at 7.3 gCOD·L^{-1}·d^{-1}. The selective pressure favoring the enrichment of the culture in PHA-storing organisms was ensured by first applying the sequential feast–famine cycles, and second by uncoupling the nitrogen source from the carbon feeding. Semiquantitative fluorescence in situ hybridization (FISH) analyses performed throughout the SBR operation confirmed the enrichment of the community in Paracoccus, which is a well-known bacterial genus with PHA-storing capacity [12].

The PHA-enriched biomasses were finally produced in accumulation assays performed in a 60 L fed-batch reactor inoculated with the MMC selected in the second stage of the process, which was pulse-wise fed with the fermented stream produced in the acidogenic reactor. After reaching maximum PHA accumulation, sulfuric acid was added until pH 2 was reached in order to quench biological activity, the biomass was then centrifuged to remove the maximum water content, and then it was freeze-dried. The process produced 1.5 kg dried biomass with a PHA content of 50.0 ± 2.4 wt%. The polymer contained 75% HB and 25% HV, as reported elsewhere [12].

3.2. Biomass Mastication

A modified masticating unit (model HR1886, Philips, Eindhoven, Netherlands) was used to process the biomass. The unit consisted of a squeezing chamber made of a conical screw (minimum and maximum external diameters D of 34.5 mm and 44 mm, respectively) embedded in a conical barrel with filter (with a gap *h* between screw and barrel varying from 4.5 to 2 mm from feeding to screw tip) and terminated by a pulp outlet. The unit operated at a constant screw speed of 80 rpm. A syringe pump (model NE-300 "Just Infusion"TM, from New Era Pump Systems Inc.) operating at 10 mL per minute was used to feed the biomass in the unit. The dried MMC PHA biomass was premixed with either a fixed amount of distilled water (different premixes ranging from 30 wt% to 70 wt% biomass were tested) or with 2 M NaOH (to reach a premix of 50 wt%) before filling in a 100 mL syringe. The masticated biomass was either collected at the pulp outlet and fed again to the syringe pump for successive passes in the unit or directly used for PHA extraction and digestion. The residence time of the biomass in the masticating unit was between 2 and 3 min for the set feeding conditions and the biomass premix containing 50 wt% distilled water.

3.3. PHA Extraction and Digestion

For NaOH digestion, a protocol optimized elsewhere for biomasses containing different amounts (from 35 to 70%) of poly(3-hydroxybutyrate-co-3-hydroxyvalerate) (PHBV) with HV content varying from 18% to 31% [28] was employed with a modification. Instead of 2 g of biomass (dried, lyophilized or fresh [28]), 20 g of biomass (dried or masticated) was mixed under agitation with 100 mL of 0.3 M NaOH at 30 °C for 4.8 h. The mixture was then centrifuged (11,000 rpm, 10 min), and the recovered PHA-rich solid was washed several times with distilled water. The PHA-rich solid was then dried overnight at 60 °C in a ventilated oven (model FD 115, Binder GmbH, Tuttlingen, Germany), weighed to compute PHA recovery and then stored in black plastic bags until PHA characterization. Figure 2 presents a scheme summarizing the workflows used for both extraction and digestion. For PHA extraction, dried biomass (sample F), wet biomass (sample E) or the masticated biomass (with 2 M NaOH, samples C1 and C5, or with water, samples D1 and D5) were mixed with 100 mL of $CHCl_3$ under agitation at 30 °C for 2 h. The PHA-rich solution was then separated from the solid biomass residues via filtration using a metal screen (1 micron). The filtrate was poured into a Petri dish and dried in a ventilated oven at 60 °C overnight. The resulting PHA films were weighed to compute PHA recovery and then stored in black plastic bags until PHA characterization.

The product recovery, *recovery*, is defined by the following ratio:

$$recovery = \frac{mPRODUCT}{mBIOMASS \times 0.5} \quad (3)$$

where *mPRODUCT* is the weight of film or pellet recovered from the downstreaming of a dried biomass with a weight of *mBIOMASS* and initial PHA content of 50.0 ± 2.4 wt%.

3.4. Characterization of the Recovered Products

A TGA Q500 apparatus (from TA Instruments New Castle, DE, USA) was used to perform the thermogravimetric analysis (TGA) of extracted PHA films and digested PHA pellets. Typically, around 10 mg of film or pellets were loaded in the sample pan and were heated from 40 °C to 600 °C at a rate of 10 °C/min using nitrogen as a purge flow at a flow rate of 200 mL/min. TGA curves were acquired in triplicate. Differential scanning calorimetry (DSC) was performed with a 200 F3 Maia® instrument (from Netzsch- GmbH, Kelb, Germany). The samples were placed in a closed aluminum pan and subjected to three scans. A first heating (25–180 °C at 10 °C/min) was performed to erase the thermal history of the sample. Then, a cooling ramp was conducted at a rate of −10 °C/min down to −25 °C, where the sample was maintained for 5 min. Finally, a second heating was applied under the same conditions as the first heating. The second heating was used to determine all DSC output gathered in Table 2. All DSC experiments were done under nitrogen purging. The FTIR spectra of film samples and powdered pellets were acquired with an FTIR Spectrometer (Spectrum100, PerkinElmer Inc., Waltham, MA, USA) equipped with an attenuated total reflectance (ATR) accessory for measuring films and a diffuse reflectance (DRIFT) accessory for measuring the pellets. For each sample, 32 scans were recorded between 4000 and 400 cm^{-1} and averaged to give spectra with a resolution of 1 cm^{-1}.

3.5. Rheology

A stress-controlled rotational rheometer (ARG2, TA Instruments, New Castle, DE, USA) was used with 25 mm parallel plates and a gap varying between 0.55 and 1.02 mm for the characterization of biomass premixes in water (slurries). All tests were performed at room temperature. Right after loading a slurry between the plates of the rheometer, a dynamic time sweep using a small amplitude oscillatory shear (SAOS) with a frequency of 1 Hz and an amplitude of 0.05% was performed for 5 min to monitor the structural recovery of the slurry after loading. A stress sweep in oscillatory mode at 1 Hz was then performed from 100 to 5000 Pa to determine the regime of linear viscoelasticity (LVE). In a second set of experiments, another sample was loaded, and the same dynamic time sweep was performed as above. This was followed by a frequency sweep from 100 Hz down to 0.01 Hz performed with a SAOS of 0.05%, which belongs to the LVE regime determined in the previous experiment. Finally, a third experiment was carried out with a third sample. After performing the same dynamic time sweep, a flow curve was measured by ramping the steady shear stress from 10 Pa to 5000 Pa, applying each stress during 10 s and reading the shear rate during the last two seconds of the applied stress to compute an average shear rate. The experimental protocol used for characterizing the PHA films and pellets at 170 °C with 40 mm parallel plates encompassed a dynamic time sweep (at 1 Hz with a SAOS varying from 0.1% to 10% depending on the LVE regime of the tested sample) applied for 8 min to monitor the thermal degradation of the film or pellets after loading between the preheated plates. Then, a frequency sweep was performed from 100 to 0.03 Hz with a SAOS varying between 0.1% and 10% depending on the PHA sample. The LVE regime was ensured by monitoring the sinusoidal traces of both dynamic stress and strain recorded online by the rheometer. The frequency sweep was repeated several times to further probe the thermal degradation of the samples.

4. Conclusions

The masticating device used in the present study was not effective in assisting the digestion of a wet mixed microbial biomass in NaOH for the recovery of more PHA. The recovered products contain significant amounts of impurities and thus may only find application as rheological modifiers or thermal stabilizers for PHA. In spite of the too-small improvement in the recovery of pure PHA with $CHCl_3$, the new tested continuous mechanical pretreatment is attractive when large amounts of wet biomass are processed. Indeed, the product recovered with mastication shows better melt stability and melt viscoelasticity. There is still ample room for exploring routes to improve the mastication of wet PHA biomasses in future studies. Mastication parameters and unit designs can be changed to increase mechanical energy and allow the processing of biomasses containing more water. It will be of interest to check whether mastication can boost both PHA recovery and thermal properties for biomasses with higher PHA content. Alternatively, testing biomasses with lower PHA content and higher water content with a modified prototype enabling biomass dewatering before mastication will be challenging, as such feeds are known to be very difficult to downstream. Finally, future studies should also screen biomasses containing other types of PHA, e.g., PHBV with different HV contents, and extend to economically and ecologically attracting solvents to confirm the results achieved here with $CHCl_3$.

Author Contributions: Conceptualization, L.H.; validation, H.K.S.S., M.M. and L.H.; formal analysis, H.K.S.S., M.M. and L.H.; investigation, H.K.S.S., M.M. and L.H.; resources, L.H., J.A.C. and M.A.M.R.; data curation, H.K.S.S., M.M. and L.H.; writing—original draft preparation, H.K.S.S. and L.H.; writing—review and editing, H.K.S.S., M.M., L.H., J.A.C. and M.A.M.R.; visualization, H.K.S.S. and L.H.; supervision, L.H., J.A.C. and M.A.M.R.; project administration, L.H. and J.A.C.; funding acquisition, L.H. All authors have read and agreed to the published version of the manuscript.

Funding: This research was funded by the Portuguese Foundation for Science and Technology (FCT), grant number PTDC/BII-BIO/5626/2020. Additional funding from the Portuguese Foundation for Science and Technology (FCT) in the framework of the Strategic Funding of the IPC (projects UIDB/05256/2020 and UIDP/05256/2020) and contract CEECINST/00156/2018 is also acknowledged.

Institutional Review Board Statement: Not applicable.

Informed Consent Statement: Not applicable.

Data Availability Statement: Data is contained within the article.

Conflicts of Interest: The authors declare no conflict of interest.

Sample Availability: Samples are no more available.

References

1. Kunasundari, B.; Sudesh, K. Isolation and recovery of microbial polyhydroxyalkanoates. *Express Polym. Lett.* **2011**, *5*, 620–634. [CrossRef]
2. Biodegradable Polymers in Various Environments According to Established Standards and Certification Schemes: Renewable Carbon Publications, November 2021. Available online: https://renewable-carbon.eu/publications/product/biodegradable-polymers-in-various-environments-according-to-established-standards-and-certification-schemes-graphic-pdf/ (accessed on 19 October 2022).
3. Baidurah, S.; Kobayashi, T. Polyhydroxyalkanoates: Production and Biodegradation—A Review. *Encycl. Mater. Plast. Polym.* **2022**, *2*, 530–540. [CrossRef]
4. Bioplastics Facts and Figure. European Bioplastics e.V. (EUBP). Available online: https://docs.european-bioplastics.org/publications/EUBP_Facts_and_figures.pdf (accessed on 22 October 2022).
5. *Global Plastics Outlook: Economic Drivers, Environmental Impacts and Policy Options*; OECD Publishing: Paris, France, 2022. [CrossRef]
6. Plackett, D.; Siró, I. Polyhydroxyalkanoates (PHAs) for food packaging. In *Multifunctional and Nanoreinforced Polymers for Food Packaging*; Lagaron, J.-M., Ed.; Woodhead Publishing Limited: Cambridge, UK, 2011; pp. 498–526.
7. Cunha, M.; Berthet, M.-A.; Pereira, R.; Covas, J.A.; Vicente, A.A.; Hilliou, L. Development of polyhydroxyalkanoate/beer spent grain fibers composites for film blowing applications. *Polym. Compos.* **2014**, *36*, 1859–1865. [CrossRef]
8. Hilliou, L.; Covas, J.A. Production and Processing of Polymer-Based Nanocomposites. In *Nanomaterials for Food Packaging*; Cerqueira, M.A., Lagaron, J.-M., Pastrana, L.M., Vicente, A.A., Eds.; Elsevier: Amsterdam, The Netherlands, 2018; pp. 111–146.
9. Teixeira, P.F.; Covas, J.A.; Suarez, M.J.; Angulo, I.; Hilliou, L. Film Blowing of PHB-Based Systems for Home Compostable Food Packaging. *Int. Polym. Process* **2020**, *35*, 440–447. [CrossRef]
10. Kumar, M.; Rathour, R.; Singh, R.; Sun, Y.; Pandey, A.; Gnansounou, E.; Lin, K.-Y.A.; Tsang, D.C.; Thakur, I.S. Bacterial polyhydroxyalkanoates: Opportunities, challenges, and prospects. *J. Clean. Prod.* **2020**, *263*, 121500. [CrossRef]
11. Samorì, C.; Abbondanzi, F.; Galletti, P.; Giorgini, L.; Mazzocchetti, L.; Torri, C.; Tagliavini, E. Extraction of polyhydroxyalkanoates from mixed microbial cultures: Impact on polymer quality and recovery. *Bioresour. Technol.* **2015**, *189*, 195–202. [CrossRef]
12. Matos, M.; Cruz, R.A.P.; Cardoso, P.; Silva, F.; Freitas, E.B.; Carvalho, G.; Reis, M.A.M. Combined Strategies to Boost Polyhydroxyalkanoate Production from Fruit Waste in a Three-Stage Pilot Plant. *ACS Sustain. Chem. Eng.* **2021**, *9*, 8270–8279. [CrossRef]

13. Fernández-Dacosta, C.; Posada, J.A.; Kleerebezem, R.; Cuellar, M.C.; Ramirez, A. Microbial community-based polyhydroxyalkanoates (PHAs) production from wastewater: Techno-economic analysis and ex-ante environmental assessment. *Bioresour. Technol.* **2015**, *185*, 368–377. [CrossRef]
14. Kurian, N.S.; Das, B. Comparative analysis of various extraction processes based on economy, eco-friendly, purity and recovery of polyhydroxyalkanoate: A review. *Int. J. Biol. Macromol.* **2021**, *183*, 1881–1890. [CrossRef]
15. Pagliano, G.; Galletti, P.; Samorì, C.; Zaghini, A.; Torri, C. Recovery of Polyhydroxyalkanoates From Single and Mixed Microbial Cultures: A Review. *Front. Bioeng. Biotechnol.* **2021**, *9*, 624021. [CrossRef]
16. Pérez-Rivero, C.; López-Gómez, J.P.; Roy, I. A sustainable approach for the downstream processing of bacterial polyhydroxyalkanoates: State-of-the-art and latest developments. *Biochem. Eng. J.* **2019**, *150*, 107283. [CrossRef]
17. Mannina, G.; Presti, D.; Montiel-Jarillo, G.; Carrera, J.; Suárez-Ojeda, M.E. Recovery of polyhydroxyalkanoates (PHAs) from wastewater: A review. *Bioresour. Technol.* **2019**, *297*, 122478. [CrossRef]
18. Koller, M.; Niebelschütz, H.; Braunegg, G. Strategies for recovery and purification of poly[(*R*)-3-hydroxyalkanoates] (PHA) biopolyesters from surrounding biomass. *Eng. Life Sci.* **2013**, *13*, 549–562. [CrossRef]
19. Byrom, D. Polyhydroxyalkanoates. In *Plastics from Microbes: Microbial Synthesis of Polymers and Polymer Precursors*; Mobley, D.P., Ed.; Carl Hanser Verlag: Munich, Germany, 1994; p. 5.
20. Patel, M.; Gapes, D.J.; Newman, R.H.; Dare, P.H. Physico-chemical properties of polyhydroxyalkanoate produced by mixed-culture nitrogen-fixing bacteria. *Appl. Microbiol. Biotechnol.* **2009**, *82*, 545–555. [CrossRef]
21. Palmieri, S.; Tittarelli, F.; Sabbatini, S.; Cespi, M.; Bonacucina, G.; Eusebi, A.; Fatone, F.; Stipa, P. Effects of different pre-treatments on the properties of polyhydroxyalkanoates extracted from sidestreams of a municipal wastewater treatment plant. *Sci. Total Environ.* **2021**, *801*, 149633. [CrossRef]
22. Bocaz-Beltrán, J.; Rocha, S.; Pinto-Ibieta, F.; Ciudad, G.; Cea, M. Novel alternative recovery of polyhydroxyalkanoates from mixed microbial cultures using microwave-assisted extraction. *J. Chem. Technol. Biotechnol.* **2021**, *96*, 2596–2603. [CrossRef]
23. Choi, J.-I.; Lee, S.Y. Efficient and economical recovery of poly(3-hydroxybutyrate) from recombinantEscherichia coli by simple digestion with chemicals. *Biotechnol. Bioeng.* **1999**, *62*, 546–553. [CrossRef]
24. Tamer, I.M.; Moo-Young, M.; Chisti, Y. Optimization of poly(β-hydroxybutyric acid) recovery from Alcaligenes latus: Combined mechanical and chemical treatments. *Bioprocess Eng.* **1998**, *19*, 459–468. [CrossRef]
25. Jiang, Y.; Mikova, G.; Kleerebezem, R.; van der Wielen, L.A.; Cuellar, M.C. Feasibility study of an alkaline-based chemical treatment for the purification of polyhydroxybutyrate produced by a mixed enriched culture. *AMB Express* **2015**, *5*, 5. [CrossRef]
26. Villano, M.; Valentino, F.; Barbetta, A.; Martino, L.; Scandola, M.; Majone, M. Polyhydroxyalkanoates production with mixed microbial cultures: From culture selection to polymer recovery in a high-rate continuous process. *New Biotechnol.* **2014**, *31*, 289–296. [CrossRef]
27. Burniol-Figols, A.; Skiadas, I.V.; Daugaard, A.E.; Gavala, H.N. Polyhydroxyalkanoate (PHA) purification through dilute aqueous ammonia digestion at elevated temperatures. *J. Chem. Technol. Biotechnol.* **2020**, *95*, 1519–1532. [CrossRef]
28. Rodrigues, A.M.; Franca, R.D.G.; Dionísio, M.; Sevrin, C.; Grandfils, C.; Reis, M.A.M.; Lourenço, N.D. Polyhydroxyalkanoates from a Mixed Microbial Culture: Extraction Optimization and Polymer Characterization. *Polymers* **2022**, *14*, 2155. [CrossRef] [PubMed]
29. Meneses, L.; Esmail, A.; Matos, M.; Sevrin, C.; Grandfils, C.; Barreiros, S.; Reis, M.A.M.; Freitas, F.; Paiva, A. Subcritical Water as a Pre-Treatment of Mixed Microbial Biomass for the Extraction of Polyhydroxyalkanoates. *Bioengineering* **2022**, *9*, 302. [CrossRef] [PubMed]
30. Colombo, B.; Pereira, J.; Martins, M.; Torres-Acosta, M.A.; Dias, A.C.; Lemos, P.C.; Ventura, S.P.; Eisele, G.; Alekseeva, A.; Adani, F.; et al. Recovering PHA from mixed microbial biomass: Using non-ionic surfactants as a pretreatment step. *Sep. Purif. Technol.* **2020**, *253*, 117521. [CrossRef]
31. Lorini, L.; Martinelli, A.; Pavan, P.; Majone, M.; Valentino, F. Downstream processing and characterization of polyhydroxyalkanoates (PHAs) produced by mixed microbial culture (MMC) and organic urban waste as substrate. *Biomass Convers. Biorefin.* **2020**, *11*, 693–703. [CrossRef]
32. Elhami, V.; van de Beek, N.; Wang, L.; Picken, S.J.; Tamis, J.; Sousa, J.A.B.; Hempenius, M.A.; Schuur, B. Extraction of low molecular weight polyhydroxyalkanoates from mixed microbial cultures using bio-based solvents. *Sep. Purif. Technol.* **2022**, *299*, 121773. [CrossRef]
33. Mongili, B.; Azim, A.A.; Garofalo, S.F.; Batuecas, E.; Re, A.; Bocchini, S.; Fino, D. Novel insights in dimethyl carbonate-based extraction of polyhydroxybutyrate (PHB). *Biotechnol. Biofuels* **2021**, *14*, 13. [CrossRef]
34. de Souza Reis, G.A.; Michels, M.H.A.; Fajardo, G.L.; Lamot, I.; De Best, J.H. Optimization of Green Extraction and Purification of PHA Produced by Mixed Microbial Cultures from Sludge. *Water* **2020**, *12*, 1185. [CrossRef]
35. Alfano, S.; Lorini, L.; Majone, M.; Sciubba, F.; Valentino, F.; Martinelli, A. Ethylic Esters as Green Solvents for the Extraction of Intracellular Polyhydroxyalkanoates Produced by Mixed Microbial Culture. *Polymers* **2021**, *13*, 2789. [CrossRef]
36. del Oso, M.S.; Mauricio-Iglesias, M.; Hospido, A. Evaluation and optimization of the environmental performance of PHA downstream processing. *Chem. Eng. J.* **2021**, *412*, 127687. [CrossRef]
37. Herschel, W.H.; Bulkley, R. Konsistenzmessungen von Gummi-Benzollösungen. *Kolloid Z.* **1926**, *39*, 291–300. [CrossRef]
38. Kohlbruger, K. *Co-Rotating Twin Screw Extruders*; Carl Hanser Verlag: Munich, Germany, 2007.

39. Galindo-Rosales, F.J.; Moldenaers, P.; Vermant, J. Assessment of the Dispersion Quality in Polymer Nanocomposites by Rheological Methods. *Macromol. Mater. Eng.* **2011**, *296*, 331–340. [CrossRef]
40. Teixeira, P.; Covas, J.; Hilliou, L. In-Line Rheo-Optical Investigation of the Dispersion of Organoclay in a Polymer Matrix during Twin-Screw Compounding. *Polymers* **2021**, *13*, 2128. [CrossRef]
41. Roja, K.; Sudhakar, D.R.; Anto, S.; Mathimani, T. Extraction and characterization of polyhydroxyalkanoates from marine green alga and cyanobacteria. *Biocatal. Agric. Biotechnol.* **2019**, *22*, 101358. [CrossRef]
42. Kansiz, M.; Billman-Jacobe, H.; McNaughton, D. Quantitative Determination of the Biodegradable Polymer Poly(β-hydroxybutyrate) in a Recombinant *Escherichia coli* Strain by Use of Mid-Infrared Spectroscopy and Multivariative Statistics. *Appl. Environ. Microbiol.* **2000**, *66*, 3415–3420. [CrossRef]
43. Sato, H.; Murakami, R.; Padermshoke, A.; Hirose, F.; Senda, K.; Noda, I.; Ozaki, Y. Infrared Spectroscopy Studies of CH···O Hydrogen Bondings and Thermal Behavior of Biodegradable Poly(hydroxyalkanoate). *Macromolecules* **2004**, *37*, 7203–7213. [CrossRef]
44. Chan, C.M.; Johansson, P.; Magnusson, P.; Vandi, L.-J.; Arcos-Hernandez, M.; Halley, P.; Laycock, B.; Pratt, S.; Werker, A. Mixed culture polyhydroxyalkanoate-rich biomass assessment and quality control using thermogravimetric measurement methods. *Polym. Degrad. Stab.* **2017**, *144*, 110–120. [CrossRef]
45. Hilliou, L.; Machado, D.; Oliveira, C.S.S.; Gouveia, A.R.; Reis, M.A.M.; Campanari, S.; Villano, M.; Majone, M. Impact of fermentation residues on the thermal, structural, and rheological properties of polyhydroxy(butyrate-*co*-valerate) produced from cheese whey and olive oil mill wastewater. *J. Appl. Polym. Sci.* **2015**, *133*, 42818. [CrossRef]
46. Larsson, M.; Markbo, O.; Jannasch, P. Melt processability and thermomechanical properties of blends based on polyhydroxyalkanoates and poly(butylene adipate-co-terephthalate). *RSC Adv.* **2016**, *6*, 44354–44363. [CrossRef]
47. Zhila, N.O.; Sapozhnikova, K.Y.; Kiselev, E.G.; Nemtsev, I.V.; Lukyanenko, A.V.; Shishatskaya, E.I.; Volova, T.G. Biosynthesis and Properties of a P(3HB-*co*-3HV-*co*-4HV) Produced by *Cupriavidus necator* B-10646. *Polymers* **2022**, *14*, 4226. [CrossRef]
48. Laycock, B.; Halley, P.; Pratt, S.; Werker, A.; Lant, P. The chemomechanical properties of microbial polyhydroxyalkanoates. *Prog. Polym. Sci.* **2013**, *38*, 536–583. [CrossRef]
49. Orts, W.J.; VanderHart, D.L.; Bluhm, T.L.; Marchessault, R.H. Cocrystallization in random copolymers of poly(β-hydroxybutyrate-*co*-β-hydroxyvalerate) and its effect on crystalline morphology. *Can. J. Chem.* **1995**, *73*, 2094–2100. [CrossRef]
50. Organ, S. Variation in melting point with molecular weight for hydroxybutyrate/hydroxyvalerate copolymers. *Polymer* **1993**, *34*, 2175–2179. [CrossRef]
51. Kim, K.J.; Doi, Y.; Abe, H. Effects of residual metal compounds and chain-end structure on thermal degradation of poly(3-hydroxybutyric acid). *Polym. Degrad. Stab.* **2006**, *91*, 769–777. [CrossRef]
52. Kopinke, F.-D.; Remmler, M.; Mackenzie, K. Thermal decomposition of biodegradable polyesters—I: Poly(β-hydroxybutyric acid). *Polym. Degrad. Stab.* **1996**, *52*, 25–38. [CrossRef]
53. Cunha, M.; Fernandes, B.; Covas, J.A.; Vicente, A.A.; Hilliou, L. Film blowing of PHBV blends and PHBV-based multilayers for the production of biodegradable packages. *J. Appl. Polym. Sci.* **2015**, *133*, 42165. [CrossRef]

molecules

MDPI

Review

Synthesis and Applications of Hybrid Polymer Networks Based on Renewable Natural Macromolecules

Dariya Getya [1,2] and Ivan Gitsov [1,2,3,*]

1 Department of Chemistry, State University of New York—ESF, Syracuse, NY 13210, USA; dgetya@syr.edu
2 The Michael M. Szwarc Polymer Research Institute, Syracuse, NY 13210, USA
3 The BioInspired Institute, Syracuse University, Syracuse, NY 13244, USA
* Correspondence: igivanov@syr.edu; Tel.: +1-315-470-6860

Abstract: Macromolecules obtained from renewable natural sources are gaining increasing attention as components for a vast variety of sustainable polymer-based materials. Natural raw materials can facilitate continuous-flow production due to their year-round availability and short replenishment period. They also open new opportunities for chemists and biologists to design and create "bioreplacement" and "bioadvantaged" polymers, where complex structures produced by nature are being modified, upgraded, and utilized to create novel materials. Bio-based macromonomers are expected not only to compete with but to replace some petroleum-based analogs, as well. The development of novel sustainable materials is an ongoing and very dynamic process. There are multiple strategies for transforming natural macromolecules into sophisticated value-added products. Some methods include chemical modification of macromolecules, while others include blending several components into one new system. One of the most promising approaches for incorporating renewable macromolecules into new products is the synthesis of hybrid networks based on one or more natural components. Each one has unique characteristics, so its incorporation into a network brings new sustainable materials with properties that can be tuned according to their end-use. This article reviews the current state-of-the-art and future potential of renewable natural macromolecules as sustainable building blocks for the synthesis and use of hybrid polymer networks. The most recent advancements and applications that involve polymers, such as cellulose, chitin, alginic acid, gellan gum, lignin, and their derivatives, are discussed.

Keywords: sustainable; renewable; natural macromolecules; hybrid networks; cellulose; chitin; alginic acid; gellan gum; lignin

1. Introduction

Humanity has been using natural polymers since day one of civilization. Cotton plant cultivation (~90% cellulose, a polysaccharide) has been verifiably practiced for 7800 years, with 6000-year-old cotton fabrics found during archaeological excavations in Peru [1]. Wool (a protein mixture with low lipid content) appeared as clothing material around the same time [2]. Wood (a composite of natural macromolecules cellulose, chemi-cellulose, and lignin) was used to make tools and provide shelter. For millennia, these natural materials have been exploited to cover various needs, and naturally, with the increase in population and the need for more reliable and stronger tools, humans started to explore ways of modifying and improving what was easily available from nature. The onset of the deliberate modification of those renewable natural polymers and macromolecules can be traced to the mid-19th century with the invention of cellulose nitrate, made by treating cotton with nitric and sulfuric acids [3]. Regrettably, the first envisioned use of the new material was for military purposes as "Schiesswolle" (gun cotton) [4]. Further adjustments to the method yielded a new material called "celluloid" (CLD), a colloidal mixture of nitrocellulose and camphor. This invention is mostly associated with the name of John

Wesley Hyatt, who introduced it for numerous everyday uses, from hair combs (Figure 1A) to billiard balls (Figure 1B), collars, cuffs, and shirtfronts [5]. CLD was also the mainstay of the film industry until the 1920s, when it was replaced with cellulose acetate (CA) as the film base. First prepared in 1865, CA turned out to be a commercial success and played an important role in material manufacturing for cinematographic and photographic films, frames for eyeglasses (Figure 1C), packaging containers, and more. CA was also combined with other polymers. An example is shown in Figure 1D.

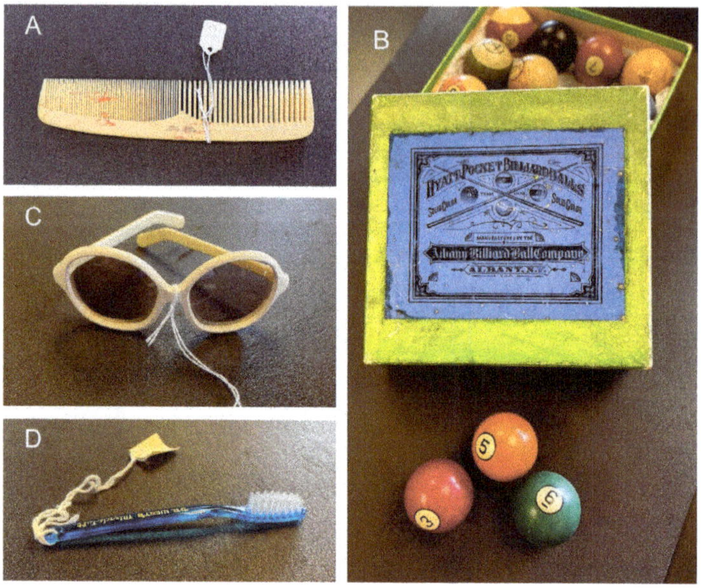

Figure 1. (**A**) Dupont comb, cellulose nitrate; (**B**) Billiard balls in original packaging, cellulose nitrate-based composite; (**C**) Sunglasses, cellulose acetate; (**D**) Dr. West's miracle-tuft toothbrush, cellulose acetate and nylon. Plastics Artifact Collection, Special Collections Research Center, Syracuse University Libraries.

The petrochemical revolution in the 20th century introduced new precursors for polymers that led to the creation of a wide variety of oil-based materials. By the year 1930, practically every branch of human activity took advantage of the remarkably versatile properties and longevity of synthetic polymers [6]. However, with the gradual depletion of nonrenewable resources and environmental concerns caused by increased non-biodegradable waste, the research focus turned again to renewable biomass resources for the development of hybrid materials [7,8]. These raw materials are beneficial due to the short period of their replenishment cycle, which leads to the continuous flow of production supplies. Renewable feedstock resources include a wide variety of materials, such as carbon fibers, clay, plant-based oils, lignin, vanillin, tannin-based materials, and even pineapple leaf fibers [9]. Hybrid materials combine natural and synthetic macromolecules into one system, making up a rapidly emerging structural class in material science [10]. These systems possess properties superior to the sum of their components. One particularly intriguing group of hybrid materials is physically or chemically cross-linked gels and hydrogels. They have been extensively studied and used in diverse applications [11]. Many of them contain natural polymers with functional groups along their backbone suitable for multi-molecular assembly. While the combinations of natural and wholly synthetic network components are numerous, this review will focus only on those that involve polysaccharides (cellulose, chitin, alginic acid, gellan gum) and polyphenols (lignin).

2. Networks: Classification and Applications

Networks are complex systems classified into three main groups: "entanglement networks" (Figure 2A), chemically cross-linked networks (Figure 2B), and physical gels (Figure 2C). In the case of carbohydrates, entanglement networks are also often considered physical gels since, when dissolved, they actively participate in various interactions, such as hydrogen bonding and static interactions. This process is commonly called gelation and is described for various biopolymers, such as gelatin [12], gellan gum [13], alginic acid [14], and starch [15]. Chemically cross-linked networks are multi-molecular systems where polymer chains are covalently bound via a cross-linker [16]. In physical gels, cross-links do not form through chemical bonds on the polymers' backbone but occur by so-called "junction zones"—a lateral aggregation of chains [17].

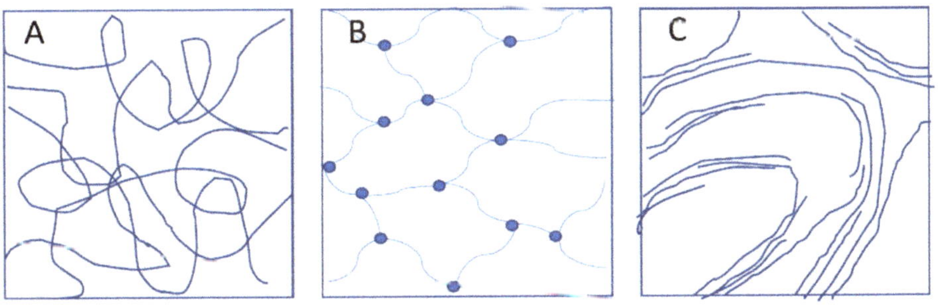

Figure 2. Classification of networks. (**A**) Entanglement networks; (**B**) Covalently cross-linked networks; (**C**) Physical gels with multiple chain junction zones.

Gels based on biopolymers are typically hydrophilic and, therefore, can bind and retain large quantities of water without dissolving. Such gels are called hydrogels, and in general, they consist of cross-linked polymers with 3-dimensional (3D) network structures. It is known that the cross-linking method affects the morphology of hydrogels and their physicochemical properties [18]. Physical hydrogels are reversible and tend to respond to stimuli rapidly, but they are often fragile and sensitive to the chemical environment. Covalently cross-linked hydrogels, on the other hand, are solid and stable, but they are more rigid and cannot rapidly respond to stimuli due to their highly cross-linked structure. Therefore, the type and conditions of synthesis must match the envisioned application of the resulting material [19]. Networks can be synthesized via various strategies and methods. When utilizing a natural macromolecule, one must take into consideration its solubility, reactivity, and processability. If any of these properties are not suitable for the method chosen, the macromolecule must undergo a chemical modification or physical treatment. The deliberate combination of the above-described cross-linked systems produces new types of polymer gels labeled as semi-interpenetrating polymer networks (SIPNs) and interpenetrating polymer networks (IPNs) [20]. In SIPNs, the first polymer is cross-linked (blue dots in Figure 3A), and the second polymer (doted lines in Figure 3A) is linear and non-crosslinked; therefore, it can be extracted and washed off [21,22]. Within the IPN, one network consists of chains of polymer 1 cross-linked with cross-link 1 (blue dots in Figure 3B). Separately, chain 2 is cross-linked with cross-link 2, shown as black dots in Figure 3B. An important characteristic of the IPN is the absence of a covalent bond between the two distinct networks [23]. Depending on the type of polymers that form the IPN or SIPN, their applications vary and range from industrial and military [24,25] uses (sound and vibrational damping materials [26,27], shape-memory foams [28]) to biomedical uses [29], such as tissue engineering [30,31], drug delivery [32], wound dressing [33], and even dentistry [34].

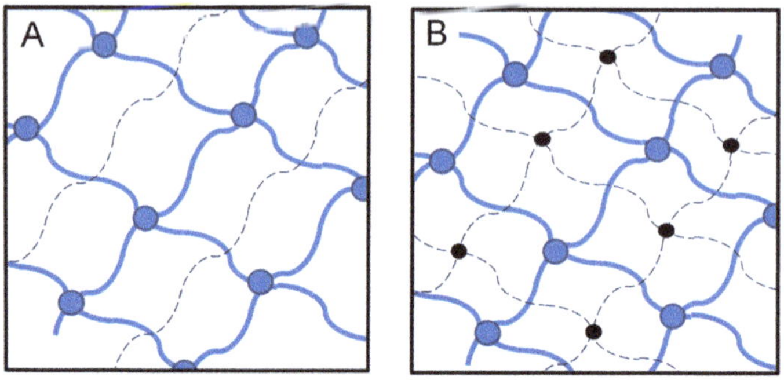

Figure 3. Schematic representation of SIPNs (**A**) and IPNs (**B**).

Natural polymers are typically limited in their reactivity and processability, and corresponding physical gels usually lack mechanical strength and thermal stability. That is why their inclusion into IPNs is often the best method to overcome those limitations, as IPNs possess reduced interfacial tension and increased adhesion between the phases that lead to more stable morphology [35]. Such new hybrid materials have advantageous chemical, physical, and biological properties. Among the most widely used sustainable IPN and SIPN building blocks are carbohydrates, such as cellulose in its various forms and states, chitin as a cellulose analog, alginic acid, and gellan gum. The second most abundant natural macromolecule, lignin (a polyphenol), is extensively incorporated, as well. This review describes hybrid networks (entanglement networks, covalently cross-linked networks, and physical gels) containing those natural polymers and the recently reported synthetic routes for achieving the synthesis of various sophisticated hybrid networks.

3. Cellulose-Based Networks

Cellulose is the most abundant polymer on Earth and is considered one of the most important structural elements in plants and other living species. Each day a tree produces about 10 g of cellulose, making the global annual production of cellulose 1.5×10^{12} tons [36]. Cellulose is an unbranched polymer-homopolysaccharide composed of β-1,4-linked glucose residues. It is linear, and every other glucose unit is rotated approximately 180° with respect to its neighbors, giving cellulose a 2-fold axis. The physical properties such as crystalline state, degree of crystallinity, and molecular weight may be highly variable and depend on the source from which it was obtained [37]. Figure 4 shows the structural formula for the β-1,4-glucan polymer chain (cellulose). The repeating unit is called cellobiose and has a length of 1.03 nm. Its n—degree of polymerization—can reach up to 15 k in natural cellulose.

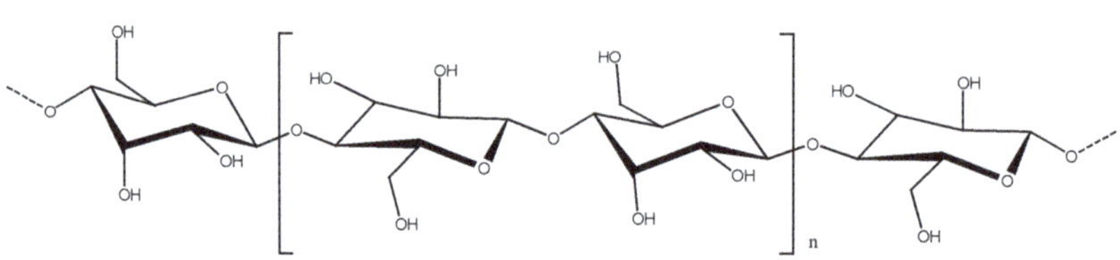

Figure 4. Structural formula of the β-1,4-glucan polymer chain. Cellobiose is shown in brackets. n—degree of polymerization.

The biggest limitation of using cellulose in organic reactions is its solubility. Typically, to create an entanglement network, cellulose must be dissolved completely to ensure its homogeneity in the system. Its dissolution is quite challenging due to the partial crystallinity and a high degree of polymerization (DP). Cellulose is not soluble in water, nor in the majority of organic solvents, so multiple other systems are used, such as ionic liquids, LiOH/urea, NaOH/urea, NaOH/thiourea, and others [38].

Physical gels of cellulose can be created using the freeze–thaw (FT) method, where growing ice crystals upon freezing create the irreversible aggregation of physically confined cellulose nanocrystals between ice domains, and upon thawing, a network of aggregated cellulose fibers is formed (Figure 5). The described FT procedure yields a gel in liquid nitrogen at relatively low concentrations (~4 wt%) either in water or in polar organic solvents, producing hydrogels and organogels, respectively [39]. The FT method works for carboxymethylcellulose (CMC), as well. CMC is readily soluble in water, so to decrease the solubility, the ionized carboxyl groups must be protonated. Treatment with sulfuric acid yields insoluble and flexible transparent CMC-based films, allowing for the creation of a water-impermeable layer. When adhered to a filter paper, the hydrogels kept the surface on top of the paper waterproof for more than 24 h [40]. When an aqueous solution of citric acid was used, edible, highly biodegradable hydrogels with high compressive recoverability could be prepared [41].

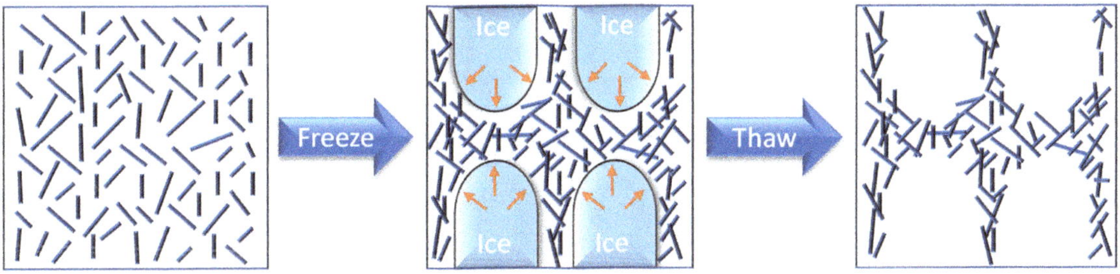

Figure 5. Preparation of physical gels of cellulose nanofibers using the freeze–thaw (FT) method.

Covalently cross-linked networks based on unsubstituted cellulose can be prepared when cross-linked via epichlorohydrin [42] or N,N′-methylene bisacrylamide [43]. An interesting case of a co-polymerized cross-linked hydrogel of microcrystalline cellulose and polyvinyl alcohol was reported by combining both physical and chemical cross-linking methods. This hydrogel was tested for the loading and release of a 5-fluorouracil drug in phosphate-buffered saline [44].

Chemical modification of cellulose fibers seems to be the most common approach to cellulose integration into gels and composites. Altering the surface of cellulose fibers helps to change the hydrophobicity, reactivity, and solubility of chains [45]. There are multiple ways to chemically modify the surface of cellulose, but grafting methods remain the most common. Figure 6 represents the two most frequently used approaches, called "grafting to" and "grafting from". These methods are employed to modify the backbone of other carbohydrates as well.

Grafting can also produce reactive cellulose-based macromonomers. The reaction between the hydroxyl groups on the surface of cellulose fibers and 4-vinylbenzyl chloride enables the formation of cellulose-based cross-linkers, called "cellu-mers". They can undergo radical copolymerization with either styrene to form SIPNs (Figure 7A) [46]. This strategy allows for the synthesis of versatile homogeneous sustainable networks, where incorporated cellulose fibers do not agglomerate, and synthesized materials have improved mechanical and thermal properties. The method was further extended with modified poly(ethylene glycol), PEG, to form hydrogels in high yields (Figure 7B) [47]. The PEG/cellu-mer hydrogels were able to bind and remove several organic dyes from an aqueous solution, showing potential for water remediation. Olad et al. described the synthesis of an amphiphilic SIPN nanocomposite where sulfonated carboxymethyl cellulose-*g*-poly(acrylic acid-*co*-acrylamide)/polyvinyl alcohol/montmorillonite was synthesized using free-radical graft copolymerization in an aqueous solution (Figure 7C). The hydrogels showed excellent swelling ability, indicating their potential application in agriculture [48].

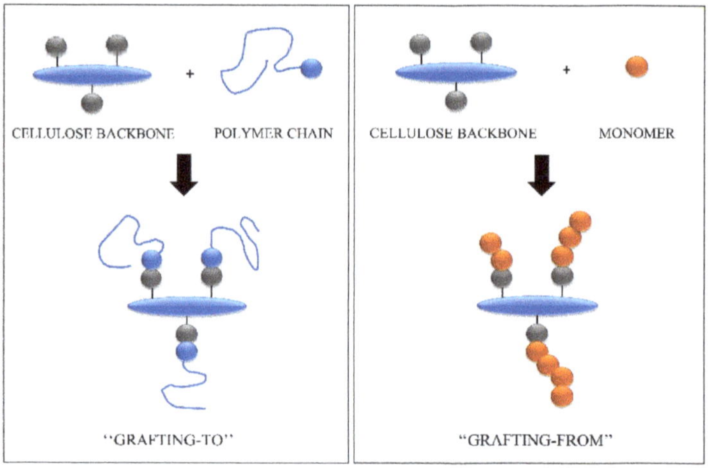

Figure 6. Grafting approaches used to modify carbohydrate's backbone.

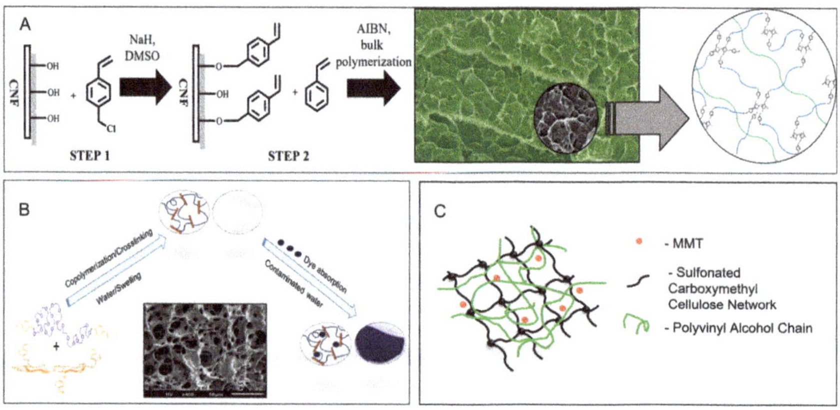

Figure 7. Various cellulose-based networks. (**A**) SIPN prepared after copolymerization of cellulose macromonomers ("cellu-mers") and styrene [46]; (**B**) Hydrogel based on cellu-mers and modified PEG [47]; (**C**) Carboxymethyl cellulose-based SIPN nanocomposite [48].

Some other examples include the preparation of a hydrogel based on quaternary cellulose via an amino-anhydride "click" reaction between maleic anhydride copolymer and poly(acrylamine hydrochloride) [49]. The synthesis was performed in a "green" solvent (water), and the SIPN was tested as a super adsorbent for wastewater remediation purposes. Another strategy to synthesize cellulose-based IPN was reported, where doubly cross-linked graphene oxide modified cellulose nanocrystal/poly(N-isopropylacrylamide) IPN was prepared using an ultrasound one-pot strategy. The material was able to adsorb and remove organic dyes such as Congo red (CR) and methylene blue (MB) from the solution. They showed excellent water-swelling properties and good mechanical properties, making them a low-cost material for the purification of dye wastewater [50].

These research reports, selected among many, prove the importance and continuous development of cellulose-based networks and gels via multiple strategies and approaches.

4. Chitin-Containing Networks

Chitin is the second most abundant polysaccharide in nature after cellulose and is considered analogous to it as a structural material [51]. It is usually obtained by extraction from arthropods, nematodes, and fungi. The polymer is insoluble in common solvents and cannot be extracted by a solvent extraction method, which makes the process tough and laborious. Chitin tends to form microfibrils (also referred to as rods or crystallites) of around 3 nm in diameter that are stabilized by H-bonds involving the N-glucoacetate residue and hydroxyl groups (Figure 8). This acetamide group in its structural repeating unit makes the H-bonding in chitin even more complex and strong than that in cellulose.

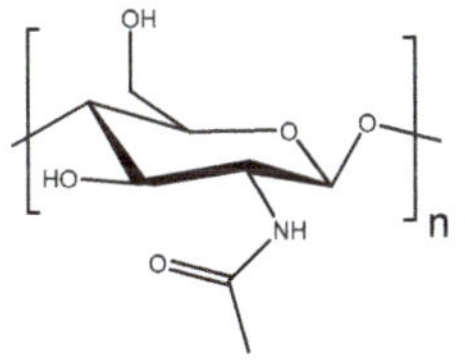

Figure 8. Chitin repeating unit. n—degree of polymerization.

Due to its limited solubility in most organic solvents, the use of chitin is significantly limited to a few applications [52]. Ionic liquids are used to dissolve chitin and prepare relatively weak gels, but they still require high concentrations to form a gel [53]. Ionic liquids such as 1-butyl-3-methylimidazole acetate (BmimAc) dissolves native chitin and, upon regeneration, form a gel-like substance [54]. There are multiple other examples of physical gel formation by dissolving chitin from different sources in ionic liquids [55]. One of them is a preparation of a tannic acid-mediated self-assembled chitin hydrogel that has improved strength (8.18 MPa) and Young's modulus (0.36 MPa) compared to pure chitin hydrogel. Due to the high noncovalent cross-linking density, the hydrogel enabled effective energy dissipation and high toughness, Figure 9 [56].

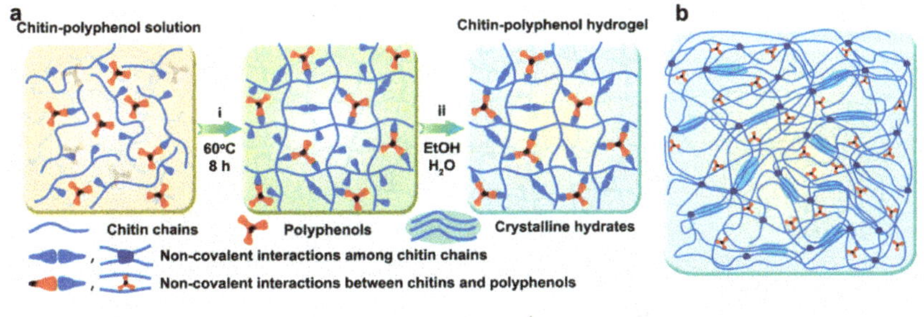

Figure 9. Preparation of the chitin–polyphenol hydrogel. (**a**) Processing principles; (i) Partial noncovalent network; (ii) Final formation of a multiple noncovalent network; (**b**) Chitin–polyphenol hydrogel. Modified with permission from [56]. Copyright 2023, published by Royal Society of Chemistry.

Kasprzak and Galiński have recently reported the synthesis of a chitin-based electrolyte with a solution-casting technique using ionic liquids. This method allowed the creation of highly efficient hydrogels in terms of specific capacitance, power density, and cyclability. In addition, this approach can also be used to synthesize chitin/cellulose-based hydrogel electrolytes. Specific capacitance values were in the range of 92–98 $F \cdot g^{-1}$, with excellent capacitance retention (97–98%) after 20,000 galvanostatic charge and discharge cycles. Such results show that chitin is a promising component to be used in the electrochemical industry to produce green electrochemical capacitors, Figure 10 [57].

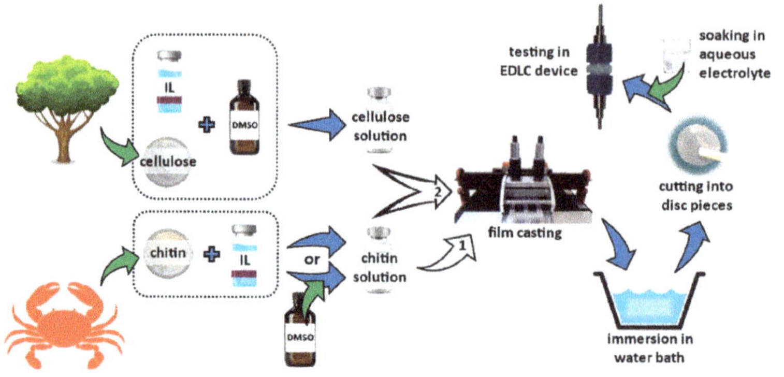

Figure 10. **(A)** The preparation of hydrogel electrolytes. Modified with permission from [57]. Copyright 2023, published by Springer Nature.

Natural chitin–polyphenol hydrogels were reported to be highly biocompatible and biodegradable, making them promising materials for future tissue engineering [58]. Typically, chitin-based physical gels have low gel stability with weak strength, and an irreversible covalent cross-linking alleviates the above-mentioned problems. In 2019, Chen et al. reported the synthesis of hydrogels based on quinone-crosslinked chitin nanofibers using amino groups. There, surface-deacetylated chitin nanofibers (S-ChNF) formed a hydrogel after a reaction in hydroquinone (HQ)/copper (Cu(II)) solutions. Compared to non-surface-deacetylated, S-ChNF-based hydrogel displayed almost 10-fold higher tensile strength because of the extended cross-linking effect between quinone and amino groups, Figure 11 [59].

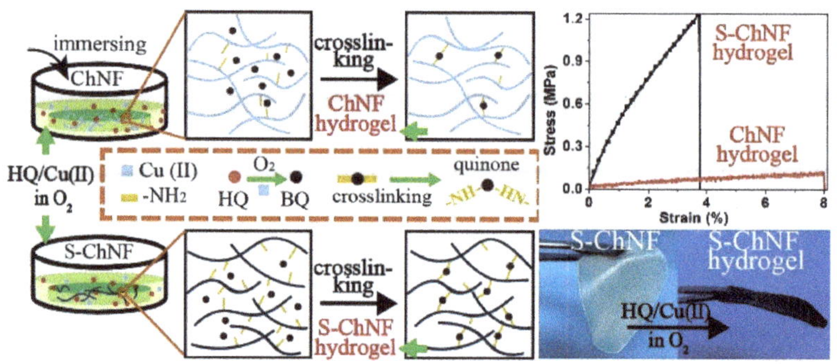

Figure 11. Comparison of chitin–based hydrogels prepared with chitin nanofibers (ChNFs) and surface–deacetylated chitin nanofibers (S-ChNF). Reproduced with permission from [59]. Copyright 2023, published by Elsevier.

Other examples of recently reported chitin-based gels are the double cross-linked chitin/KH560 networks [60]. Such systems have two types of cross-links: physical and chemical. They can be synthesized when dissolving chitin in an aqueous solution of KOH/urea with a freeze–thaw process followed by using KH560 (an epoxy-functional silane) as a cross-linking agent and coagulating in ethanol solution at low temperature. The material was biocompatible, with excellent mechanical properties and a high swelling ratio. The presence of two types of cross-links makes this hydrogel an excellent damping material since cross-linking bonds can serve as "sacrificial bonds", effectively dispersing the stress and withstanding large deformations, Figure 12.

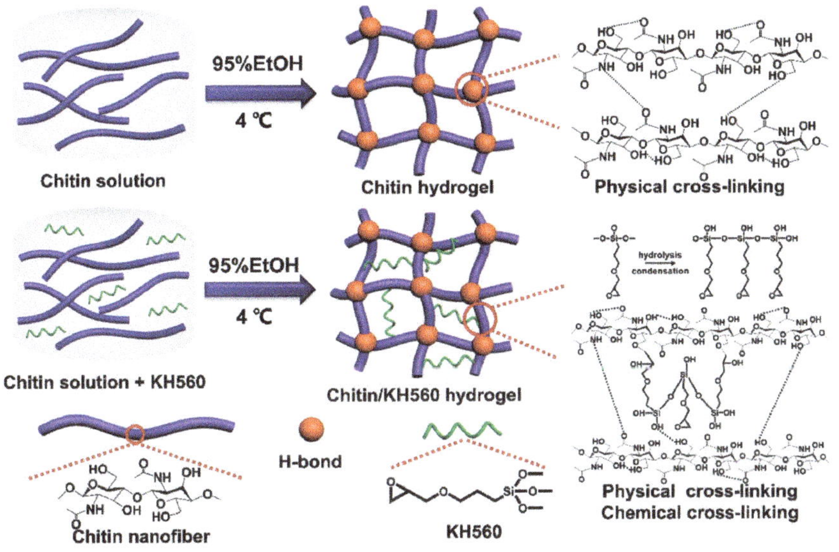

Figure 12. Preparation of double cross-linked chitin/KH560 hydrogel. Reproduced with permission from [60]. Copyright 2023, published by Elsevier.

A similar approach was used for the synthesis of injectable thermosensitive hydrogels. Double cross-linked furyl-modified hydroxypropyl chitin polymer was synthesized homogeneously in an aqueous solution. Firstly, physically cross-linked hydrogel based on furyl-modified hydroxypropyl chitin was formed upon cooling to 4 °C. Then, irreversible cross-linking with maleimide-terminated PEG 2k Da using the Diels–Alder reaction was achieved under physiological conditions. Potentially, during the second step, there was sufficient time to stir-in various therapeutic molecules and/or cells, making these gels an excellent material for drug delivery, three-dimensional cell culture, and tissue repair. Additionally, dually cross-linked hydrogels have enhanced mechanical strength compared to physically cross-linked ones [61].

A SIPN hydrogel could also be prepared via radical polymerization. Unsaturated poly(vinyl alcohol)/maleic anhydride-based macromonomer was cross-linked with acrylic acid in the presence of water-soluble chitin to generate a SIPN. Linear chains of chitin were homogeneously dispersed throughout the network. Such a completely sustainable synthetic method allowed for the preparation of non-toxic, biocompatible, and robust hydrogels [62]. Along with other natural polymers, chitin composite hydrogels were evaluated for the effective adsorptive removal of textile dyes from water [63,64].

Chemical or enzymatic deacetylation of chitin leads to the creation of a new biopolymer—chitosan, Figure 13 [65]. When the degree of deacetylation (DA) is higher than 50 mol%, the product is soluble in dilute acid solutions. This approach is frequently used to create advanced products based on chitin and chitosan. Their biological activity and antimicrobial properties make them perfect candidates for various applications, such as cell culture media [66] and antimicrobial materials [67].

Networks based on chitosan are being investigated for various other purposes. Yang et al. in 2023 described the development of a robust, anti-freezing, and conductive double-network for a stable-performance flexible electronic device. Various chitosan–polyacrylamide double-network ionic hydrogels were chemically cross-linked with different substrates. The chitosan physical network enabled effective energy dissipation, while the covalent linkage between the polyacrylamide network and the substrate surface provided interfacial adhesion. The hydrogel–substrate combinations were reported to maintain high interfacial toughness at low temperatures. In addition, the high conductivity of the hydrogel–metal interface makes it a promising material for developing a flexible sensor to detect strain and pressure within a broad temperature range, Figure 14 [68].

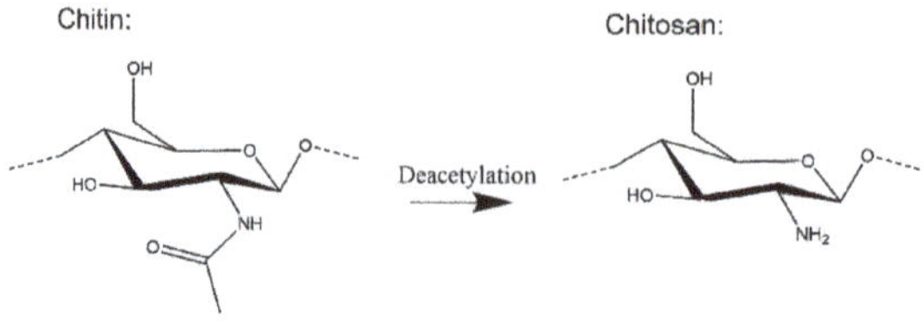

Figure 13. Transformation of Chitin to Chitosan.

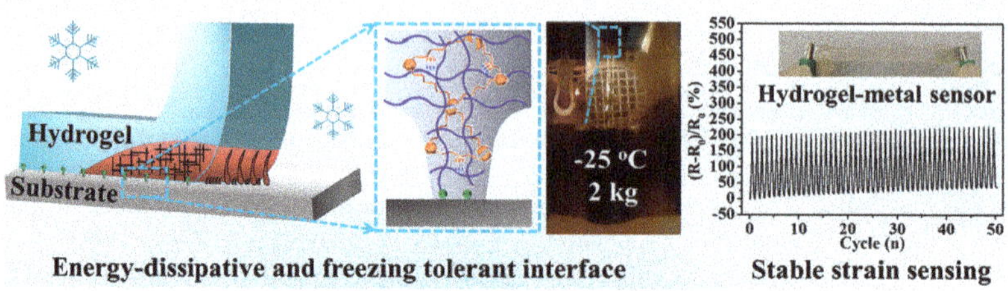

Figure 14. Chitosan–polyacrylamide double-network ionic hydrogel for flexible electronics. Modified with permission from [68]. Copyright 2023, published by Elsevier.

Several other interesting examples of sophisticated chitosan-based systems were recently reported. Wan et al. reported a synthesis of poly(acrylic acid)/chitosan-*g*-poly (vinylamine) composite hydrogel for efficient adsorption of methylene blue dye from aqueous solutions. It was reported that the hydrogel achieved 85.24% removal of dye even after five adsorption cycles [69]. Another report announced the synthesis of chitosan-based hydrogel containing in situ bio-reduced silver nanoparticles for accelerated healing of infected full-thickness skin defects. This material could enable a ground-breaking sustainable anti-infection, anti-inflammation, collagen-stimulating treatment that promotes the formation of epithelia and blood vessels [70].

5. Alginic Acid Networks

Alginic acid (AA) is a natural polysaccharide produced by brown seaweeds such as kelp and macroalgae or produced by microbial fermentation using special bacteria [71]. Alginate consists of two basic building blocks, α-l-guluronic acid (G) and β-d-mannuronic acid (M) residues, linearly linked together with 1–4 linkages (Figure 15A). The composition and distribution of these M/G residues impact the properties and functionalities of alginate. Due to the nontoxicity and biocompatibility of this natural polymer, the most common use of AA-based networks is in biomedical, agricultural, and food-packaging industries [72]. Sodium alginate (SA) is the sodium salt form of alginic acid, frequently used for the synthesis of gels and networks (Figure 15B).

The "egg-box" model is a classic model describing the gelation mechanism of alginates. The most recent studies show that the chains pair into dimers cooperatively when the ratio of Ca^{2+} ions to G units in alginate $R_{(Ca/G)}$ is 0.25 for higher G alginate and 0.50 for lower G alginate. At higher R, the "egg-box" dimers act as a structural basis for further lateral aggregation by successively doubling dimers into tetramers, octamers, and hexadecamers (Figure 15C) [73,74].

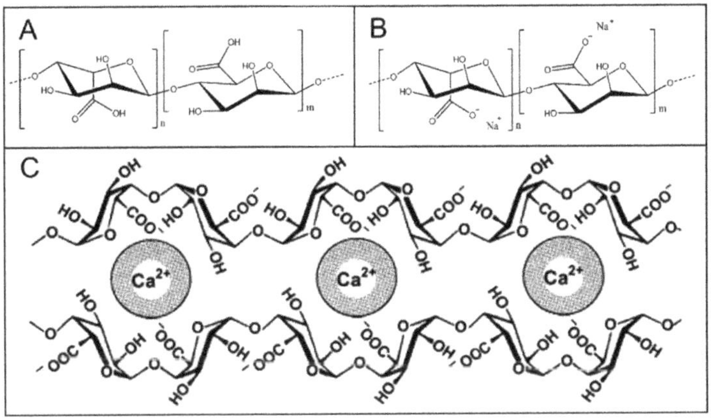

Figure 15. (**A**) Structural formula of alginic acid; (**B**) Structural formula of sodium alginate; (**C**) The "egg–box" model for alginate gelation with calcium ions. Ca^{2+} ions bind to the carboxyl groups of the guluronic acid residues of two neighboring alginate chains. Reproduced with permission from [74]. Copyright 2023, published by John Wiley and Sons.

"Entanglement networks" based on alginates can be formed at lower pH when the chains associate. However, the gel is weak, soft, and soluble in alkaline solutions. Therefore, it does not find much practical use. Casting aqueous AA solutions leads to the creation of transparent and flexible films, but these films are water-soluble unless they are immersed in a multivalent ion salt solution [75]. In the presence of divalent cations like calcium (Ca^{2+}) and magnesium (Mg^{2+}), alginate chains cross-link and form water-insoluble physical gels [76]. Most recent reports include various other metals, such as strontium (Sr^{2+}), zinc (Zn^{2+}), and even aluminum (Al^{3+}). Depending on the metal, hydrogels exhibit different properties that lead to various applications [77], but the drawback of the use of these materials is their low dimensional stability. Pure alginate gels formed with calcium ions have high porosity and are known for their healing potential and prospects in anti-tumor applications. However, they are unstable at physiological pH or in environments with a high concentration of phosphate or citrate ions because Ca^{2+} ions can be extracted from the hydrogel structures, leading to their collapse [78].

Covalently cross-linked alginate gels are more mechanically stable compared to physically cross-linked ones [79,80]. If the material is intended to be applied for biomedical purposes, biocompatible cross-linkers should be used. For example, glutaraldehyde was utilized as a cross-linking agent in the preparation of sodium alginate/poly(vinyl alcohol) IPN [81]. Although tests have indicated that the prepared IPN hydrogel membranes were safe for drug delivery applications, some concerns related to glutaraldehyde cytotoxicity remain, as it is a contact irritant and dermal sensitizer [82].

Biocompatible chemically cross-linked gels based on modified sodium alginate/poly (vinyl alcohol) using poly(ethylene glycol) diacrylate as a cross-linking agent were recently reported. The authors have compared the mechanical strength of chemically cross-linked hydrogels with those prepared via a physical cross-linking mechanism using Ca^{2+} (Figure 16) [18]. They also studied the drug release profile using *Echinacea purpurea* (*EP*) extract as a model compound. Results indicated that hydrogels prepared via ionic cross-linking had better swelling than those prepared via chemical cross-linking. However, chemically cross-linked ones had enhanced mechanical strength, and the surface of the hydrogels was more regular and denser than in physical gels. Analysis of release profiles of the model drug showed that *EP* could be delivered in a controlled manner for an extended time, but only with the alginate-based hydrogels synthesized using chemical cross-linking. In the case of physical gels, an undesirable phenomenon called a "burst effect" was observed, which is a very rapid release of the active substance due to rapid surface desorption instead of slow and steady release.

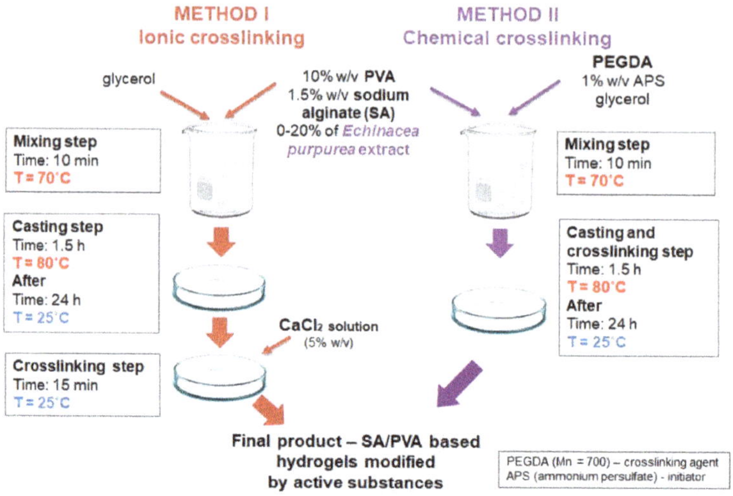

Figure 16. Preparation of physically– (Method 1) and chemically (Method 2) crosslinked gels. Reproduced with permission from [18]. Copyright 2023, published by MDPI.

Chemically cross-linked alginate hydrogels were reported to recover gold nanoparticles from wastewater effectively. Particles of sericin and alginate chemically cross-linked by proanthocyanidins were synthesized and characterized. Sericin is usually discharged in wastewater as a silk manufacturing by-product; therefore, the synthesized hydrogel is interesting not only as a system able to recover gold from aqueous media but also as a possible solution to problems with sericin disposal [83].

Kim et al. proved that the formation of IPN or SIPN not only improves the mechanical and thermal stability of alginate-based gels, but this improvement happens without losing the advantages of polysaccharides. A series of SPNs were prepared with 75:25, 50:50, and 25:75 ratios of alginate to poly(vinyl) alcohol, where there is only one cross-linked network inside—alginate. Results revealed that by utilizing the same components but altering the method of hydrogel preparation, pH-responsive behavior, stability, and water retention properties can be controlled and altered according to the desired uses [19].

Alginate IPNs and SIPNs can be utilized for other interesting applications. For example, Wang et al. synthesized reusable porous alginate/poly(vinyl alcohol) SIPN hydrogel to remove Pb(II) from aqueous solutions [84]. They used a natural biopolymer sodium alginate (SA) as the main chain, sodium acrylate (NaA) as the monomer, and poly(vinyl alcohol) (PVA). NaA monomers were grafted onto the SA chains, while PVA chains were interpenetrated and entangled within the cross-linked network. This facile grafting polymerization in an aqueous medium created hydrogels with excellent adsorption properties and removal efficiency toward Pb(II). Incorporated PVA polymer chains improved the regularity of the network and, therefore, increased the mechanical strength and hydrophilicity. More importantly, ~94% of the bound Pb(II) can be de-adsorbed and recycled using an acid-leaching process. The system can be regenerated five times, with the possibility to de-adsorb and recycle about 89.9% of the initially sequestered Pb(II). In addition, the adsorption capacity for Pb(II) could reach up to 502.5 mg/g after being reused for five cycles. Figure 17 shows the structure of the SIPN and its adsorption actions for Pb(II).

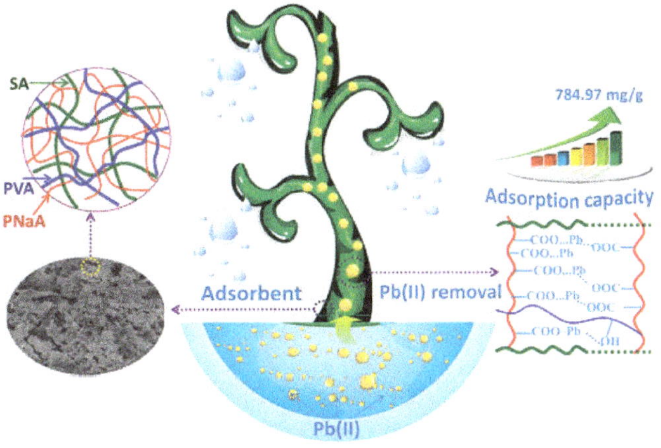

Figure 17. Structure of the network and its adsorption actions for Pb(II). Reproduced with permission from [84]. Copyright 2023, published by Frontiers in Chemistry.

Notably, the most common use of alginate-based IPNs is for biomedical purposes. When they are used for controlled drug-release, the loaded substance can be retained to the maximum extent without negatively affecting its pharmacological effect due to the excellent biocompatibility and stable porous structure of the network [85,86]. For example, controlled delivery of aspirin using IPN hydrogels based on sodium alginate and nanocellulose was reported recently [87]. The first framework network made with sodium alginate and nanocellulose was built after acetic acid coagulation, followed by $CaCl_2$ chelation. Ions of Ca^{2+} can exchange ions with sodium alginate and chelate with carboxyl groups on the surface of cellulose to form a uniform interpenetrating network structure. The resulting hydrogel showed good cytocompatibility and nontoxicity, was mechanically robust, and had a prolonged aspirin release time of ~ 80 h, making it an ideal system for prolonged aspirin release (Figure 18).

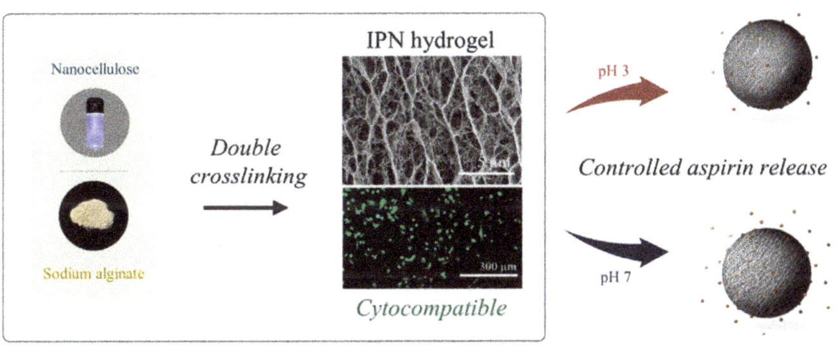

Figure 18. Synthesis of sodium alginate/nanocellulose-based IPN for controlled aspirin release. Reproduced with permission from [87]. Copyright 2023, published by Elsevier.

Alginate-based hydrogels have gained increased interest as an alternative to the existing clinical methods for treating cartilage injuries [88]. Injectable in situ forming hydrogels as carriers for drug delivery or scaffolds for tissue engineering applications offer multiple benefits as they require a minimally invasive procedure for implantation. The cross-linkers used for this application should be carefully selected, as they should show no cytotoxicity towards encapsulated cells and should provide an appropriate gelation time after the injection. An example of a suitable cross-linker was recently reported describing a synthesis of an injectable IPN based on modified cartilage extracellular matrix (ECM) and alginate [89]. Horseradish peroxidase (HRP) in the presence of hydrogen peroxide was used as a cross-linking mediator after the extracellular matrix was functionalized with phenol groups (ECM-ph). The synthesized injectable hydrogel was then impregnated with silk fibroin nanofibers (SFN). It was evaluated for articular cartilage tissue engineering applications. The material provided a suitable microenvironment for cell growth and viability. Collagen content measurements showed that this hydrogel has great potential and capability to mimic natural cartilage. Mechanical strength, however, needed further improvement [89].

6. Gellan Gum-Derived Networks

Gellan Gum (GG) is a linear electronegative exopolysaccharide produced by *Pseudomonas elodea*. Its main chain consists of four repeating carbohydrate units: two D-glucose, one L-rhamnose, and one D-glucuronic acid (Figure 19A). On average, about 25% of repeating units contain an o-acetyl group linked to the C-6 position of one of the δ-D-glucopyranosyl units [90,91]. GG is in coil form at elevated temperatures in solution, but upon cooling, the coil transforms into a thermally-reversible double helix. These anti-parallel double helices self-assemble and form oriented bundles that are connected via untwined regions of polysaccharide chains (Figure 19B) [92]. GG is biodegradable and nonpathogenic (approved by the FDA) and allowed for use as a food additive [93]. However, it is investigated for other applications, as well [94]. The most investigated use of GG hydrogels for biomedical purposes is for drug delivery [95]. Physical hydrogels based on GG that can bind and release glucose can be easily prepared by dispersing the hydrocolloid powder in water at 90 °C. Recently, hydrogels based on high acyl (HA) and a modified version—low acyl (LA)—GG have been studied and compared [96]. HA GG has acetyl and glyceryl substitutions on the first glucose of the repeating unit at the O-2 and O-6 positions, while in LA, GG acyl groups are removed (Figure 19(A1,A2)). Synthesized HA hydrogels had a much greater swelling ratio than LA hydrogels, and both had more significant swelling in DI water than in 50 mM KCl solution. Swelling slows the release of glucose by decreasing the diffusion flux, which must be considered when developing GG-based hydrogels. However, the hydrogels were brittle; to improve the mechanical strength, chemical cross-linking must be considered as an alternative synthesis. An example of this approach was recently published. Covalently cross-linked redox-responsive implantable hydrogels were synthesized and used as paclitaxel carriers for HER2-positive breast cancer therapy [97]. These systems can be prepared using two types of buffers: acetate buffer and phosphate buffer, cross-linked with different concentrations of l-cysteine (Figure 20). The hydrogels were loaded with paclitaxel:β-cyclodextrin, and in vitro tests indicated that they expressed great anti-tumor activity. Rheological analysis showed that samples had characteristic strong gel behavior and could withstand the load over a broad range of frequencies. Potentially, they could be used in the post-surgical treatment of HER2-overexpressing breast tumors [97].

Three-dimensional (3D) printing is increasingly used to produce hydrogel structures with increased complexity and recoverable mechanical properties [98]. GG-based hydrogels are increasingly popular as novel biocompatible inks for printing, as evidenced by a recent review [99]. Here, only a few newly published articles will be discussed in detail.

Duffy et al. reported using poly-ε-lysine/GG hydrogels for the 3D reactive inkjet printing of corneal constructs [100]. It is possible to modify the structure of the printed hydrogels to include pores if desired. The printed hydrogels were 80% transparent, cytocompatible with corneal epithelial and endothelial cells, and facilitated cell attachment to their surface (Figure 21).

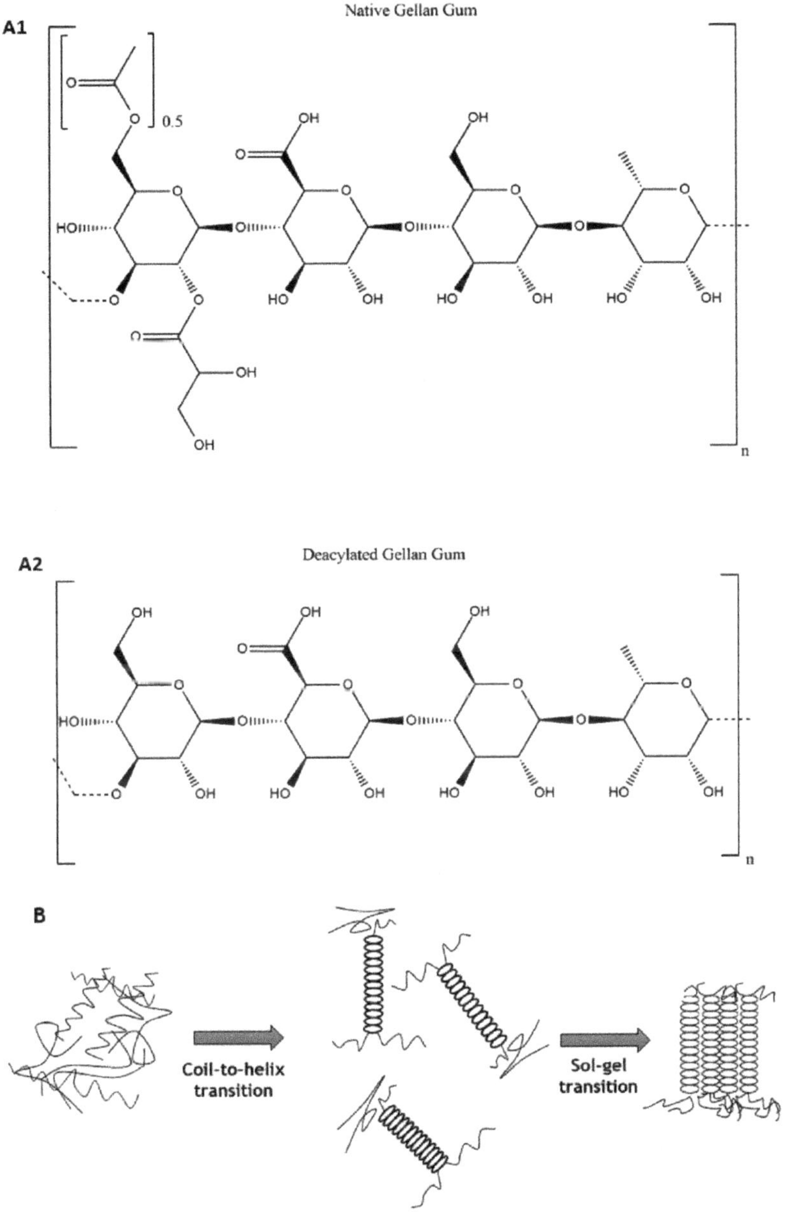

Figure 19. Chemical structure of high acyl (**A1**) and low acyl (**A2**) gellan gum; Gelation process of gellan gum (**B**). Reproduced with permission from [91]. Copyright 2023, published by Elsevier.

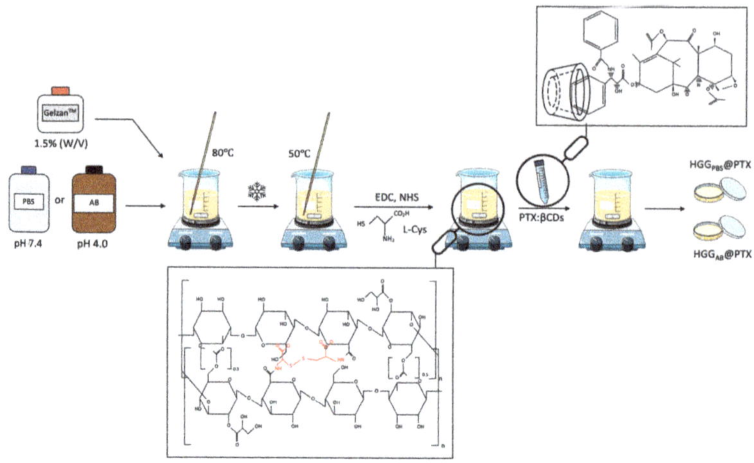

Figure 20. Preparation of GG-based hydrogels loaded with paclitaxel for local breast cancer therapy. Reproduced with permission from [97]. Copyright 2023, published by Elsevier.

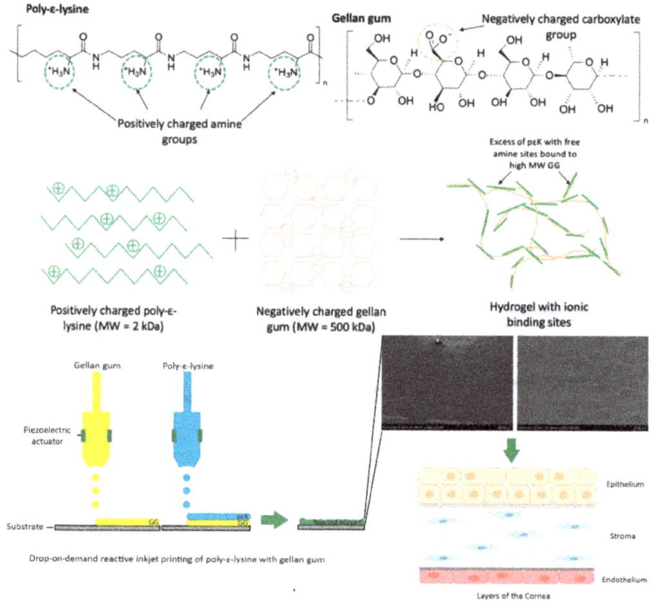

Figure 21. Chemical structures of poly(ε-lysine), gellan gum, their hydrogel, and reactive inkjet 3D printing. Reproduced with permission from [100]. Copyright 2023, published by Elsevier.

Three-dimensional printing was also used to enhance the mechanical strength of the final IPNs product. For example, a starch-based excipient known as PREGEFLO®PI10 was used to reinforce a GG–collagen IPN hydrogel [101]. A geometric addition method was developed for a new biocompatible bio-ink by incorporating PREGEFLO®PI10 into the IPN loaded with adipose-derived stem cells. A multi-printhead pneumatic extrusion-based 3D bioprinter was used to bio-print the improved IPN ink to demonstrate the potential for on-site fabrication (Figure 22).

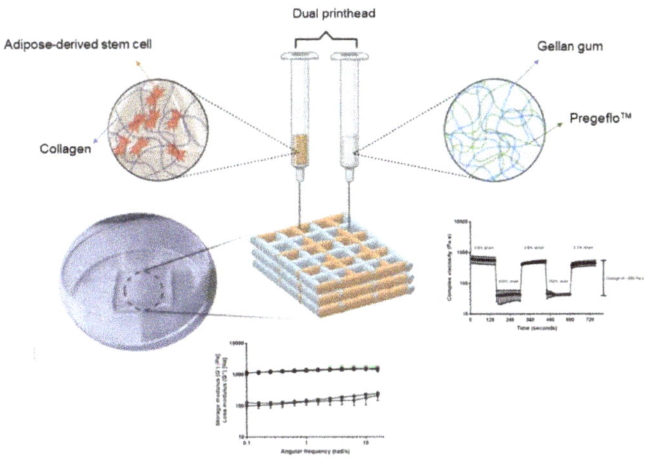

Figure 22. Schematic representation of printing the improved IPN hydrogel ink formula using a multi-printhead 3D bioprinter. Reproduced with permission from [101]. Copyright 2023, published by Elsevier.

Environmental cleanup is another emerging field where GG-based IPNs were used. The failure to treat industrial waste before it is discharged into water streams is a significant problem for the industrial, agricultural, and health sectors. The removal of various organic dyes used industrially in food, pharmaceuticals, paper, and textiles is one of the most demanding water remediation tasks [102]. The presence of dyes in wastewater poses a great threat to living organisms as it prevents the passage of sunlight into the depths of water bodies, but also it is harmful to humans and animals who consume such water for their life needs [103,104].

GG is a great material for wastewater remediation [105]. A multifunctional cross-linker with increased surface hydrophobicity was synthesized by grafting styrenic moieties onto the surface of GG [106]. These modified GG polymers could be co-polymerized with styrene, forming a SIPN network (Figure 23). These networks were mechanically robust, and their binding ability towards several organic dyes was high, showing that they could be used as potential materials in environmental cleanup [105].

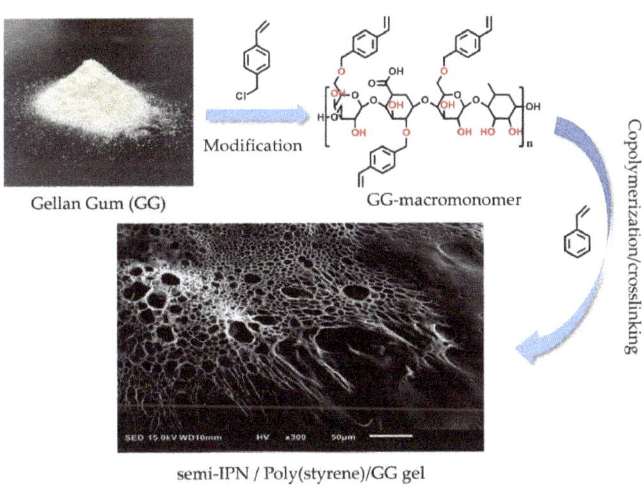

Figure 23. Synthesis of GG/PSt SIPN [106].

7. Lignin-Based Networks

Lignin accounts for approximately 30% of all organic carbon and plays three main roles in plants: (a) it provides structural integrity to the cell wall and strength to the stem and root; (b) it waterproofs the cell wall to make the transport of water and solutes through the vascular system possible; (c) it protects plants against pathogens [107]. This complex, irregular, three-dimensional, amorphous, and interlinked biomacromolecule consists of phenylpropane units linked by C-C and C-O-C bonds. There are three monolignol moieties that exist as p-hydroxyphenyl (H), guaiacyl (G), and syringyl (S) units, Figure 24 [108]. Lignins in softwoods (SWs) are mainly composed of P_G-units (with minor amounts of H-units), while hardwood (HW) lignins are mainly composed of G- and S-units. H-units' content may be slightly higher in grasses [109].

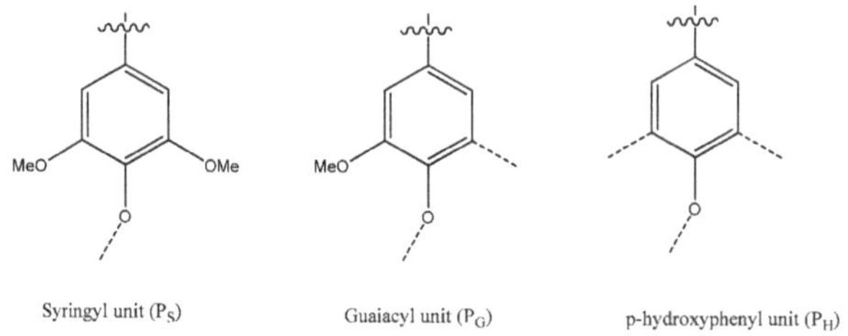

Figure 24. Lignin methoxyl substituted S-, G-, and H- moieties as generic P_S, P_G, and P_H units. See also reference [110].

The most frequent lignin linkage is the β–O–4-(β-aryl ether), and it is one of the most easily chemically cleaved (Figure 25A). The other linkages presented in that figure are more resistant to chemical degradation.

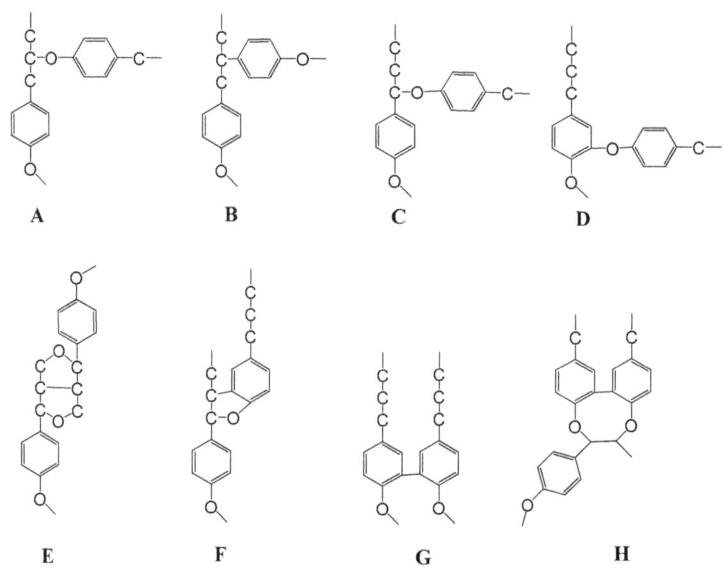

Figure 25. Various C–C and C–O–C bonds found in lignin structure. The main linkages are (**A**) β-aryl ether β–O–4; (**B**) diphenyl ethane β–1; (**C**) α-aryl ether α–O–4; (**D**) diphenylether 4–O–5 (from oligomer–oligomer couplings); (**E**) pinoresinol β–β; (**F**) phenylocumaran β–5 (from monomer–oligomer couplings); (**G**) biphenyl 5–5; (**H**) dibenzodioxicine 5–5.

Lignin is the main by-product of the pulp and paper industry, and interest in its reutilization has dramatically increased. Two methods are generally used to repurpose technical lignin. The first one is the depolymerization of lignin, breaking it down into monomeric units with further modification to form functional materials. The second approach is a chemical modification of the existing polymer to obtain a functionalized polymer. The first method is more common since lignin has excellent potential as a renewable source of aromatic compounds in lieu of petroleum. Therefore, recently, many reports have related to the cleavage of the C–O and C–C linkages. For example, the most recent one is on the cleavage of C–C/C–O bonds in lignin using dye-sensitized photoelectrosynthetic solar cells. This method presents a promising direction for the sustainable production of chemicals based on lignin [111,112].

The incorporation of lignin in functional hydrogels is another direction for the upcycling of this natural macromolecule [113]. Physical networks are typically prepared by blending/mixing two or more polymers to immobilize lignin within the existing matrix of the polymer [114]. This leads to the creation of an IPN or a SIPN, where lignin provides mechanical strength, processability, and toughness to the network, just like it does for plants. These systems are being used for various applications, including but not limited to agriculture [115], wastewater treatment [116], biomedical [117], and energy storage fields [118]. A supercapacitor energy storage IPN device was produced using a one-pot electrosynthesis of polyaniline and sulfonated lignin [119]. The nanocomposite formulation was prepared on an electrochemically etched carbon fiber electrode (Figure 26). The constructed device displayed great energy/power performance and flexibility. This approach allows for the synthesis of metal-free electrode active materials from green and cost-effective sources and marks an essential step toward green energy technology.

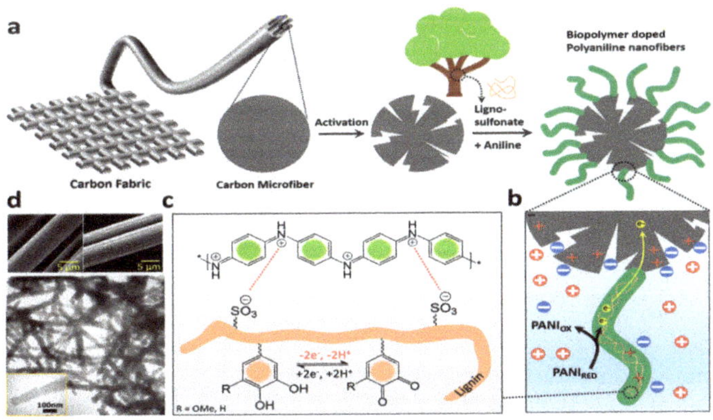

Figure 26. A supercapacitor energy storage device based on polyaniline and sulfonated lignin; (**a**) The preparation of biopolymer doped polyaniline nanofiber; (**b**) Faradaic charge transfer and electrical double layer charge accumulation at the electrode/electrolyte interface; (**c**) Structure and the interaction within the components in the nanocomposite; (**d**) SEM images of the carbon fiber substrate before (left) and after (right) electrochemical etching along with TEM image of the resulting PANI-LS nanocomposite. Reproduced with permission from [119]. Copyright 2023, published by American Chemical Society.

Lignin can also be used for environmental remediation purposes, for example, as "super-swelling" hydrogels for dye removal. In this case, technical lignins were cross-linked with poly(methyl vinyl ether-*co*-maleic acid), PMVE/MA, Figure 27 [120]. The results indicated that the adsorption of methylene blue from solutions was comparable to that reported for activated charcoals. An interesting and important fact is that these renewable and biocompatible hydrogels were prepared via a "green" approach, with no organic solvents or potentially toxic reagents used, and water was the main reaction by-product. This strategy produced a system that was not only prepared under conditions with minimal harm to the environment but was also able to remove contaminants from it [120].

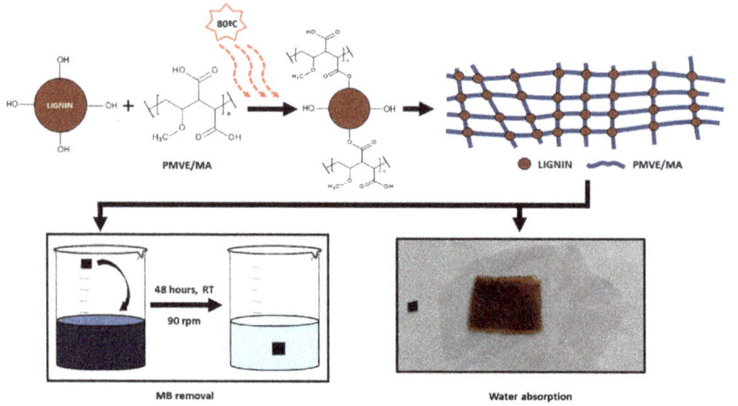

Figure 27. Proposed chemical reactions taking place during the cross-linking of different technical lignins with PMVE/MA via ester linkages. Reproduced with permission from [120]. Copyright 2023, published by Elsevier.

Multiple polar sites in lignin's framework can be used for the physical cross-linking of hydrophilic polymers via H-bonding. Wang et al. utilized lignin in lieu of traditional bisphenol A to prepare lignin-based epoxy resin adhesives with double-interpenetrating network structures [121]. In their work, lignin was modified with methacryloyl chloride and epichlorohydrin, introducing double bonds and oxirane rings into the macromolecule to form the network (Figure 28). In that way, the free volume inside the gel can be reduced and prevent the infiltration of external water molecules. The final product could be used as a high-quality water-resistant adhesive.

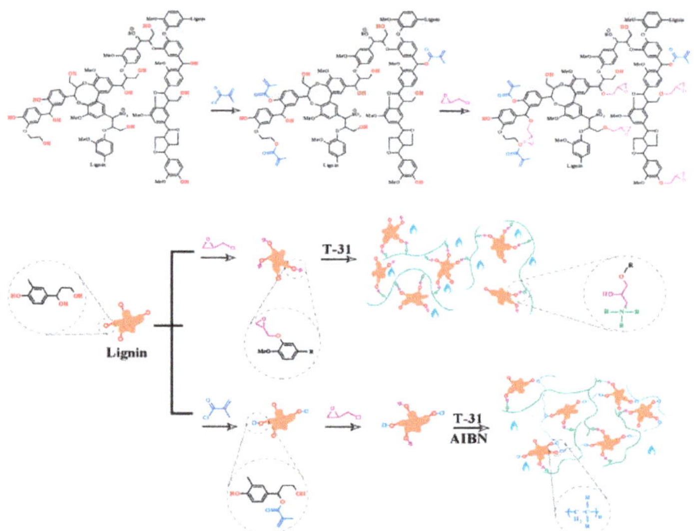

Figure 28. Schematic preparation of the double-IPN lignin-based epoxy resin adhesives resistant to extree environments. Reproduced with permission from [121]. Copyright 2023, published by American Chemical Society.

8. Other Hybrid Networks

In recent decades, metal-coordinated networks (MOFs) and frameworks have been rapidly developing. MOFs are porous coordination polymers that have both organic and inorganic components. Metal cations such as Ag^+, Zn^{2+}, Co^{2+}, Cu^{2+}, and Mo^{6+} can be easily introduced to the frameworks, and their antibacterial activity is proven [122]. Metal–phenolic networks (MPNs), on the other hand, are non-toxic and more environmentally friendly due to the milder conditions of their synthesis. Polyphenols and metal ions readily dissociate at low pH, enabling the controlled release of a metal ion, with Fe being the most common metal. MPNs have great potential in chemotherapy, as they can produce highly toxic hydroxyl radicals (•OH) that effectively help the drug kill tumor cells [123].

The few examples discussed so far clearly indicate that work on sustainable hybrid networks is rapidly evolving and growing. Various renewable natural macromolecules can serve as sustainable building blocks for these materials. New complex systems are being developed and synthesized with properties designed and adjusted for specific applications. Reports that were not mentioned in the main text are listed in Table 1.

Table 1. Additional examples of sustainable networks, their synthesis, and applications.

Natural Precursor	Potential Use	Synthesis Method	Reference
Cellulose and locust bean gum	3D printing	Suspensions of locust bean gum and cellulose microfibers are mixed and homogenized. Prepared hydrogel is inserted into Luer Lock syringes and refrigerated for complete gelation. Inks are printed using a 3D bioprinter and stored at ambient temperature.	[124]
Chitin	Potential tissue engineering	Chitin is dissolved in an 8% NaOH/6% urea solution using the freezing/thawing method and then blended with sodium alginate solution, attapulgite powder, and epichlorohydrin.	[125]
Chitin and alginate	Wound healing	Self-healing hydrogels are prepared by cross-linking acrylamide-modified β-chitin (Am-β-Chn) with alginate dialdehyde (AD). Chitin solution (2%) is prepared, and a fixed volume of various AD concentrations are added to yield pre-hydrogel mixtures. Solutions are left to gel completely.	[126]
Gellan gum (GG)	Tumor therapy and bone reconstruction	GG is dissolved in deionized water at 80 °C. Graphene oxide is dispersed in deionized water using ultrasonication. Bio-inks are formed by mixing GG and GO aqueous solutions in different concentrations. Prepared GG/GO 3D scaffolds are then soaked in $CaCl_2$ solution. Final hydrogels are obtained after freeze-drying.	[127]
Gellan gum	Sustained drug release	Spherical diclofenac sodium-loaded carboxymethyl tamarind gum-GG IPN microbeads are prepared using $AlCl_3$ as an ionic cross-linker. Aqueous GG and CTG solutions are mixed, and the diclofenac sodium is added. The homogeneous drug–polymer mixtures are added dropwise into 3% w/v $AlCl_3$ solution, resulting in the formation of rigid spherical microbeads.	[128]

Table 1. *Cont.*

Natural Precursor	Potential Use	Synthesis Method	Reference
Gellan gum and cellulose	3D bioprinting of skin cells	Hydrogel formulations are prepared by mixing GG water solution and a suspension of nano-fibrillated cellulose. Hydrogels are fully cross-linked in 30 min by immersion in a $CaCl_2$ 1% (w/v) bath for mechanical characterization.	[129]
Polylactic acid, chitosan, cellulose nanocrystals, and ethyl lauryl alginate	Antibacterial food packaging material	An antibacterial food packaging material with a trilayer structure is obtained through the combination of extrusion, coaxial electrospinning, and coating techniques.	[130]
Cellulose nanofiber/polyvinyl alcohol with aldehyde cellulose nanofiber as cross-linker	Highly temperature-resistant hydrogels	Cellulose nanofibers and polyvinyl alcohol are dispersed into water and stirred. HNO_3 catalyst is added to the mixture, and the hydrogel precursor is transferred into a sealed transparent mold. Hydrogel is placed in an oven for 24 h.	[131]
Sodium alginate and carboxymethyl cellulose	Material with excellent water-induced shape memory properties	A dual-crosslinked gel is formed by sodium alginate reaction with glutaraldehyde (GA) and Ca^{2+}. Carboxymethyl cellulose (CMC) is introduced into the cross-linked network.	[132]
Lignin and polyurethane	Tough, thin, printable hydrogels	Lignin is used as a toughening component for polyether-based polyurethane (HPU) hydrogels. Various concentrations (0, 0.5, 1, and 2.5 wt%) are prepared from lignin stock solutions and added to HPU at pH 7–8. Samples are cast and dried at 60 °C for 3 days.	[133]
Lignin, poly(ethylene glycol) methyl ether methacrylate (PEGMA)-grafted lignin	Mechanically responsive and self-healing hydrogels	Hyperbranched copolymers are prepared by atom transfer radical polymerization to form supramolecular hydrogels with a very low critical gelation concentration of 1 wt% in the presence of α-cyclodextrin.	[134]

Advantages of Hybrid Networks and Remaining Challenges

Renewable natural macromolecules used for the synthesis of hybrid polymer networks offer a range of advantages. They are non-toxic, biocompatible, biodegradable, and, more importantly, available all year round [135]. In addition, they vary in molecular mass, chemical structure, and reactive groups available. Those reactive groups often serve as helpful anchors for direct cross-linking, chemical modification, or chelation. Thus, natural macromolecules are good candidates to substitute or aid petroleum-based polymers, leading to minimizing the ecological footprint related to the extraction and processing of oil itself. Important future use might be in the biomedical field for vital cases such as tissue regeneration or the restoration of physiological functions [136].

There are, however, certain limitations related to the commercial use of natural polymers and their hybrid networks. A few of them are discussed here. The first problem is the lack of mass production due to modest efficiency in extraction and purification. This increases the final price of the natural polymers and macromolecules, making the materials based on them not economically practical in several key markets [137]. To alleviate this problem, many countries are trying to implement environmental laws that aim to increase mass production and use of sustainable raw materials [138]. The second problem stems from limitations in compatibility with synthetic polymers [139] and a lack of stability, reactivity, and processability [21]. Chemical modification is one of the most effective methods to modify the surface properties of natural materials and make them more compatible and disperse better within other polymers [140–142]. However, on an industrial scale, such modifications might not be cost-sustainable. Not only the modification agents can be harmful, but the solvents, as well. The third problem is associated with the characterization of the natural polymers and macromolecules due to the great dispersity in molecular mass, diversity in repeating units, and solubility before and after modification [143]. Nuclear Magnetic Resonance (NMR) is a powerful spectroscopic technique for the structure elucidation of small and large molecules but has certain limitations when used to characterize polysaccharides due to their poor solubility in most common NMR solvents and signal overlap from structurally similar moieties and fragments. That is why two- and three-dimensional techniques are increasingly used [144]. Since NMR is a colligative method, it cannot differentiate between macromolecules with different degrees of modification and shows merely the total content of newly incorporated reactive functionalities. Matrix-assisted laser desorption and ionization time-of-flight (MALDI-TOF) is a precise non-destructive mass spectrometry method that is increasingly paired with NMR to associate the degree of modification within a specific population (fraction) or even individual macromolecules [145]. However, it also has limitations associated with the choice of matrix and the inability of the high molecular mass fractions to fly. Thus, despite recent advances, more research is needed to improve the analytical precision and characterization efficiency of those natural polymers and macromolecules incorporated in the hybrid networks.

9. Conclusions

Significant research efforts to reduce humans' ecological footprint have produced an increasing number of hybrid networks containing one or more natural polymers and macromolecules. Despite the dispersity in size and molecular mass of the natural components (a fundamental limitation), they display new or improved properties that render them attractive for a broad array of applications, including but not limited to environmental cleanup, biotechnology, the food industry, and medicine. Two new directions are currently emerging in this area of study. The first one is the search for new bioderived building blocks. While most studies utilize the polymers discussed in this review—cellulose, chitin, alginic acid, gellan gum, and lignin—more efforts are being devoted to the exploration of other natural macromolecules such as starch [146], hyaluronic acid [147], and spider silk [148,149]. Many of those studies employ environmentally friendly solvents, low energy, and biocatalysis, which constitute the second direction for the utilization of "green" cross-linking processes. The work by Kaplan's group and others is a good illustrative example of the attempts in this area [150,151]. Additive manufacturing is the latest innovative technology that will greatly improve the synthesis of multifunctional polymer networks. Due to increasing speed and printing resolution, a new generation of bioinks is expected to emerge shortly [152]. Three-dimensional printing will be broadly used to generate hybrid networks in three-dimensional form as analogs or semi-artificial substitutes for human organs and tissues [153].

All these recent developments trace a tendency of ever-increasing usage of natural polymers and macromolecules and their incorporation in new versatile networks through innovative, environmentally friendly technologies.

Author Contributions: Writing—original draft preparation, D.G.; writing—review and editing, I.G.; visualization, D.G. and I.G.; supervision, I.G.; project administration, I.G.; funding acquisition, I.G. All authors have read and agreed to the published version of the manuscript.

Funding: Partial financial support of this study by Syracuse BioInspired Institute at Syracuse University (Grant 1168699-1-8709) is acknowledged with thanks.

Institutional Review Board Statement: Not applicable.

Informed Consent Statement: Not applicable.

Data Availability Statement: No new data were created or analyzed in this study. Data sharing is not applicable to this article.

Conflicts of Interest: The authors declare no conflict of interest.

Sample Availability: No samples were created or analyzed in this study. Sample availability is not applicable to this article.

References

1. Splitstoser, J.C.; Dillehay, T.D.; Wouters, J.; Claro, A. Early pre-Hispanic use of indigo blue in Peru. *Sci. Adv.* **2016**, *2*, e1501623. [CrossRef]
2. Beker, C.; Benecke, N.; Grabundžija, A.; Küchelmann, H.-C.; Pollock, S.; Schier, W.; Schoch, C.; Schrakamp, I.; Schütt, B.; Schumacher, M. The Textile Revolution. Research into the Origin and Spread of Wool Production. *eTopoi J. Anc. Stud.* **2016**, *6*, 102–151.
3. Schönbein, C.F. Notitz über eine Veränderung der Pflanzenfaser und einiger andern organischen Substanzen. In *Bericht Über Die Verhandlungen der Naturforschenden Gesellschaft in Basel*; 1846; Volume 7, pp. 26–27. Available online: https://books.google.com/books?id=TSVJAAAAcAAJ&pg=PA26 (accessed on 13 July 2023).
4. Schönbein, C.F. Ueber Schiesswolle. In *Bericht über die Verhandlungen der Naturforschenden Gesellschaft in Basel*; 1846; Volume 7, p. 27. Available online: https://books.google.com/books?id=TSVJAAAAcAAJ&pg=PA27 (accessed on 13 July 2023).
5. Hyatt, J.W. Address of Acceptance. *Ind. Eng. Chem.* **1914**, *6*, 158–161. [CrossRef]
6. Porta, R. Anthropocene, the plastic age and future perspectives. *FEBS Open Bio* **2021**, *11*, 948–953. [CrossRef] [PubMed]
7. Olivetti, E.A.; Cullen, J.M. Toward a sustainable materials system. *Science* **2018**, *360*, 1396–1398. [CrossRef]
8. Mori, R. Replacing all petroleum-based chemical products with natural biomass-based chemical products: A tutorial review. *RSC Sustain.* **2023**, *1*, 179–212. [CrossRef]
9. Mohanty, A.K.; Misra, M.; Drzal, L.T. Sustainable Bio-Composites from Renewable Resources: Opportunities and Challenges in the Green Materials World. *J. Polym. Environ.* **2002**, *10*, 19–26. [CrossRef]
10. Kwak, M.; Herrmann, A. Nucleic Acid/Organic Polymer Hybrid Materials: Synthesis, Superstructures, and Applications. *Angew. Chem.* **2010**, *49*, 8574–8587. [CrossRef]
11. El Sayed, M.M. Production of Polymer Hydrogel Composites and Their Applications. *J. Polym. Environ.* **2023**, *31*, 2855–2879. [CrossRef]
12. Djabourov, M.; Leblond, L.; Papon, P. Gelation of aqueous gelatin solutions. II. Rheology of the sol-gel transition. *J. Phys.* **1988**, *49*, 333–343. [CrossRef]
13. Quinn, F.X.; Hatakeyama, T.; Yoshida, H.; Takahashi, M.; Hatakeyama, H. The Conformational Properties of Gellan Gum Hydrogels. *Polym. Gels Net.* **1993**, *1*, 93–114. [CrossRef]
14. Draget, K.I.; Skjåk Bræk, G.; Smidsrød, O. Alginic acid gels: The effect of alginate chemical composition and molecular weight. *Carbohydr. Polym.* **1994**, *25*, 31–38. [CrossRef]
15. Ai, Y.; Jane, J.I. Gelatinization and rheological properties of starch. *Starch-Stärke* **2015**, *67*, 213–224. [CrossRef]
16. Akhtar, M.F.; Hanif, M.; Ranjha, N.M. Methods of synthesis of hydrogels. A review. *Saudi Pharm. J.* **2016**, *24*, 554–559. [CrossRef] [PubMed]
17. DeRossi, D.; Kajiwara, K.; Osada, Y.; Yamauchi, A. Polymer gels. In *Fundamentals and Biomedical Applications*; Springer: New York, NY, USA, 1991; p. 354. [CrossRef]
18. Bialik-Was, K.; Królicka, E.; Malina, D. Impact of the Type of Crosslinking Agents on the Properties of Modified Sodium Alginate/Poly(vinyl Alcohol) Hydrogels. *Molecules* **2021**, *26*, 2381. [CrossRef]
19. Kim, Y.J.; Min, J. Property modulation of the alginate-based hydrogel via semi-interpenetrating polymer network (semi-IPN) with poly(vinyl alcohol). *Int. J. Biol. Macromol.* **2021**, *193*, 1068–1077. [CrossRef]
20. IUPAC. Compendium of Chemical Terminology, 2nd ed. In *Gold Book*; McNaught, A.D., Wilkinson, A., Eds.; Blackwell Scientific Publications: Oxford, UK, 1997.
21. Shivashankar, M.; Mandal, B.K. A Review on Interpenetrating Polymer Network. *Int. J. Pharm. Pharm. Sci.* **2012**, *4*, 1–7.
22. Zeng, M.; Fang, Z. Preparation of sub-micrometer porous membrane from chitosan/polyethylene glycol semi-IPN. *J. Membr. Sci.* **2004**, *245*, 95–102. [CrossRef]
23. Dragan, E.S. Design and applications of interpenetrating polymer network hydrogels. A review. *Chem. Eng. J.* **2014**, *243*, 572–590. [CrossRef]
24. Comstock, M.J. Sound and Vibration Damping with Polymers. In *Sound and Vibration Damping with Polymers*; Corsaro, R.D., Sperling, L.H., Eds.; ACS Symposium Series; American Chemical Society: Washington DC, USA, 1991; pp. 1–4. [CrossRef]
25. Mouritz, A.P.; Gellert, E.; Burchill, P.; Challis, K. Review of advanced composite structures for naval ships and submarines. *Compos. Str.* **2001**, *53*, 21–42. [CrossRef]
26. Moradi, G.; Nassiri, P.; Ershad-Langroudi, A.; Monazzam, M.R. Preparation of Sound Absorption Material Based on Interpenetrating Polymer Network (PU/PMMA IPN). *Health Scope* **2019**, 8. [CrossRef]
27. Culin, J. Interpenetrating polymer network composites containing polyurethanes designed for vibration damping. *Polimery/Polymers* **2016**, *61*, 159–165. [CrossRef]
28. Yao, Y.; Zhou, T.; Yang, C.; Liu, Y.; Leng, J. Preparation and characterization of shape memory composite foams with interpenetrating polymer networks. *Smart Mat. Struct.* **2006**, *25*, 035002. [CrossRef]
29. Vasile, C.; Pamfil, D.; Stoleru, E.; Baican, M. New developments in medical applications of hybrid hydrogels containing natural polymers. *Molecules* **2020**, *25*, 1539. [CrossRef]

30. Boyer, C.; Figueiredo, L.; Pace, R.; Lesoeur, J.; Rouillon, T.; Visage, C.L.; Tassin, J.F.; Weiss, P.; Guicheux, J.; Rethore, G. Laponite nanoparticle-associated silated hydroxypropylmethylcellulose as an injectable reinforced interpenetrating network hydrogel for cartilage tissue engineering. *Acta Biomater.* **2018**, *65*, 112–122. [CrossRef] [PubMed]
31. Naseri, N.; Deepa, B.; Mathew, A.P.; Oksman, K.; Girandon, L. Nanocellulose-Based Interpenetrating Polymer Network (IPN) Hydrogels for Cartilage Applications. *Biomacromolecules* **2016**, *17*, 3714–3723. [CrossRef]
32. Raina, N.; Rani, R.; Khan, A.; Nagpal, K.; Gupta, M. Interpenetrating polymer network as a pioneer drug delivery system: A review. *Polym. Bull.* **2020**, *77*, 5027–5050. [CrossRef]
33. Yang, C.; Xue, R.; Zhang, Q.; Yang, S.; Liu, P.; Chen, L.; Wang, K.; Zhang, X.; Wei, Y. Nanoclay cross-linked semi-IPN silk sericin/poly(NIPAm/LMSH) nanocomposite hydrogel: An outstanding antibacterial wound dressing. *Mater. Sci. Eng.* **2017**, *81*, 303–313. [CrossRef]
34. Vallittu, P.K. Interpenetrating Polymer Networks (IPNs) in Dental Polymers and Composites. *J. Adhes. Sci. Technol.* **2009**, *23*, 961–972. [CrossRef]
35. Liu, C.; Ding, J.; Zhou, L.; Chen, S. Mechanical properties, water-swelling behavior, and morphology of water-swellable rubber prepared using crosslinked sodium polyacrylate. *J. Appl. Polym. Sci.* **2006**, *102*, 1489–1496. [CrossRef]
36. Gan, P.G.; Sam, S.T.; bin Abdullah, M.F.; Omar, M.F. Thermal properties of nanocellulose-reinforced composites: A review. *J. Appl. Polym. Sci.* **2020**, *137*, 48544. [CrossRef]
37. Saxena, I.M.; Broun, R.M. Cellulose Biosynthesis: Current Views and Evolving Concepts. *Ann. Bot.* **2005**, *96*, 9–21. [CrossRef]
38. Zainal, S.H.; Mohd, N.H.; Suhaili, N.; Anuar, F.H.; Lazim, A.M.; Othaman, R. Preparation of cellulose-based hydrogel: A review. *J. Mater. Res. Technol.* **2021**, *10*, 935–952. [CrossRef]
39. Lewis, L.; Hatzikiriakos, S.G.; Hamad, W.Y.; Maclachlan, M.J. Freeze-Thaw Gelation of Cellulose Nanocrystals. *ACS Macro Lett.* **2019**, *8*, 486–491. [CrossRef]
40. Wang, M.; Jia, X.; Liu, W.; Lin, X. Water insoluble and flexible transparent film based on carboxymethyl cellulose. *Carbohyd. Pol.* **2021**, *255*, 117353. [CrossRef] [PubMed]
41. Sekine, Y.; Nankawa, T.; Yunoki, S.; Sugita, T.; Nakagawa, H.; Yamada, T. Eco-friendly Carboxymethyl Cellulose Nanofiber Hydrogels Prepared via Freeze Cross-Linking and Their Applications. *ACS Appl. Polym. Mater.* **2020**, *2*, 5482–5491. [CrossRef]
42. Zhou, J.; Chang, C.; Zhang, R.; Zhang, L. Hydrogels prepared from unsubstituted cellulose in NaOH/urea aqueous solution. *Macromol. Biosci.* **2007**, *7*, 804–809. [CrossRef]
43. Geng, H. A one-step approach to make cellulose-based hydrogels of various transparency and swelling degrees. *Carbohydr. Polym.* **2018**, *186*, 208–216. [CrossRef]
44. Seera, S.D.K.; Kundu, D.; Banerjee, T. Physical and chemical crosslinked microcrystalline cellulose-polyvinyl alcohol hydrogel: Freeze–thaw mediated synthesis, characterization and in vitro delivery of 5-fluorouracil. *Cellulose* **2020**, *27*, 6521–6535. [CrossRef]
45. Gomri, C.; Cretin, M.; Semsarilar, M. Recent progress on chemical modification of cellulose nanocrystal (CNC) and its application in nanocomposite films and membranes—A comprehensive review. *Carbohydr. Polym.* **2022**, *294*, 119790. [CrossRef]
46. Getya, D.; Gitsov, I. Reactive Cellu-mers—A Novel Approach to Improved Cellulose/Polymer Composites. *Polymers* **2022**, *14*, 1670. [CrossRef]
47. Getya, D.; Lucas, A.; Gitsov, I. Composite Hydrogels Based on Poly(Ethylene Glycol) and Cellulose Macromonomers as Fortified Materials for Environmental Cleanup and Clean Water Safeguarding. *Int. J. Mol. Sci.* **2023**, *24*, 7558. [CrossRef]
48. Olad, A.; Aghayan, J.; Gharekhani, H.; Akbarzadeh, M. Carboxymethyl cellulose-based semi-IPN hydrogel nanocomposite with improved physicochemical and mechanical properties. *J. Polym. Res.* **2022**, *29*, 460. [CrossRef]
49. Liu, Y.; Wei, H.; Li, S.; Wang, G.; Guo, T.; Han, H. Facile fabrication of semi-IPN hydrogel adsorbent based on quaternary cellulose via amino-anhydride click reaction in water. *Int. J. Biol. Macromol.* **2022**, *207*, 622–634. [CrossRef]
50. Zhang, L.; Zhao, D.; Lu, Y.; Chen, J.; Li, H.; Xie, J.; Xu, Y.; Yuan, H.; Liu, X.; Zhu, X.; et al. A graphene oxide modified cellulose nanocrystal/PNIPAAm IPN hydrogel for the adsorption of Congo red and methylene blue. *New J. Chem.* **2021**, *45*, 16679–16688. [CrossRef]
51. Lee, S.; Hao, L.T.; Park, J.; Oh, D.X.; Hwang, D.S. Nanochitin and Nanochitosan: Chitin Nanostructure Engineering with Multiscale Properties for Biomedical and Environmental Applications. *Adv. Mater.* **2023**, *35*, 2203325. [CrossRef] [PubMed]
52. Maddaloni, M.; Vassalini, I. Green Routes for the Development of Chitin/Chitosan. *Sustain. Chem.* **2020**, *1*, 325–344. [CrossRef]
53. Prasad, K.; Murakami, M.; Kaneko, Y.; Takada, A.; Nakamura, Y.; Kadokawa, J. Weak gel of chitin with ionic liquid, 1-allyl-3-methylimidazolium bromide. *Int. J. Biol. Macromol.* **2009**, *45*, 221–225. [CrossRef]
54. Wu, Y.; Sasaki, T.; Irie, S.; Sakurai, K. A novel biomass-ionic liquid platform for the utilization of native chitin. *Polymer* **2008**, *49*, 2321–2327. [CrossRef]
55. Shamshina, J.L.; Berton, P.; Rogers, R.D. Advances in Functional Chitin Materials: A Review. *ACS Sustain. Chem. Eng.* **2019**, *7*, 6444–6457. [CrossRef]
56. Lin, X.; Zhang, L.; Duan, B. Polyphenol-mediated chitin self-assembly for constructing a fully naturally resourced hydrogel with high strength and toughness. *Mater. Horiz.* **2021**, *8*, 2503–2512. [CrossRef]
57. Kasprzak, D.; Galiński, M. Chitin and chitin-cellulose composite hydrogels prepared by ionic liquid-based process as the novel electrolytes for electrochemical capacitors. *J. Solid State Electrochem.* **2021**, *25*, 2549–2563. [CrossRef]
58. Liao, J.; Hou, B.; Huang, H. Preparation, properties and drug controlled release of chitin-based hydrogels: An updated review. *Carbohydr. Polym.* **2022**, *283*, 119177. [CrossRef]

59. Chen, C.; Li, D.; Yano, H.; Abe, K. Bioinspired hydrogels: Quinone crosslinking reaction for chitin nanofibers with enhanced mechanical strength via surface deacetylation. *Carbohydr. Polym.* **2019**, *207*, 411–417. [CrossRef] [PubMed]
60. Chen, B.; Wu, S.; Ye, Q. Fabrication and characterization of biodegradable KH560 crosslinked chitin hydrogels with high toughness and good biocompatibility. *Carbohydr. Polym.* **2021**, *259*, 117707. [CrossRef]
61. Bi, B.; Ma, M.; Lv, S.; Zhuo, R.; Jiang, X. In-situ forming thermosensitive hydroxypropyl chitin-based hydrogel crosslinked by Diels-Alder reaction for three dimensional cell culture. *Carbohydr. Polym.* **2019**, *212*, 368–377. [CrossRef]
62. Liu, X.; Yang, W.; Xiao, C. Self-healable and pH-sensitive high-strength water-soluble chitosan/chemically cross-linked polyvinyl alcohol semi-IPN hydrogel. *Int. J. Biol. Macromol.* **2019**, *138*, 667–672. [CrossRef] [PubMed]
63. Chen, X.; Hong, L. Preparation of chitin composite hydrogel for dye-contaminated water treatment. *Pol. J. Environ. Stud.* **2022**, *31*, 15–23. [CrossRef] [PubMed]
64. Sirajudheen, P.; Poovathumkuzhi, N.C.; Vigneshwaran, S.; Chelaveettil, B.M.; Meenakshi, S. Applications of chitin and chitosan-based biomaterials for the adsorptive removal of textile dyes from water—A comprehensive review. *Carbohydr. Polym.* **2021**, *273*, 118604. [CrossRef]
65. Desai, N.; Rana, D.; Salave, S.; Gupta, R.; Patel, P.; Karunakaran, B.; Sharma, A.; Giri, J.; Benival, D.; Kommineni, N. Chitosan: A Potential Biopolymer in Drug Delivery and Biomedical Applications. *Pharmaceutics* **2023**, *15*, 1313. [CrossRef] [PubMed]
66. Xu, H.; Zhang, L.; Zhang, H.; Luo, J.; Gao, X. Green Fabrication of Chitin/Chitosan Composite Hydrogels and Their Potential Applications. *Macromol. Biosci.* **2021**, *21*, 2000389. [CrossRef]
67. Thirupathi, K.; Raorane, C.J.; Ramkumar, V.; Ulagesan, S.; Santhamoorthy, M.; Raj, V.; Krishnakumar, G.S.; Phan, T.T.V.; Kim, S.C. Update on Chitosan-Based Hydrogels: Preparation, Characterization, and Its Antimicrobial and Antibiofilm Applications. *Gels* **2023**, *9*, 35. [CrossRef] [PubMed]
68. Yang, Y.; Zhou, M.; Peng, J.; Wang, X.; Liu, Y.; Wang, W.; Wu, D. Robust, anti-freezing and conductive bonding of chitosan-based double-network hydrogels for stable-performance flexible electronic. *Carbohydr. Polym.* **2023**, *276*, 118753. [CrossRef] [PubMed]
69. Wan, X.; Rong, Z.; Zhu, K.; Wu, Y. Chitosan-based dual network composite hydrogel for efficient adsorption of methylene blue dye. *Int. J. Biol. Macromol.* **2022**, *222*, 725–735. [CrossRef] [PubMed]
70. Zhao, F.; Liu, Y.; Song, T.; Zhang, B.; Li, D.; Xiao, Y.; Zhanga, X. A chitosan-based multifunctional hydrogel containing in situ rapidly bioreduced silver nanoparticles for accelerating infected wound healing. *J. Mater. Chem. B* **2022**, *10*, 2135–2147. [CrossRef]
71. Goh, C.H.; Heng, P.W.S.; Chan, L.W. Alginates as a useful natural polymer for microencapsulation and therapeutic applications. *Carbohydr. Polym.* **2012**, *88*, 1–12. [CrossRef]
72. Ramdhan, T.; Ching, S.H.; Prakash, S.; Bhandari, B. Physical and mechanical properties of alginate based composite gels. *Trends Food Sci. Technol.* **2020**, *106*, 150–159. [CrossRef]
73. Wang, Y.; Zhao, Y.; He, J.; Sun, C.; Lu, W.; Zhang, Y.; Fang, Y. Doubling growth of egg-box structure during Calcium-mediated molecular assembly of alginate. *J. Colloid Interface Sci.* **2023**, *634*, 747–756. [CrossRef]
74. Yin, L.; Fei, L.; Tang, C.; Yin, C. Synthesis, characterization, mechanical properties and biocompatibility of interpenetrating polymer network–superporous hydrogel containing sodium alginate. *Polym. Int.* **2007**, *56*, 1563–1571. [CrossRef]
75. Pavlath, A.E.; Gossett, C.; Camirand, W.; Robertson, G.H. Ionomeric Films of Alginic Acid. *J. Food Sci.* **2006**, *64*, 61–63. [CrossRef]
76. Bhattacharyya, A.; Mukhopadhyay, P.; Pramanik, N.; Kundu, P.P. Effect of polyethylene glycol on bis(2-hydroxyethyl) terephthalate-based polyurethane/alginate Ph-sensitive blend for oral protein delivery. *Adv. Pol. Technol.* **2016**, *35*, 21525. [CrossRef]
77. Hu, C.; Lu, W.; Mata, A.; Nishinari, K.; Fang, Y. Ions-induced gelation of alginate: Mechanisms and applications. *Int. J. Biol. Macromol.* **2021**, *177*, 578–588. [CrossRef]
78. Siqueira, P.; Siqueira, É.; de Lima, A.E.; Siqueira, G.; Pinzón-Garcia, A.D.; Lopes, A.P.; Segura, M.E.C.; Isaac, A.; Pereira, F.V.; Botaro, V.R. Three-dimensional stable alginate-nanocellulose gels for biomedical applications: Towards tunable mechanical properties and cell growth. *Nanomaterials* **2019**, *9*, 78. [CrossRef]
79. Jungst, T.; Smolan, W.; Schacht, K.; Scheibel, T.; Groll, J. Strategies and molecular design criteria for 3D printable hydrogels. *Chem. Rev.* **2016**, *116*, 1496–1539. [CrossRef]
80. Hashemnejad, S.M.; Kundu, S. Rheological properties and failure of alginate hydrogels with ionic and covalent crosslinks. *Soft Matter.* **2019**, *15*, 7852–7862. [CrossRef]
81. Kulkarni, R.V.; Sreedhar, V.; Mutalik, S.; Setty, C.M.; Sa, B. Interpenetrating network hydrogel membranes of sodium alginate and poly(vinyl alcohol) for controlled release of prazosin hydrochloride through skin. *Int. J. Biol. Macromol.* **2010**, *47*, 520–527. [CrossRef] [PubMed]
82. Centers for Disease Control and Prevention. Toxicological Profile for Glutaraldehyde. Available online: https://www.atsdr.cdc.gov/ToxProfiles/tp208-c2.pdf (accessed on 4 April 2023).
83. Santos, N.T.; Das, G.; Moraes, L.F.; da Silva, M.G.C.; Vieira, M.G.A. Recovery of gold through adsorption onto sericin and alginate particles chemically crosslinked by proanthocyanidins. *J. Clean. Prod.* **2020**, *253*, 119925. [CrossRef]
84. Wang, W.; Liu, X.; Wang, X.; Zong, L.; Kang, Y.; Wang, A. Fast and highly efficient adsorption removal of toxic Pb(II) by a reusable porous semi-IPN hydrogel based on alginate and poly(vinyl alcohol). *Front. Chem.* **2021**, *9*, 662482. [CrossRef] [PubMed]
85. Zou, Z.; Zhang, B.; Nie, X.; Cheng, Y.; Hu, Z.; Liao, M.; Li, S. A sodium alginate-based sustained-release IPN hydrogel and its applications. *RSC Adv.* **2020**, *10*, 39722–39730. [CrossRef]

86. Kadri, R.; Elkhoury, K.; Ben Messaoud, G.; Kahn, C.; Tamayol, A.; Mano, J.F.; Arab-Tehrany, E.; Sánchez-González, L. Physicochemical Interactions in Nanofunctionalized Alginate/GelMA IPN Hydrogels. *Nanomaterials* **2021**, *11*, 2256. [CrossRef]
87. Ma, H.; Zhao, J.; Liu, Y.; Liu, L.; Yu, J.; Fan, Y. Controlled delivery of aspirin from nanocellulose-sodium alginate interpenetrating network hydrogels. *Ind. Crops Prod.* **2023**, *192*, 116081. [CrossRef]
88. Lee, K.Y.; Mooney, D.J. Alginate: Properties and biomedical applications. *Prog. Polym. Sci.* **2012**, *37*, 106–126. [CrossRef]
89. Shojarazavi, N.; Mashayekhan, S.; Pazooki, H.; Mohsenifard, S.; Baniasadi, H. Alginate/cartilage extracellular matrix-based injectable interpenetrating polymer network hydrogel for cartilage tissue engineering. *J. Biomater. Appl.* **2021**, *36*, 803–817. [CrossRef] [PubMed]
90. Ansson, P.; Lindberg, B. Structural studies of gellan gum, an extracellular polysaccharide elaborated by pseudomonas elodea. *Carbohydr. Res.* **1983**, *124*, 135–139. [CrossRef]
91. Gomes, D.; Batista-Silva, J.P.; Sousa, A.; Passarinha, L.A. Progress and opportunities in Gellan gum-based materials: A review of preparation, characterization and emerging applications. *Carbohydr. Pol.* **2023**, *311*, 120782. [CrossRef]
92. Bemiller, J.N. Structure-Property Relationships of Water-Soluble Polysaccharides. *J. Appl. Glycosci.* **1996**, *43*, 377–384.
93. Title 21-Food and Drugs. Chapter I-Food and Drug Administration. Department of Health and Human Services. Subchapter B -Food for Human Consumption. Available online: https://www.accessdata.fda.gov/scripts/cdrh/cfdocs/cfcfr/cfrsearch.cfm?fr=172.665 (accessed on 4 April 2023).
94. Vilela, C.A.; Correia, C.; da Silva Morais, A.; Santos, T.C.; Gertrudes, A.C.; Moreira, E.S.; Frias, A.M.; Learmonth, D.A.; Oliveira, P.; Oliveira, J.M. In vitro and in vivo performance of methacrylated gellan gum hydrogel formulations for cartilage repair. *J. Biomed. Mater. Res. A* **2018**, *106*, 1987–1996. [CrossRef]
95. D'Arrigo, G.; Navarro, G.; Di Meo, C.; Matricardi, P.; Torchilin, V. Gellan gum nanohydrogel containing anti-inflammatory and anti-cancer drugs: A multi-drug delivery system for a combination therapy in cancer treatment. *Eur. J. Pharm. Biopharm.* **2014**, *87*, 208–216. [CrossRef]
96. Kanyuck, K.M.; Mills, T.B.; Norton, I.T.; Norton-Welch, A.B. Swelling of high acyl gellan gum hydrogel: Characterization of network strengthening and slower release. *Carbohydr. Polym.* **2021**, *259*, 117758. [CrossRef]
97. Nieto, C.; Vega, M.A.; Rodríguez, V.; Pérez-Esteban, P.; Martín del Valle, E.M. Biodegradable gellan gum hydrogels loaded with paclitaxel for HER2+ breast cancer local therapy. *Carbohydr. Polym.* **2022**, *294*, 119732. [CrossRef]
98. Kunwar, P.; Andrade, B.L.; Poudel, A.; Xiong, Z.; Aryal, U.; Geffert, Z.J.; Poudel, S.; Fougnier, D.; Gitsov, I.; Soman, P. Printing Double-Network Tough Hydrogels Using Temperature-Controlled Projection Stereolithography (TOPS). *ACS Appl. Mater. Interf.* **2023**, *15*, 30780–30792. [CrossRef]
99. Cernencu, A.I.; Ioniță, M. The current state of the art in gellan-based printing inks in tissue engineering. *Carbohydr. Polym.* **2023**, *309*, 120676. [CrossRef]
100. Duffy, G.L.; Liang, H.; Williams, R.L.; Wellings, D.A.; Black, K. 3D reactive inkjet printing of poly-ε-lysine/gellan gum hydrogels for potential corneal constructs. *Mater. Sci. Eng. C* **2021**, *131*, 112476. [CrossRef] [PubMed]
101. Ng, J.Y.; Yu, P.; Murali, D.M.; Liu, Y.S.; Gokhale, R.; Ee, P.L.R. The influence of pregelatinized starch on the rheology of a gellan gum-collagen IPN hydrogel for 3D bioprinting. *Chem. Eng. Res. Des.* **2023**, *192*, 477–486. [CrossRef]
102. Ibrahim, A.G.; Elkony, A.M.; El-Bahy, S.M. Methylene blue uptake by gum arabic/acrylic amide/3-allyloxy-2-hydroxy-1-propanesulfonic acid sodium salt semi-IPN hydrogel. *Int. J. Biol. Macromol.* **2021**, *186*, 268–277. [CrossRef]
103. Piaskowski, K.; Świderska-Dąbrowska, R.; Zarzycki, P.K. Dye removal from water and wastewater using various physical, chemical, and biological processes. *J. AOAC Int.* **2018**, *101*, 1371–1384. [CrossRef] [PubMed]
104. Mishra, S.; Cheng, L.; Maiti, A. The utilization of agro-biomass/byproducts for effective bio-removal of dyes from dyeing wastewater: A comprehensive review. *J. Environ. Chem. Eng.* **2021**, *9*, 104901. [CrossRef]
105. Karthika, J.S.; Vishalakshi, B. Novel stimuli responsive gellan gum-graft-poly (DMAEMA) hydrogel as adsorbent for anionic dye. *Int. J. Biol. Macromol.* **2015**, *81*, 648–655. [CrossRef]
106. Getya, D.; Gitsov, I. Stronger together. Poly(styrene) gels reinforced by soft gellan gum. *Gels* **2022**, *8*, 607. [CrossRef]
107. Humphreys, J.M.; Chapple, C. Rewriting the lignin roadmap. *Curr. Opin. Plant Biol.* **2002**, *5*, 224–229. [CrossRef]
108. Vanholme, R.; Demedts, B.; Morreel, K.; Ralph, J.; Boerjan, W. Lignin Biosynthesis and Structure. *Plant Physiol.* **2010**, *153*, 895–905. [CrossRef]
109. Christensen, C.S.L.; Rasmussen, S.K. Lignin Biosynthesis and Structure. *Agronomy* **2019**, *9*, 256. [CrossRef]
110. Ralph, J.; Lundquist, K.; Brunow, P.G.; Lu, F.; Kim, H.; Schatz, P.; Boerjan, W. Lignins: Natural polymer from oxidative coupling of 4-hydroxyphenylpropanoids. *Phytochem. Rev.* **2004**, *3*, 29–60. [CrossRef]
111. Li, S.; Park, S.; Sherman, B.D.; Yoo, C.G.; Leem, G. Photoelectrochemical approaches for the conversion of lignin at room temperature. *Chem. Commun.* **2022**, *59*, 401–413. [CrossRef] [PubMed]
112. Kim, S.; Kang, H.C.; Chu, C.; Li, S.; Yoo, K.; Wijethunga, U.K.; Zheng, W.; Yoo, C.G.; Sherman, B.D.; Lee, J.-J.; et al. Solar Energy Driven C-C Bond Cleavage of Lignin with a D–π–A Organic Dye-Sensitized Photoanode. *Sustain. Energy Fuels* **2023**, *7*, 2339–2348. [CrossRef]
113. Meng, Y.; Lu, J.; Cheng, Y.; Li, Q.; Wang, H. Lignin-based hydrogels: A review of preparation, properties, and application. *Int. J. Biol. Macromol.* **2019**, *135*, 1006–1019. [CrossRef]
114. Gujjala, L.K.S.; Kim, J.; Won, W. Technical lignin to hydrogels: An Eclectic review on suitability, synthesis, applications, challenges and future prospects. *J. Clean. Prod.* **2022**, *363*, 132585. [CrossRef]

115. Kaur, R.; Sharma, R.; Chahal, G.K. Synthesis of lignin-based hydrogels and their applications in agriculture: A review. *Chem. Papers* **2021**, *75*, 4465–4478. [CrossRef]
116. Liu, Y.; Jin, C.; Zhaozhe, Y.; Wu, G.; Liu, G.; Kong, Z. Recent advances in lignin-based porous materials for pollutants removal from wastewater. *Int. J. Biol. Macromol.* **2021**, *187*, 880–891. [CrossRef]
117. Sugiarto, S.; Leow, Y.; Tan, C.L.; Wang, G.; Kai, D. How far is lignin from being a biomedical material? *Bioact. Mater.* **2022**, *8*, 71–94. [CrossRef]
118. Liu, C.; Li, Y.; Zhuang, J.; Xiang, Z.; Jiang, W.; He, S.; Xiao, H. Conductive hydrogels based on industrial lignin: Opportunities and challenges. *Polymers* **2022**, *14*, 3739. [CrossRef]
119. Dianat, N.; Rahmanifar, M.S.; Noori, A.; El-Kady, M.F.; Chang, X.; Kaner, R.B.; Mousavi, M.F. Polyaniline-lignin interpenetrating network for supercapacitive energy storage. *Nano Lett.* **2021**, *21*, 9485–9493. [CrossRef]
120. Domínguez-Robles, J.; Soledad Peresin, M.; Tamminen, T.; Rodríguez, A.; Larrañeta, E.; Jääskeläinen, A. Lignin-based hydrogels with "super-swelling" capacities for dye removal. *Int. J. Biol. Macromol.* **2018**, *115*, 1249–1259. [CrossRef] [PubMed]
121. Wang, W.; Li, Y.; Zhang, H.; Chen, T.; Sun, G.; Han, Y.; Li, J. Double-interpenetrating-network lignin-based epoxy resin adhesives for resistance to extreme environment. *Biomacromolecules* **2022**, *23*, 779–788. [CrossRef]
122. Zhang, W.; Ye, G.; Liao, D.; Chen, X.; Lu, C.; Nezamzadeh-Ejhieh, A.; Khan, M.S.; Liu, J.; Pan, Y.; Dai, Z. Recent Advances of Silver-Based Coordination Polymers on Antibacterial Applications. *Molecules* **2022**, *27*, 7166. [CrossRef]
123. Chen, W.; Liu, M.; Yang, H.; Nezamzadeh-Ejhieh, A.; Lu, C.; Pan, Y.; Liu, J.; Bai, Z. Recent Advances of Fe(III)/Fe(II)-MPNs in Biomedical Applications. *Pharmaceutics* **2023**, *15*, 1323. [CrossRef]
124. Adedeji, O.E.; Choi, J.Y.; Park, G.E.; Kang, H.J.; Aminu, M.O.; Min, J.H.; Chinma, C.E.; Moon, K.D.; Jung, Y.H. Formulation and characterization of an interpenetrating network hydrogel of locust bean gum and cellulose microfibrils for 3D printing. *Innov. Food Sci. Emerg. Technol.* **2022**, *80*, 103086. [CrossRef]
125. He, M.; Wang, Z.; Cao, Y.; Zhao, Y.; Duan, B.; Chen, Y.; Xu, M.; Zhang, L. Construction of Chitin/PVA Composite Hydrogels with Jellyfish Gel-Like Structure and Their Biocompatibility. *Biomacromolecules* **2014**, *15*, 3358–3365. [CrossRef]
126. Balitaan, J.N.I.; Hsiao, C.D.; Yeh, J.M.; Santiago, K.S. Innovation inspired by nature: Biocompatible self-healing injectable hydrogels based on modified-β-chitin for wound healing. *Int. J. Biol. Macromol.* **2020**, *162*, 723–736. [CrossRef] [PubMed]
127. Zhu, S.; Yao, L.; Pan, C.; Tian, J.; Li, L.; Luo, B.; Zhou, C.; Lu, L. 3D printed gellan gum/graphene oxide scaffold for tumor therapy and bone reconstruction. *Compos. Sci. Technol.* **2021**, *208*, 108763. [CrossRef]
128. Jana, S.; Pramanik, R.; Kumar, A.; Kumar, K. Gellan gum (GG)-based IPN microbeads for sustained drug release. *J. Drug Deliv. Sci. Technol.* **2022**, *69*, 103034. [CrossRef]
129. Lameirinhas, N.S.; Teixeira, M.C.; Carvalho, J.P.F.; Valente, B.F.A.; Pinto, R.J.B.; Oliveira, H.; Luís, J.L.; Pires, L.; Oliveira, J.M.; Vilela, C.; et al. Nanofibrillated cellulose/gellan gum hydrogel-based bioinks for 3D bioprinting of skin cells. *Int. J. Biol. Macromol.* **2023**, *229*, 849–860. [CrossRef] [PubMed]
130. Luzi, F.; Puglia, D.; Gracia, L. Development of a sustainable and antibacterial food packaging material based in a biopolymeric multilayer system composed by polylactic acid, chitosan, cellulose nanocrystals and ethyl lauroyl arginate. *Food Packag. Shelf Life* **2023**, *36*, 101050. [CrossRef]
131. Zhu, L.; Liu, Y.; Jiang, Z.; Sakai, E.; Qiu, J.; Zhu, P. Highly temperature resistant cellulose nanofiber/polyvinyl alcohol hydrogel using aldehyde cellulose nanofiber as cross-linker. *Cellulose* **2019**, *26*, 5291–5303. [CrossRef]
132. He, X.; Zeng, L.; Cheng, X.; Yang, C.; Chen, J.; Chen, H.; Ni, H.; Bai, Y.; Yu, W.; Zhao, K.; et al. Shape memory composite hydrogel based on sodium alginate dual crosslinked network with carboxymethyl cellulose. *Eur. Polym. J.* **2021**, *156*, 110592. [CrossRef]
133. Oveissi, F.; Naficy, S.; Le, T.Y.L.; Fletcher, D.F.; Dehghani, F. Tough and Processable Hydrogels Based on Lignin and Hydrophilic Polyurethane. *ACS Appl. Bio Mater.* **2018**, *1*, 2073–2081. [CrossRef] [PubMed]
134. Rico-García, D.; Ruiz-Rubio, L.; Pérez-Alvarez, L.; Hernández-Olmos, S.L.; Guerrero-Ramírez, G.L.; Vilas-Vilela, J.L. Lignin-based hydrogels: Synthesis and applications. *Polymers* **2020**, *12*, 81. [CrossRef]
135. Kulkarni, V.; Butte, K.; Rathod, S. Natural Polymers—A Comprehensive Review. *Int. J. Res. Pharm. Biomed. Sci.* **2012**, *3*, 1597–1613.
136. Yaşayan, G.; Alarçin, E.; Bal-Öztürk, A.; Avci-Adali, M. Natural polymers for wound dressing applications. In *Studies in Natural Products Chemistry*; Elsevier: Amsterdam, The Netherlands, 2022; Volume 74. [CrossRef]
137. Nagalakshmaiah, M.; Afrin, S.; Malladi, R.P.; Elkoun, S.; Robert, M.; Ansari, M.A.; Svedberg, A.; Karim, Z. Biocomposites: Present trends and challenges for the future. *Green Compos. Automot. Appl.* **2018**, *9*, 197–215. [CrossRef]
138. Environment Policy: General Principles and Basic Framework. Available online: https://www.europarl.europa.eu/factsheets/en/sheet/71/environment-policy-general-principles-and-basic-framework (accessed on 24 July 2023).
139. Chen, S.; Zhang, Z.L.; Song, F.; Wang, X.L.; Wang, Y.Z. Rapid Synthesis of Polymer-Grafted Cellulose Nanofiber Nanocomposite via Surface-Initiated Cu(0)-Mediated Reversible Deactivation Radical Polymerization. *Macromolecules* **2021**, *54*, 7409–7420. [CrossRef]
140. Kalia, S.; Boufi, S.; Celli, A.; Kango, S. Nanofibrillated cellulose: Surface modification and potential applications. *Colloid Polym. Sci.* **2014**, *292*, 5–31. [CrossRef]
141. Littunen, K.; Hippi, U.; Johansson, L.S.; Österberg, M.; Tammelin, T.; Laine, J.; Seppälä, J. Free radical graft copolymerization of nanofibrillated cellulose with acrylic monomers. *Carbohydr. Polym.* **2011**, *84*, 1039–1047. [CrossRef]
142. Yu, L.; Dean, K.; Li, L. Polymer Blends and Composites from Renewable Resources. *Prog. Polym. Sci.* **2006**, *31*, 576–602. [CrossRef]

143. John, M.J.; Anandjiwala, R.D. Recent Developments in Chemical Modificationand Characterization of Natural Fiber-ReinforcedComposites. *Polym. Compos.* **2008**, *29*, 119–237. [CrossRef]
144. Cheng, H.N.; Neiss, T.G. Solution NMR Spectroscopy of Food Polysaccharides. *Polym. Rev.* **2012**, *52*, 81–114. [CrossRef]
145. Wang, J.; Zhao, J.; Nie, S.; Xie, M.; Li, S. MALDI mass spectrometry in food carbohydrates analysis: A review of recent researches. *Food Chem.* **2023**, *399*, 133968. [CrossRef]
146. Tanan, W.; Panichpakdee, J.; Saengsuwan, S. Novel biodegradable hydrogel based on natural polymers: Synthesis, characterization, swelling/reswelling and biodegradability. *Eur. Polym. J.* **2019**, *112*, 678–687. [CrossRef]
147. Chu, C.K.; Joseph, A.J.; Limjoco, M.D.; Yang, J.; Bose, S.; Thapa, L.S.; Langer, R.; Anderson, D.G. Chemical Tuning of Fibers Drawn from Extensible Hyaluronic Acid Networks. *J. Am. Chem. Soc.* **2020**, *142*, 19715–19721. [CrossRef]
148. Kim, U.-J.; Park, J.; Li, C.; Jin, H.-J.; Valluzzi, R.; Kaplan, D.L. Structure and Properties of Silk Hydrogels. *Biomacromolecules* **2004**, *5*, 786–792. [CrossRef] [PubMed]
149. Rammensee, S.; Huemmerich, D.; Hermanson, K.D.; Scheibel, T.; Bausch, A.R. Rheological characterization of hydrogels formed by recombinantly produced spider silk. *Appl. Phys. A* **2006**, *82*, 261–264. [CrossRef]
150. McGill, M.; Grant, J.M.; Kaplan, D.L. Enzyme-Mediated Conjugation of Peptides to Silk Fibroin for Facile Hydrogel Functionalization. *Ann. Biomed. Eng.* **2020**, *48*, 1905–1915. [CrossRef] [PubMed]
151. Li, C.; Wang, L.; Chen, Z.; Li, Y.; Li, J. Facile and green preparation of diverse arabinoxylan hydrogels from wheat bran by combining subcritical water and enzymatic crosslinking. *Carbohydr. Polym.* **2020**, *241*, 116317. [CrossRef] [PubMed]
152. Chimene, D.; Lennox, K.K.; Kaunas, R.R.; Gaharwar, A.K. Advanced Bioinks for 3D Printing: A Materials Science Perspective. *Ann. Biomed. Eng.* **2016**, *44*, 2090–2102. [CrossRef] [PubMed]
153. Anand, R.; Amoli, M.S.; Huysecom, A.-S.; Amorim, P.A.; Agten, H.; Geris, L.; Bloemen, V. A tunable gelatin-hyaluronan dialdehyde/methacryloyl gelatin interpenetrating polymer network hydrogel for additive tissue manufacturing. *Biomed. Mater.* **2022**, *17*, 045027. [CrossRef]

molecules

MDPI

Article

Smart and Efficient Synthesis of Cyclic Poly(*N*-isopropylacrylamide)s by Ring Expansion RAFT (RE-RAFT) Polymerization and Analysis of Their Unique Temperature-Responsive Properties

Jin Motoyanagi *, Kenichi Bessho and Masahiko Minoda *

Faculty of Molecular Chemistry and Engineering, Graduate School of Science and Technology, Kyoto Institute of Technology, Matsugasaki, Sakyo-ku, Kyoto 606-8585, Japan

* Correspondence: jinmoto@kit.ac.jp (J.M.); minoda@kit.ac.jp (M.M.); Tel.: +81-75-724-7537 (J.M.); +81-75-724-7513 (M.M.)

Abstract: Cyclic polymers have many interesting properties compared to their linear analogs, but there are very few examples of their synthesis. This is because most cyclic polymers have been synthesized by stepwise processes, including synthesizing homo- or hetero-telechelic end-functionalized precursor polymers and consecutive intramolecularly coupling of both ends of the polymers. This requires a complicated synthesis, and the product yields are very low because the target cyclic polymers are usually synthesized under highly dilute conditions, consequently, making it difficult to systematically analyze the properties of cyclic polymers. In the present research, we have synthesized cyclic polymers using a ring expansion polymerization method. Particularly, the ring expansion RAFT polymerization (RE-RAFT polymerization) that we have developed using a cyclic chain transfer agent is a smart method that can synthesize cyclic polymers very efficiently. In this paper, we successfully synthesized cyclic-poly(*N*-isopropylacrylamide), which is widely known as a thermo-responsive polymer, by RE-RAFT polymerization. Furthermore, we have compared the thermo-responsive properties of the cyclic-poly(*N*-isopropylacrylamide)s with those of their linear analogs.

Keywords: RAFT polymerization; cyclic polymer; poly(*N*-isopropylacrylamide); ring-expansion RAFT polymerization; cyclic trithiocarbonate derivative; thermo-responsive property

1. Introduction

In recent years, cyclic polymers with no chain ends have exhibited many interesting properties compared to their linear analogs, such as distinguished glass transition temperature, viscosity, diffusion behavior, and hydrodynamic volume [1–5]. However, the synthesis of cyclic polymers with external stimuli-responsive polymers as the backbone moiety has been very limited because the synthesis is only possible through stepwise methods, resulting in very low yields [6,7]. The present study focuses on the thermo-responsive polymer among external stimulus-responsive polymers and aims to develop an efficient synthetic method for cyclic temperature-responsive polymers. In previous research, linear and cyclic poly(*N*-isopropylacrylamide) (polyNIPAM), which is one of the representative thermo-responsive polymers, have been demonstrated to display disparate aggregation and phase transition behaviors [8–10]. Because the phase transition temperature of polyNIPAM is proximate to body temperature [11,12], polyNIPAM and its derivatives have been well recognized as promising materials in biomedical and biotechnology fields [13–15]. Moreover, Tezuka has demonstrated that the thermal stability of self-assembling polymeric micelles is significantly influenced by topological factors [16–18]. The designed amphiphilic linear triblock copolymers and their cyclization products were observed to self-assemble into flower-like micelles. However, the cloud point (CP) of the cyclic polymers was found to increase by over 40 °C in comparison to the linear polymers, despite no significant change in chemical composition [17]. Thus, cyclic polymers exhibit distinctive properties that diverge from those of linear counterparts in response to external stimuli, and they are highly anticipated as new functional materials.

Over the past few decades, numerous researchers have endeavored to synthesize cyclic polymers. The most prevalent synthetic methodology is the end coupling method. First, homo- or hetero-telechelic polymers are synthesized by introducing reactive groups at both ends of a linear polymer using living polymerization, and then cyclic polymers are synthesized by intramolecular coupling of the terminal functional groups [3]. Therefore, this synthetic method requires a multi-step synthetic process and the coupling reactions usually need to be carried out under highly dilute conditions to favor intramolecular reactions, resulting in very low product yields. It was therefore a cumbersome task to systematically synthesize cyclic polymers and investigate the correlation between their structure and properties. On the other hand, ring-expansion polymerization (REP) has been developed as an alternative method for the synthesis of cyclic polymers. REP is superior to the end-coupling methods mentioned above because it is a smarter, more straightforward synthetic approach and allows for the yielding of large amounts of cyclic polymers [19,20]. In REP, the utilization of designed cyclic initiators that can generate a couple of active sites within a molecule allows the synthesis of cyclic polymers; therefore, the cyclic topology is maintained throughout the polymerization by successive insertion of monomers. The earliest examples of REP work include the synthesis of cyclic polycaprolactones using cyclic alkoxytin derivatives as initiators by Kricheldorf [21]. Additionally, Grubbs has demonstrated the successful synthesis of cyclic polyolefins by ring-opening metathesis polymerization employing designed cyclic carbene–ruthenium complexes as initiators [22,23]. In addition, Ouchi used a cyclic hemiacetal ester compound as an initiator for living cationic polymerization and succeeded in the synthesis of cyclic polyvinyl ethers [24,25]. Moreover, the synthesis of cyclic polylactides by ring-opening polymerization of lactides using nitrogen-containing heterocyclic carbenes as activators has also been reported [26]. Thus, several methods for the synthesis of cyclic polymers using REP have been reported, but the types of monomers that can be applied are limited, and there are very few examples of the synthesis of cyclic polymers from commonly used vinyl monomers.

To establish a more versatile strategy for the synthesis of cyclic polymers using REP, we focused on reversible addition-fragmentation chain transfer (RAFT) polymerization [27–32], which is applicable to a wide range of vinyl monomers such as (meth)acrylates, (meth)acrylamides, and styrene derivatives and has the potential to control molecular weight and repeating unit arrangement (e.g., block polymer structures). In RAFT polymerization, monomers are inserted between the growing ends of the polymer chain and the RAFT agent residues. Therefore, we have conceived of an efficient synthesis of cyclic polymers by ring-expansion RAFT (RE-RAFT) polymerization using a cyclic RAFT agent with two potential reaction sites in the molecule. RE-RAFT polymerization using cyclic chain transfer agents has been previously reported by Pan, but under very limited conditions, i.e., under γ-irradiation with ^{60}Co and at -30 °C [33]. We have already succeeded in the efficient synthesis of cyclic-poly(*tert*-butyl acrylate) for the first time by performing RE-RAFT polymerization of *tert*-butyl acrylate (TBA) using cyclic trithiocarbonate (4,7-diphenyl-[1,3]dithiepane-2-thione, hereinafter referred to as CTTC) as a RAFT agent [34]. The structural analyses of the obtained polymers using size exclusion chromatography (SEC), ^{1}H NMR, and matrix-assisted laser desorption ionization-time of flight mass spectrometry (MALDI-TOF-MS) revealed the formation of cyclic-polyTBA. Furthermore, we found that the "ring-fusion" of the generated cyclic polymers due to the cross-propagation occurred during the RE-RAFT polymerization, resulting in the production of a mixture of cyclic polymers of various sizes [34]. In this study, with the aim of synthesizing functional cyclic polymers, we investigated the synthesis of cyclic-polyNIPAM (Scheme 1a) by the RE-RAFT polymerization of *N*-isopropylacrylamide. The major difference between this study and the synthesis of cyclic-polyNIPAMs in previous research [8–10] is that the synthesis of cyclic-polyNIPAM by RE-RAFT polymerization can be performed in a simple one-pot synthesis with high yield. The obtained polymers were characterized in detail by ^{1}H NMR, SEC, and MALDI-TOF-MS. Furthermore, to investigate the unique thermo-responsive properties of the cyclic-polyNIPAMs, we synthesized cyclic-polyNIPAMs with different molecular weights and compared their thermo-responsive properties with those of the linear analogs. This RE-RAFT polymerization is a smart synthetic method that can efficiently obtain cyclic polymers by simply adding a cyclic RAFT agent to a conventional free radical polymerization system. For example, it is possible to synthesize multi-component cyclic polymers by copolymerization of *N*-isopropylacrylamide with other acrylamide monomers, so it is expected that it can be applied to the control of thermo-responsive properties.

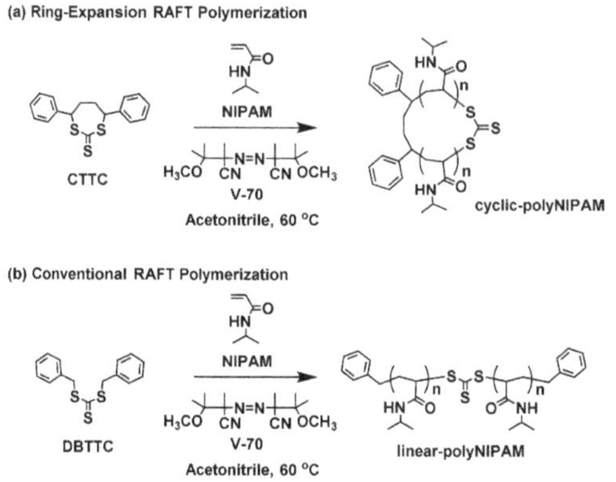

Scheme 1. RAFT polymerization of NIPAM (**a**) with CTTC as a cyclic RAFT agent and (**b**) with DBTTC as an acyclic RAFT agent.

2. Results

2.1. Ring-Expansion RAFT (RE-RAFT) Polymerization of NIPAM with CTTC

We investigated RE-RAFT polymerization of NIPAM at 60 °C in acetonitrile using 2,2′-azobis(4-methoxy-2,4-dimethylvaleronitrile) (V-70) as a radical initiator and CTTC as a cyclic RAFT reagent. ($[NIPAM]_0/[CTTC]_0/[V\text{-}70]_0 = 200/1/0.4$, $[NIPAM]_0 = 20$ wt%) (Scheme 1a). In Figure 1, the time-conversion curve obtained for the polymerization of NIPAM showed an induction phase of approximately 0.5 h, reaching a conversion of about 70% within 2 h. After the induction phase, the pseudo-first-order kinetic plot of NIPAM polymerization showed a linear relationship, indicating that the concentration of the growing active species remained constant throughout the polymerization, suggesting that a controlled polymerization process occurred. However, SEC analysis showed peculiar polymerization behavior, similar to the previously reported RE-RAFT polymerization of TBA [34]. In fact, from the very early stages of polymerization (conv.~15%), the chromatogram showed bimodal peaks, including a new peak appearing in the high molecular weight region (Figure 2, black line) (Table 1). This phenomenon is due to the fact that two pathways occur simultaneously in this RE-RAFT polymerization [23,34]: the "ring-expansion pathway", in which cyclic polymers are synthesized by ring expansion polymerization with monomer insertion into the cyclic species, and the "ring-fusion pathway", by intermolecular chain transfer reactions between the formed cyclic polymers containing one or more trithiocarbonate moieties (Scheme 2). The latter path resulted in generating higher molecular weight cyclic polymers. Considering the reaction mechanism of the "ring-fusion pathway", the resulting high-molecular-weight polymers are expected to have a ring structure in which multiple polyNIPAM segments are linked to each other via multiple trithiocarbonate (TC) moieties. Therefore, cleavage of the TC moieties yields individual polyNIPAM segments, and a detailed investigation of the resulting polyNIPAM segment polymers provides useful information about the polymerization behavior of the RE-RAFT polymerization of NIPAM. As previously reported [34], the TC moieties connecting the polyNIPAM segments are expected to be hydrolyzed by primary amines to generate thiol end groups, so by adding an excess amount of TBA and performing a Michael addition reaction to prevent the formation of inter- and intramolecular disulfide bonds, polyNIPAM polymers capped at both ends with TBA residues can be obtained. The polymers obtained by RE-RAFT polymerization were hydrolyzed with hexylamine in THF followed by the reaction with TBA (Scheme 3). After 1 h of reaction at room temperature, the UV–vis spectrum of the reaction mixture showed the disappearance of the 310 nm peak assignable to the TC moieties, indicating that the quantitative hydrolysis had progressed. The obtained polyNIPAM from the cleavage reaction was then characterized by SEC to study the growth of polyNIPAM segments during RE-RAFT polymerization (Figure 2, red line). Regardless of the reaction time, the SEC curves showed a unimodal peak, and as monomer conversion increased, the peak top shifted to the high molecular weight region, while the molecular weight distribution (M_w/M_n) remained relatively narrow, around 1.2. The M_n values calculated from the SEC curves were found to be smaller than the theoretical molecular weights. This discrepancy is likely due to the interaction between the SEC column and polyNIPAM caused by the use of THF as the eluent in the SEC measurement. These results indicate that the polymerization of NIPAM in the RE-RAFT polymerization conditions using the cyclic RAFT reagent CTTC proceeded in a controlled manner, producing cyclic-polyNIPAM consisting of multiple polyNIPAM segments with uniform chain lengths linked with TC groups.

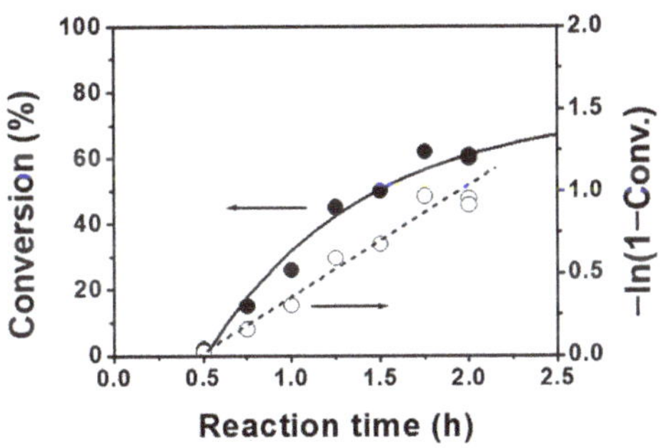

Figure 1. A time-conversion curve (black circles) and pseudo-first-order kinetic plot (white circles) for the polymerization of NIPAM. Polymerization was conducted with CTTC/V-70 in acetonitrile at 60 °C ($[NIPAM]_0/[CTTC]_0/[V\text{-}70]_0 = 200/1/0.4$, $[NIPAM]_0 = 20$ wt%).

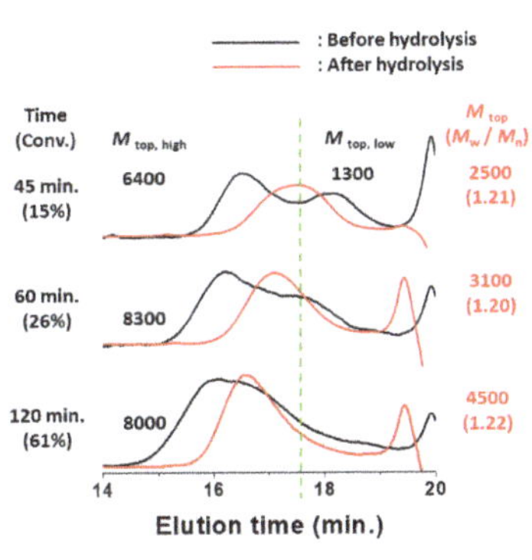

Figure 2. SEC curves of polyNIPAMs measured in THF. Polymerization was conducted with CTTC/V-70 in acetonitrile at 60 °C ($[NIPAM]_0/[CTTC]_0/[V\text{-}70]_0 = 200/1/0.4$, $[NIPAM]_0 = 20$ wt%) (Black line). SEC curves of the cleaved segment polyNIPAM from the generated polymers by RE-RAFT polymerization of NIPAM with CTTC obtained at various NIPAM conversions (Red line).

Table 1. Molecular weights and molecular weight distributions of the cyclic-polyNIPAMs obtained by RE-RAFT polymerization of NIPAM with a RAFT reagent CTA.

	CTA	Reaction Time (h)	NIPAM Conversion (%) [a]	$M_{n,\,theory}$ [b] (g mol^{-1})	$M_{n,\,top}$ [c,d] (g mol^{-1})	M_w/M_n [e]
cyclic-polyNIPAM$_{30}$	CTTC	0.75	15	3400	6400 and 1300	2.15
cyclic-polyNIPAM$_{52}$	CTTC	1	26	5900	8300	1.68
cyclic-polyNIPAM$_{122}$	CTTC	2	61	14,000	8000	1.78
cyclic-polyNIPAM$_{360}$	CTTC	2 [e]	72	41,000	28,000	1.53
linear-polyNIPAM$_{146}$	DBTTC	2	73	16,000	6600	1.21
linear-polyNIPAM$_{450}$	DBTTC	2 [e]	90	51,000	36,000	1.61

[a] Determined by ^{1}H NMR. [b] Calculated from $M_{n,\,theory}$ = ([NIPAM]$_0$/[CTA]$_0$ × (molecular weight of NIPAM) × (conversion of NIPAM). [c] Estimated by PSt-calibrated SEC in THF. [d] Estimated from peak top value of SEC curve. [e] Polymerization conditions; [NIPAM]$_0$/[CTA]$_0$/[V-70]$_0$ = 500/1/0.4, [NIPAM]$_0$ = 20 wt%.

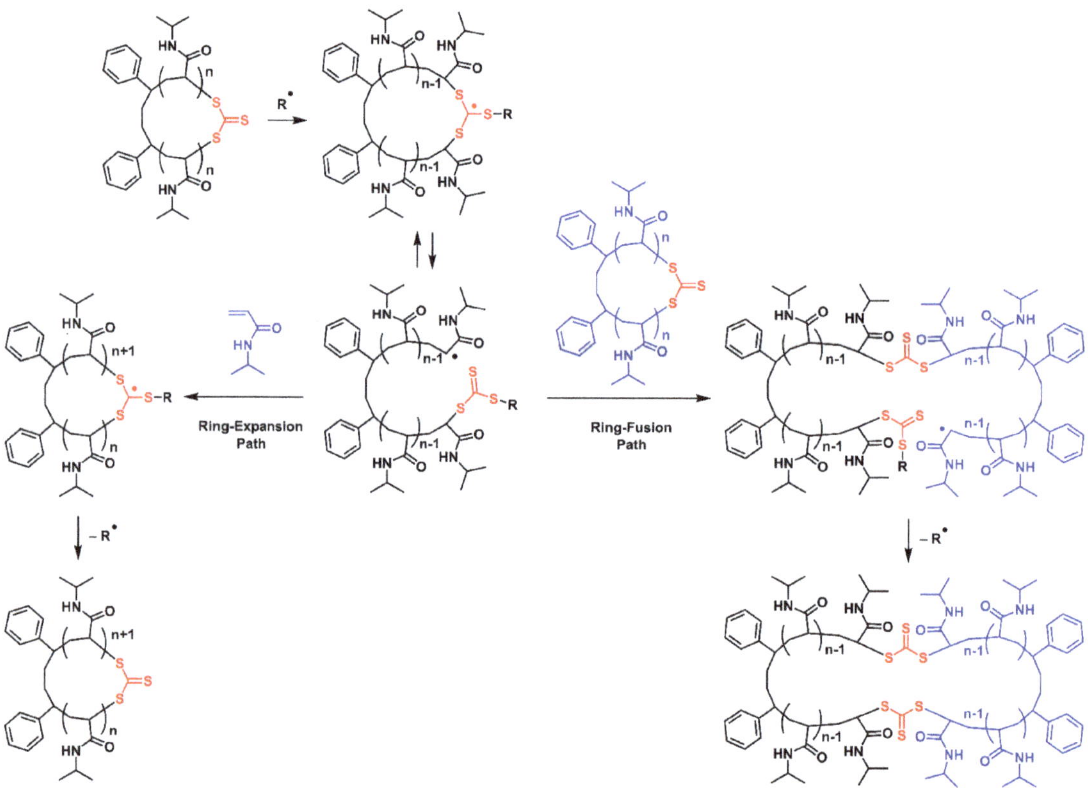

Scheme 2. Schematic of ring-expansion RAFT polymerization of NIPAM with a cyclic RAFT reagent CTTC containing "ring-expansion paths" and "ring-fusion paths".

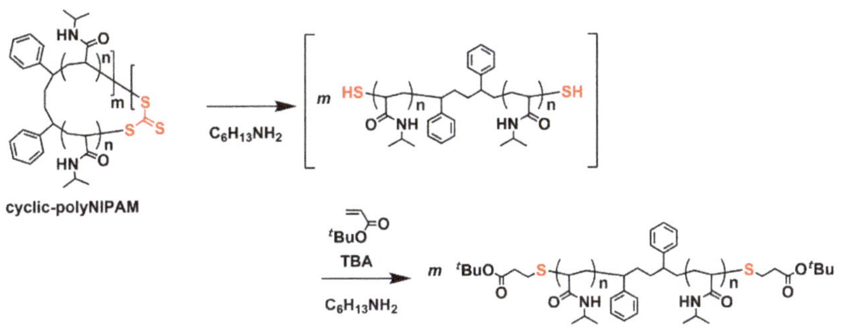

Scheme 3. The hydrolysis of the cyclic-polyNIPAM with hexylamine followed by the Michael addition reaction with TBA.

2.2. Characterization of the Obtained Polymers by RE-RAFT Polymerization with CTTC

After reprecipitation of the obtained cyclic-polyNIPAM (good solvent; acetone, poor solvent; hexane), the structure of the isolated cyclic-polyNIPAM$_{122}$ (under the same polymerization conditions as in Figures 1 and 2, NIPAM conversion = 61%) was analyzed in detail by ^{1}H NMR spectroscopy (Figure 3A). As a reference polymer, linear-polyNIPAM$_{146}$ was synthesized separately by conventional RAFT polymerization using *S*,*S*-dibenzyl trithiocarbonate (DBTTC) as an acyclic RAFT reagent at 60 °C in acetonitrile ([NIPAM]$_0$/[DBTTC]$_0$/[V-70]$_0$ = 200/1/0.4, [NIPAM]$_0$ = 20 wt%, NIPAM conversion = 73%, linear-polyNIPAM: M_n = 6600, M_w/M_n = 1.21; estimated by polystyrene calibrated SEC in THF) (Scheme 1b). The ^{1}H NMR spectra of the linear-polyNIPAM$_{146}$ are shown in Figure 3B. The key protons for the structural analysis are the phenyl proton peak (peaks d′) arising from the end of the linear-polyNIPAM$_{146}$, the pair of methine protons (peak c′) adjacent to the TC moieties, and the methine peak of the pendant isopropyl groups (peak b′). In DMSO-d_6, peak c′ was clearly observed at 4.3–4.5 ppm, but peaks d′ at 6.9–7.3 ppm overlapped with the NH proton of the amide group, so their integral ratio cannot be directly estimated. On the other hand, in methanol-d_4, peaks d′ appeared at 6.9–7.3 ppm, but peak c′ overlapped with the residual solvent, so a direct comparison of their integral ratio is not possible. Therefore, the isopropyl methine protons (peak b′), which was clearly observed at 3.8–4.1 ppm in both deuterated solvents, was used as the standard peak, and the integral ratio of peak d′ to peak c′ was estimated from the ratio of peak c′/peak b′ in DMSO-d_6 and the ratio of peak d′/peak b′ in methanol-d_4. As a result, the peak intensity ratio of the terminal phenyl protons (peak d′) and the methine protons adjacent to the TC group (peak c′) was determined to be 10.0:1.96, which is in good agreement with the expected telechelic structure (Figure 3B). Based on the peak intensity ratio of peak d′/peak b′ in methanol-d_4, the number-average degree of polymerization (DP_n) of the linear-polyNIPAM$_{146}$ was evaluated to be 147. This DP_n value was in good agreement with the theoretical value of 146, which was calculated based on the feed molar ratio of NIPAM to DBTTC and the monomer conversion. By comparison with the ^{1}H NMR data of the linear-polyNIPAM$_{146}$, we identified the chemical structure of the obtained polyNIPAM based on RE-RAFT polymerization (Figure 3A). The phenyl proton peaks d derived from the CTTC residue appeared at 6.9–7.4 ppm in methanol-d_4 and the characteristic peak c at 4.4 ppm of the methine protons adjacent to the TC moieties was also observed in DMSO-d_6. The ratio of the resonance intensities of the phenyl protons versus methine protons [Figure 3A, peaks d and peak c, respectively] was 10.0:1.96, which indicates that NIPAM monomers were inserted between the C–S bond of the TC moiety and the benzyl carbon (–S–C(=S) –*S*–*C*H(Ph)CH$_2$CH$_2$–) in the initiation process of the RE-RAFT polymerization and consecutive insertions during the propagation. This result indicates that the molecular structure thoroughly retained the cyclic structure of CTTC and the resulting polyNIPAMs consist of a cyclic structure without termini. Furthermore, DP_n of the polyNIPAM segments was evaluated to be 137 from the intensity ratio of the peak b and peak c. This DP_n value is nearly identical to the theoretical one (Theoretical DP_n = $[NIPAM]_0/[CTTC]_0 \times$ (conversion of NIPAM)). If acyclic polyNIPAM were included in the product polymers, these ratios (peak area ratios of peak d/peak c and peak b/peak c) would deviate from the theoretical values, clearly indicating that the target cyclic polyNIPAM was formed based on RE-RAFT polymerization.

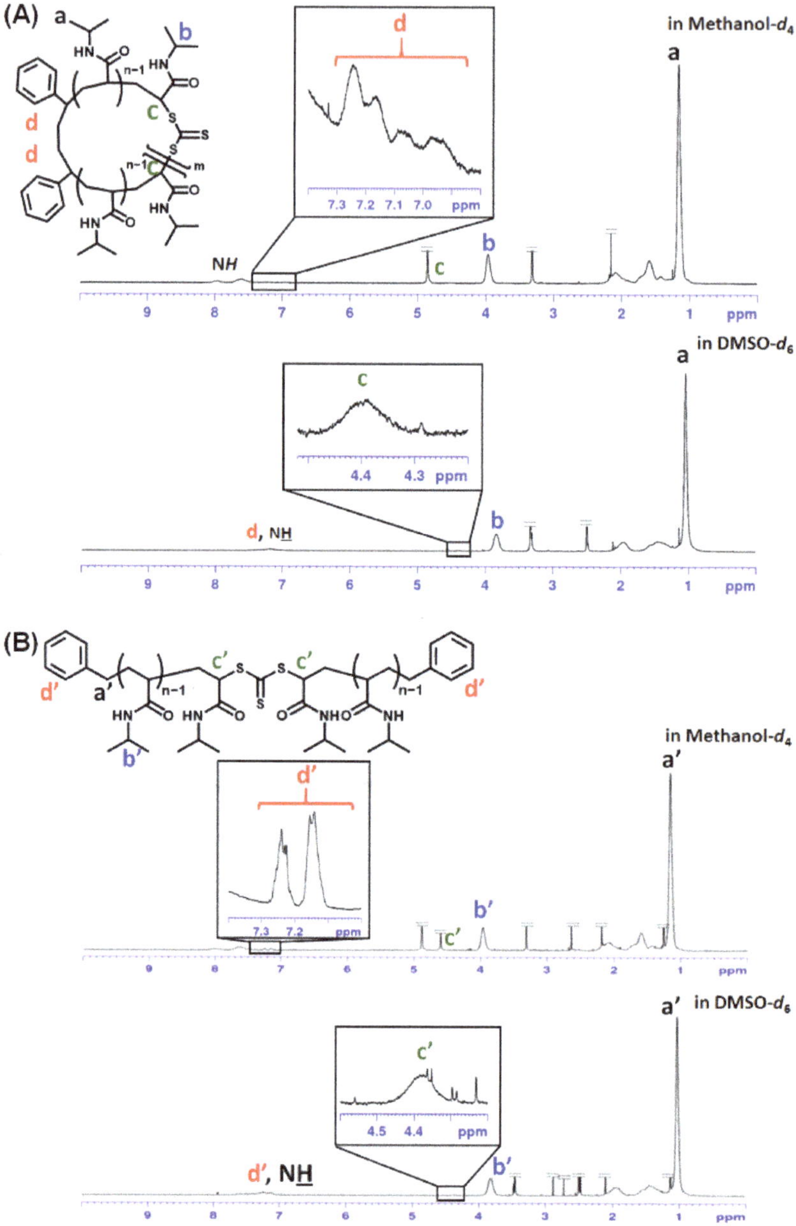

Figure 3. ^{1}H NMR spectra of (**A**) cyclic-polyNIPAM$_{122}$ by CTTC/V-70 and (**B**) linear-polyNIPAM$_{146}$ by DBTTC/V-70 in methanol-d_4 and DMSO-d_6.

To further verify that the polyNIPAM synthesized by the RE-RAFT polymerization has a cyclic structure, MALDI-TOF-MS measurements (matrix: 9-nitroanthracene, cationizing agent: sodium trifluoroacetate) were performed. With the aim of obtaining data with higher resolution, we employed the polymer sample produced at the early stage of the RE-RAFT polymerization. Figure 4 shows the MALDI-TOF-MS of polyNIPAM (cyclic-polyNIPAM$_{30}$) obtained at Conv. = 15% under the same polymerization conditions as in Figures 1 and 2. Figure 4 depicts a series of the peaks attributed to the cyclic-polyNIPAM of various DP_n along with a summary comparing the observed and calculated m/z values. The calculated m/z values were based on the cyclic-polyNIPAM having a CTTC moiety and repeating NIPAM units, ionized with Na^+ [CTTC-(NIPAM)$_n$ + Na]$^+$ (M_n = 316 + (113 × n) + 23). The observed and calculated m/z values are nearly identical, suggesting that the obtained polyNIPAMs possess no terminal structure and consist only of a cyclic structure. And the minor series of peaks are probably attributable to the structure in which the trithiocarbonate group in cyclic-polyNIPAM is hydrolyzed to thiol groups under ionization conditions (calculated m/z values; M_n = 316 + (113 × n) − 44).

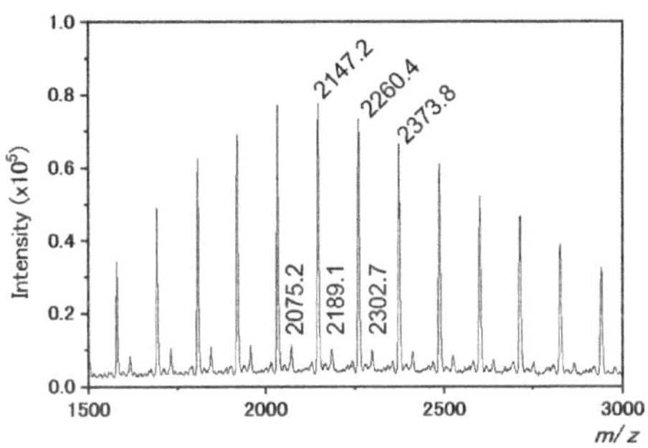

Major Peaks

Cyclic-polyNIPAM + Na$^+$	m/z found	m/z calculated
DP_n = 16	2147.2	2148.4
DP_n = 17	2260.4	2261.5
DP_n = 18	2373.8	2374.5

Minor Peaks

Hydrolysis Cyclic-polyNIPAM	m/z found	m/z calculated
DP_n = 16	2075.2	2079.4
DP_n = 17	2189.1	2192.5
DP_n = 18	2302.7	2305.5

Figure 4. MALDI-TOF MS of the cyclic-polyNIPAM$_{30}$ (matrix: 9-nitroanthracene; cationizing agent: sodium trifluoroacetate).

2.3. *Temperature-Responsive Properties of the Cyclic-polyNIPAM and Linear-polyNIPAM*

One of the fundamental physical properties of temperature-responsive water-soluble polymers is the thermal phase transition at a specific temperature that is specific to the polymer structure, expressed as the lower critical solution temperature (LCST) and/or cloud point (CP) temperature. Liu et al. compared the thermal phase transition temperatures of cyclic and linear polyNIPAM [8]. In their study, they measured the LCST and CP temperatures of cyclic and linear polyNIPAM in aqueous solution using a temperature-dependent turbidity measurement method. Here, the LCST and CP temperatures are defined as the temperatures at which the transmittance decreases by 1% and 50%, respectively. They found that the LCST and CP temperature of the linear polymer were 34 °C and 37 °C, respectively, while these values for the cyclic polymer were 27 °C and 35 °C, respectively, indicating that the LCST and CP temperature of the cyclic polymer are lower [8]. CP is a macroscopic parameter that defines the temperature at which a solution becomes turbid and can be confirmed by visual inspection. On the other hand, the LCST is defined as a point on the phase diagram and is measured by critical temperature measurement using DCS measurement [11,12]. Here, the onset temperature at which the transmittance decreases by 1% can be regarded as the transition temperature from coil-to-sphere, so the values at which the transmittance changes by 1% and 50% were used in this study to compare with the results reported by Liu et al. [8]. Since it is known that the temperature response of polyNIPAM depends on the molecular weight of the polymer [35], we synthesized cyclic-polyNIPAM$_{360}$ (M_n = 28,000, M_w/M_n = 1.53) and linear-polyNIPAM$_{450}$ (M_n = 36,000,

M_w/M_n = 1.61) with relatively similar molecular weight and molecular weight distribution. Due to the broad molecular weight distribution of the cyclic-polyNIPAM$_{360}$ as prepared by RE-RAFT polymerization, the polymer sample was fractionated by preparative SEC.

Figure 5 shows the temperature dependence of the transmittance of aqueous solutions (concentration of 1.0 g/L) of cyclic-polyNIPAM$_{360}$ (Figure 5a) and linear-polyNIPAM$_{450}$ (Figure 5b) at a wavelength of 500 nm upon heating and cooling at a rate of 0.5 °C/min. Upon heating the aqueous solutions, the linear-polyNIPAM$_{450}$ solution showed a rapid decrease in transmittance, while the cyclic-polyNIPAM$_{360}$ solution showed a more gradual decrease. Based on the point at which the transmittance of the solutions began to decrease, the onset temperature for the cyclic-polyNIPAM$_{360}$ and the linear-polyNIPAM$_{450}$ were determined to be 29 °C and 32 °C, respectively, indicating that, above these temperatures, interchain interactions began to occur, forming aggregates, which resulted in a decrease in transmittance due to scattering of the incident light. Furthermore, the CP temperatures were determined from the temperature at which transmittance reaches 50%, which were 31 °C and 33 °C for the cyclic-polyNIPAM$_{360}$ and the linear-polyNIPAM$_{450}$, respectively. For these temperatures, the cyclic-polyNIPAM$_{360}$ exhibited lower temperatures than the linear-polyNIPAM$_{450}$. Thus, during the temperature increase process, the water molecules that were hydrated to the polyNIPAM segments dehydrated at a relatively low temperature to form intramolecular hydrogen bonds, resulting in a lower onset temperature for the cyclic-polyNIPAM$_{360}$. This result is similar to the trend previously reported for the thermo-responsive properties of the cyclic- and linear-polyNIPAMs by Liu [8]. On the other hand, as the temperature decreased, the linear-polyNIPAM$_{450}$ began to rapidly increase in transmittance, while the cyclic-polyNIPAM$_{360}$ showed a clearly slower rise in transmittance and remained opaque until around 30 °C. It is speculated that these phenomena are because the cyclic-polyNIPAM has no terminal structure, which makes the molecular chains more constrained compared to the linear analog, allowing the formation of enhanced intramolecular hydrogen bonds. That is, during the temperature decrease process, the intramolecular hydrogen bonds were formed relatively strongly in the cyclic-polyNIPAM, making it difficult for redissolution due to hydration to occur.

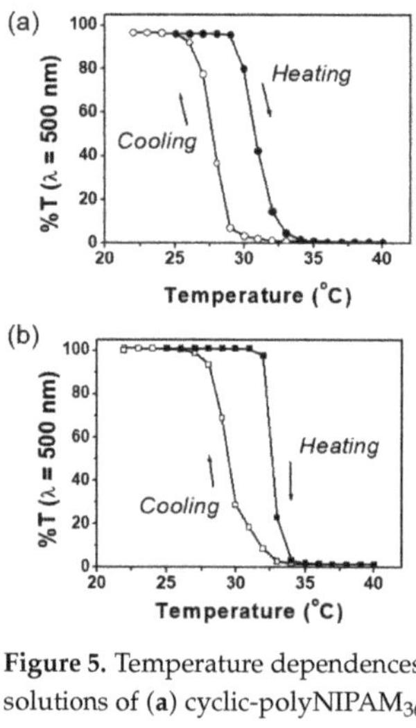

Figure 5. Temperature dependences of optical transmittance at 500 nm obtained for 1.0 g/L aqueous solutions of (**a**) cyclic-polyNIPAM$_{360}$ and (**b**) linear-polyNIPAM$_{450}$.

3. Materials and Methods

3.1. Chemicals and Reagents

Unless otherwise stated, all commercial reagents including 2,2′-azobis(4-methoxy-2,4-dimethylvaleronitrile) (V-70; FUJIFILM Wako Pure Chemical Corporation, Osaka, Japan, 95%) and *S*,*S*-dibenzyl trithiocarbonate (DBTTC; FUJIFILM Wako Pure Chemical Corporation, Osaka, Japan, 98%) were used as received. *N*-isopropylacrylamide (NIPAM; FUJIFILM Wako Pure Chemical Corporation, Osaka, Japan, 98%) was recrystallized from a mixture of toluene/*n*-hexane (1/1, *v*/*v*) prior to use. A cyclic trithiocarbonate derivative, 4,7-diphenyl-[1,3]dithiepane-2-thione (CTTC) was synthesized according to the literature [36,37].

3.2. Methods

^{1}H NMR spectra were assessed in methanol-d_4 or DMSO-d_6 at 25 °C on a Bruker model AC-500 spectrometer (Bruker, Billerica, MA, USA), operating at 500 MHz, where chemical shifts (δ in ppm) were determined with respect to non-deuterated solvent residues as internal standards. Analytical size exclusion chromatography (SEC) was performed at 40 °C, using 8.0 mm × 300 mm polystyrene gel columns (TOSOH-TSKgel H type × 2) on a TOSOH GPC-8320 system (TOSOH, Tokyo, Japan), equipped with a UV-8000 variable-wavelength UV–vis detector (TOSOH, Tokyo, Japan). The number-average molecular weight (M_n) and molecular weight distribution (M_w/M_n) were calculated from the chromatographs with respect to 15 polystyrene standards (Scientific Polymer Products, Inc., Ontario, NY, USA; M_n = 580–670,000 g mol^{-1}, M_w/M_n = 1.01–1.07). Preparative size exclusion chromatography (SEC) was performed at 25 °C by using 21.5 mm × 300 mm polystyrene gel columns (TOSOH TSKgel G2000H, G2500H, and G3000H) on a TOSOH model CCPE equipped with a RI-8022 RI detector (TOSOH, Tokyo, Japan). Matrix-assisted laser desorption/ionization time-of-flight mass spectrometry (MALDI-TOF-MS) was performed on a Bruker model amaZon SL spectrometer (Bruker, Billerica, MA, USA) using THF as a solvent, 9-nitroanthracene as a matrix, and sodium trifluoroacetate as a cationizing agent. UV–vis spectra were recorded using a quartz cell of 1 cm path length on a JASCO Type V-730 spectrometer equipped with a Peltier thermostatted cell holder ETCS-761 (JASCO, Hachioji, Japan).

3.3. Ring-Expansion RAFT Polymerization of NIPAM with CTTC (Synthesis of Cyclic-polyNIPAM$_{30}$, Cyclic-polyNIPAM$_{52}$ and Cyclic-polyNIPAM$_{122}$)

Ring-expansion RAFT polymerization of NIPAM was carried out with CTTC as a cyclic RAFT agent and V-70 as an initiator. To a solution of NIPAM (321 mg, 2.8 mmol) and CTTC (4.5 mg, 14 µmol) in acetonitrile (1.28 g) was added V-70 (1.9 mg, 6.0 µmol) in a glass tube ($[NIPAM]_0/[CTTC]_0/[V\text{-}70]_0 = 200/1/0.4$). The resulting solution was degassed by three freeze-pump-thaw cycles, and then the glass tube was sealed under vacuum, heated at 60 °C for 0.5–2 h, and quenched by rapid cooling. The reaction mixture was analyzed by SEC and ^{1}H NMR spectroscopy. By the ^{1}H NMR spectra of reaction mixtures, the NIPAM conversions were calculated based on the peak intensity ratio of the isopropyl methyl protons and the vinyl protons of NIPAM. The acetonitrile solution of the reaction mixture was poured into a large amount of *n*-hexane to precipitate the polymers and remove the unreacted monomers. The resultant polymer was collected by centrifugation and dried under reduced pressure. The isolated polymer structure of cyclic-polyNIPAM$_{122}$ was analyzed by ^{1}H NMR spectroscopy and the isolated polymer of cyclic-polyNIPAM$_{30}$ was analyzed by MALDI-TOF-MS.

3.4. Conventional RAFT Polymerization of NIPAM with DBTTC (Synthesis of Linear-polyNIPAM$_{146}$)

Conventional RAFT polymerization of NIPAM was carried out with DBTTC as a bifunctional acyclic RAFT agent and V-70 as an initiator. To a solution of NIPAM (161 mg, 1.4 mmol) and DBTTC (2.1 mg, 7.2 µmol) in acetonitrile (642 mg) was added V-70 (0.86 mg, 2.8 µmol) in a glass tube ($[NIPAM]_0/[DBTTC]_0/[V\text{-}70]_0 = 200/1/0.4$). The resulting solution was degassed by three freeze-pump-thaw cycles, and then the glass tube was sealed under vacuum, heated at 60 °C for 0.5–2 h, and quenched by rapid cooling. The reaction mixture was analyzed by SEC and ^{1}H NMR spectroscopy. The acetonitrile solution of the reaction mixture was poured into a large amount of *n*-hexane to precipitate the polymers and remove the unreacted monomers. The resultant polymer was collected by centrifugation and dried under reduced pressure. The isolated polymer structure was analyzed by ^1H NMR spectroscopy and MALDI-TOF-MS.

3.5. Synthesis of Cyclic-polyNIPAM$_{360}$ and Linear-polyNIPAM$_{450}$ for Analysis of Temperature-Responsive Properties

For the synthesis of cyclic-polyNIPAM$_{360}$, V-70 (0.76 mg, 2.4 μmol) was added to a solution of NIPAM (321 mg, 2.8 mmol) and CTTC (1.8 mg, 5.6 μmol) in acetonitrile (1.28 g) in a glass tube ($[NIPAM]_0/[CTTC]_0/[V\text{-}70]_0 = 500/1/0.4$). The resulting solution was degassed by three freeze-pump-thaw cycles, and then the glass tube was sealed under vacuum, heated at 60 °C for 2 h, and quenched by rapid cooling. The acetonitrile solution of the reaction mixture was poured into a large amount of *n*-hexane to precipitate the polymers and remove the unreacted monomers. The resultant polymer was collected by centrifugation and dried under reduced pressure. Furthermore, the obtained polymer sample was fractionated by preparative SEC.

For the synthesis of linear-polyNIPAM$_{450}$, V-70 (0.34 mg, 1.1 μmol) was added to a solution of NIPAM (161 mg, 1.4 mmol) and DBTTC (0.84 mg, 2.8 μmol) in acetonitrile (642 mg) in a glass tube ($[NIPAM]_0/[DBTTC]_0/[V\text{-}70]_0 = 500/1/0.4$). The resulting solution was degassed by three freeze-pump-thaw cycles, and then the glass tube was sealed under vacuum, heated at 60 °C for 2 h, and quenched by rapid cooling. The acetonitrile solution of the reaction mixture was poured into a large amount of *n*-hexane to precipitate the polymers and remove the unreacted monomers. The resultant polymer was collected by centrifugation and dried under reduced pressure.

4. Conclusions

In this study, we demonstrated that thermo-responsive cyclic-polyNIPAM can be successfully synthesized by ring expansion RAFT polymerization of NIPAM using a cyclic chain transfer agent CTTC. This RE-RAFT polymerization is a smart synthetic method that can efficiently obtain the desired cyclic polymers by simply adding a cyclic RAFT agent to a conventional free radical polymerization system. Since cyclic-polyNIPAM does not possess chain ends, it exhibits a different thermal phase transition behavior compared to linear-polyNIPAM with almost the same molecular weight and molecular weight distribution, exhibiting lower onset and CP temperatures. It is speculated that the cyclic topology facilitates intramolecular hydrogen bond formation in cyclic-polyNIPAM. By using the RE-RAFT polymerization, it is possible to synthesize multi-component cyclic polymers by copolymerization of NIPAM with other acrylamide monomers, so it is expected that it can be applied to the control of thermo-responsive properties. Large-scale synthesis of cyclic polymers enable us to elucidate the relationships between the physical and functional properties of cyclic polymers and their structures in detail.

Author Contributions: Conceptualization, J.M. and M.M.; methodology, J.M. and M.M.; validation, J.M.; formal analysis, K.B.; investigation, J.M.; resources, J.M. and M.M.; data curation, K.B. and J.M.; writing—original draft preparation, J.M.; visualization, J.M.; supervision, M.M.; project administration, J.M.; funding acquisition, J.M. and M.M. All authors have read and agreed to the published version of the manuscript.

Funding: This research was funded by the JSPS KAKENHI, Grant-in-Aid for Scientific Research (C), grant numbers 20K05605 and 22K05216.

Institutional Review Board Statement: Not applicable.

Informed Consent Statement: Not applicable.

Data Availability Statement: Data are contained within the article.

Conflicts of Interest: The authors declare no conflicts of interest.

References

1. Endo, K. Synthesis and Properties of Cyclic Polymers. *Adv. Polym. Sci.* **2008**, *217*, 121–183.
2. Laurent, B.A.; Grayson, S.M. Synthetic approaches for the preparation of cyclic polymers. *Chem. Soc. Rev.* **2009**, *38*, 2202–2213. [CrossRef] [PubMed]
3. Kricheldorf, H.R. Cyclic Polymers: Synthetic Strategies and Physical Properties. *J. Polym. Sci. Part A Polym. Chem.* **2010**, *48*, 251–284. [CrossRef]
4. Jia, Z.; Monteiro, M.J. Cyclic Polymers: Methods and Strategies. *J. Polym. Sci. Part A Polym. Chem.* **2012**, *50*, 2085–2097. [CrossRef]
5. Haque, F.M.; Grayson, S.M. The synthesis, properties and potential applications of cyclic polymers. *Nat. Chem.* **2020**, *12*, 433–444. [CrossRef]
6. An, X.; Tang, Q.; Zhu, W.; Zhang, K.; Zhao, Y. Synthesis, Thermal Properties, and Thermoresponsive Behaviors of Cyclic Poly(2-(dimethylamino)ethyl Methacrylate)s. *Macromol. Rapid Commun.* **2016**, *37*, 980–986. [CrossRef]
7. Zhang, H.; Wu, W.; Zhao, X.; Zhao, Y. Synthesis and Thermoresponsive Behaviors of Thermo-, pH-, CO_2-, and Oxidation-Responsive Linear and Cyclic Graft Copolymers. *Macromolecules* **2017**, *50*, 3411–3423. [CrossRef]
8. Xu, J.; Ye, J.; Liu, S. Synthesis of Well-Defined Cyclic Poly(*N*-isopropylacrylamide) via Click Chemistry and Its Unique Thermal Phase Transition Behavior. *Macromolecules* **2007**, *40*, 9103–9110. [CrossRef]
9. Qiu, X.-P.; Winnik, F.M. Effect of Topology on the Properties of Poly(*N*-isopropylacrylamide) in Water and in Bulk. *Macromol. Symp.* **2009**, *278*, 10–13. [CrossRef]
10. Daneshyan, S.; Sodeifian, G. A new approach for synthesis of cyclic poly(*N*-isopropylacrylamide), for applying in biomaterial applications. *Polym. Bull.* **2024**, *81*, 929–949. [CrossRef]
11. Halperin, A.; Kröger, M.; Winnik, F.M. Poly(N-isopropylacrylamide) Phase Diagrams: Fifty Years of Research. *Angew. Chem. Int. Ed.* **2015**, *54*, 15342–15367. [CrossRef] [PubMed]
12. Osváth, Z.; Iván, B. The Dependence of the Cloud Point, Clearing Point, and Hysteresis of Poly(N-isopropylacrylamide) on Experimental Conditions: The Need for Standardization of Thermoresponsive Transition Determinations. *Macromol. Chem. Phys.* **2017**, *218*, 1600470. [CrossRef]
13. Gibson, M.I.; O'Reilly, R.K. To aggregate, or not to aggregate? considerations in the design and application of polymeric thermally-responsive nanoparticles. *Chem. Soc. Rev.* **2013**, *42*, 7204–7213. [CrossRef] [PubMed]
14. Lanzalaco, S.; Armelin, E. Poly(N-isopropylacrylamide) and Copolymers: A Review on Recent Progresses in Biomedical Applications. *Gels* **2017**, *3*, 36. [CrossRef] [PubMed]
15. Lanzalaco, S.; Mingot, J.; Torras, J.; Alemán, C.; Armelin, E. Recent Advances in Poly(N-isopropylacrylamide) Hydrogels and Derivatives as Promising Materials for Biomedical and Engineering Emerging Applications. *Adv. Eng. Mater.* **2023**, *25*, 2201303. [CrossRef]
16. Adachi, K.; Honda, S.; Hayashi, S.; Tezuka, Y. ATRP-RCM Synthesis of Cyclic Diblock Copolymers. *Macromolecules* **2008**, *41*, 7898–7903. [CrossRef]
17. Honda, S.; Yamamoto, T.; Tezuka, Y. Topology-Directed Control on Thermal Stability: Micelles Formed from Linear and Cyclized Amphiphilic Block Copolymers. *J. Am. Chem. Soc.* **2010**, *132*, 10251–10253. [CrossRef]
18. Honda, S.; Yamamoto, T.; Tezuka, Y. Tuneable enhancement of the salt and thermal stability of polymeric micelles by cyclized amphiphiles. *Nat. Commun.* **2013**, *4*, 1574. [CrossRef]
19. Ouchi, M.; Kammiyada, H.; Sawamoto, M. Ring-expansion cationic polymerization of vinyl ethers. *Polym. Chem.* **2017**, *8*, 4970–4977. [CrossRef]
20. Chang, Y.A.; Waymouth, R.M. Recent Progress on the Synthesis of Cyclic Polymers via Ring-Expansion Strategies. *J. Polym. Sci. Part A Polym. Chem.* **2017**, *55*, 2892–2902. [CrossRef]
21. Kricheldorf, H.R.; Lee, S.R. Polylactones. 35. Macrocyclic and Stereoselective Polymerization of β-D, L-Butyrolactone with Cyclic Dibutyltin Initiators. *Macromolecules* **1995**, *28*, 6718–6725. [CrossRef]
22. Bielawski, C.W.; Benitez, D.; Grubbs, R.H. An "Endless" Route to Cyclic Polymers. *Science* **2002**, *297*, 2041–2044. [CrossRef] [PubMed]
23. Xia, Y.; Boydston, A.J.; Yao, Y.; Kornfield, J.A.; Gorodetskaya, I.A.; Spiess, H.W.; Grubbs, R.H. Ring-Expansion Metathesis Polymerization: Catalyst-Dependent Polymerization Profiles. *J. Am. Chem. Soc.* **2009**, *131*, 2670–2677. [CrossRef] [PubMed]
24. Kaitz, J.A.; Diesendruck, C.E.; Moore, J.S. End Group Characterization of Poly(phthalaldehyde): Surprising Discovery of a Reversible, Cationic Macrocyclization Mechanism. *J. Am. Chem. Soc.* **2013**, *135*, 12755–12761. [CrossRef] [PubMed]
25. Kammiyada, H.; Konishi, A.; Ouchi, M.; Sawamoto, M. Ring-Expansion Living Cationic Polymerization via Reversible Activation of a Hemiacetal Ester Bond. *ACS Macro Lett.* **2013**, *2*, 531–534. [CrossRef]
26. Culkin, D.A.; Jeong, W.; Csihony, S.; Gomez, E.D.; Balsara, N.P.; Hedrick, J.L.; Waymouth, R.M. Zwitterionic Polymerization of Lactide to Cyclic Poly(Lactide) by Using N-Heterocyclic Carbene Organocatalysts. *Angew. Chem. Int. Ed.* **2007**, *46*, 2627–2630. [CrossRef]
27. Chiefari, J.; Chong, Y.K.; Ercole, F.; Krstina, J.; Jeffery, J.; Le, T.P.L.; Mayadunne, R.T.A.; Meijs, G.F.; Moad, C.L.; Moad, G.; et al. Living Free-Radical Polymerization by Reversible Addition–Fragmentation Chain Transfer: The RAFT Process. *Macromolecules* **1998**, *31*, 5559–5562. [CrossRef]
28. Perrier, S.; Takolpuckdee, P. Macromolecular Design via Reversible Addition-Fragmentation Chain Transfer (RAFT)/Xanthates (MADIX) Polymerization. *J. Polym. Sci. Part A Polym. Chem.* **2005**, *43*, 5347–5393. [CrossRef]

29. Moad, G.; Rizzardo, E.; Thang, S.H. Radical addition-fragmentation chemistry in polymer synthesis. *Polymer* **2008**, *49*, 1079–1131. [CrossRef]
30. Chen, M.; Zhong, M.; Johnson, J.A. Light-Controlled Radical Polymerization: Mechanisms, Methods, and Applications. *Chem. Rev.* **2016**, *116*, 10167–10211. [CrossRef]
31. Minoda, M.; Otsubo, T.; Yamamoto, Y.; Zhao, J.; Honda, Y.; Tanaka, T.; Motoyanagi, J. The First Synthesis of Periodic and Alternating Glycopolymers by RAFT Polymerization: A Novel Synthetic Pathway for Glycosaminoglycan Mimics. *Polymers* **2019**, *11*, 70. [CrossRef] [PubMed]
32. Motoyanagi, J.; Oguri, A.; Minoda, M. Synthesis of Well-Defined Alternating Copolymer Composed of Ethylmaleimide and Hydroxy-Functionalized Vinyl Ether by RAFT Polymerization and Their Thermoresponsive Properties. *Polymers* **2020**, *12*, 2255. [CrossRef] [PubMed]
33. He, T.; Zheng, G.H.; Pan, C. Synthesis of Cyclic Polymers and Block Copolymers by Monomer Insertion into Cyclic Initiator by a Radical Mechanism. *Macromolecules* **2003**, *36*, 5960–5966. [CrossRef]
34. Motoyanagi, J.; Fujii, H.; Minoda, M. Development of Ring-Expansion RAFT Polymerization of *tert*-Butyl Acrylate with a Cyclic Trithiocarbonate Derivative toward the Facile Synthesis of Cyclic Polymers. *Molecules* **2024**, *29*, 1839. [CrossRef] [PubMed]
35. Xia, Y.; Yin, X.; Burke, N.A.D.; Stöver, H.D.H. Thermal Response of Narrow-Disperse Poly(*N*-isopropylacrylamide) Prepared by Atom Transfer Radical Polymerization. *Macromolecules* **2005**, *38*, 5937–5943. [CrossRef]
36. Hong, J.; Fan, Z. Synthesis of Multiblock Polymer Containing Narrow Polydispersity Blocks. *Macromol. Rapid Commun.* **2006**, *27*, 57–62. [CrossRef]
37. Lee, A.W.M.; Chan, W.H.; Wong, H.C. One Pot Phase Transfer Synthesis of Trithiocarbonates from Carbon Bisulphide and Alkyl Halides. *Synth. Commun.* **1988**, *18*, 1531–1536. [CrossRef]

molecules

MDPI

Article

Graphene Modification by Curcuminoids as an Effective Method to Improve the Dispersion and Stability of PVC/Graphene Nanocomposites

Sławomir Wilczewski [1,*], Katarzyna Skórczewska [1,*], Jolanta Tomaszewska [1], Magdalena Osial [2], Agnieszka Dąbrowska [3,4], Kostiantyn Nikiforow [5], Piotr Jenczyk [2] and Hubert Grzywacz [2]

1 Faculty of Chemical Technology and Engineering, Bydgoszcz University of Science and Technology, Seminaryjna 3 Street, 85-326 Bydgoszcz, Poland; jolanta.tomaszewska@pbs.edu.pl
2 Institute of Fundamental Technological Research, Polish Academy of Sciences, Pawińskiego 5B Street, 02-106 Warsaw, Poland; mosial@ippt.pan.pl (M.O.)
3 Faculty of Chemistry, University of Warsaw, Pasteura 1 Street, 02-093 Warsaw, Poland
4 Biological and Chemical Research Centre, University of Warsaw, Żwirki i Wigury 101, 02-089 Warsaw, Poland
5 Institute of Physical Chemistry, Polish Academy of Sciences, Kasprzaka 44/52, 01-224 Warsaw, Poland
* Correspondence: slawomir.wilczewski@pbs.edu.pl (S.W.); katarzyna.skorczewska@pbs.edu.pl (K.S.)

Abstract: A large amount of graphene-related research is its use as a filler for polymer composites, including thin nanocomposite films. However, its use is limited by the need for large-scale methods to obtain high–quality filler, as well as its poor dispersion in the polymer matrix. This work presents polymer thin-film composites based on poly(vinyl chloride) (PVC) and graphene, whose surfaces were modified by curcuminoids. TGA, UV–vis, Raman spectroscopy, XPS, TEM, and SEM methods have confirmed the effectiveness of the graphene modification due to π–π interactions. The dispersion of graphene in the PVC solution was investigated by the turbidimetric method. SEM, AFM, and Raman spectroscopy methods evaluated the thin-film composite's structure. The research showed significant improvements in terms of graphene's dispersion (in solutions and PVC composites) following the application of curcuminoids. The best results were obtained for materials modified with compounds obtained from the extraction of the rhizome of *Curcuma longa* L. Modification of the graphene's surface with these compounds also increased the thermal and chemical stability of PVC/graphene nanocomposites.

Keywords: graphene; curcuminoids; poly(vinyl chloride); nanocomposites stability; polymer films

1. Introduction

Research concerning the application of graphene materials is carried out in many areas, such as medicine, chemistry, materials engineering, energetics, and electronics. A large area of graphene–related research regards its use as a filler for polymer nanocomposites [1–5]. Graphene materials (graphene, multilayer graphene, graphene oxide) show a considerable variation regarding their dispersibility in the polymer matrix, depending on the number of defects and layers and the presence of oxygen–containing functional groups. Furthermore, studies have shown that adding graphene to the polymer matrix can positively change the physicochemical properties of plastics [6–10].

An interesting and still relatively unexplored group of polymer nanocomposites with graphene are poly(vinyl chloride)–based materials (PVC/GN). Analysis of the literature has shown that in these materials with relatively well-dispersed graphene, regardless of the method of preparation (melt mixing, solvent casting, in situ polymerization method), result in an increase in tensile strength and modulus of elasticity [11–16], as well as in increased impact strength [16,17]. Such materials are also characterized by higher thermal stability [13,18,19], swelling agent resistance [20–22] and glass transition temperature (T_g) [13–15,23,24], lower volume and surface resistivity compared to matrix material [14,25,26]. However, the improvement of poly(vinyl chloride) properties using graphene has limitations, mainly due to the poor dispersibility of the filler in the matrix and insufficient interactions at the filler–polymer interface. Depending on the method of filler introduction, maximum properties are obtained in materials containing between 0.1 and 0.5 wt.% graphene. Filler agglomeration affects the deterioration of all the properties mentioned above of PVC/GN polymer nanocomposites.

The agglomerative behavior of graphene is due to strong π–π interactions and van der Waals forces [27–30]. The main preventive methods are based on covalent modification of chemical reactions with graphene's carbon atoms [2,27,29] and non–covalent modification based on the adsorption of particles onto the material's surface. Due to the functionalization mechanism, non-covalent methods to counteract graphene aggregation are divided into four main groups. The first includes methods related to π–π interactions [28,31–34]. The second stabilization mechanism is based on cation–π interactions [35,36]. Other methods apply hydrophilic–hydrophobic effects [2,27,29,37] and surface modifications of graphene with other nanoparticles [2,38,39]. Improving graphene's dispersion in the PVC matrix is also based on the above methods. However, the covalent solutions presented so far [40–44] use reagents harmful to health and the environment, such as aniline, N–methylpyrrolidone, hydrazine, and terephthalaldehyde. Non–covalent methods, on the other hand, are mainly based on the use of plasticizers [14,25], which alter the properties of poly(vinyl chloride). Soft varieties of PVC in terms of properties can be considered a different group of materials compared to rigid PVC, so they also have entirely different practical applications [45].

The π–π interactions refer to non–covalent interactions (overlapping orbitals) between the π bonds of aromatic rings. However, this is a misleading description, because the direct sandwich overlap is electrostatically repulsive. Instead, we are dealing with staggered stacking (parallel displaced) or π–teeing (perpendicular T–shaped), which makes the described interactions electrostatically attractive. The displacement occurs due to the superposition of carbon atoms with partially negative charges over hydrogen atoms with partially positive charges. Similarly, T–type perpendicular overlap is electrostatically attractive because here, too, there is an overlap of different charges [32,33,46].

A promising method of graphene modification using π–π stacking is the modification of its surface with curcuminoids. The interaction between the aromatic rings of curcumin and the GN surface was used to obtain graphene by reducing graphene oxide, through ultrasonic exfoliation of graphite [31,34,47]. However, the most common use of the graphene–curcumin system is in drug delivery systems [48,49]. Curcuminoids, thanks to their high chemical and thermal stability (in combination with graphene, even up to 220 °C), are also attractive for applications in the modification of polymers, including PVC [50–52].

Therefore, the present work aimed to modify surface graphene with curcuminoids and to investigate the potential of this modification to improve graphene dispersion in poly(vinyl chloride)/graphene nanocomposites obtained by the solvent-casting method. In this work, we used a commercially available mixture of three types of curcumin (CU) (curcumin, demethoxycurcumin, and bisdemethoxycurcumin) and an extract of *Curcuma longa* L. (CE) obtained by extraction of the powdered rhizome of the plant, which was presented in our earlier work [22]. In addition to the aforementioned main ingredients, CE also contained Ar—curcumene, (-)—zingiberene, β—sesquiphellandrene, Ar—turmerone, α—turmerone, β—turmerone, (6R, 7R)—bisabolene and (E)—atlantone.

2. Results and Discussion

2.1. UV-Vis Spectroscopy

Before the morphology studies were conducted, the UV-vis studies were performed, in order to determine if the GN was successfully modified with curcuminoids. Figure 1 shows the UV–Vis spectra of graphene, curcumin, *Curcuma longa* L. extract, and GN–CU and GN–CE colloidal suspensions. The analysis showed that the maximum absorption for graphene was observed at 272 nm, corresponding to π–π^* transitions of C=C bonds [53,54]. Two absorption bands were observed in the spectrum of curcumin; at 263 nm, with π–π^* and C=C bonds, and at 422 nm, which corresponds to n–π^* transitions of C=O bonds [54,55]. In the spectrum of the *Curcuma* extract, the described transitions were observed at 233 nm and 422 nm, respectively. The shift of the absorption maximum (233 nm) toward the ultraviolet spectrum is due to the presence of the other CE components. A maximum wavelength of 233 nm is characteristic of ar–turmerone [56,57]. Two absorption bands were observed on the GN–CU and GN–CE spectra at 269 nm and 414 nm, respectively, which were attributed to the π–π^* transitions of C=C bonds and n–π^* bonds of C=O. The presence of a band at 414 nm, characteristic of curcumin, indicates the effective functionalization of graphene [54,58]. In addition, the change in band position at 269 nm confirms the interaction of π–π graphene and curcuminoid–containing modifiers [21,55,59].

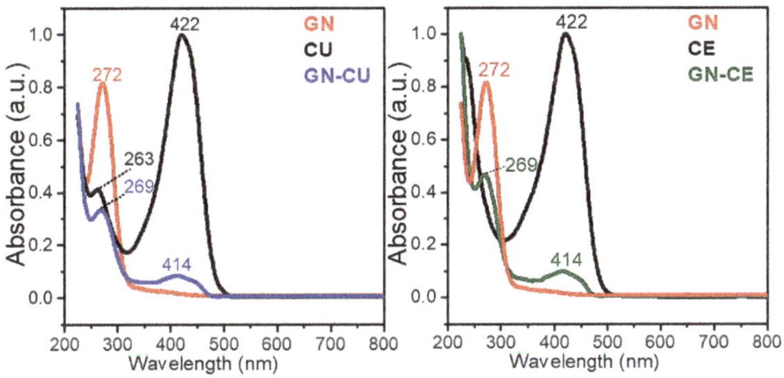

Figure 1. UV–Vis spectra of the GN, CE, CU and GN–CU, GN–CE dispersion in methanol.

2.2. XPS Analysis

Next, X-ray photoelectron spectroscopy (XPS) was used to characterize graphene surfaces' functionalization accurately. Figure 2 shows the XPS spectra of GN and curcuminoid–modified materials. In the image, one can see an increase in the intensity of the O1s band in materials after modification. It indicates the effective deposition of curcuminoids (which contain oxygen functional groups) onto the graphene surface [4,47]. Finally, the deconvolution of O1s and C1s peaks and the XPS characterization of GN, GN–CU, and GN–CE sheets are presented in Figure 3 and Table 1. Since graphene materials were modified, mainly sp^2 hybridization was expected. Unfortunately, it was impossible to match sp^2 in any way, so an additional sp^3 hybridization peak was used. This fact may be due to the partial defecting of the graphene used and its partial oxidation (according to the manufacturer, about 2.5% was oxidized) [31,47,60].

The distance between the C sp^2 and C sp^3 peaks at 0.8 eV, especially for the starting sample, agrees very well with literature data on peak distances of thin carbon films, which are, respectively, 0.7 eV for graphene oxide and reduced graphene oxide [61], 0.8 eV for diamond-like carbon films [62,63], and 0.9 eV for graphite and graphite–like materials [64–66]. Furthermore, the position of the C sp^2 peak around 284.5 eV is also characteristic of this type of material [64] and generally corresponds to values found in the literature (284.25–285 eV), including graphene materials [4,31,61,67].

The rest of the smaller carbon peaks (Figure 3, Table 1) of the carbon functional groups are characterized by the expected distance from the main peak C sp^2 [68]. Considering the literature data, the chemical compounds of turmeric are composed of carbon–oxygen groups [69–71]. The obtained spectra for the samples with CU and CE confirm an increase in the proportion of oxygen groups (O1s), which was 0.6 at.% for GN and 5.3 and 3.3 at.% for GN–CU and GN–CE, respectively, which confirms the deposition of curcuminoids on the graphene's surface. However, the changes in the π electrons (Table 1 C1s spectra) indicate the physical nature of their interactions as a result of non-covalent π–π stacking modification [63,72].

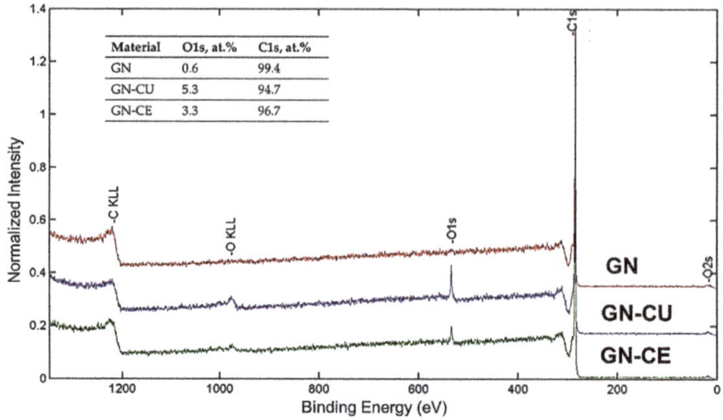

Figure 2. XPS spectra of the GN, GN–CU and GN–CE.

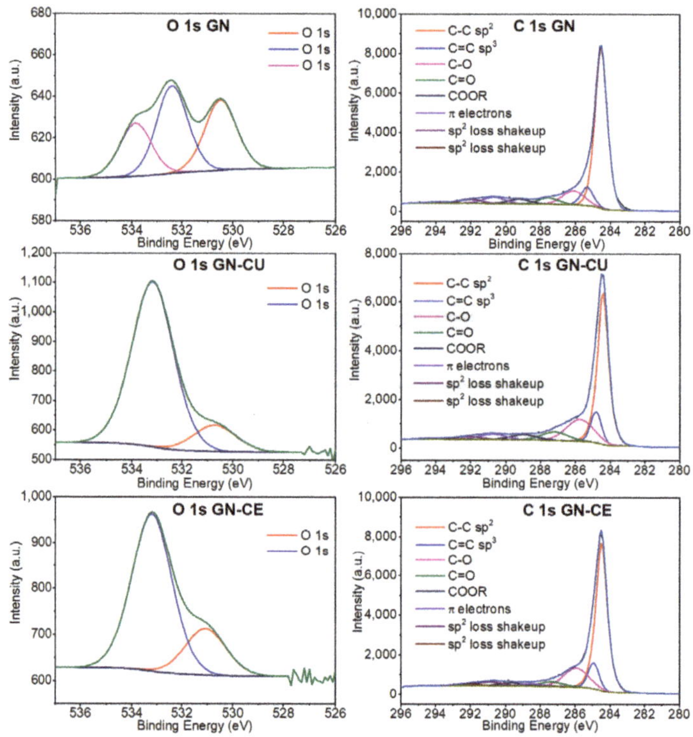

Figure 3. Peaks deconvolution of O1s and C1s of the GN, GN–CU and GN–CE.

Table 1. XPS characterization of GN, GN–CU and GN–CE.

Binding Energy, eV	Chemical Bonds	FWHM	Atomic, %
	GRAPHENE		
	O1s		
530.47	O1s	1.47	0.2
532.38	O1s	1.47	0.2
533.84	O1s	1.47	0.2
			0.6
	C1s		
284.53	C-C sp^2	0.82	65.3
285.33	C=C sp^3	0.82	7.5
286.11	C-O	1.49	10.0
287.56	C=O	1.49	4.5
289.21	COOR	1.49	3.5
290.68	π electrons	1.49	4.6
292.03	sp^2 loss shakeup	1.49	2.8
293.87	sp^2 loss shakeup	1.49	1.1
			99.4
			100.0
	GRAPHENE–CURCUMIN		
	O1s		
530.7	O1s	1.95	0.7
533.17	O1s	1.95	4.6
			5.3
	C1s		
284.42	C-C sp^2	0.73	50.9
284.81	C=C sp^3	0.73	9.6
285.72	C-O	1.82	16.5
287.18	C=O	1.82	6.4
288.99	COOR	1.82	4.0
290.73	π electrons	1.82	4.2
292.14	sp^2 loss shakeup	1.82	2.1
294.15	sp^2 loss shakeup	1.82	1.0
			94.7
			100
	GRAPHENE–*Curcuma Longa* L. EXTRACT		
	O1s		
531.06	O1s	1.87	0.8
533.19	O1s	1.87	2.5
			3.3

Table 1. *Cont.*

Binding Energy, eV	Chemical Bonds	FWHM	Atomic, %
	C1s		
284.48	C-C sp^2	0.76	59.4
284.94	C=C sp^3	0.76	9.4
285.92	C-O	1.65	15.2
287.35	C=O	1.65	3.7
289.66	COOR	1.65	2.4
288.45	π electrons	1.65	2.0
290.88	sp^2 loss shakeup	1.65	3.2
292.16	sp^2 loss shakeup	1.65	1.5
			96.7
			100.0

2.3. Thermogravimetric Analysis (TGA)

A thermogravimetric analysis was performed to determine the modifier's content on the graphene surface. The test was carried out in the temperature range from 30 to 700 °C. The analysis results (Figure 4) showed that the graphene was thermally stable over the entire temperature range, in good agreement with the literature data [34,73]. The extract from *Curcuma longa* L was stable up to about 175 °C, and its degradation occurred between 175 °C and about 450 °C (Figure 4A) while the curcumin was stable up to 250 °C, and its decomposition occurred at temperatures ranging from 250 to 550 °C (Figure 4B). The differences in the thermal stability of these modifiers are due to the presence of additional components in the extract. Their decomposition, up to a temperature of about 320 °C, is related to the degradation of the -OH and -OCH_3 groups. Furthermore, the curcuminoids are decomposed [74–76].

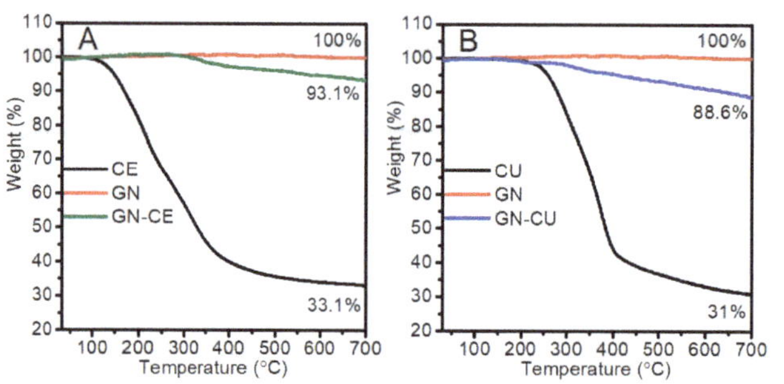

Figure 4. TGA of curcumin, GN, GN–CU (**A**), and *Curcuma longa* L extract and GN, GN–CE (**B**).

Both GN–CU and GN–CE showed thermal stability up to about 340 °C, and their decomposition was associated with the degradation of the organic modifier [34,73]. It is worth noting that the thermal stability of the modified graphene is well above the processing temperature of PVC (below 200 °C) [77–79]. This indicates that the proposed modification can also be used to obtain poly(vinyl chloride)/graphene nanocomposites by the melt mixing method. From the mass differences at 700 °C, the modifier's content on the GN surface was determined by comparing the reference GN sample with the mass loss for GN–CU and GN–CE, respectively. The calculations showed that the mass of curcumin on the graphene surface was 11.4% (Figure 4B), while the mass of the extract was 6.9% (Figure 4A).

2.4. Morphology Characterization

The morphology of the materials was observed by transmission electron microscopy (Figure 5A–C) and scanning electron microscopy (Figure 5(A1–C1)). The observations showed that the modified graphene was a material composed of several layers, which was characterized by a crumpled morphology. This is typical of types of graphene with several layers and relatively large, flake diameters (according to the manufacturer, three-layered graphene with a diameter of up to 10 μm) [1,60,80].

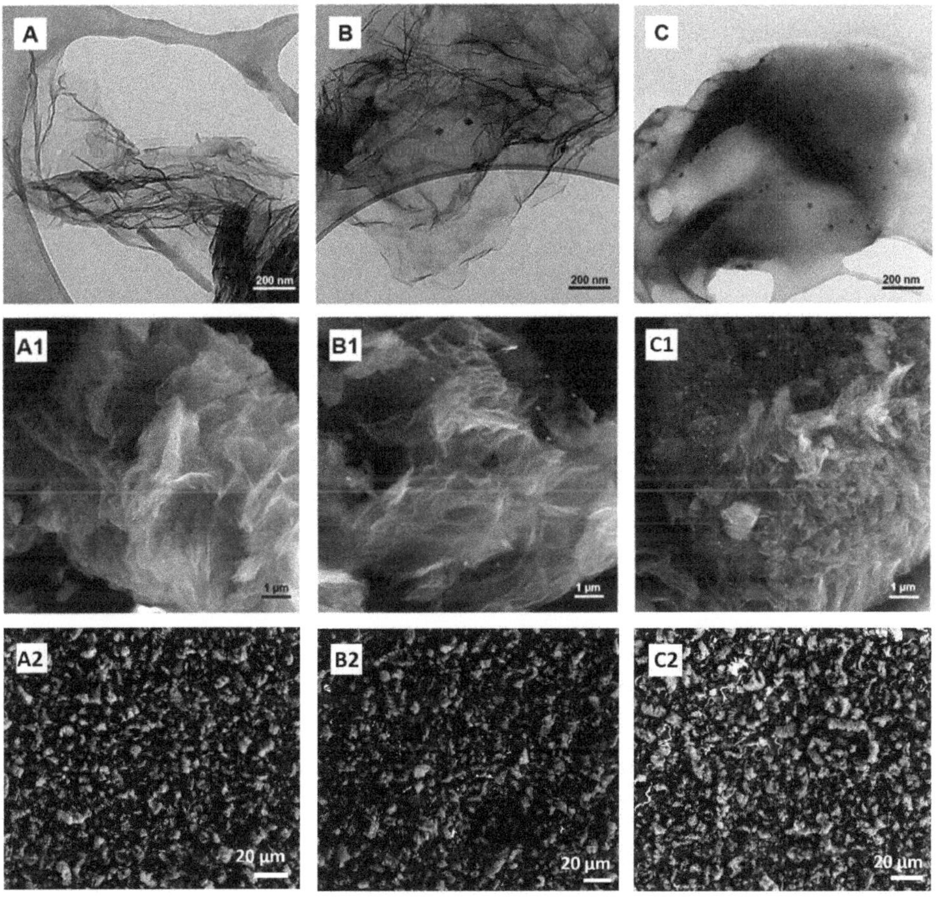

Figure 5. TEM images of GN, GN–CU, GN–CE (**A–C**), SEM images of GN, GN–CU, GN–CE (**A1**,**B1**,**C1**) with scale bar 1 μm, and GN, GN–CU, GN–CE (**A2**,**B2**,**C2**) with scale bar 20 μm (survey images).

Spherical curcuminoid particles (brighter white dots) can be seen in the SEM images (Figure 5(B1,C1)) of the modified graphene [81,82]. Simultaneously, they have a smaller diameter than those in GN–CU (Figure 5(B1)). On the other hand, in TEM images, it can be seen that both modifiers, CU and CE, on the graphene's surface have a floral–spherical shape (Figure 5B,C) [83,84]. As in the SEM images, the curcuminoids from the extract (Figure 5C) were much better deposited on the graphene surface. They are well-dispersed, as present in all SEM pictures of the specimen, and have a smaller diameter. Figure 5(A2–C2) shows the uniform distribution of the flakes in the whole volume as a survey image. Based on the SEM images, it can be argued that CU and CE mainly cover the edges of the graphene flakes, but the TEM images, especially that of GN–CE, indicate that the modifier can cover the surface of graphene to a greater extent.

2.5. Raman Spectroscopy

The following step was the evaluation of the structure of graphene materials using Raman spectroscopy, an effective method for studying carbon materials [85]. Recorded spectra presented in Figure 6 for GN, GN–CU and GN–CE reveal the presence of the D-band, which is characteristic of graphene with a disordered carbon structure, and the G-band, typical for graphite materials, which, in the case of graphene, is also associated with the number of layers [5,86]. To estimate the size of graphene domains, the D- and G-band intensity (ID/IG) ratio was used, where this quantity is inversely proportional to the size of the crystallites [15,27,86,87]. In our materials, it was 1.5–1.6 (depending on the background treatment) for all fillers, indicating a high level of defects in curcuminoid-modified materials, which can increase their wettability by the polymer matrix. Furthermore, the intensity ratio 2D: G < 1 confirms the presence of multilayered, and not structuralized, material. The 2D band is broad and weak, and is slightly shifted towards higher frequencies (~2700 cm^{-1}), which indicates the residual compression stress in the material. The RBS parameter (Raman Band Separtion) is relatively high and equals 287 (GN–CE) to 293 (GN–CU), whereas non modified filler has a value of ~291. This confirms the different mechanisms of carbon structure modification by both curcuminoids.

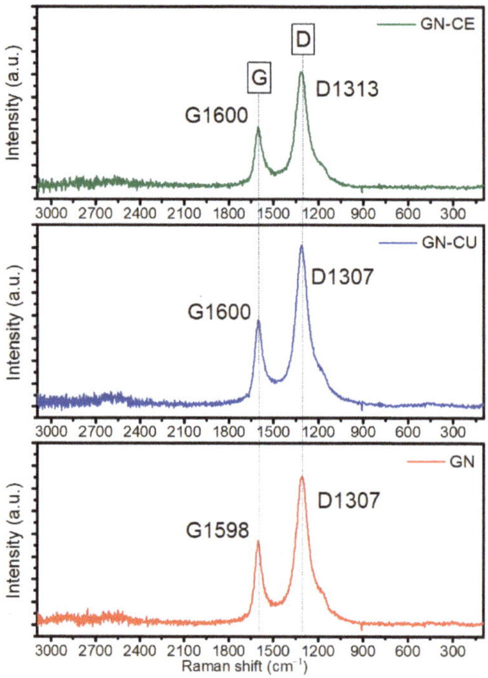

Figure 6. Raman spectra of GN, GN–CU and GN–CE.

In paper [31], it was shown that the stacking of more graphene layers causes a shift of the G-band peak towards higher frequencies, and the Rs for the monolayer is generally higher than for a bilayer. In curcuminoid-modified materials, a shift toward higher wavenumbers was observed, from 1598 cm^{-1} for GN to 1600 cm^{-1} for graphene modified with curcumin and *Curcuma longa* L. extract, further confirming the better foliation of the material after surface functionalization with curcuminoids. Bands typical for disordered carbons are not visible (360 cm^{-1}, 1620 cm^{-1}).

2.6. Dispersion Stability Analysis

As we pointed out in our earlier work [21], in nanocomposites obtained by solvent casting, the critical step in determining the homogeneity of graphene dispersion in the PVC matrix is obtaining a stable filler dispersion in the polymer solution. Therefore, the stability of GN, GN–CU, and GN–CE dispersions in a 3% poly(vinyl chloride) solution was evaluated, from which thin polymer films were obtained after THF evaporation. The test was carried out using the multiple light-scattering method (used in the Turbiscan Lab Toulouse, France). Figure 7 shows the change in the percentage of backscattered and transmitted light as a function of the height of the measuring cell over 24 h. This time is sufficient to obtain a polymer film from a PVC solution at room temperature.

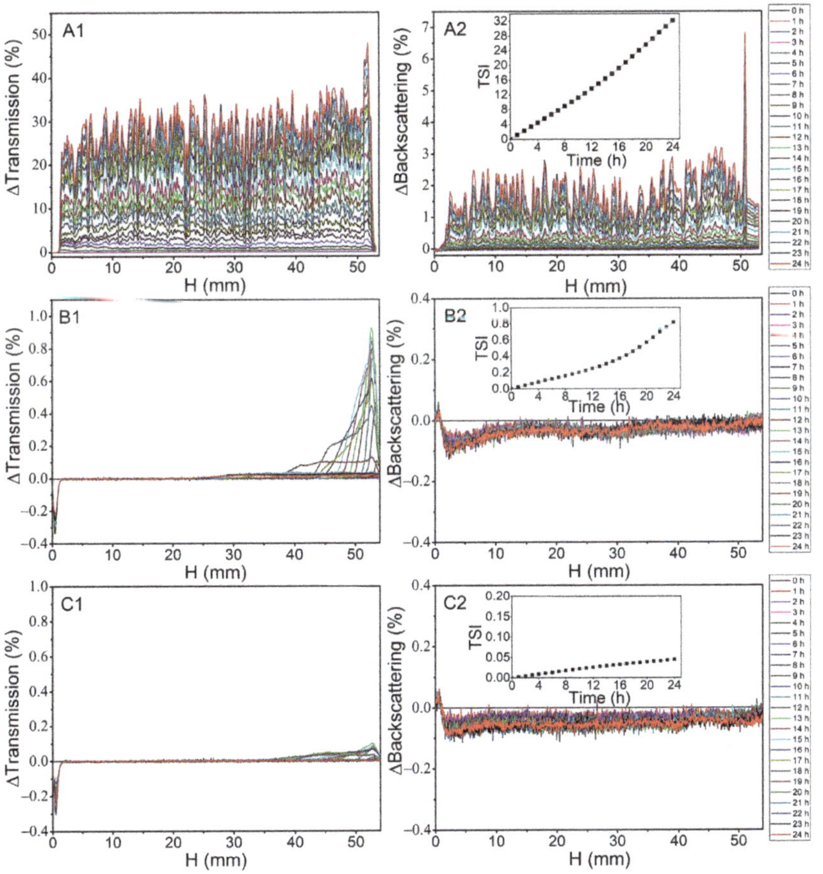

Figure 7. ΔTransmission and ΔBackscattering of GN (**A1**,**A2**), GN–CU (**B1**,**B2**) and GN–CE (**C1**,**C2**) dispersion.

The analysis shows that the destabilization and stratification of graphene dispersion occur due to re-agglomeration and sedimentation of the filler [88,89]. The study demonstrated that the unmodified graphene does not form stable dispersions in PVC solution (Figure 7(A1,A2)), as evidenced by changes in transmitted and backscattered light [88,90]. Functionalization of the GN's surface with curcuminoids improved its stability in PVC solution. Neither GN–CU nor GN–CE re-agglomerated, as evidenced by the absence of significant changes in backscattered light (Figure 7(B2,C2)) [89]. However, in Figure 7(B1,C1), we can see an increase in the solution's transparency after about 16 h, especially for GN-CU. This is due to the gravitational descent of the filler particles.

The Turbiscan Stability Index (TSI) (a specific Turbiscan parameter) was determined to compare the destabilization kinetics of the prepared systems, in particular, the differences between the CU and CE effect. TSI is an efficient method of comparing several systems, where the higher its value is, the less stable the system is [91,92]. Other sources [21,28] distinguish a more precise division of TSI values, indicating that the dispersion is stable when TSI is less than five. When its value is between 5 and 20, we are dealing with a partially stable system; above 20, the dispersion is sedimentary. The results presented in Figure 7(A2–C2) show that the GN dispersion is stable for only 4 h, while modifying its surface resulted in extending this time to 24 h. TSI analysis proved that the most stable systems are formed by GN–CE, for which the discussed value, even after 24 h, was below 0.05.

2.7. Graphene Dispersion in Nanocomposites

The structure of nanocomposite polymer thin films obtained by solvent casting was studied using Raman spectroscopy and atomic force microscopy. In addition, the morphologies of the obtained materials were also observed on a macro- and micrometer scale, using images from digital photography and SEM. As mentioned in the introduction, the problem with graphene aggregation occurs in materials containing about 0.5 wt.% filler or more, so the structure evaluation was carried out for materials with 1 wt.% GN.

Regarding the Raman spectra of the nanocomposites (Figure 8), D– and G–bands typical of graphene and peaks originating from poly(vinyl chloride) were observed. Signals were assigned to the following chemical groups: a band at about 360 cm^{-1} to C–Cl in the trans configuration of the polymer [93], bands at 635 and 695 cm^{-1} to C–Cl stretching [93,94], a band at approximately 1113 cm^{-1} to C–C stretching [93], a band at ~1430 and 1498 cm^{-1} to C–H symmetrical stretching in CH_2 group [21] and a band at approximately 2916 cm^{-1} to C–H asymmetrical stretching in CH_2 group [21,93]. In addition, a peak was observed in PVC/1% GN–CU and PVC/1% GN-CE at about 1185 cm^{-1}. It was attributed to the C–O bonds of curcuminoids [95]. The study did not reveal any new chemical bonds in PVC, which indicates the physical nature of filler–polymer interactions. Analysis of the graphene–derived bands showed changes in the fillers' structures, due to their dispersion in the poly(vinyl chloride) matrix. Firstly, the RBS decreased for all materials, with 243 (GN–CE), 247 GN, and 249 (GN–CU), but it increased in the same way from GN–CE to GN–CU. Raman shifts of G bands decreased in all samples below 1600 cm^{-1} (1592 cm^{-1}, 1597 cm^{-1}, 1598 cm^{-1}), being the lowest in GN–CE and the highest in GN–CU. This can be explained by the layer-stacking enhancement in the matrix's presence. Thus, the best filler dispersion was observed in GN–CU. Contrarily, the D-bands have been shifted towards higher Rs (1349 cm^{-1} in modified materials and 1350 cm^{-1} in GN). The D (first–order peak) moves towards lower frequencies in the case of tensile stress and towards higher ones in the presence of compressive stress, as observed in nanocomposites. Additionally, the I_D to I_G ratio decreases, especially in the case of the GN–CU, indicating better structural ordering. However, the sheer increase in the intensity of the PVC-derived band (2916 cm^{-1}) in the materials after its functionalization may indicate better dispersion of these materials and deposition inside the polymer matrix [93] (the spectrum is collected from the polymer, not from the agglomerate on the surface of the thin polymer film).

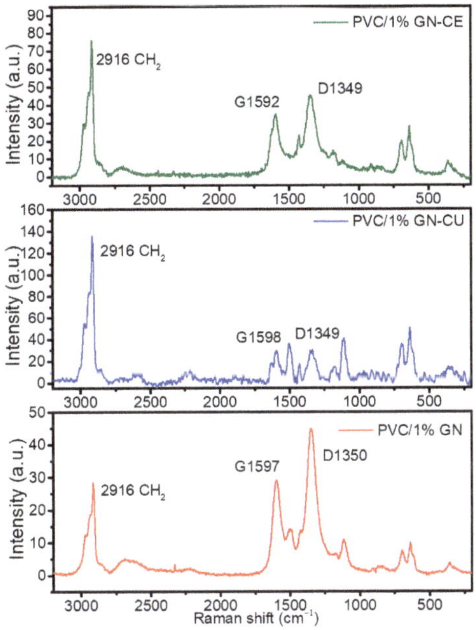

Figure 8. Raman spectra of PVC/1%GN, PVC/1%GN–CU and PVC/1%GN–CE.

AFM also investigated the surface structure of the obtained nanocomposites in contact with a surface (the upper side of the surface that was not in contact with glass during THF evaporation, forming a polymer film). AFM allows for the direct evaluation of the dispersion of graphene in the PVC matrix and the determination of roughness [94, 96]. Figure 9 shows 2D and 3D images of nanocomposites containing modified and unmodified graphene.

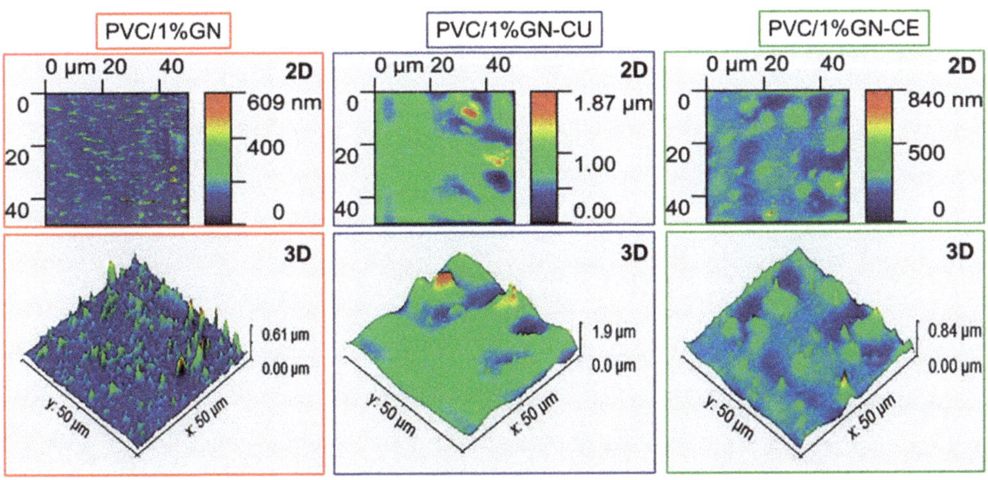

Figure 9. AFM 2D and 3D images of PVC/1% GN, PVC/1%GN–CU and PVC/1% GN–CE.

The image of the surface topography of the nanocomposites shows some corrugation of the samples due to the evaporation of solvent during the formation of the polymer film [22]. AFM PVC/1%GN shows extreme inhomogeneity of the material, and the 3D image shows a large number of pointed tops, resulting from the presence of agglomerates. Additionally, the 2D image shows a lack of homogeneity of the material; the filler is distributed point-wise, and the sample is flat outside the place of its presence. In PVC/1%GN-CU nanocomposites, an increase in the homogeneity of the material was observed (no pointed vertices, no significant differences in roughness). At the same time, the roughness of the whole material increased significantly, which is a normal phenomenon in materials containing large amounts of dispersed filler [15,94,97]. However, the surface topography analysis showed the best dispersion in PVC films containing GN–CE; its material is homogeneous, and at the same time, its roughness is reduced compared to PVC/1%GN–CU.

The observations made earlier were confirmed by analyzing the homogeneity of the polymer films on a macroscopic scale. The presented digital images of nanocomposites containing 1 wt.% GN, GN–CU, and GN–CE Figure 10A–C, respectively, showed that the obtained materials had no visual defects in terms of their structure or the presence of pores. At the same time, it can be noted that the composite with unmodified GN is heterogeneous, and the filler is present in the form of agglomerates. Modifying the GN's surface with curcuminoids significantly enhances the homogeneity of the nanocomposites, where the PVC/1% GN–CE material is characterized by the best graphene dispersion (the two phases in the material cannot be distinguished).

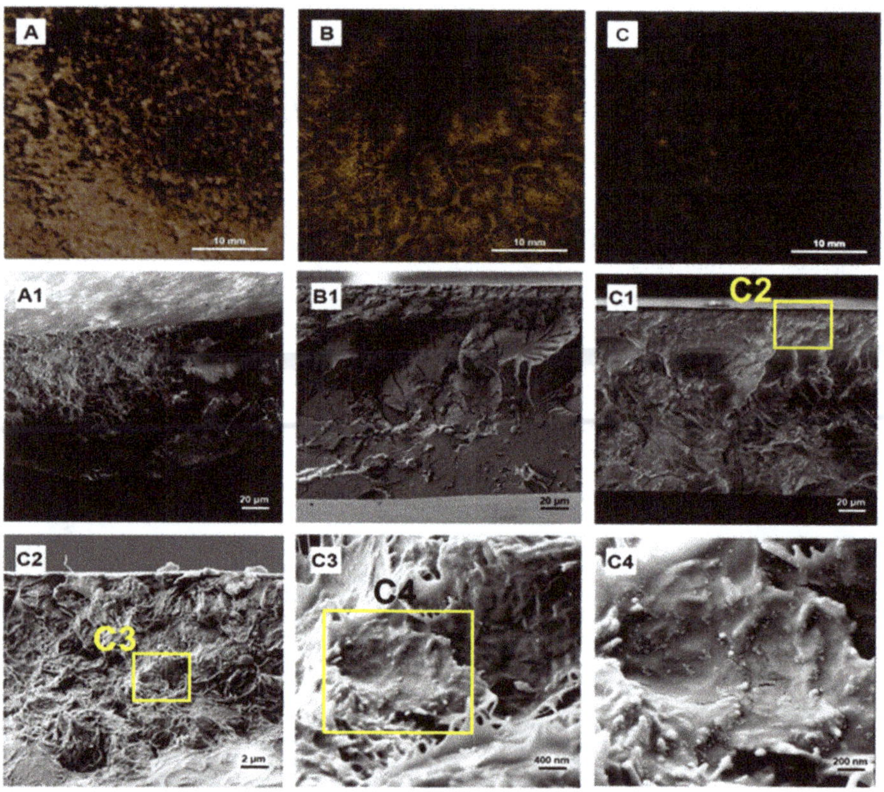

Figure 10. Digital photos of PVC/1% GN, PVC/1%GN–CU and PVC/1% GN–CE (**A–C**), SEM images of those materials (**A1**,**B1**,**C1**) and zoomed in structure of PVC/1%GN–CE (**C2–C4**).

The same observations were confirmed by SEM imaging on the surface of the PVC/1%GN cryogenic breakthrough (Figure 10(A1)); we observed two phases in the material, and a graphene agglomerate is surrounded by a polymer matrix. A significant improvement in GN dispersion characterizes the PVC/1%GN–CU nanocomposites, but areas without filler can be observed on the surface of the cryogenic breakthrough (Figure 10(B1)). The image of the PVC/1% GN–CE breakthrough (Figure 10(C1)) is homogeneous throughout, and no areas of unmodified PVC can be seen, indicating the best filler dispersion [16,17,21,24]. At the same time, we can see that the structure of the nanocomposite differs significantly from that typically observed in unmodified poly(vinyl chloride) [20–22]. Its image may resemble balls of wool (Figure 10(C2)). Further zooming in on the structure (Figure 10(C3,C4)) shows the polymer matrix's thorough coverage of individual graphene flakes, which may translate into a good filler–polymer interfacial interaction. At the same time, we can see that the modifier in the form of CE remains on the graphene flakes and does not pass into the PVC in significant amounts. The Figure 10(C4) image shows curcuminoid particles having spherical shapes [81,82] on the surface of the flakes. These spheres were not observed in the expanded ends of the polymer. The jaggedness of the PVC/GN–CE confirms that good filler dispersion further affects the fracture ductility of modified PVC.

Observations of the structure demonstrated the best improvement in the dispersion of graphene in materials modified by *Curcuma longa* L. extract. Therefore, the CE-modified filler was used to study the effect of the applied modification on the properties of poly(vinyl chloride)/graphene nanocomposites. Materials containing 0.01, 0.1, 0.5, and 1 wt% GN were produced.

2.8. Thermal and Mechanical Properties of Nanocomposites

Thermal properties were determined by the thermogravimetric method and the Congo Red test, with which the thermal stability time was measured. The results obtained are summarized in Table 2 and are shown in Figure 11. The TGA thermograms (Figure 11A) show mass losses at temperatures up to 170 °C, which are related to the evaporation of residual THF [20,23]. The content of solvent was about 6% (Table 2), despite the applied evaporation and drying methods (reduced pressure at 50 °C for 2 weeks). Earlier studies have shown that higher evaporation temperature would lead to the degradation of the material [20].

Thermal degradation of PVC and PVC/GN nanocomposites takes place in two stages related to the decomposition of the polymer. The first step occurs at 200 °C to 375 °C, and is related to the dehydrochlorination of PVC and the formation of conjugated polyene structures. The second step in the range from 375 °C to 600 °C corresponds to thermal cracking of the carbonaceous, conjugated polyene sequences and the formation of residual chars [11,23].

Table 2. Thermal properties of PVC/GN and PVC/GN–CE nanocomposites.

Material	Cont. of THF, %	Max. DTG I, °C	Max. DTG II, °C	Residual Mass, %	Congo Red Test, min
PVC	5.2 (0.4)	269.8 (2.1)	452.4 (1.5)	7.7 (0.2)	3.5 (0.02)
PVC/0.01%GN	5.8 (0.1)	269.0 (0.6)	450.9 (0.6)	7.8 (0.4)	3.1 (0.04)
PVC/0.01%GN–CE	5.5 (0.1)	273.5 (0.3)	449.9 (1.1)	8.2 (0.1)	3.3 (0.08)
PVC/0.1%GN	5.8 (0.4)	273.0 (0.3)	454.2 (0.4)	7.5 (0.2)	3.1 (0.04)
PVC/0.1%GN–CE	6.4 (0.1)	274.9 (0.8)	452.9 (1.4)	8.3 (0.2)	2.8 (0.03)
PVC/0.5%GN	5.7 (0.6)	274.7 (2.0)	446.8 (0.6)	7.2 (0.4)	2.8 (0.04)
PVC/0.5%GN–CE	6.1 (0.2)	278.3 (0.8)	451.4 (1.5)	8.6 (0.1)	2.4 (0.03)
PVC/1%GN	6.8 (0.5)	270.6 (0.5)	438.1 (2.9)	8.0 (0.7)	2.6 (0.08)
PVC/1%GN–CE	6.6 (0.2)	278.5 (0.7)	453.5 (1.3)	9.9 (0.3)	2.3 (0.03)

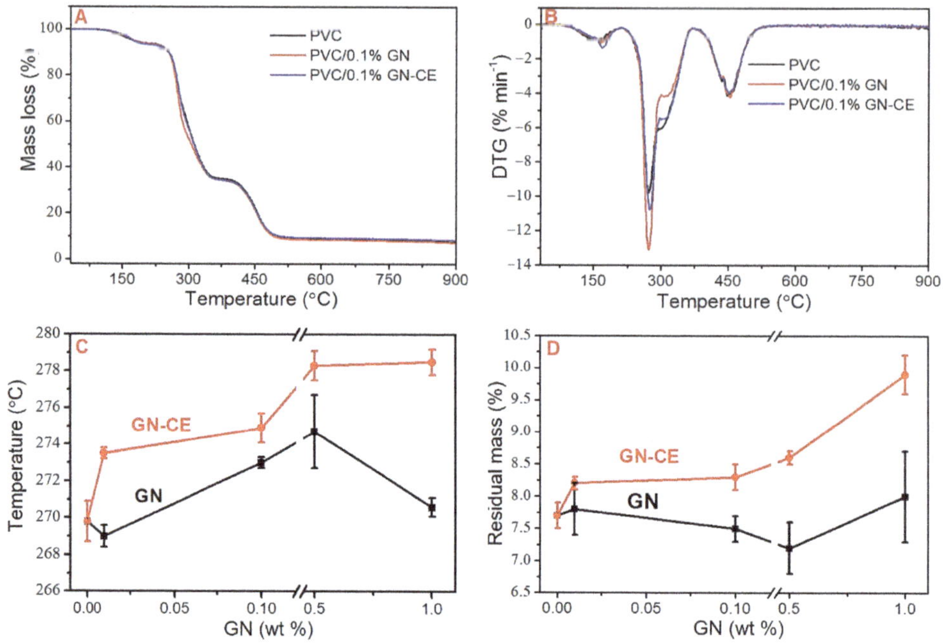

Figure 11. (**A**) Example TGA thermograms, (**B**) Example DTG curves, (**C**) Max. DTG I analysis, (**D**) Residual mass analysis.

Due to the presence of residual THF in PVC, the nanocomposite films' thermal stability was evaluated based on the differential thermogravimetry (DTG) [13,19]. DTG maxima of the first (max. DTG I) and second (max. DTG II) stages of decomposition, as well as the residual mass after the heating process, were used to analyze the thermal stability of the PVC/GN nanocomposites. The examples of thermogravimetric and DTG curves are presented in Figure 11A,B; all results are summarized in Table 2 (standard deviation of the obtained mean values in brackets). The max. DTG I analysis (Figure 11C) showed an increase in the thermal stability of graphene–CE modified materials, where the maximum temperature observed in the case of PVC/0.5%GN–CE was 8.5 °C higher compared to the unmodified matrix material. The introduction of GN–CE at a higher concentration does not cause a statistically significant change in thermal stability. In the case of nanocomposites with unmodified graphene, a significant improvement in stability was observed for materials with 0.1% GN; the temperature of the maximal degradation rate was in this case 3.2 °C higher than for poly(vinyl chloride). PVC/0.5%GN showed no statistically significant increase in this temperature, while an increase in standard deviation was observed. A further increase in the concentration of graphene filler to 1% had the effect of lowering the thermal stability of the nanocomposite materials. It can be assumed that the increase in standard deviation and decrease in thermal stability is due to the agglomeration of nanoparticles in the matrix material.

The max. DTG II analysis (Table 2) showed no statistically significant change in the temperature of the maximal degradation rate for materials with GN–CE. In contrast, a decrease in thermal stability was observed for materials with the unmodified graphene for concentrations of 0.5% and 1%, which may be related to the non-uniform distribution of the filler in the matrix. Residual mass analysis (Figure 11D) showed increases of these values in graphene–CE modified materials, and no statistically significant change in nanocomposites with GN. The increase in residual mass in materials with GN–CE indicates that a well-dispersed filler can act protectively against the carbon chain of the polymer, by forming an insulating layer [24,98].

A commonly used method for assessing the thermal stability of PVC materials is the Congo Red test. The results presented in Table 2 show that the introduction of graphene into polymer matrix results accelerates the dehydrochlorination process, which can take place according to various mechanisms [77]. The authors [19] indicated that the lower stability of PVC/GN composites may be due to the fact that GN nano-flakes act as reinforcing particulate fillers that attract chloride from the PVC. Obtained nanocomposites are sensitive to heat so they can undergo thermomechanical degradation processes under processing. Therefore, the use of thermal stabilizers will be required in the manufacture of these materials by conventional processing methods.

At the same time, it should be emphasized that discrepancies in the results of thermal stability measured by two methods are not unusual, due to the conditions under which the tests were conducted. Thermal stability by the Congo Red method is determined at a constant temperature that is close to the upper temperature range of PVC processing. In the TGA method, the sample is heated under a nitrogen atmosphere at a specified heating rate, over a wide temperature range, usually up to about 900 °C.

Figure 12 shows exemplary stress–strain curves for PVC with the nanocomposites on its matrix, and the determined tensile strength (TS) vs. GN and GN–CE concentration. The course of the stress–strain curves shows no yield point regardless of the material composition, which is as expected. The PVC/GN–CE nanocomposites had, compared to PVC/GN materials, a slightly higher mean TS value and a lower standard deviation, which is due to the better dispersion of the modified filler in the PVC matrix. However, the introduction of graphene into the PVC matrix, regardless of the graphene modification, did not resulted in a statistically significant change in the tensile strength of PVC nanocomposites.

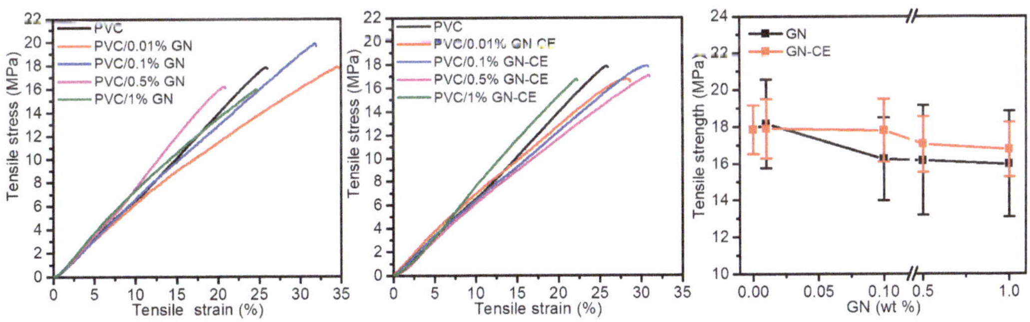

Figure 12. Mechanical properties of PVC, PVC/GN and PVC/GN-CE nanocomposites.

An improvement of the mechanical properties of PVC/GN nanocomposites depends on the homogeneous dispersion of graphene in the polymer matrix and the structure of the filler. The lack of defects on the GN surface results in its high integrity and good mechanical properties, which directly determines the improvement of the mechanical properties of nanocomposites [26,99,100]. On the other hand, the dense GN network limits the penetration of poly(vinyl chloride) macromolecules between its layers, which makes the matrix discontinuous [14,24,100]. This explains the lack of a significant improvement in the mechanical properties of materials with GN–CE, where filler dispersion was improved.

2.9. Electrical Properties of Nanocomposites

Figure 13 shows the results of the resistivity tests of the obtained materials. The surface resistivity of poly(vinyl chloride) was determined to be 4.7×10^{14} Ω, and significant change occurs when the GN concentration in the composites is 0.5% and 1%. The lowest resistivity, which equals 1.4×10^{7} Ω, is characteristic of the PVC/1%GN sample, which makes it possible to conclude that it is an antistatic material [25,101]. Nanocomposites containing

GN–CE did not show a decrease in surface resistivity, and the lowest value (1.5×10^{14} Ω) was shown by the sample PVC/1%GN–CE.

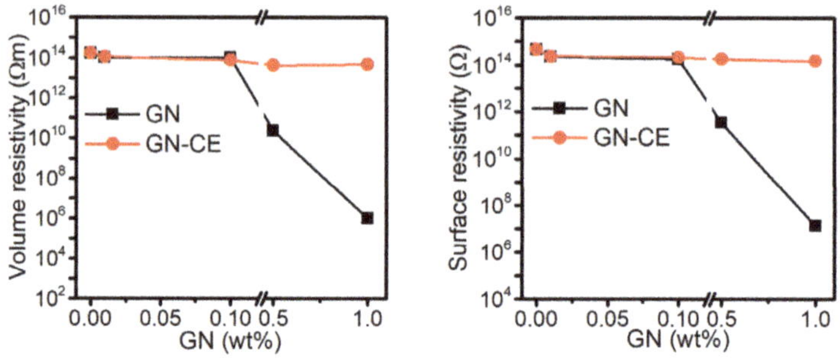

Figure 13. Surface and volume resistivity of PVC, PVC/GN and PVC/GN–CE nanocomposites.

The volume resistivity of PVC/GN composites also decreases with the increase in graphene concentration in the matrix. The materials containing 1% of GN were characterized by the lowest resistivity, and equaled 9.8×10^5 Ωm. Such a large change in resistivity (for PVC was 1.7×10^{14} Ωm) indicates that the dispersion of the GN was sufficient to create conduction paths. The PVC/GN–CE composites, as in the case of surface resistivity, did not show large changes in volume resistivity as the filler content in the matrix increased.

References [14,101] present that the influence of GN on the resistivity or conductivity of PVC/GN nanocomposites is significantly related to the dispersion of the filler in the polymer's matrix. Although the modification of graphene with turmeric significantly improved the homogeneity of the structure of PVC/GN–CE nanocomposites, the expected improvement in electrical properties was not found. Graphene surface modification with curcuma extract is possible by π–π interactions. Some studies have shown the disturbances in the displacement of π electrons on the graphene's surface, leading to the significant deterioration of its electrical properties [20,29]. This explains the lack of change regarding the resistivity of GN–CE-containing materials, despite a significant improvement of the graphene's dispersion in the PVC matrix.

2.10. Swelling Behaviour of Nanocomposites

The numerous applications of poly(vinyl chloride) are due to its resistance to solvents [19,102,103]. However, PVC dissolves completely in THF and cyclohexanone, and it undergoes limited swelling when contacted with acetone. In this work, the chemical resistance of the proposed PVC/GN nanocomposites was tested by analyzing the swelling process in acetone. Swelling curves, i.e., the dependence of the degree of swelling as a function of time, are shown in Figure 14.

It was found that all the obtained materials undergo limited swelling, while the dependence of swelling degree on the exposure time to the swelling agent has a shape of a sigmoid function. Therefore, Equation (1) was used to approximate the swelling curves [20,22].

$$S_d = \frac{S_E}{1 + 10^{(t_M - t)p}} \quad (1)$$

where:

S_d is swelling degree, %,
S_E is equilibrium swelling, upper asymptote, %,
t_M is time in which the swelling occurs with a maximum rate, s,
t is time of exposure to the swelling agent, s,
p is comparison parameter, 1 s^{-1}.

The parameters of the equation and the coefficient of determination R^2 are summarized in Table 3. The proposed model describes the experimental results with high accuracy, as evidenced by the high values of the coefficient of determination.

Table 3. Parameters of the model describing the swelling process.

Material	S_E, %	t_M, s	p, s^{-1}	R^2
PVC	48.2 (0.3)	348 (4)	0.007 (0.0004)	0.994
PVC/0.01% GN	43.8 (0.4)	477 (7)	0.004 (0.0002)	0.994
PVC/0.01% GN–CE	42.9 (0.3)	570 (7)	0.003 (0.0001)	0.996
PVC/0.1% GN	43.9 (0.4)	307 (6)	0.006 (0.0004)	0.990
PVC/0.1% GN–CE	42.1 (0.2)	262 (3)	0.010 (0.0005)	0.996
PVC/0.5% GN	49.5 (0.2)	262 (2)	0.009 (0.0003)	0.998
PVC/0.5% GN–CE	32.1 (0.3)	498 (8)	0.004 (0.0003)	0.992
PVC/1% GN	43.5 (0.1)	267 (1)	0.010 (0.0002)	0.999
PVC/1% GN–CE	27.3 (0.2)	443 (4)	0.007 (0.0003)	0.997

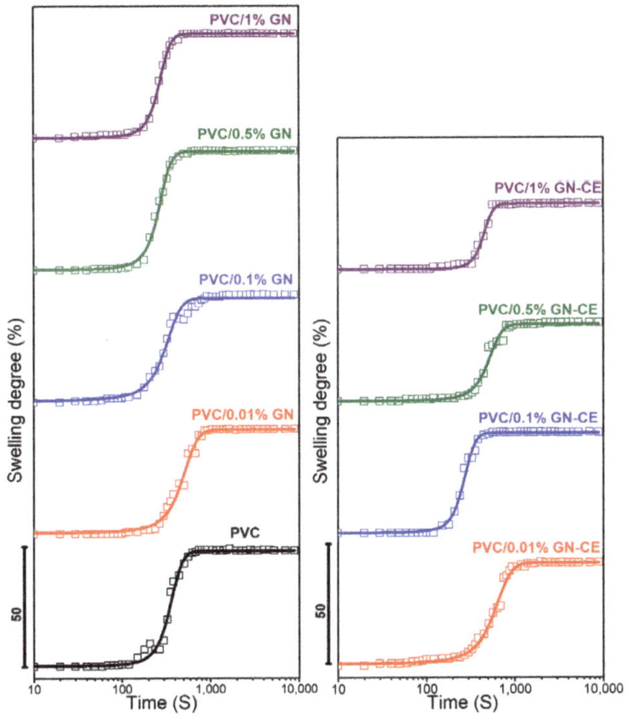

Figure 14. Swelling degree of PVC, PVC/GN and PVC/GN–CE nanocomposites vs. time.

PVC/GN nanocomposites were found to have a lower equilibrium swelling ratio compared to unmodified PVC; the lowest SE value observed for PVC/1%GN, i.e., 43.5%, shows an improvement of about 5% in terms of the chemical resistance on the polymer matrix. Materials obtained with the addition of extract-modified graphene were characterized by a lower value of SE compared to corresponding materials with unmodified graphene. This effect was attributed to the better dispersion of the filler in the polymer matrix, as evidenced by significantly lower SE values for materials with 0.5 and 1% GN–CE. They were 32.1 and 27.3%, respectively, which represents a reduction in the equilibrium degree of swelling relative to PVC of 14.1%, in the case of PVC/0.5%GN–CE, and of 20.9%, in PVC/1%GN–CE. In addition, it should be noted that the agglomeration of the filler in the polymer matrix can lead to lower resistance to solvents in the form of acetone, which was observed in PVC/0.5%GN, where the SE was 1.3% higher relative to PVC. As the GN–CE content of the nanocomposites increases, the time in which the swelling occurs with a maximum rate (t_M) also increases, which further confirms the correlation between acetone resistance of the materials and filler surface modifications. In materials with unmodified GN, a decrease in t_M values was observed with an increase in its content in the PVC matrix.

An increase in the chemical resistance of PVC, by using carbon fillers, occurs as a result of a decrease in its free volume. Consequently, this leads to a reduced access of solvents to the polymer chain [22,103]. In the case of the materials in question, this mechanism is very likely, as confirmed by the higher resistance of nanocomposites containing GN–CE, characterized by a better dispersion of graphene in the polymer matrix.

3. Materials and Methods

3.1. Materials

Graphene–based nanopowder with a flake thickness of 1.6 nm (maximum of 3 atomic monolayers), a flake length of 10 μm, and a specific surface area of 400 ÷ 800 m^2g^{-1} was purchased from USA Graphene Laboratories Inc. Graphene was dispersed in a poly(vinyl chloride) solution using unmodified suspensive poly(vinyl chloride) Neralit 601 (Czech Republic, Spolana s.r.o. Anwil S.A. group), with a K number of 59–61, a bulk density of 0.56–0.63 g cm^{-3}, a specific density of 1.39 g cm^{-3}, and 97% purity. As a solvent for PVC and an environment used for graphene modification, tetrahydrofuran (THF) (Chempur, Piekary Śląskie, Poland) was used. Curcumin from *Curcuma longa* L. turmeric powder (Sigma-Aldrich, St. Louis, MO, USA) and *Curcuma longa* L. rhizome extract (made in-house [22]) was used to modify the surface of graphene. Methanol (Chempur, Piekary Śląskie, Poland) was used to wash modified graphene.

3.2. Graphene Modification

In order to functionalize the graphene's surface with curcuminoids in the first step, 100 cm^3 solutions of CU and CE in THF with a concentration of 2.25 mg cm^{-3} were prepared by ultrasound (frequency of 20 kHz, 40% amplitude, time 10 min, temperature 23 °C) using a SONOPULS rod-shaped probe homogenizer from Bandelin. Graphene was then introduced into the solution, which was prepared so that the dispersion concentration was 1.5 mg cm^{-3}. Next, GN was sonicated under the same conditions as curcuminoids for 2 h. It was then stirred with a magnetic stirrer for 15 h at 200 rpm. After this time, the material was centrifuged 4 times at 12,000 rpm for 4 min and washed after each step with methanol, in order to remove the undeposited modifier. After centrifugation, the graphene was dried at 45 °C for 65 h to evaporate residual solvents.

3.3. Preparation of PVC/GN Dispersions and Nanocomposites

In the first step, poly(vinyl chloride) was dissolved in THF at 25 °C for about 48 h, yielding a solution of 3 wt%. Then, the dry graphene was added to the solution and dispersed using ultrasound (conditions as in Section 2.2) for 60 min. The amount of graphene in the prepared dispersions for dispersion stability testing was 1 wt% per polymer mass.

Thin films of PVC/GN, PVC/GN–CU, and PVC/GN–CE nanocomposites containing 1% filler for structure test were obtained by solvent evaporation, where the dispersions were poured onto Petri dishes with a diameter of 7 cm, and the solvent was evaporated at 50 °C for 24 h. Nanocomposites for testing properties were dried in a vacuum drier under a reduced pressure (max 20 mbar absolute) at 50 °C for 2 weeks, in order to remove the THF residue from the polymer films. All samples were coded to take into account the graphene content and the presence of CE, e.g., a sample containing GN at a concentration of 0.01 wt.% was coded as PVC/1%GN, whereas a sample containing the same graphene modified by *Curcuma longa* L. rhizome extract was coded as PVC/1%GN–CE, and that modified by Curcumin from *Curcuma longa* L. turmeric powder was consequently coded as PVC/1%GN–CU.

3.4. Characterization

Absorption studies were conducted using a UV-vis/NIR spectrometer (Perkin Elmer Lambda 1050+) in a quartz cuvette. For this purpose, solutions of CU, CE in methanol with a concentration of 0.008 mg cm^{-3} and dispersions of GN, GN–CU, and GN–CE with a concentration of 0.03 mg cm^{-3} were prepared. The study was conducted in the 210–800 nm range. The experimental results were normalized in the range 0–1 using the OriginLab software.

X–ray photoelectron spectroscopy: a PHI 5000 VersaProbe (ULVAC–PHI Inc., Hagisono, Japan) spectrometer was used to conduct XPS measurements under the following conditions: monochromatic Al Kα radiation ($h\nu$ = 1486.6 eV), an X-ray source operating at 25 W, 15 kV, 100 µm spot, pass energy of 23.5 eV, energy step of 0.1 eV. Obtained XPS spectra were analyzed with CasaXPS software, using the set of the sensitivity factors native for the hardware. Shirley background and Gaussian–Lorentzian peak shape were used for deconvolution of all spectra.

The thermal stability of the fillers and nanocomposites was assessed by the thermogravimetric method using a TG 209 F3 Tarsus apparatus (Netzsch). The heating rate was about 10 °C min^{-1} in an open ceramic crucible, under a nitrogen atmosphere, at 30 to 700 °C for fillers and at 30 to 900 °C for nanocomposites. In addition, static thermal stability tests were determined for the nanocomposite materials using the Congo Red test. The test was carried out at 200 °C, measuring the thermal stability time, i.e., the time during which the sample shows no signs of degradation in the form of hydrogen chloride release and color change of the indicator paper.

Raman spectra were collected with a Raman DXR microscope (Thermo Fisher Scientific, Waltham, MA, USA) using two laser lines, 532 nm and 780 nm, where the shorter wavelength was used for polymer composites, and fillers were studied using the longer wavelength line. The specimens were measured with 50 repetitions of 50 s each (20 in successive control measurements). The aperture was 50 µm, the lenses were 10 and 50 mm, and the laser beam power was 0.1 mW for 780 nm and 1 mW for 532 nm. Each sample was measured at least dozens of times (from 80 to >300) at different locations, and the final signal was obtained as an average and statistically representative one. The morphology of the samples was studied using scanning electron microscopy (SEM), and a Zeiss Crossbeam 350 microscope; in the case of polymer nanocomposites, cryogenic breakthroughs with a sputtered gold layer were observed. As a second technique, a Zeiss Libra 120 Plus transmission electron microscope (TEM) (Stuttgart, Germany) operating at 120 kV was used, where the aqueous suspension was placed on a copper grid coated with a formvar polymer layer. Finally, the sample was left in the air (under a fume hood) to dry.

Stability testing of the obtained graphene dispersions in PVC solution was carried out by multiple light scattering using the Turbiscan Lab apparatus (Formulaction SA, Toulouse, France), where suspensions of GN, GN–CU and GN–CE in 3% PVC solution in THF were placed in glass cylindrical test tubes, with a working height of 54 mm. The samples were then placed in a Turbiscan Lab instrument and scanned with light at approximately 880 nm. Scanning was carried out every 1 h for 24 h at room temperature (23 °C). The Turbiscan Stability Index (TSI) was determined from the results.

Filler dispersion in the matrix was also determined using an atomic force microscope Nanosurf (Liestal, Switzerland) in contact mode with the PPP–XYCONTR probe (NANOSENSORS, Neuchatel, Switzerland), where the radius of the aluminum-coated blade was 7 nm, while the line angle was 20 degrees. 2D and 3D images were taken within a 50 × 50 µm area. To determine the surface topography of the area, 500 lines were measured (at 20 nm intervals), and the number of measurement points per line was 5000. The scanning speed of the sample was 5 µm s^{-1}.

A study of tensile properties was carried out using an in-house-built tensile tester, with the parameters presented in our earlier work [22]. The tests were carried out on samples with a thickness of 0.2 ± 0.07 mm and a width of 2 ± 0.1 mm, while the length of the measuring section L0 was 1.5 mm. The materials were stretched at a constant speed of 0.1 mm s^{-1}.

The electrical properties of nanocomposites were determined by measuring system consisting of a 6517A electrometer and a 8009 measuring chamber (Keithley Instrument Inc., Cleveland, OH, USA). The surface and volume resistivity were measured. The tests were carried out on film samples with a diameter of 70 mm in air, at a temperature of 23 °C and humidity of 50%, and at a voltage of 10 V.

The resistance of the obtained materials to swelling in acetone was tested in accordance with the method proposed in our earlier study [20]. The change in the swelling degree (S_d) was determined (see Equation (2)), depending on the immersion time in the swelling agent. The changes in the samples' diameters were set on the basis of photos using the NIS Elements 4.0 software. The frequency picture-taking depended on the exposure time to the swelling agent. The measurement temperature was 20 °C and the initial diameter of the samples h_0 = 10 mm.

$$S_d = \frac{h - h_0}{h_0} \times 100\% \quad (2)$$

where:

h is sample diameter after time t (mm),
h_0 is initial sample diameter (mm).

In order to analyse the obtained results, Origin 8.6 Pro software with implemented statistical analysis modules was used. ANOVA with Tukey's post–hoc test was used to compare the significance of the difference for the mean values of the obtained results. The normal distribution was confirmed by the Shapiro–Wilk test, and the homogeneity of variance by the Levene's test. All analyses were performed assuming a significance level below 0.05.

4. Conclusions

Graphene is one of the most promising nanofillers of nanocomposites, improving their performance. However, its uniform distribution in the polymer matrix is challenging. Several studies refer the need to reduce the agglomeration of GN and its derivatives during nanocomposite fabrication, in order to achieve improved properties [104–107]. In the present work, an environmentally safe method of graphene surface modification is presented, using curcumin and an extract derived from the powdered rhizome of *Curcuma longa* L.

Observations of the modified materials' morphology showed better CE dispersion on the graphene's surface. Spectroscopic studies have shown that modification occurs via a π–π stacking mechanism. TGA analysis revealed a higher amount of deposited curcumin on GN flakes i.e., 11.4 wt.%, while for the extract it was 6.9 wt.%. In addition, it is worth noting that the modified materials have a high thermal stability of about 340 °C. The analysis of the stability of GN, GN–CU, and GN–CE dispersions, as well as observations of the polymer films structure obtained after solvent evaporation from the dispersions, showed a significant improvement in the homogeneity of the nanocomposites after modification by curcuminoids. At the same time, extract-modified graphene showed better dispersion improvement in PVC. That indicates that not only the amount of modifier deposited on the graphene surface, but also its size and dispersion important in improving PVC/GN homogeneity. The analysis of the properties of nanocomposites with GN-CE has shown a better thermal and chemical stability of these materials. On the other hand, based on the results of the Congo Red test, it was stated that the proposed nanocomposites will require thermal stabilization at the processing stage by traditional methods, which is a well-known phenomenon in the field of the PVC technology.

Summarizing, the proposed naturally derived compounds in the form of CE can be an alternative to previously proposed modifications of graphene for use in PVC and other polymer composites. In addition, the high thermal stability of the proposed modification, as well as the relatively low cost and possibility of obtaining the modifier on a large scale, provide opportunities for the technological use of CE in the future production of graphene polymer nanocomposites, through traditional processing methods at high temperatures.

Author Contributions: Conceptualization, S.W., K.S. and J.T.; methodology, S.W., K.S., J.T., M.O., A.D., K.N., P.J. and H.G.; validation, S.W., K.S. and J.T.; formal analysis, S.W. and K.S.; investigation, S.W., K.S., M.O., A.D., K.N., P.J. and H.G.; resources, J.T.; data curation, S.W., K.N. and A.D.; writing—original draft preparation, S.W., K.S., J.T., M.O., A.D. and K.N.; writing—review and editing, S.W., K.S., J.T., M.O., A.D. and K.N.; visualization, S.W.; supervision, J.T. and K.S.; project administration, S.W. All authors have read and agreed to the published version of the manuscript.

Funding: This work was supported by the Ministry of Education and Science of Poland.

Institutional Review Board Statement: Not applicable.

Informed Consent Statement: Not applicable.

Data Availability Statement: The data that support the findings of this study are available from the corresponding author upon reasonable request.

Conflicts of Interest: The authors declare that they have no known competing financial interest or personal relationship that could have appeared to influence the work reported in this paper.

References

1. Saleem, H.; Haneef, M.; Abbasi, H.Y. Synthesis route of reduced graphene oxide via thermal reduction of chemically exfoliated graphene oxide. *Mater. Chem. Phys.* **2018**, *204*, 1–7. [CrossRef]
2. Johnson, D.W.; Dobson, B.P.; Coleman, K.S. A manufacturing perspective on graphene dispersions. *Curr. Opin. Colloid Interface Sci.* **2015**, *20*, 367–382. [CrossRef]
3. Abdel Ghany, N.A.; Elsherif, S.A.; Handal, H.T. Revolution of Graphene for different applications: State-of-the-art. *Surf. Interfaces* **2017**, *9*, 93–106. [CrossRef]
4. Drewniak, S.; Muzyka, R.; Stolarczyk, A.; Pustelny, T.; Kotyczka-Morańska, M.; Setkiewicz, M. Studies of reduced graphene oxide and graphite oxide in the aspect of their possible application in gas sensors. *Sensors* **2016**, *16*, 103. [CrossRef] [PubMed]
5. Mikhailov, S. *Physics and Applications of Graphene-Experiments*; InTech: Rijeka, Croatia, 2011; ISBN 9789533071527.
6. Punetha, V.D.; Rana, S.; Yoo, H.J.; Chaurasia, A.; McLeskey, J.T.; Ramasamy, M.S.; Sahoo, N.G.; Cho, J.W. Functionalization of carbon nanomaterials for advanced polymer nanocomposites: A comparison study between CNT and graphene. *Prog. Polym. Sci.* **2017**, *67*, 1–47. [CrossRef]
7. Galpaya, D.; Wang, M.; Liu, M.; Motta, N.; Waclawik, E.; Yan, C. Recent Advances in Fabrication and Characterization of Graphene-Polymer Nanocomposites. *Graphene* **2012**, *1*, 30–49. [CrossRef]
8. Sreenivasulu, B.; Ramji, B.R.; Nagaral, M. A Review on Graphene Reinforced Polymer Matrix Composites. *Mater. Today Proc.* **2018**, *5*, 2419–2428. [CrossRef]
9. Phiri, J.; Gane, P.; Maloney, T.C. General overview of graphene: Production, properties and application in polymer composites. *Mater. Sci. Eng. B Solid-State Mater. Adv. Technol.* **2017**, *215*, 9–28. [CrossRef]
10. Mohan, V.B.; Lau, K.; Hui, D.; Bhattacharyya, D. Graphene-based materials and their composites: A review on production, applications and product limitations. *Compos. Part B Eng.* **2018**, *142*, 200–220. [CrossRef]
11. Deshmukh, K.; Joshi, G.M. Thermo-mechanical properties of poly (vinyl chloride)/graphene oxide as high performance nanocomposites. *Polym. Test.* **2014**, *34*, 211–219. [CrossRef]
12. Nawaz, K.; Ayub, M.; Ul-Haq, N.; Khan, M.B.; Niazi, M.B.K.; Hussain, A. The Effect of Graphene Nanosheets on the Mechanical Properties of Polyvinylchloride. *Polym. Compos.* **2016**, *37*, 1572–1576. [CrossRef]
13. Mindivan, F.; Göktaş, M. Preparation of new PVC composite using green reduced graphene oxide and its effects in thermal and mechanical properties. *Polym. Bull.* **2020**, *77*, 1929–1949. [CrossRef]
14. Li, Q.; Shen, F.; Zhang, Y.; Huang, Z.; Muhammad, Y.; Hu, H.; Zhu, Y.; Yu, C.; Qin, Y. Graphene incorporated poly(vinyl chloride) composites prepared by mechanical activation with enhanced electrical and thermo–mechanical properties. *J. Appl. Polym. Sci.* **2020**, *137*, 21–23. [CrossRef]
15. Akhina, H.; Gopinathan Nair, M.R.; Kalarikkal, N.; Pramoda, K.P.; Hui Ru, T.; Kailas, L.; Thomas, S. Plasticized PVC graphene nanocomposites: Morphology, mechanical, and dynamic mechanical properties. *Polym. Eng. Sci.* **2018**, *58*, E104–E113. [CrossRef]
16. Wang, H.; Xie, G.; Fang, M.; Ying, Z.; Tong, Y.; Zeng, Y. Mechanical reinforcement of graphene/poly(vinyl chloride) composites prepared by combining the in-situ suspension polymerization and melt-mixing methods. *Compos. Part B Eng.* **2017**, *113*, 278–284. [CrossRef]

17. Wang, H.; Xie, G.; Yang, C.; Zheng, Y.; Ying, Z.; Ren, W.; Zeng, Y. Enhanced Toughness of Multilayer Graphene-Filled Poly(vinyl chloride) Composites Prepared Using Melt-Mixing Method. *Polym. Compos.* **2017**, *38*, 138–146. [CrossRef]
18. Abd El-Kader, M.F.H.; Awwad, N.S.; Ibrahium, H.A.; Ahmed, M.K. Graphene oxide fillers through polymeric blends of PVC/PVDF using laser ablation technique: Electrical behavior, cell viability, and thermal stability. *J. Mater. Res. Technol.* **2021**, *13*, 1878–1886. [CrossRef]
19. Mindivan, F.; Göktaş, M.; Dike, A.S. Mechanical, thermal, and micro- and nanostructural properties of polyvinyl chloride/graphene nanoplatelets nanocomposites. *Polym. Compos.* **2020**, *41*, 3707–3716. [CrossRef]
20. Wilczewski, S.; Skórczewska, K.; Tomaszewska, J.; Lewandowski, K. Structure and properties of poly (vinyl chloride)/ graphene nanocomposites. *Polym. Test.* **2020**, *81*, 106282. [CrossRef]
21. Wilczewski, S.; Skórczewska, K.; Tomaszewska, J.; Lewandowski, K.; Szulc, J.; Runka, T. Manufacturing homogenous PVC/graphene nanocomposites using a novel dispersion agent. *Polym. Test.* **2020**, *91*, 106868. [CrossRef]
22. Wilczewski, S.; Skórczewska, K.; Tomaszewska, J.; Lewandowski, K.; Studziński, W.; Osial, M.; Jenczyk, P.; Grzywacz, H.; Domańska, A. *Curcuma longa* L. Rhizome Extract as a Poly(vinyl chloride)/Graphene Nanocomposite Green Modifier. *Molecules* **2022**, *27*, 8081. [CrossRef]
23. Hasan, M.; Banerjee, A.N.; Lee, M. Enhanced thermo-optical performance and high BET surface area of graphene@PVC nanocomposite fibers prepared by simple facile deposition technique: N2 adsorption study. *J. Ind. Eng. Chem.* **2015**, *21*, 828–834. [CrossRef]
24. Wang, L.; Wei, X.; Wang, G.; Zhao, S.; Cui, J.; Gao, A.; Zhang, G.; Yan, Y. A facile and industrially feasible one-pot approach to prepare graphene-decorated PVC particles and their application in multifunctional PVC/graphene composites with segregated structure. *Compos. Part B Eng.* **2020**, *185*, 107775. [CrossRef]
25. Wei, Z.B.; Zhao, Y.; Wang, C.; Kuga, S.; Huang, Y.; Wu, M. Antistatic PVC-graphene Composite through Plasticizer-mediated Exfoliation of Graphite. *Chin. J. Polym. Sci.* **2018**, *36*, 1361–1367. [CrossRef]
26. Ahmed, R.M.; Ibrahiem, A.A.; El-Bayoumi, A.S.; Atta, M.M. Structural, mechanical, and dielectric properties of polyvinylchloride/graphene nano platelets composites. *Int. J. Polym. Anal. Charact.* **2021**, *26*, 68–83. [CrossRef]
27. Ma, J.; Liu, J.; Zhu, W.; Qin, W. Solubility study on the surfactants functionalized reduced graphene oxide. *Colloids Surf. A Physicochem. Eng. Asp.* **2018**, *538*, 79–85. [CrossRef]
28. Dai, J.; Wang, G.; Ma, L.; Wu, C. Study on the surface energies and dispersibility of graphene oxide and its derivatives. *J. Mater. Sci.* **2015**, *50*, 3895–3907. [CrossRef]
29. Tkalya, E.E.; Ghislandi, M.; de With, G.; Koning, C.E. The use of surfactants for dispersing carbon nanotubes and graphene to make conductive nanocomposites. *Curr. Opin. Colloid Interface Sci.* **2012**, *17*, 225–232. [CrossRef]
30. Kuila, T.; Bose, S.; Mishra, A.K.; Khanra, P.; Kim, N.H.; Lee, J.H. Chemical functionalization of graphene and its applications. *Prog. Mater. Sci.* **2012**, *57*, 1061–1105. [CrossRef]
31. Hatamie, S.; Akhavan, O.; Sadrnezhaad, S.K.; Ahadian, M.M.; Shirolkar, M.M.; Wang, H.Q. Curcumin-reduced graphene oxide sheets and their effects on human breast cancer cells. *Mater. Sci. Eng. C* **2015**, *55*, 482–489. [CrossRef]
32. Grimme, S. Do special noncovalent π-π stacking interactions really exist? *Angew. Chemie Int. Ed.* **2008**, *47*, 3430–3434. [CrossRef] [PubMed]
33. Wang, W.; Zhang, Y.; Wang, Y.B. Noncovalent π···π interaction between graphene and aromatic molecule: Structure, energy, and nature. *J. Chem. Phys.* **2014**, *140*, 094302. [CrossRef]
34. Navik, R.; Gai, Y.; Wang, W.; Zhao, Y. Curcumin-assisted ultrasound exfoliation of graphite to graphene in ethanol. *Ultrason. Sonochem.* **2018**, *48*, 96–102. [CrossRef]
35. Yang, Y.K.; He, C.E.; Peng, R.G.; Baji, A.; Du, X.S.; Huang, Y.L.; Xie, X.L.; Mai, Y.W. Non-covalently modified graphene sheets by imidazolium ionic liquids for multifunctional polymer nanocomposites. *J. Mater. Chem.* **2012**, *22*, 5666–5675. [CrossRef]
36. Jeong, S.Y.; Kim, S.H.; Han, J.T.; Jeong, H.J.; Jeong, S.Y.; Lee, G.W. Highly concentrated and conductive reduced graphene oxide nanosheets by monovalent cation-π interaction: Toward printed electronics. *Adv. Funct. Mater.* **2012**, *22*, 3307–3314. [CrossRef]
37. Texter, J. Graphene dispersions. *Curr. Opin. Colloid Interface Sci.* **2014**, *19*, 163–174. [CrossRef]
38. Achari, A.; Datta, K.; De, M.; Dravid, V.; Eswaramoorthy, M. Amphiphilic aminoclay–RGO hybrids: A simple strategy to disperse a high concentration of RGO in water. *Nanoscale* **2013**, *5*, 5316–5320. [CrossRef] [PubMed]
39. Kou, L.; Gao, C. Making silica nanoparticle-covered graphene oxide nanohybrids as general building blocks for large-area superhydrophilic coatings. *Nanoscale* **2011**, *3*, 519–528. [CrossRef]
40. Ma, F.; Yuan, N.; Ding, J. The conductive network made up by the reduced graphene nanosheet/polyaniline/polyvinyl chloride. *J. Appl. Polym. Sci.* **2013**, *128*, 3870–3875. [CrossRef]
41. Li, P.; Chen, X.; Zeng, J.-B.; Gan, L.; Wang, M. Enhancement of Interfacial Interaction between Poly(vinyl chloride) and Zinc Oxide Modified Reduced Graphene Oxide. *RSC Adv.* **2016**, *6*, 5784–5791. [CrossRef]
42. Khaleghi, M.; Didehban, K.; Shabanian, M. Simple and fast preparation of graphene oxide@ melamine terephthaldehyde and its PVC nanocomposite via ultrasonic irradiation: Chemical and thermal resistance study. *Ultrason. Sonochem.* **2018**, *43*, 275–284. [CrossRef] [PubMed]
43. Hu, J.; Jia, X.; Li, C.; Ma, Z.; Zhang, G.; Sheng, W.; Zhang, X.; Wei, Z. Effect of interfacial interaction between graphene oxide derivatives and poly(vinyl chloride) upon the mechanical properties of their nanocomposites. *J. Mater. Sci.* **2014**, *49*, 2943–2951. [CrossRef]

44. Khaleghi, M.; Didehban, K.; Shabanian, M. Effect of new melamine-terephthaldehyde resin modified graphene oxide on thermal and mechanical properties of PVC. *Polym. Test.* **2017**, *63*, 382–391. [CrossRef]
45. Piszczek, K. *Żelowanie Suspensyjnego, Nieplastyfikowanego Poli(chlorku winylu)*; Wydawnictwa Uczelniane Uniwersytetu Technologiczno-Przyrodniczego: Bydgoszcz, Poland, 2009.
46. Thakuria, R.; Nath, N.K.; Saha, B.K. The Nature and Applications of π-π Interactions: A Perspective. *Cryst. Growth Des.* **2019**, *19*, 523–528. [CrossRef]
47. Hatamie, S.; Ahadian, M.M.; Iraji zad, A.; Akhavan, O.; Jokar, E. Photoluminescence and electrochemical investigation of curcumin-reduced graphene oxide sheets. *J. Iran. Chem. Soc.* **2018**, *15*, 351–357. [CrossRef]
48. Alibolandi, M.; Mohammadi, M.; Taghdisi, S.M.; Ramezani, M.; Abnous, K. Fabrication of aptamer decorated dextran coated nano-graphene oxide for targeted drug delivery. *Carbohydr. Polym.* **2017**, *155*, 218–229. [CrossRef] [PubMed]
49. Bani, F.; Adeli, M.; Movahedi, S.; Sadeghizadeh, M. Graphene-polyglycerol-curcumin hybrid as a near-infrared (NIR) laser stimuli-responsive system for chemo-photothermal cancer therapy. *RSC Adv.* **2016**, *6*, 61141–61149. [CrossRef]
50. Saltos, J.A.; Shi, W.; Mancuso, A.; Sun, C.; Park, T.; Averick, N.; Punia, K.; Fatab, J.; Raja, K. Curcumin-derived green plasticizers for Poly(vinyl) chloride. *RSC Adv.* **2014**, *4*, 54725–54728. [CrossRef]
51. Larrañeta, E.; Imízcoz, M.; Toh, J.X.; Irwin, N.J.; Ripolin, A.; Perminova, A.; Domínguez-Robles, J.; Rodríguez, A.; Donnelly, R.F. Synthesis and Characterization of Lignin Hydrogels for Potential Applications as Drug Eluting Antimicrobial Coatings for Medical Materials. *ACS Sustain. Chem. Eng.* **2018**, *6*, 9037–9046. [CrossRef]
52. Velho, S.; Brum, L.; Petter, C.; dos Santos, J.H.; Šimunić, Š.; Kappa, W.H. Development of structured natural dyes for use into plastics. *Dye. Pigment.* **2017**, *136*, 248–254. [CrossRef]
53. Thema, F.T.; Moloto, M.J.; Dikio, E.D.; Nyangiwe, N.N.; Kotsedi, L.; Maaza, M.; Khenfouch, M. Synthesis and characterization of graphene thin films by chemical reduction of exfoliated and intercalated graphite oxide. *J. Chem.* **2013**, *3*, 150536. [CrossRef]
54. Marković, Z.M.; Kepić, D.P.; Matijašević, D.M.; Pavlović, V.B.; Jovanović, S.P.; Stanković, N.K.; Milivojević, D.D.; Spitalsky, Z.; Holclajtner-Antunović, I.D.; Bajuk-Bogdanović, D.V.; et al. Ambient light induced antibacterial action of curcumin/graphene nanomesh hybrids. *RSC Adv.* **2017**, *7*, 36081–36092. [CrossRef]
55. Barua, S.; Karak, N.; Chattopadhyay, P.; Islam, J.; Phukan, M.M.; Konwar, B.K. Biocompatible hyperbranched epoxy/silver-reduced graphene oxide-curcumin nanocomposite as an advanced antimicrobial material. *RSC Adv.* **2014**, *4*, 47797–47805. [CrossRef]
56. Husein, S.G.; Hamdani, S.; Rinaldi, K. Optimization of ar-Turmerone Isolation Method from Turmeric Oleoresin using Silica Gel Adsorbent. *Lett. Appl. NanoBioSci.* **2023**, *12*, 53. [CrossRef]
57. Kadam, P.V.; Yadav, K.N.; Bhingare, C.L.; Patil, M.J. Standardization and quantification of curcumin from *Curcuma longa* extract using UV visible spectroscopy and HPLC. *J. Pharmacogn. Phytochem.* **2018**, *7*, 1913–1918.
58. Pourhajibagher, M.; Parker, S.; Chiniforush, N.; Bahador, A. Photoexcitation triggering via semiconductor Graphene Quantum Dots by photochemical doping with Curcumin versus perio-pathogens mixed biofilms. *Photodiagn. Photodyn. Ther.* **2019**, *28*, 125–131. [CrossRef]
59. Malekmohammadi, S.; Hadadzadeh, H.; Farrokhpour, H.; Amirghofran, Z. Immobilization of Gold Nanoparticles on Folate-Conjugated Dendritic Mesoporous Silica-Coated Reduced Graphene Oxide Nanosheets: A New Nanoplatform for Curcumin pH-Controlled and Targeted Delivery. *Soft Matter* **2018**, *14*, 2400–2410. [CrossRef]
60. Fu, C.; Zhao, G.; Zhang, H.; Li, S. Evaluation and characterization of reduced graphene oxide nanosheets as anode materials for lithium-ion batteries. *Int. J. Electrochem. Sci.* **2013**, *8*, 6269–6280.
61. Stobinski, L.; Lesiak, B.; Malolepszy, A.; Mazurkiewicz, M.; Mierzwa, B.; Zemek, J.; Jiricek, P.; Bieloshapka, I. Graphene oxide and reduced graphene oxide studied by the XRD, TEM and electron spectroscopy methods. *J. Electron Spectros. Relat. Phenomena* **2014**, *195*, 145–154. [CrossRef]
62. Henderson, A.E.; Fitzgerald, A.G.; Rickerby, D.S.; Starey, B.E. Electron spectroscopy and electron microscopy of a a-C:H films. *J. Electron Spectros. Relat. Phenomena* **1990**, *52*, 475–483. [CrossRef]
63. Mezzi, A.; Kaciulis, S. Surface investigation of carbon films: From diamond to graphite. *Surf. Interface Anal.* **2010**, *42*, 1082–1084. [CrossRef]
64. Jackson, S.T.; Nuzzo, R.G. Determining hybridization differences for amorphous carbon from the XPS C 1s envelope. *Appl. Surf. Sci.* **1995**, *90*, 195–203. [CrossRef]
65. Díaz, J.; Paolicelli, G.; Ferrer, S.; Comin, F. Separation of thesp3 and sp2 components in the C1s photoemission spectra of amorphous carbon films. *Phys. Rev. B Condens. Matter Mater. Phys.* **1996**, *54*, 8064–8069. [CrossRef] [PubMed]
66. Haerle, R.; Riedo, E.; Pasquarello, A.; Baldereschi, A. sp2/sp3 hybridization ratio in amorphous carbon from C 1s core-level shifts: X-ray photoelectron spectroscopy and first-principles calculation. *Phys. Rev. B Condens. Matter Mater. Phys.* **2001**, *65*, 1–9. [CrossRef]
67. Fujimoto, A.; Yamada, Y.; Koinuma, M.; Sato, S. Origins of sp3C peaks in C1s X-ray Photoelectron Spectra of Carbon Materials. *Anal. Chem.* **2016**, *88*, 6110–6114. [CrossRef] [PubMed]
68. Butenko, Y.V.; Krishnamurthy, S.; Chakraborty, A.K.; Kuznetsov, V.L.; Dhanak, V.R.; Hunt, M.R.C.; Šiller, L. Photoemission study of onionlike carbons produced by annealing nanodiamonds. *Phys. Rev. B Condens. Matter Mater. Phys.* **2005**, *71*, 075420. [CrossRef]
69. Dosoky, N.S.; Setzer, W.N. Chemical composition and biological activities of essential oils of *Curcuma* species. *Nutrients* **2018**, *10*, 1196. [CrossRef]

70. Li, S. Chemical Composition and Product Quality Control of Turmeric (*Curcuma longa* L.). *Pharm. Crop.* **2011**, *5*, 28–54. [CrossRef]
71. Jantan, I.; Saputri, F.C.; Qaisar, M.N.; Buang, F. Correlation between chemical composition of *Curcuma domestica* and *Curcuma xanthorrhiza* and their antioxidant effect on human low-density lipoprotein oxidation. Evidence-based Complement. *Altern. Med.* **2012**, *2012*, 438356. [CrossRef]
72. Qu, S.; Li, M.; Xie, L.; Huang, X.; Yang, J.; Wang, N.; Yang, S. Noncovalent functionalization of graphene attaching [6,6]-phenyl-C61- butyric acid methyl ester (PCBM) and application as electron extraction layer of polymer solar cells. *ACS Nano* **2013**, *7*, 4070–4081. [CrossRef]
73. George, G.; Sisupal, S.B.; Tomy, T.; Kumaran, A.; Vadivelu, P.; Suvekbala, V.; Sivaram, S.; Ragupathy, L. Facile, environmentally benign and scalable approach to produce pristine few layers graphene suitable for preparing biocompatible polymer nanocomposites. *Sci. Rep.* **2018**, *8*, 11228. [CrossRef]
74. Refat, M.S. Synthesis and characterization of ligational behavior of curcumin drug towards some transition metal ions: Chelation effect on their thermal stability and biological activity. Spectrochim. *Acta Part A Mol. Biomol. Spectrosc.* **2013**, *105*, 326–337. [CrossRef] [PubMed]
75. Mohan, P.R.K.; Sreelakshmi, G.; Muraleedharan, C.V.; Joseph, R. Water soluble complexes of curcumin with cyclodextrins: Characterization by FT-Raman spectroscopy. *Vib. Spectrosc.* **2012**, *62*, 77–84. [CrossRef]
76. Lestari, M.; Indrayanto, G. Chapter Three—Curcumin. *Profiles Drug Subst. Excip. Relat. Methodol.* **2014**, *39*, 113–204. [CrossRef] [PubMed]
77. Klapiszewski, Ł.; Tomaszewska, J.; Skórczewska, K.; Jesionowski, T. Preparation and characterization of eco-friendly Mg(OH)2/lignin hybrid material and its use as a functional filler for poly(vinyl chloride). *Polymers* **2017**, *9*, 258. [CrossRef]
78. Tomaszewska, J.; Sterzyński, T.; Piszczek, K. Rigid Poly(vinyl chloride) Gelation in a Brabender Measuring Mixer. III. Transformation in the Torque Maximum. *J. Appl. Polym. Sci.* **2007**, *106*, 3158–3164. [CrossRef]
79. Lewandowski, K.; Skórczewska, K.; Piszczek, K.; Urbaniak, W. Recycled glass fibres from wind turbines as a filler for poly(Vinyl chloride). *Adv. Polym. Technol.* **2019**, *2019*, 8960503. [CrossRef]
80. Yang, G.; Li, L.; Lee, W.B.; Ng, M.C. Structure of graphene and its disorders: A review. *Sci. Technol. Adv. Mater.* **2018**, *19*, 613–648. [CrossRef] [PubMed]
81. Huang, X.; Huang, X.; Gong, Y.; Xiao, H.; McClements, D.J.; Hu, K. Enhancement of curcumin water dispersibility and antioxidant activity using core-shell protein-polysaccharide nanoparticles. *Food Res. Int.* **2016**, *87*, 1–9. [CrossRef] [PubMed]
82. Meabed, O.M.; Shamaa, A.; Abdelrahman, I.Y.; El-Sayyed, G.S.; Mohammed, S.S. The Effect of Nano-chitosan and Nano-curcumin on Radiated Parotid Glands of Albino Rats: Comparative Study. *J. Clust. Sci.* **2022**, *34*, 977–989. [CrossRef]
83. Gupta, V.; Aseh, A.; Ríos, C.N.; Aggarwal, B.B.; Mathur, A.B. Fabrication and characterization of silk fibroin-derived curcumin nanoparticles for cancer therapy. *Int. J. Nanomed.* **2009**, *4*, 115–122. [CrossRef] [PubMed]
84. Taha, S.; El-Sherbiny, I.; Enomoto, T.; Salem, A.; Nagai, E.; Askar, A.; Abady, G.; Abdel-Hamid, M. Improving the functional activities of curcumin using milk proteins as nanocarriers. *Foods* **2020**, *9*, 986. [CrossRef]
85. Gedde, U.W. *Polymer Physics*; Chapman & Hall: London, UK; Glasgow, UK; Weinheim, Germany; New York, NY, USA; Tokyo, Japan; Melbourne, Australia; Madras, India, 1995.
86. Lee, J.W.; Kim, S.W.; Cho, Y.L.; Jeong, H.Y.; Song, Y.I.; Suh, S.J. Characterization and dispersion stabilities of multi-layer graphene-coated copper prepared by electrical wire-explosion method. *J. Nanosci. Nanotechnol.* **2016**, *16*, 11286–11291. [CrossRef]
87. Ferralis, N. Probing mechanical properties of graphene with Raman spectroscopy. *J. Mater. Sci.* **2010**, *45*, 5135–5149. [CrossRef]
88. Delforce, L.; Hofmann, E.; Nardello-Rataj, V.; Aubry, J.M. TiO2 nanoparticle dispersions in water and nonaqueous solvents studied by gravitational sedimentation analysis: Complementarity of Hansen Parameters and DLVO interpretations. *Colloids Surf. A Physicochem. Eng. Asp.* **2021**, *628*, 127333. [CrossRef]
89. Navarrete-Segado, P.; Tourbin, M.; Frances, C.; Grossin, D. Masked stereolithography of hydroxyapatite bioceramic scaffolds: From powder tailoring to evaluation of 3D printed parts properties. *Open Ceram.* **2022**, *9*, 100235. [CrossRef]
90. Yue, M.; Huang, M.; Zhu, Z.; Huang, T.; Huang, M. Effect of ultrasound assisted emulsification in the production of Pickering emulsion formulated with chitosan self-assembled particles: Stability, macro, and micro rheological properties. *LWT* **2022**, *154*, 112595. [CrossRef]
91. Santos, J.; Trujillo-Cayado, L.A.; Carrillo, F.; López-Castejón, M.L.; Alfaro-Rodríguez, M.C. Relation between Droplet Size Distributions and Physical Stability for Zein Microfluidized Emulsions. *Polymers* **2022**, *14*, 2195. [CrossRef] [PubMed]
92. Pascaud, K.; Mercé, M.; Roucher, A.; Destribats, M.; Backov, R.; Schmitt, V.; Sescousse, R.; Brouillet, F.; Sarda, S.; Ré, M.I. Pickering emulsion as template for porous bioceramics in the perspective of bone regeneration. *Colloids Surf. A Physicochem. Eng. Asp.* **2022**, *643*, 128748. [CrossRef]
93. Baibarac, M.; Stingescu, L.; Stroe, M.; Negrila, C.; Matei, E.; Cotet, L.C.; Anghel, I.; Şofran, I.E.; Baia, L. Poly(vinyl chloride) spheres coated with graphene oxide sheets: From synthesis to optical properties and their applications as flame-retardant agents. *Polymers* **2021**, *13*, 565. [CrossRef]
94. Godínez-García, A.; Vallejo-Arenas, D.D.; Salinas-Rodríguez, E.; Gómez-Torres, S.A.; Ruíz, J.C. Spraying synthesis and ion permeation in polyvinyl chloride/graphene oxide membranes. *Appl. Surf. Sci.* **2019**, *489*, 962–975. [CrossRef]
95. Dhakal, S.; Chao, K.; Schmidt, W.; Qin, J.; Kim, M.; Chan, D. Evaluation of Turmeric Powder Adulterated with Metanil Yellow Using FT-Raman and FT-IR Spectroscopy. *Foods* **2016**, *5*, 36. [CrossRef]

96. Kugler, S.; Spychaj, T. Nanostruktury węglowe i błony lub powłoki polimerowe z ich udziałem: Cz. II. Błony i powłoki polimerowe z udziałem nanostruktur węglowych. *Polimery/Polymers* **2013**, *58*, 177–180. [CrossRef]
97. Deshmukh, K.; Khatake, S.M.; Joshi, G.M. Surface properties of graphene oxide reinforced polyvinyl chloride nanocomposites. *J. Polym. Res.* **2013**, *20*, 286. [CrossRef]
98. Xiao, Y.; Xin, B.; Chen, Z.; Lin, L.; Liu, Y.; Hu, Z. Enhanced thermal properties of graphene-based poly (vinyl chloride) composites. *J. Ind. Text.* **2019**, *48*, 1348–1363. [CrossRef]
99. Wang, H.; Xie, G.; Zhu, Z.; Ying, Z.; Zeng, Y. Enhanced tribological performance of the multi-layer graphene filled poly(vinyl chloride) composites. *Compos. Part A Appl. Sci. Manuf.* **2014**, *67*, 268–273. [CrossRef]
100. Vadukumpully, S.; Paul, J.; Mahanta, N.; Valiyaveettil, S. Flexible conductive graphene/poly(vinyl chloride) composite thin films with high mechanical strength and thermal stability. *Carbon* **2011**, *49*, 198–205. [CrossRef]
101. Wang, H.; Xie, G.; Fang, M.; Ying, Z.; Tong, Y.; Zeng, Y. Electrical and mechanical properties of antistatic PVC films containing multi-layer graphene. *Compos. Part B Eng.* **2015**, *79*, 444–450. [CrossRef]
102. Brostow, W.; Hagg Lobland, H.E. *Materials: Introduction and Applications*; John Wiley & Sons: Hoboken, NJ, USA, 2017; ISBN 978-0-470-52379-7.
103. Broza, G.; Piszczek, K.; Schulte, K.; Sterzynski, T. Nanocomposites of poly(vinyl chloride) with carbon nanotubes (CNT). *Compos. Sci. Technol.* **2007**, *67*, 890–894. [CrossRef]
104. Sun, X.; Huang, C.; Wang, L.; Liang, L.; Cheng, Y.; Fei, W.; Li, Y. Recent Progress in Graphene/Polymer Nanocomposites. *Adv. Mater.* **2021**, *33*, 2001105. [CrossRef] [PubMed]
105. Govindaraj, P.; Sokolova, A.; Salim, N.; Juodkazis, S.; Fuss, F.K.; Fox, B.; Hameed, N. Distribution states of graphene in polymer nanocomposites: A review. *Compos. Part B Eng.* **2021**, *226*, 109353. [CrossRef]
106. Reis, T.M.C.; Castro, V.G.; Amurin, L.G.; Silva, G.G. Graphene oxide dispersion in epoxy resin prepared by direct phase transfer from ethanol: Rheology and aging. *Compos. Part C Open Access* **2023**, *10*, 100340. [CrossRef]
107. Sharma, A.; Sharma, S. Graphene-based polymer coatings for preventing marine corrosion: A review. *J. Coatings Technol. Res.* **2023**, *20*, 413–432. [CrossRef]

molecules

MDPI

Article

Synthesis of Novel Polymer-Assisted Organic-Inorganic Hybrid Nanoflowers and Their Application in Cascade Biocatalysis

Joana F. Braz [1], Nadya V. Dencheva [1,*], Marc Malfois [2] and Zlatan Z. Denchev [1,*]

[1] IPC—Institute for Polymers and Composites, University of Minho, 4800-056 Guimarães, Portugal
[2] ALBA Synchrotron Facility, Cerdanyola del Valés, 0890 Barcelona, Spain
* Correspondence: nadiad@dep.uminho.pt (N.V.D.); denchev@dep.uminho.pt (Z.Z.D.)

Abstract: This study reports on the synthesis of novel bienzyme polymer-assisted nanoflower complexes (PANF), their morphological and structural characterization, and their effectiveness as cascade biocatalysts. First, highly porous polyamide 6 microparticles (PA6 MP) are synthesized by activated anionic polymerization in solution. Second, the PA6 MP are used as carriers for hybrid bienzyme assemblies comprising glucose oxidase (GOx) and horseradish peroxidase (HRP). Thus, four PANF complexes with different co-localization and compartmentalization of the two enzymes are prepared. In samples NF GH/PA and NF GH@PA, both enzymes are localized within the same hybrid flowerlike organic-inorganic nanostructures (NF), the difference being in the way the PA6 MP are assembled with NF. In samples NF G/PAiH and NF G@PAiH, only GOx is located in the NF, while HRP is preliminary immobilized on PA6 MP. The morphology and the structure of the four PANF complexes have been studied by microscopy, spectroscopy, and synchrotron X-ray techniques. The catalytic activity of the four PANF was assessed by a two-step cascade reaction of glucose oxidation. The PANF complexes are up to 2–3 times more active than the free GOx/HRP dyad. They also display enhanced kinetic parameters, superior thermal stability in the 40–60 °C range, optimum performance at pH 4–6, and excellent storage stability. All PANF complexes are active for up to 6 consecutive operational cycles.

Keywords: hybrid nanoflowers; polyamide microparticles; synchrotron WAXS/SAXS; polymer-assisted biocatalysts; enzyme kinetics; GOx/HRP enzyme cascade

1. Introduction

In the last decades, cascade biocatalysis has attracted significant interest from both academia and industry as a promising technology for green and sustainable chemical production. Unlike the step-by-step synthesis, cascade biocatalysts accomplish two or more consecutive reactions in one-pot, thus avoiding isolation and purification of intermediates. Such procedure has become more efficient due to reduced reaction time and production costs, meanwhile not creating toxic wastes [1]. Nowadays, numerous drugs and fine chemicals [2,3], including chiral alcohols and amines [4,5], carbohydrates [6], and polymers [7] are synthesized through enzyme cascade catalysis. However, the industrial use of cascade biocatalysts based on free enzymes is still problematic, due to their poor stability and difficult reuse [8].

A possible way to overcome these limitations is the application of immobilization technologies, in which two or more enzymes are co-localized close to one another on a suitable support. Various inorganic materials e.g., metal oxides, minerals, carbon allotropes, and also organic materials such as synthetic and natural biopolymers have been widely used as supports for enzyme immobilization [9]. Immobilization of enzymes results in a twofold gain: on the one hand, biocatalysts become more resistant to the harmful effects of the environment, e.g., temperature, pH, and solvents; on the other hand, enzymes can be made easily removable from the reaction medium, especially if the carrier applied is with magnetic properties [10]. It has been also demonstrated [1] that in many cases, the co-immobilization boosts the activity of the multienzyme complex. The nature of this effect is still debated, being attributed either to reduction of mass transfer limitations (substrate channeling) [11–13] or to a complete change in the microenvironment [14–16].

Three main methods are widely used for enzyme immobilization, namely (i) binding to a support, (ii) encapsulation or (iii) cross-linking [10]. Each immobilization technique has its advantages and disadvantages. The selection of the method to apply depends on the specific objectives and applications of the target enzyme complex [17,18].

The multienzyme cascade complexes can be assembled through: (i) simple fusion of enzymes combining their polypeptide chains by short cross-linkers [19]; (ii) attaching of enzymes on desired docking sites of a scaffold biomolecule, e.g., DNA [20], or (iii) co-immobilization in which the enzymes are bonded to the same support in close physical proximity, mimicking the multienzyme complexes in living cells [21]. The last method is the most frequently used to design cascade catalysts with industrial applications.

Several approaches have been developed for multienzyme co-immobilization, e.g., random or positional co-immobilization [22–24] and compartmentalization [25]. The latter approach has been developed to mimic natural enzyme organization in cells, wherein the enzymes are spatially separated in different patterns and ratios on the same support. Several carriers such as polymer capsules [25], polymersomes [26], liposome vesicles [27], and nano/microsized particles [28] have been used to integrate enzymes via encapsulation or layer-by-layer surface attachment.

Along with their enhanced stability and reusability, however, the immobilized enzymes display reduced catalytic activity in comparison to the native counterparts, due to decreased mobility, unfavorable conformation, and mass transfer limitations related to the attachment to a solid support [8–10]. A notable exception has recently been reported by Zare et al.: organic-inorganic hybrid nanoflowers (NF) [29]. The authors prepare flower-like hierarchical nanostructures comprising $Cu_3(PO_4)_2$ as the inorganic component and different enzymes as the organic component (e.g., laccase, carbonic anhydrase, and lipase) by adding copper Cu^{2+} to protein solutions in saline phosphate buffer (PBS). Later studies [30,31] confirm that this new immobilization method results in higher catalytic activity of the enzyme complexes being at the same time: simple, sustainable, and of low energy consumption. The NF synthesis that could be considered a particular case of enzyme immobilization involves three steps [29–31]: nucleation, growth, and completion. In the first stage, the protein molecules form complexes with the copper ions, mainly through coordination with amine or amide groups from the protein backbone, which become the nuclei of primary crystals of copper phosphate. Then, the metal-protein primary crystals aggregate into large agglomerates in the shape of flower petals. In the final stage, protein(s) bind the petals together, thus forming 3D flower-like protein-inorganic structures.

In subsequent studies summarized in comprehensive review articles, different metal ions, for instance, Ca^{2+}, Zn^{2+}, Mn^{2+}, Mg^{2+}, Fe^{2+}, Co^{2+}, etc. and numerous single enzymes have been used to design NF-based biocatalysts and biosensors applicable in biotechnology, biomedicine, and environmental sciences [30–34]. Since a nitrogen atom is involved in the complexation during the NF formation [29,30], besides enzymes and other nitrogen-containing biomolecules such as aptamers, natural DNA, amino acids, and chitosan, elastin-like polypeptides have also been used as the organic element to construct self-assembled NF [34]. Multienzyme NF assemblies capable to catalyze cascade reactions have also been developed. Thus, Sun et al. [35] prepare a bienzyme NF complex by random co-immobilization of glucose oxidase (GOx) and horseradish peroxidase (HRP) used for glucose detection. At the same time, Li et al. [36] develop a compartmentalized NF system by spatial co-localization of the same GOx/HRP enzyme dyad. This NF shows higher catalytic efficiency in comparison to the free bienzyme complex or to the randomly distributed NF assembles containing single GOx or HRP.

In spite of their undisputed qualities as efficient biocatalysts, hybrid NF have shown some drawbacks that impede their wide industrial application. Due to multiple centrifugation/filtration cycles during the NF collection [31–34], or in extremely acidic media [37], the delicate hierarchically ordered petals are easily broken. Along with their mechanical fragility and sensitivity to pH, conventional NF particles are too small to be filtered off easily, cannot be reused after the first reaction cycle, and their synthesis could be difficult to scale up. Consequently, there appears a demand for innovative, highly robust, and recyclable hybrid NF systems useful for enzyme-based applications such as bioremediation, biosensing, and biocatalysis [34]. In this context, chemical treatment and incorporation of other nano or microsized materials have been tried to make NF more efficient for industrial applications [38–41]. For example, cross-linking with glutaraldehyde is applied to enhance the catalytic activity and robustness of NF comprising lipase [38,39] and laccase [40]. Entrapment of preliminary synthesized α-acetolactate decarboxylase/ $Ca_3(PO_4)_2$ NF into alginate hydrogel beads improves the recyclability and stability of the NF biocatalyst applied in beer brewing industry [41]. Very recently, Luo et al. has prepared bienzyme hybrid NF in-situ grown on PVA-*co*-PE, a fibrous strip that has tested with good results as a stable and efficient colorimetric biosensor for glucose detection [42]. To the best of our knowledge, this is the first attempt to combine NF-based enzyme dyad with a synthetic polymer support.

Based on the existing research on multienzyme NF biocatalysts, it can be hypothesized that other properly structured and shaped synthetic polymer supports could also be used for more stable biocatalysts with various co-localization possibilities. Polyamides are among the few synthetic chemical analogues of proteins permitting extensive non-covalent interaction through H-bond formation. Moreover, polyamides are biocompatible semicrystalline polymers with significant mechanical strength and chemical and thermal resistance. Polyamides contain high concentrations of amide and some amine N atoms that, as referred to previously [29,30], play a key role in the hierarchical NF auto-assembly by complexation with the Me^{2+} of the inorganic component.

In this study we report on the synthesis of novel polymer-assisted bienzyme nanoflower complexes (PANF) and their application as cascade biocatalysts. The PANF complexes contain highly porous PA6 MP used as carriers for hybrid organic-inorganic NF assemblies. These NF can contain both GOx and HRP enzymes or only GOx, the HRP being preliminarily immobilized on PA6 MP. Four PANF complexes are produced with different co-localization and compartmentalization of the two enzymes. Their catalytic activity has been assessed in a cascade reaction for glucose detection. The kinetic parameters, i.e., the Michaelis constant K_m and the maximum velocity V_{max}, have also been determined. The influence of the temperature and the pH of the reaction medium on PANF activity, as well as their reusability and the storage stability, are studied.

2. Results and Discussion

2.1. Synthesis of PANF

As a first step of PANF preparation, PA6 MP are synthesized by activated anionic ring-opening polymerization (AAROP) of ε-caprolactam in solution, as explained in Section 3.3 and previous reports [43,44]. The AAROP is presented schematically in Figure S1 of the Supplementary Materials. Then, the resulting highly porous microparticles are used as supports in the process of the preparation of conventional mono and bienzyme NF. In other words, PANF represent mono or bienzyme NF assemblies grown on PA6 MP.

The PANF building up is presented schematically in Figure 1. This is a facile and sustainable auto-assembly process carried out in aqueous medium at room temperature. Altogether four PANF samples are produced, they are expected to provide different enzyme co-localization and compartmentalization.

Samples NF GH/PA and NF GH&PA are obtained with PA6 MP, decorating them with mixed GOx/HRP nanoflowers. In the first sample, the bienzyme NF are formed in the presence of the MP carrier. In the second sample, the MP carrier is added 24 h after the start of the bienzyme NF formation. In samples NF G/PAiH and NF G&PAiH, the NF contains only GOx, while the HRP enzyme is preliminary immobilized on PA6 MP by physical adsorption. The difference between these two samples is the same as in the previous sample set.

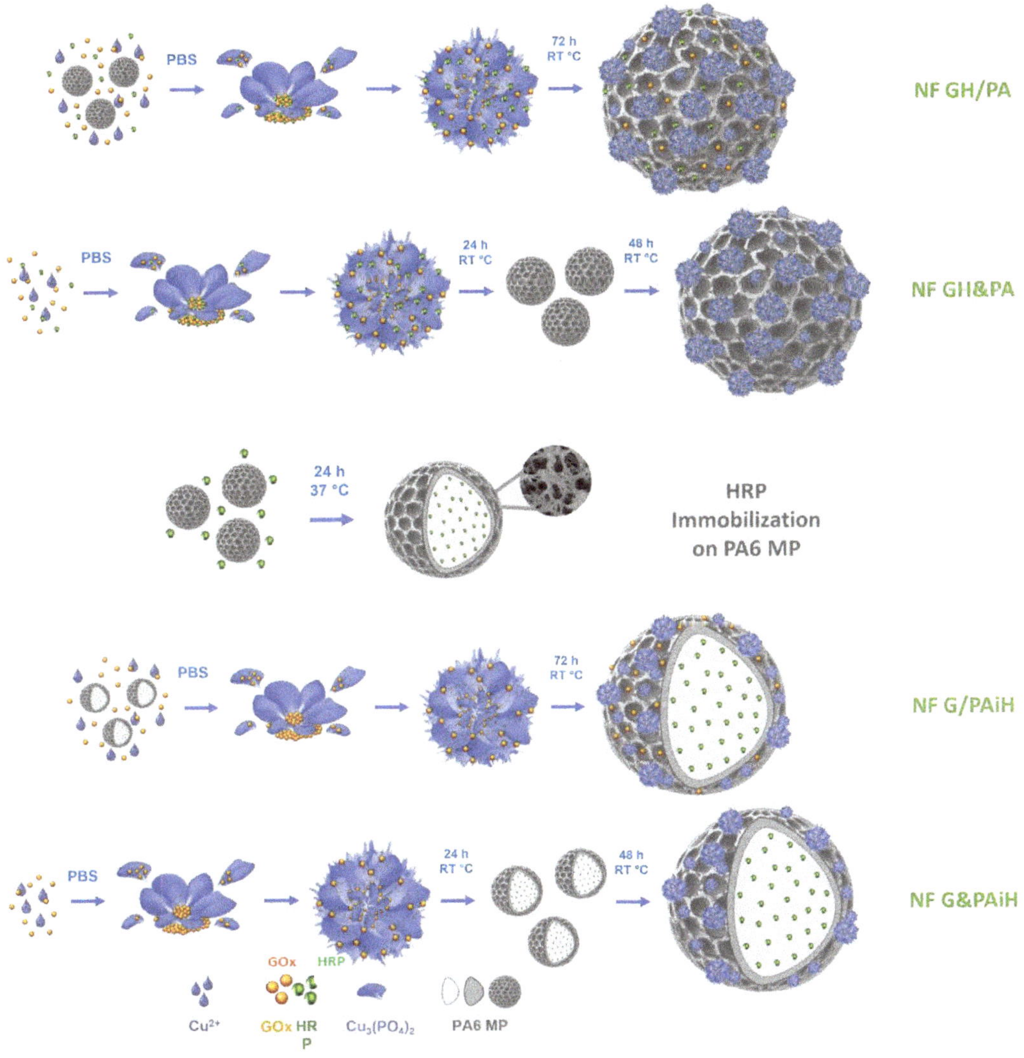

Figure 1. Schematic presentation of the design strategy for assembling NF-decorated porous PA6 microparticles carrying the GOx/HRP enzyme dyad. PBS = phosphate buffer saline; RT = room temperature (20–23 °C).

The PANF systems of this study combine the facile preparation and catalytic effectiveness of the NF hybrids with the robustness of the PA6 MP that are also proven to be good carriers for free enzymes [45]. Thus, PANF are expected to demonstrate appropriate activity and stability at different reaction conditions, as well as reutilization in various consecutive operation cycles.

2.2. SEM Morphological Studeis

As seen from the SEM micrographs in Figure 2a–f, the GOx- and the HRP-based NF display almost spherical morphology with average visible diameters of 5–15 μm and differently shaped nanostructured petals. The GOx NF (Figure 2a–c) have open, well-defined petals similar to those of a carnation. The HRP-based NF (Figure 2d–f) form denser and more closed flowers, with smaller petals as in a ranunculus. In the NF containing both enzymes, a flower-like complex with mixed petals' morphology is formed (Figure 2g–i). These three NF morphologies were easily distinguishable by direct observation.

Figure 2. SEM micrographs of: (**a–c**) GOx NF; (**d–f**) HRP NF; (**g–i**) mixed GH NF; (**j**,**k**) PA6 microparticles. For sample designation and idealized structure see Figure 1.

The preliminarily synthesized PA6 MP (Figure 2j,k) represents highly porous, scaffold-like ovoid aggregates with sizes in the range of 100–200 μm. The aggregation is sought on purpose by changing the conditions of the AAROP. According to Figure 1, the pretended PANF designs require attaching of many NF on a single PA6 microparticle. NF sizes vary in the narrow range of 5–10 μm. Therefore, the PA6 support particles should be larger than the implanted NF structures.

Selected SEM micrographs of PANF samples are presented in Figure 3.

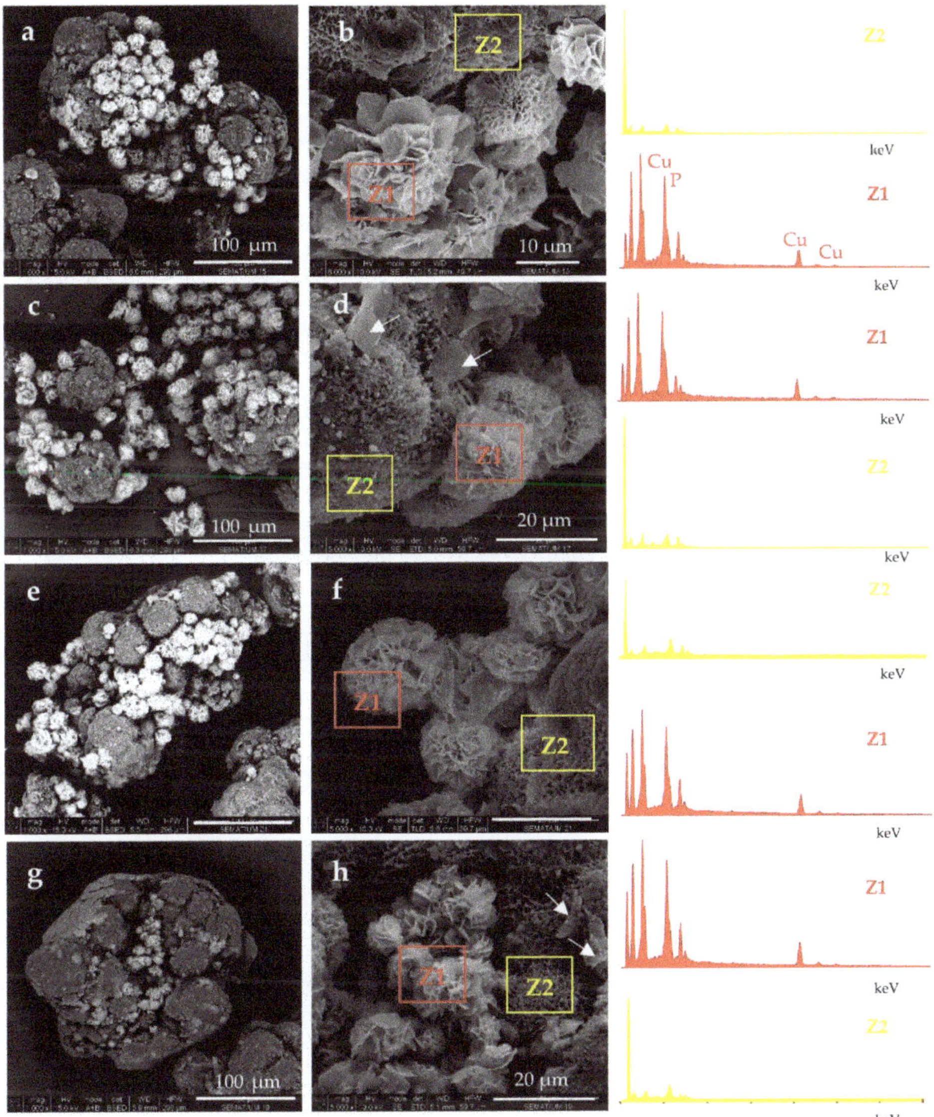

Figure 3. SEM micrographs of polymer-assisted hybrid nanoflowers (PANF): (**a**,**b**) NF GH/PA; (**c**,**d**) NF GH&PA; (**e**,**f**) NF G/PAiH; (**g**,**h**) NG G&PAiH. For sample designation see Figure 1.

The attachment of NF structures to the porous surface of the PA6 MP is clearly observable. At lower magnification (the left-hand column), the mode of the detector employed allows the observation of the NF as brighter objects sitting on the darker porous PA6 MP. This allows distinction between the NF and MP components. The higher magnification (right-hand column) shows that NF GH/PA (Figure 3a,b) and NF GH&PA (Figure 3c,d) samples contain nanoflowers of mixed morphology, while in NF G/PAiH (Figure 3e,f) and NF G&PAiH (Figure 3g,h) the NF entities are closer to the pure carnation shape of the single GOx nanoflowers in Figure 2a–c. Along with the well-formed NF, all four PANF samples comprise separate petals scattered upon the MP porous surface. They are best observed in the NF GH&PA and NF G&PAiH, i.e., the samples in which the NF formation starts before the introduction of the PA6 MP (see the arrow-indicated areas in Figure 3d,h).

More information about the composition of the PANF systems has been obtained by combining SEM with Energy Dispersive X-ray spectroscopy (EDX). For each PANF sample, EDX spectra are obtained in the NF (Z1) and PA6 (Z2) zones (see the insets to Figure 3). The numerical data of the ZAF corrected elemental analysis for Cu and P in the NF- and PA6 zones of the PANF samples are presented in Table S1 of Supplementary Materials. The weight percentage of Cu and P in the NF zones varies in the 27–36% and 9.5–11.6% respectively, being the highest in NF GH&PA and the lowest in the NF G&PAiH. At the same time, in the PA6 zones, the Cu content is between 0.0–2.5 wt.%, and that of P varies between 1.2–1.8 wt.%. Therefore, it seems that certain small amounts of the $Cu_3(PO_4)_2$/enzyme hybrid complexes are not included in the observable by SEM nanoflower morphologies, but reside as single petals in the MP pores.

According to the accepted mechanism, the self-assembly of NF takes place with the participation of amide and amine groups from the protein [29,30]. Since PA6 MP contain large concentrations of these groups, it has been decided to investigate if MP themselves are involved in the formation of NF structures. For this purpose, PA6 MP are incubated with $CuSO_4$ in PBS, in the absence of enzymes, following the procedure for PANF synthesis. The appearance of the resulting mixed morphology is presented in Figure 4.

The central SEM micrograph in Figure 4 presents clearly observable platelets with nanometric thickness ("petals") attached to the PA6 MP surface, but no nanoflower-shaped assemblies are observed. Moreover, the upper images in the same figure present the close view of these inorganic morphologies whose EDX shows strong Cu and P peaks. The bottom micrographs focus on the unchanged porous and scaffold-like morphology of the PA6 MP support. Apparently, the N-atoms from PA6 MP could induce the formation of inorganic "petals" whose EDX signals are identical to those of NF in Figure 3d. However, in the absence of proteins, these "petals" could not assemble the flower-shaped morphology. Apparently, the presence of PA6 MP has an important contribution to the morphology of the NF component in PANF. This contribution is more pronounced whenever the PA6 MP are introduced in the beginning of the PANF formation.

2.3. FTIR Structural Studies

Apart from direct observation of the typical PANF morphology by SEM, the structure of the new catalytic complexes is investigated by spectral and X-ray techniques. Figure 5 presents the FTIR spectra (ATR mode, sample weights of ca. 15 mg) of some selected samples. Curve 1 shows the spectrum of PA6 MP that contains all the peaks characteristic of aliphatic secondary amides, namely: at 3293 and 3075 cm^{-1} (associated N-H groups), a doublet at 2935 and 2863 cm^{-1} (symmetric and asymmetric C-H stretching vibrations of the CH_2 groups), 1633 cm^{-1} (Amide I band, trans CO-NH group), and 1535 cm^{-1} (Amide II band, trans CO-NH group, highly associated). The peak at 710–685 cm^{-1} is assigned to the rocking vibrations of the -$(CH_2)_5$- link. The weak broad band at 3490 cm^{-1} is ascribed to some small amounts of non-associated N-H groups or associated OH groups from adsorbed H_2O. Notably, the weak peak at 1716–1730 cm^{-1} belonging to terminal -COOH group, normally found in hydrolytic PA6, was not present in PA6 MP. This can be expected, having in mind the mechanism of AAROP (Figure S1 of the Supplementary Materials) not suggesting the formation of terminal -COOH groups.

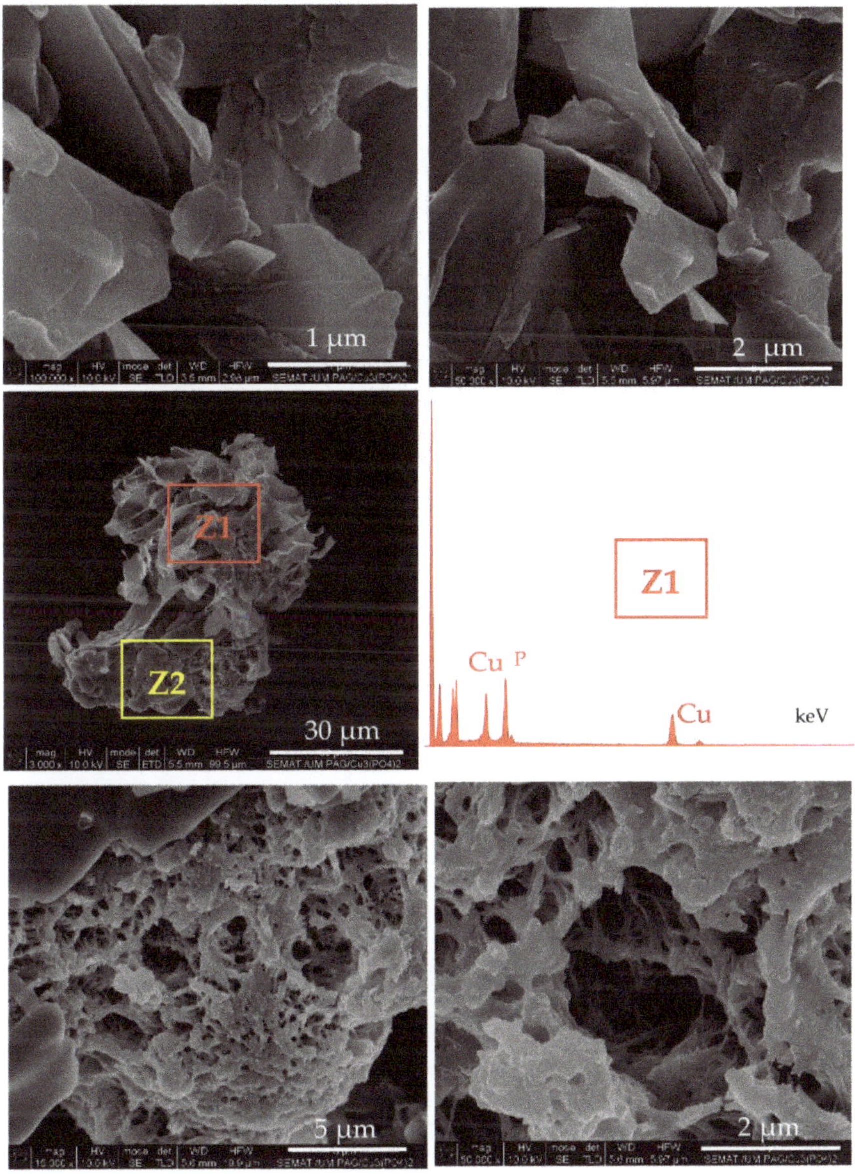

Figure 4. SEM images of micro and nanostructures obtained after incubation of PA6 MP with $CuSO_4.5H_2O$ in PBS, in the absence of enzymes. Central image: overall appearance; Top images: close view of the petal-shaped platelets originating from the inorganic component (Z1); Bottom images: close view of the porous scaffold-like morphology originating from the PA6 MP component (Z2). The EDX data of Z1 localization is presented as inset.

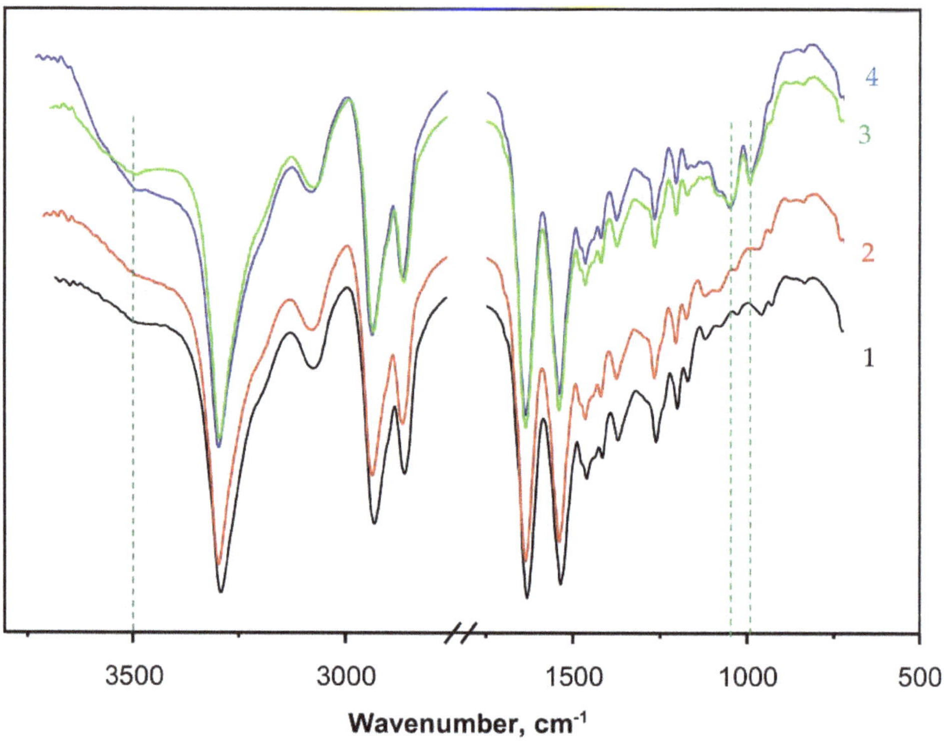

Figure 5. FTIR spectra with ATR of: 1—PA6 MP; 2—PAiH; 3—NF GH/PA; 4—NF G/PaiH. PaiH = HPR adsorbed on PA6 MP; ATR = Attenuated Total Reflection. The dashed lines indicate the appearance of peaks originating from the NF component. For more details, see the text.

Curve 2 in Figure 5 that presents the spectrum of PaiH sample (HRP immobilized by adsorption on PA6 MP) is identical to that of the neat PA6 MP. This could mean that in the former, most of the HRP have penetrated into the pores of the MP carrier and cannot be hit by the IR radiation. Curves 3 and 4 display the FTIR spectra of NF GH/PA and NF G/PAiH samples, respectively, both of them carrying nanoflowers on the surface of the PA6 MP. Apart from all FTIR bands of PA6, weak but clearly observable new bands of the $Cu_3(PO_4)_2$ from NF at 1070–1045 cm^{-1} (stretching mode $\nu_3$of PO_4^{3-}) and 987 cm^{-1} (PO_4^{3-} stretching mode ν_1) appear in these two samples [46,47]. The NF-containing samples (curves 3 and 4) also display a broad band centered close to 3500 cm^{-1} being somehow different in its shape and intensity from that of the PA6 MP. This could be due to the presence of small amounts of NH_2 groups originating from the peptide enzyme matrix overlaid with associated OH group stretching vibration peak due to humidity. Possible formation of the hydrated Cu_2PO_4OH specie would also result in a FTIR band in this region [47].

The FTIR spectra confirms the formation of NF structures on the PA6 MP in the PANF complexes, but provides insufficient direct structural information about the NF-PA6 interactions in them.

2.4. Synchrotron X-ray Studies

To shed more light on the crystalline structure of the new PANF complexes, WAXS and SAXS studies using synchrotron radiation have been performed. To the best of our knowledge, this is the first synchrotron structural characterization of hybrid inorganic-organic bienzyme NF complexes. Figure 6 shows the comparison between the linear WAXS patterns of samples representing neat PA6 MP, NF, GH and all four PANF complexes. Notably, WAXS only probes the ordered crystalline regions of PANF.

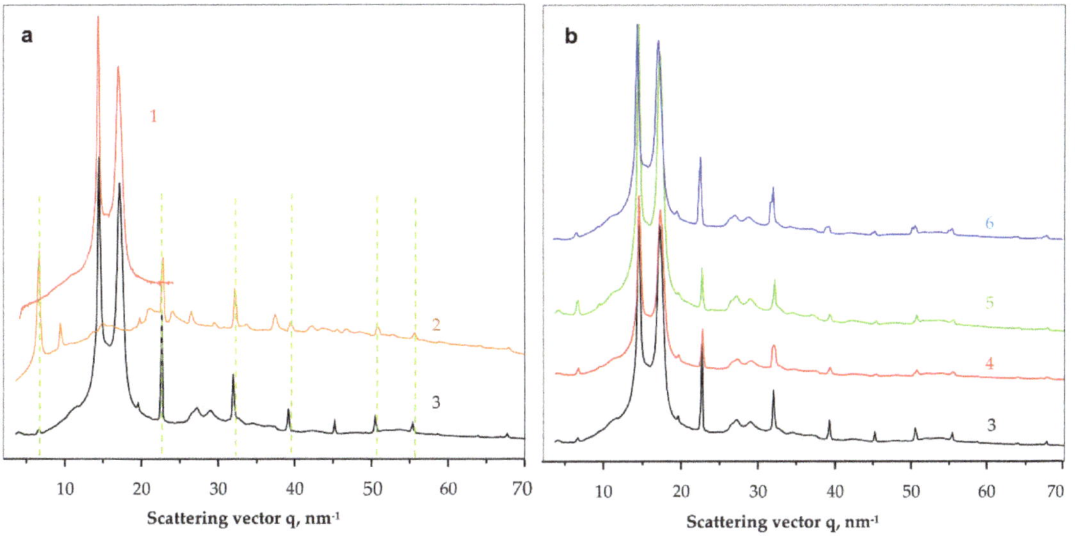

Figure 6. Comparative synchrotron WAXS patterns of NF and PANF. (**a**)—identification of the PA6 and NF building blocks; (**b**)—comparison of the scattering patterns of all PANF samples. 1—PA6 MP; 2—NF GH; 3—NF GH/PA; 4—NF GH&PA; 5—NF G&PAiH; 6—NF G/PAiH. The dahesd lines in Figure 6a relate the coinciding phospahe peaks in curves 2 and 3.

As seen from Figure 6a, the neat PA6 MP (curve 1) and the NF GH/PA complex (curve 3) present the well-known pattern of PA6 with its two strong reflections at q = 14.4 nm^{-1} and 17.2 nm^{-1} ascribed to the monoclinic unit cell of α-PA6 with $d_{[200]}$ = 4.36 Å and $d_{[002/202]}$ = 3.65 Å. These two peaks of PA6 appear in all PANF samples in Figure 6 with the same intensity ratio and angular positions. Therefore, it seems that the formation of NF in the presence of PA6 MP does not affect the crystalline structure of the MP carrier.

Comparing the WAXS pattern of the conventional bienzyme GH NF (Figure 6a, curve 2) to those of the PANF complexes (Figure 6b, curves 3–6) shows that the latter contains most all of the NF GH reflections (the dashed lines). These reflections belong to the triclinic $Cu_3(PO_4)_2$ crystalline phase since the two enzyme components are amorphous in WAXS (See Figure S2, GOx + HRP curve in the Supplementary Materials). The appearance or disappearance of some WAXS peaks of NF between 35–45 nm^{-1} or the change in their intensity means that the presence of the MP carrier influences the crystalline structure of the implanted NF. For example, the strong phosphate reflection at q = 6.7 nm^{-1} in NF GH (Figure 6a, curve 2) is extremely weak in the PANF samples (Figure 6b, curve 3–6), whereas the two reflections found in NF at q = 9.4 nm^{-1} and 37.4 nm^{-1} in NF GH are missing in the PANF patterns. This means that the structure of the protein/phosphate crystalline NF petals formed in the presence of PA6 MP differ from those that grow unperturbed in the absence of the MP support. Along with the SEM studies in Figure 4, we contemplate the WAXS results in Figure 6a as an additional proof that N-atoms of the PA6 MP could also participate in the enzyme/copper phosphate auto assembly process, thus modifying, although slightly, the crystalline structure of the incrusted NF.

The linear WAXS patterns in Figure 6b are normalized to the PA6 reflections and therefore allow conclusions about the samples′ protein content. Since the intensity of the main phosphate peaks in PANF will be proportional to the amount of NF, i.e., to that of the protein component(s), it can be inferred that sample NF GH/PA contains almost twice as much protein as NF GH@PA. A similar relation is valid for the NF G/PAiH and NF G&PAiH pair. In this case, the phosphate peak intensity should be related only to the GOx concentration.

The use of synchrotron SAXS (Figure 7) allows further clarification of the structure of the PANF samples. This method probes density periodicities with dimensions in the 20–250 angstroms range, which includes the sizes of the crystalline lamellae typically found in semi-crystalline polymers. The appearance of SAXS peaks depends on the phase contrast between the crystalline and amorphous domains of the lamellar systems. Figure 7a displays the background- and Lorentz-corrected SAXS linear profiles of the four PANF samples. To enable comparison, the SAXS pattern of the NF GH sample (conventional GOx/HRP nanoflowers without PA6 MP) are also presented.

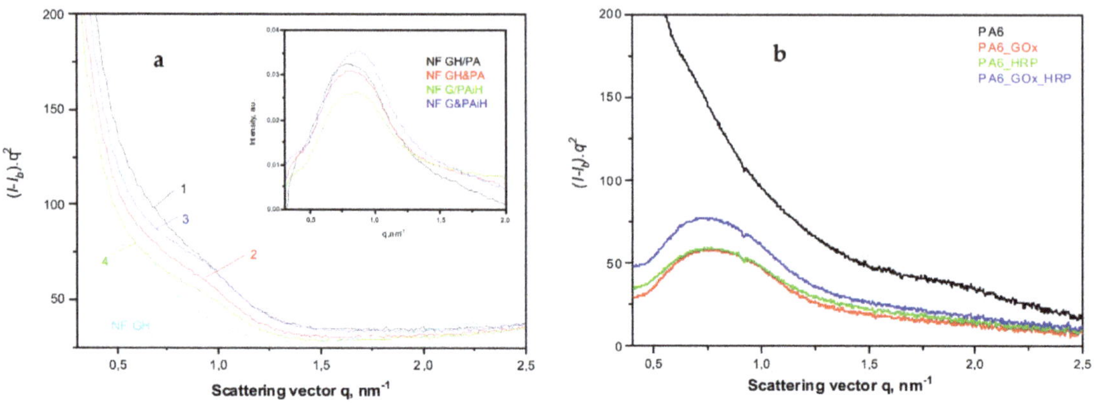

Figure 7. Comparative synchrotron SAXS patterns of: (**a**) NF and PANF; 1—NF GH/PA; 2—NF GH&PA; 3—NF G/PAiH; 4—NF G&PAiH; (**b**) PA6 MP with immobilized GOx, HRP or a mixed GOx/HRP. The inset represents the SAXS peaks obtained after subtraction of the signal of the NF GH sample from those of the four PANF samples. For more details, see the text.

All four PANF samples in Figure 7a present unresolved but clear SAXS periodicity peaks, whereas the pure PA6 MP does not show such periodicity (Figure 7b). The WAXS pattern of PA6 MP (Figure 6a, curve 1) with an overall crystallinity index above 40% (Figure S3a,b of the Supplementary Materials) suggests the presence of predominantly crystalline lamellar stacks in the PA6 MP. Hence, the absence of periodicity peak in the SAXS pattern of PA6 MP could be explained with the high porosity of the PA6 microparticles that eliminate the phase contrast between the amorphous and crystalline domains. In support of this assumption is the fact that whenever GOx or HRP are immobilized on PA6 MP by adsorption, SAXS peaks are registered with long spacings L = 80 Å (PA6 + individual GOx or HRP) or L = 86 Å (PA6 + GOx/HRP dyad) (Figure 7b). This means that the immobilized enzymes really fill the pores on the surface and in the volume of the PA6 MP, thus improving significantly the phase contrast. Similar effects are observed in the SAXS patterns of PA6 MP containing adsorption-immobilized laccase [44]. It can be therefore concluded that in all PANF complexes, some adsorption of free enzymes within the pores of the MP supports takes place, apart from their presence in the NF nanostructures.

The inset of Figure 7a shows the SAXS profiles of PANF samples after subtraction of the SAXS profile of the bienzyme nanoflowers (NF GH). This procedure is expected to remove the contribution of the NF to the SAXS intensities of PANF, revealing better the PA6 lamellar periodicity peaks. As a result, SAXS peaks are revealed with L = 82 Å for the NF GH/PA sample and 76–78 Å for the three other PANF samples. This means that in the NF GH/PA complex, certain parts of both GOx and HRP get immobilized within the PA6 MP, not being included into the NF assemblies. This causes increased L values of PA6, just like in the PA6 MP with adsorbed both GOx and HRP (Figure 7b). At the same time, the SAXS data from the inset of Figure 7a and the whole Figure 7b suggests that in NF GH&PA (PA6 MP added after the initial formation of NF), and especially in the complexes that contain only one of the enzymes (HRP) in the MP pores, these samples produce smaller SAXS long spacings. Summarizing, the synchrotron WAXS and SAXS data confirm the expected crystalline structure of the PANF complexes and validate the different localization of the GOx and HRP enzymes.

2.5. Description of the GOx-HRP Cascade Reaction

To test the activities of the PANF in this work, the known cascade reaction of glucose oxidation by the GOx/HRP enzyme dyad in the presence of the TMB chromophore (a typical substrate for HRP) has been selected (Figure 8).

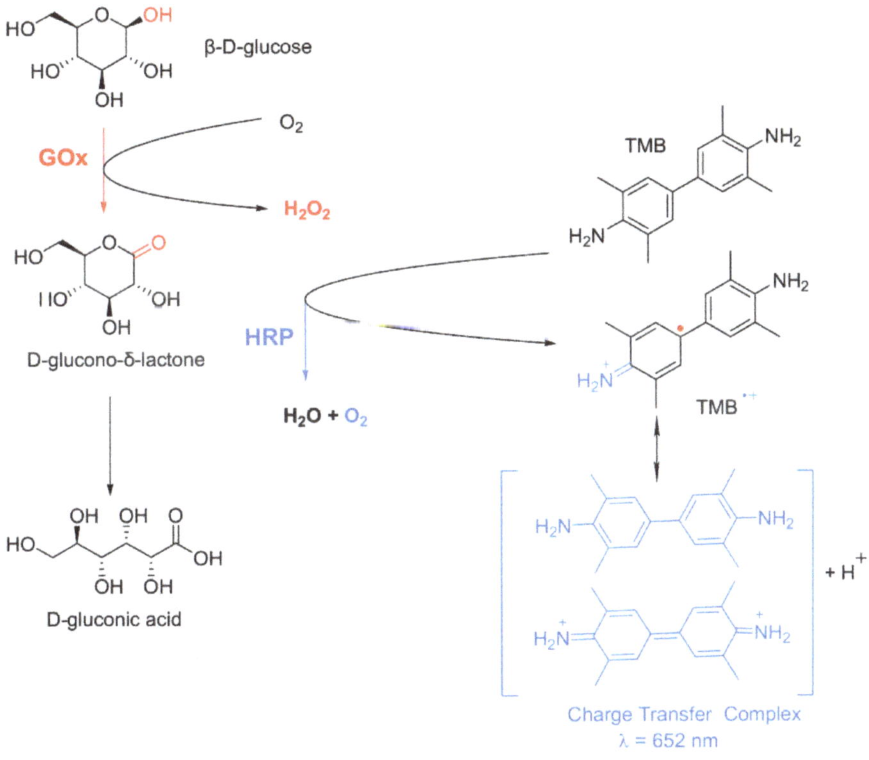

Figure 8. Schematic representation of the bienzyme cascade reaction used for β-D-glucose colorimetric determination. GOx = glucose oxidase; HRP = horseradish peroxidase; TMB = 3,3′,5,5′-tetramethlbenzidine.

In this cascade reaction, Gox catalyzes the oxidation of β-D-glucose to release D-glucono-δ-lactone and H_2O_2. The latter is used by HRP to oxidize the native diamine TMB, forming a colorless cation-radical ($TMB^{+\bullet}$) that is in rapid chemical equilibrium with a blue-green colored charge-transfer complex detectable by UV/VIS spectroscopy at λ = 652 nm. Since both H_2O_2 formation and TMB oxidation have similar reaction rates, the overall reaction rate of this cascade process will depend on how fast and effective H_2O_2 will reach HRP. Therefore, the overall reaction rate will depend on the distance between Gox, HRP, and their special orientation, i.e., on the architecture of co-immobilization of this bienzyme complex [48].

2.6 Comparative Activity Studies

The overall catalytic activity of all PANF complexes is assessed and compared to that of the free GOx/HRP system—either in the form of free enzymes or bienzyme NF without PA6 MP. For this purpose, the enzymatic cascade for colorimetric determination of β-D-glucose in Figure 8 is used. Figure 9 displays the linear part of the time dependence of the absorbance at λ = 652 nm characteristic for the blue-green charge transfer complex that TMB forms during the GOx/HRP catalyzed redox cascade. Table 1 allows the comparison between the activity parameters of all biocatalysts of this study determined at the same standard conditions.

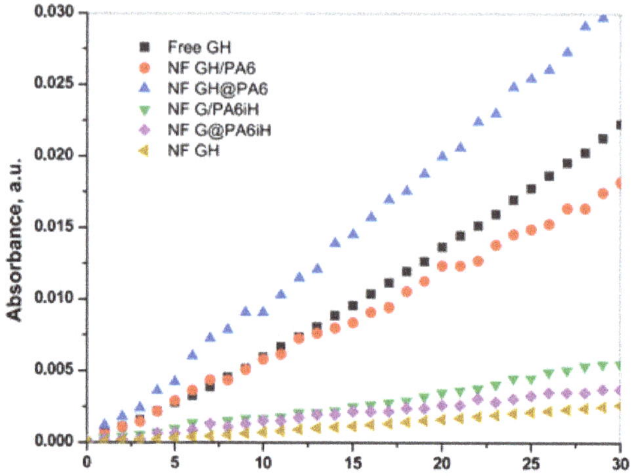

Figure 9. Overall catalytic activity of PANF complexes in comparison to the free GOx/HRP dyad (free GH) and the same incorporated in NF without PA6 MP (NF GH). For more details, see the text.

Table 1. Activity parameters of the bienzyme PANF complexes compared to the free GOx/HRP enzymes and to bienzyme standard NF without PA6 MP. Assay conditions: GOx/HRP activity toward β-D-Glucose (40 μM), in the presence of TMB (10 mM), in 0.2 M acetate buffer (pH 4, 23 °C).

System	Initial Rate $V_0 \times 10^{-4}$ [Abs s^{-1}]	Specific Initial Rate *, [Abs s^{-1} mg^{-1}]	Specific Activity, [μkat L^{-1}]	Relative Activity [%]
Free GH	8.735	0.0442	1.1340	100.0
NF GH/PA	2.336	0.0426	1.0919	96.3
NF GH@PA	2.781	0.0842	2.1597	190.4
NF G/PAiH	3.063	0.0258	0.6605	58.2
NF G@PAiH	2.092	0.0238	0.6092	53.7
NF GH	6.890	0.0115	0.2943	26.0

* Absorbance normalized per mg total protein content. The data in this table are average values of three syntheses per sample, the standard deviations being in the range of 3–5%.

The first column in Table 1 displays the slope of the linear dependencies in Figure 9 that corresponds to the initial rate V_0. For a correct interpretation of these data, they are normalized by the GOx/HRP total protein content that each sample carries. For this purpose, the residual protein in the supernatant after each PANF synthesis has been determined by UV spectroscopy using appropriate calibration plots. The protein content is calculated according to Equation (1) (Section 3.6 of Materials and Methods) and used for determination of the specific initial rate V_0 (column 2). Column 3 shows the specific activity in μkat. L^{-1}. The last column 4 presents the relative activity of PANF using that of the free GOx/HRP system (free GH) as 100%.

Table 1 shows that the catalytic activity of the samples decreases in the following order: NF GH@PA > NF GH/PA ≡ Free GH > NF G/PAiH > NF G@PAiH > NF GH

The apparent superior initial rate of the free GH and NF GH samples, in fact, is due to the elevated content of enzymes in them. The NF GH@PA complex (PA6 MP added after initial NF formation) is almost two times more active than the free GOx/HRP enzymes. If the PA6 MP are added in the beginning of the NF formation (sample NF GH/PA), the resulting activity is close to that of the free GOx/HRP. This result can be explained by the different morphologies of these two PANF samples (see the SEM micrographs in Figures 3b,d and 4). During the one-pot synthesis of NF GH/PA, some amounts of GOx and HRP enter into the pores of the microparticles, proved by the SAXS results. Meanwhile, in the NF GH@PA counterpart, GOx and HRP are predominantly in the NF domains formed during the first 24 h of the synthesis, i.e., in the absence of PA6 MP. Apparently, in this case, less amounts of the GOx and HRP get into the MP pores. Hence, the NF GH@PA morphology provides better a microenvironment and/or more favorable conditions for substrate channeling, resulting in higher catalytic activity.

The two systems that contain HRP immobilized in PA6 MP prior to the formation of the GOx-based NF are nearly two times less active than the free Gox/HRP system. This could be related to the fact that HRP (found on the MP surface and in its pores) and Gox (located predominantly in the NF assemblies) are well separated from each other, and that limits the substrate/product mass transfer. PANF complexes are also prepared, in which, instead of HRP, Gox has previously immobilized by adsorption onto PA6 MP. The results are not presented in this work due to their extremely low enzymatic activity. This result is understandable since GOx is the first enzyme to enter the cascade reaction, so if the access to it is limited, the overall velocity of the cascade reaction will be low.

Table 1 shows also that the conventional NF containing GOx/HRP display much lower activity than the free system, which is surprising and quite different from the data previously reported [29]. Most probably, this is due to the acidic medium (pH 4) used for the activity measurements in our study.

2.7. Kinetics Studies

The effect of the different enzyme co-localization in the four PANF samples on the kinetics of the GOx/HRP catalyzed cascade reaction is studied at constant TMB substrate concentration and varying the glucose substrate concentration between 50–175 μM. The initial rates are determined in the 0–30 s range. Thereafter, the Michaelis-Menten kinetic curves are constructed by plotting the initial rate, V_0, vs. glucose substrate concentration. The double reciprocal Lineweaver-Burk plot in Figure 10 is used to linearize the data of the Michaelis-Menten saturation curves and for more reliable determination of the basic kinetic parameters—the Michaelis-Menten constant K_m constant and the maximum velocity V_{max}. Kinetics studies using the same approach have been reported earlier by other authors [49–51].

Table 2. Kinetic parameters of the bienzyme PANF complexes extracted from the Michaelis-Menten equation by the double reciprocal Lineweaver-Burk plot. Data for the free GOx/HRP are presented for comparison.

System	K_m, [mM]	V_{max}, [μM s^{-1}]	V_{max}/[TP], [μM s^{-1} mg^{1}]	CE, [mg^{-1} s^{-1}]
Free GH	1.198	0.693	35.085	0.029
NF GH/PA	0.177	0.081	61.157	0.390
NF GH@PA	0.506	0.087	461.076	0.910
NF G/PAiH	1.688	0.129	77.850	0.046
NF G@PAiH	1.835	0.075	26.262	0.014

Note: The V_0 values used in the Lineweaver-Burk plot are based on the initial reaction rate between 0–30 s. $K_{cat} = V_{max}/[TP]$; $CE = K_{cat}/K_m$.

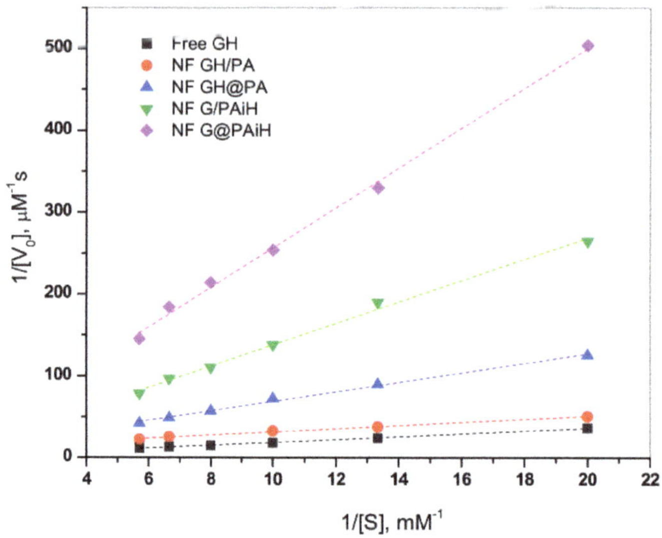

Figure 10. Double reciprocal Lineweaver-Burk plots for the PANF complexes in comparison to the free enzyme dyad. The kinetic data extracted from these plots are presented in Table 2. For sample designation and details, see Table 2 and the text.

According to Equation (2) (Section 3.8. of Materials and Methods), V_{max} is determined from the intercept, and K_m from the slope of the curves in Figure 10. K_m represents the substrate concentration at which the enzyme reaction rate is half of V_{max} and characterizes the affinity between the substrate and enzyme. Low K_m values correspond to higher affinity and vice versa. Since V_{max} depends on the enzyme concentration, in comparative studies it is more reliable to use the catalytic rate constant, K_{cat}, and the catalytic efficiency, CE. As usual, K_{cat} is determined by dividing V_{max} by the total protein content TP, while CE is defined as the ratio K_{cat}/K_m. These kinetic parameters for all four PANF systems and those of the free GOx/HRP dyad are presented in Table 2.

As seen from Figure 10 and Table 2, the NF GH/PA and NF GH&PA complexes designed to contain both GOx and HRP predominantly in the NF component display the highest affinity between the enzymes and the substrates in this cascade reaction, better than that of the free enzymes. Meanwhile, the two systems in which only GOx is located in NF and HRP is immobilized in the PA6 MP show higher K_m, i.e., lower affinity toward the substrate. The V_{max} values (Table 2, column 3) after normalization by the total protein content produce the catalytic rate constant K_{cat} (column 4). The affinity and the rate parameters are united in the so-called catalytic efficiency, CE (column 5), giving the best idea about the kinetic performance of the PANF systems, as compared to the free GOx/HRP dyad. Evidently, the adjacent co-localization of GOx and HRP within the same NF and the NF morphology itself of NF GH/PA and mostly of NF GH&PA results in the highest CE values being 13 and 30 times higher than the free enzyme dyad. The NF G/PAiH system in which NF contains GOx only and HRP has been preliminarily immobilized in the MP pores show higher K_m, but K_{cat} is still superior than with the free enzymes. Hence, in this specific PANF, the GOx active center is more difficult to be retrieved by the substrate. Despite this diffusional limitation, H_2O_2 (i.e., the product of GOx reaction and the substrate for HRP) is still able to diffuse within the PA6 MP pores. As a whole, this PANF catalyst remains active during the entire reaction duration, but the complete glucose conversion takes more time. The last NF G&PAiH system show CE values being only the half of the free enzymes. Most probably, in this case, part of the enzymes remains inaccessible for the substrate so in general, this system works far away from its optimum conditions.

Notably, the two PANF that contain both enzymes in the NF localization (NF GH&PA and NF GH/PA) show the best catalytic performance and display kinetic parameters being in times better than those of the free GOx/HRP system. This phenomenon is not very common for conventional immobilized enzymes. Evidently, the improved kinetic parameters of these two systems should be related to the close proximity of the two enzymes in a single nanoflower and the unique nanoflowers′ morphology that improves the reaction microenvironment and reduces the substrate/products diffusion limitations.

2.8. PANF Reusability Tests

One of the biggest advantages of the immobilized enzymes, considered crucial for possible large-scale use, is their reuse in various reaction cycles. Hence, the activity of the new PANF complexes is studied in six consecutive redox cycles at pH = 4 and temperature of 20–23 °C. The results are presented in Figure 11 in comparison to those of the NF system without PA6 MP and to the free GOx/HRP dyad.

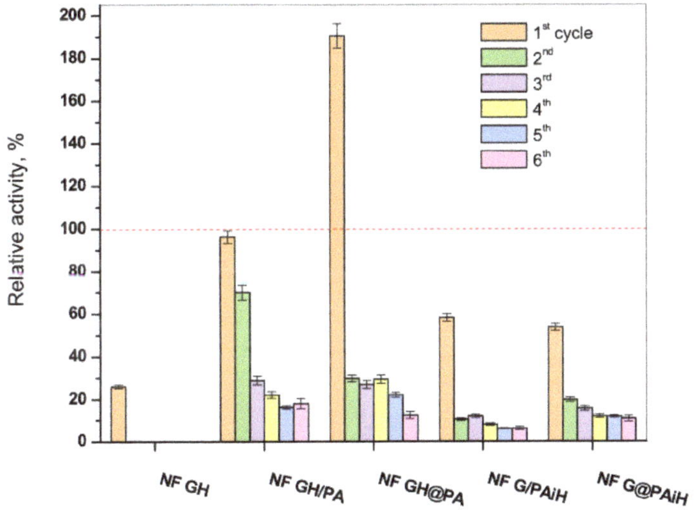

Figure 11. Relative activity of the PANF complexes measured after 1–6 cycles of utilization. The activity of the free GH dyad during its only cycle of use at 20–23 °C is taken as 100% (the dashed line). The NF GH system (mixed GOx/HPR NF without PA6 MP) survives only one cycle of use as well.

In agreement with previous studies, the classical NF system comprising GOx and HRP disintegrates after the first cycle, making impossible its recuperation for more uses, which is also the case with the free enzyme dyad. In its first reaction cycle, the conventional bienzyme NF complex shows only up to 27% of the activity of the free enzymes. The latter is indicated with the dashed line in Figure 11.

All PANF systems endure, although with decreasing activities, up to six redox cycles. The best performing during the first cycle samples NF GH&PA (almost two times more active than the free dyad) and NF GH/PA (95% of the free enzyme activity) reach in cycles 2–6 almost constant values between 20–25%. This is close to the values of the conventional bienzyme NF. For the samples with HRP located predominantly in the PA6 MP (i.e., NF G/PAiH and NF G&PAiH), these values do not exceed 20%, but their first cycle activity is also lower, in the range of 50–55%.

As seen from Figure S4 of the Supplementary Materials containing the SEM micrographs of PANF complexes after the sixth operating cycle, the drop of the catalytic activity after repeated use is to be attributed to disintegration of the NF morphologies, most probably due to their washing away and leaching of most of the enzymes out of the NF complexes.

As can be seen from Figure 11, after the second cycle, PANF retained a residual activity of about 10–30%, which changes very little in the next cycles. One of the systems, NF GH/PA, is tested even in 10 working cycles and the residual activity continues to be within the 18–20%. We attribute this retention of PANF activity to some part of GOx/HRP enzymes being adsorbed in the PA6 MP pores. Therefore, regardless of the design method used in this work, the two enzymes could take part in both processes, i.e., the NF construction and adsorption in the pores of the polymer particles, which is also confirmed by our SEM and WAXS/SAXS studies.

In summary, the conjugation of PA6 MP and NF carrying GOx and HRP in different localization of PANF allows multiple uses of these complexes in the cascade reaction of glucose oxidation.

2.9. PANF Stability at Different Temperatures and pH

The conditions for the above activity and kinetic tests are selected having in mind possible applications of PANF catalytic systems in the food industry, where room temperature and slightly acidic environments are very common. To make sure that the new complexes operate at their optimum conditions, activity tests at room temperature, 40, 50, and 60 °C and pH between 3–8 are performed and comparatively analyzed. The evolution of the relative activities of PANF as a function of temperature at pH = 4 is presented in Figure 12.

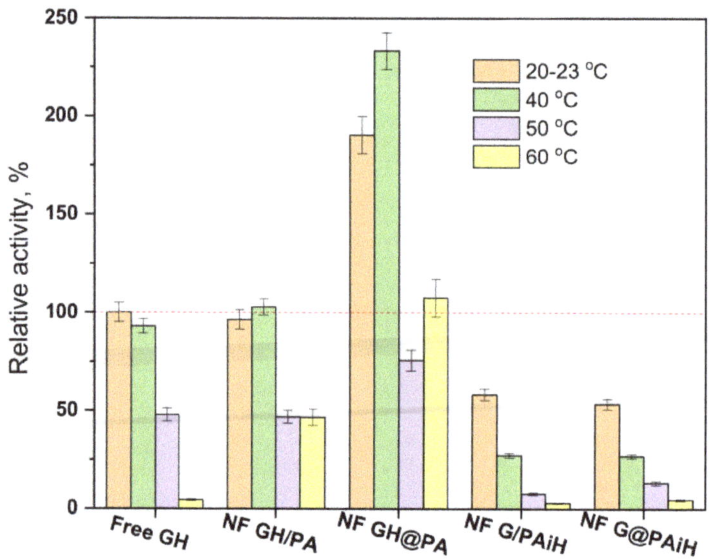

Figure 12. Relative activity of the PANF complexes measured at different temperatures of the reaction medium at pH = 4. The activity of the free GH systems at 20–23 °C is taken as 100%.

The free Gox/HRP system exhibits high activity at room temperature and at 40 °C (the activity at room temperature and pH = 4 is taken as 100%), after which the relative activity drops, and at 60°C is 4% only. The NF GH/PA sample performs similarly, with the important difference that at 60°C this system displays 46% relative activity, i.e., is 10 times more active than the free enzyme dyad. Once again, the NF GH/@PA sample shows the best performance at 40 °C of 230% relative activity that at 60 °C, was still over 100%. The other two samples containing GOx in the NF localization with HRP immobilized on PA6 MP show the highest activity of about 60% at room temperature. Increasing the reaction temperature results in a well-expressed decrease of the relative activity whereby NF G/PAiH and NF G@PAiH are outperformed by the free GH dyad at all temperatures studied.

Figure 13 represents the relative activity of the PANF complexes at different pH values compared to the activity of the free GH enzyme dyad at 20–23 °C and pH = 4 (the horizontal dashed line corresponds to 100% activity). Notably, the free enzymes at pH = 3 do not function at all; however, increasing the pH from 4 to 8 shows a clear activity optimum of 210% at pH = 6, followed by smooth decrease at higher pH values. As a whole, the profiles of activity as a function of pH are close to a normal Gaussian curve. The same relates to the performance of the NF GH/PA system, in which the maximum relative activity values were in the 4–6 pH range, being close or slightly above 150%.

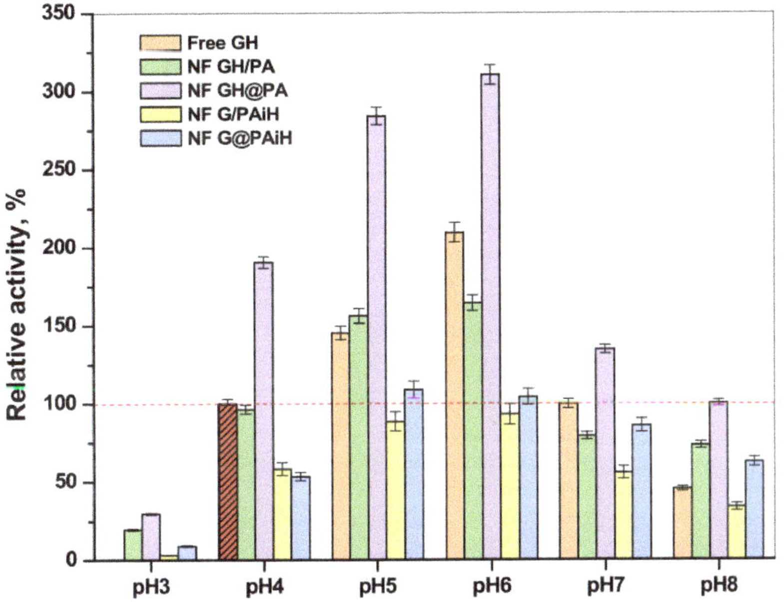

Figure 13. Relative activity of the PANF complexes measured at different pH values of the reaction medium. The activity of the free GOx/HRP systems at pH = 4 is taken as 100%.

The NF GH&PA system displays its optimum activity in the same pH range; however, the relative activities reached 275% and 325% for pH values of 4 and 6, respectively. As in most of the previous tests, the systems with HRP immobilized in PA6 MP are less active than the free GOx/HRP system, whose highest values are of ca. 100% at pH = 4. The samples NF G/PAiH and NF G@PAiH, however, show 25% activity at pH = 3, at which the free enzymes become completely deactivated. Moreover, the NF G@PAiH system at pH = 8 performs even better than the free enzyme dyad. Apparently, at extreme pH values that are out of the optimum range for both enzymes, the protection of HRP within the PA6 MP pores becomes a dominant factor.

The results from the comparative temperature and pH tests show that all PANF possess a broader window of application than the free GOx/HRP enzymes and justify the combination of the NF and PA6 MP components in the new catalytic systems.

2.10. *Shelf Life of PANF*

The storage stability (or shelf life) of the PANF systems is another important parameter to be accessed in validation of their potential for large-scale application. Figure 14 presents the retention of the relative activities of all systems after various storage periods of up to 60 days at 4 °C in PBS, pH 7. It should be noted that the starting activities of all systems at day 0 are different (see Table 1, column 4) and are taken as 100% in the storage stability study.

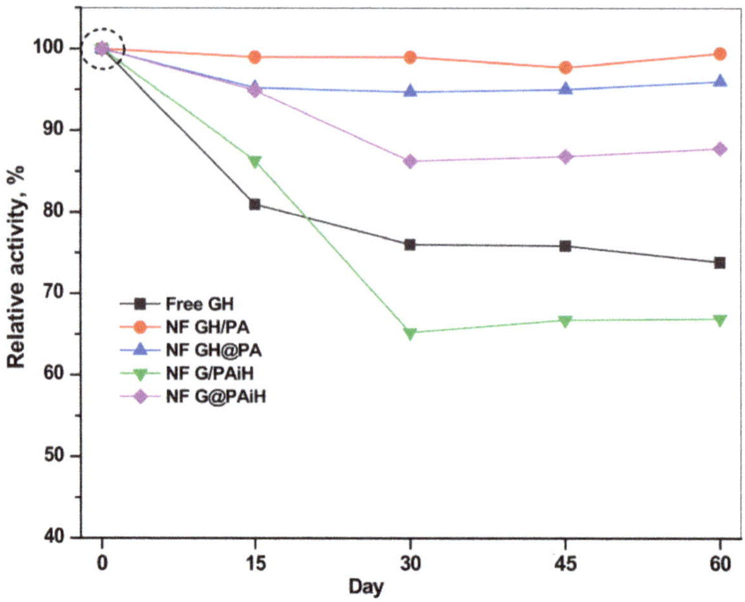

Figure 14. Storage stability of the PANF and free GH complexes at 4 °C, PBS. For each catalytic system the initial activity of the freshly prepared system is taken as 100%. For the absolute values of the initial activities see Table 1, column 4 (specific activity in μkatal·L^{-1}).

The free GH system with starting specific activity of 1.134 μkatal·L^{-1} has lost at day 60 about 25% of this value, the loss being best pronounced during the first 15 days. Meanwhile, all PANF systems are more stable than the free dyad until day 15 and retain up to 85% of their starting activity. The two PANF samples NF GH/PA and NF GH@PA lose only 2–3% until the end of the storage period. The only PANF sample that is less stable than the free enzymes after 15 days is NF G/PAiH. Its loss is up to 33% of its initial activity of 0.66 μkatal.L^{-1}.

As a whole, using PA6 MP as a support for NF enhances the storage stability, especially if both enzymes are localized in the NF component.

3. Materials and Methods

3.1. Materials and Reagents

The ε-caprolactam (ECL) monomer with reduced moisture content for anionic polymerization (AP-Nylon®) was delivered from Brüggemann Chemical (Heilbronn, Germany). Before use, it was kept under vacuum for 1 h at 23 °C. A polymerization activator, Brüggolen C20® (C20), from the same company, was used. The initiator sodium dicaprolactamato-bis-(2-methoxy-ethoxo)-aluminate (Dilactamate®, DL, 85% solution in toluene) was purchased from Katchem (Prague, Czech Republic) and applied without further treatment. Toluene, xylene, methanol, and other solvents were all of analytic grade, purchased from Merck (Lisbon, Portugal) and used as received. Gox from *Aspergillus niger* type VIII, D−(+) glucose (purum p.a.), $NaH_2PO_4 \cdot 2H_2O$, 99% (purum p.a.), $Na_2HPO_4 \cdot 2H_2O$, (98%), NaCl (99%), and CH_3COOH (99%) were purchased from Merck/Sigma Aldrich, Lisbon, Portugal. HRP from Amoracia rusticana and CH_3COONa (99%) were purchased from Alfa Aesar (Lancashire, UK). 3,3′,5,5′-tetramethylbenzidine (TMB, 99%) and $CuSO_4.5H_2O$ (99%) were purchased from Acros Organic (Porto Salvo, Portugal).

3.2. Characterization Methods

The scanning electron microscopy (SEM) studies were performed in a NanoSEM-200 apparatus of FEI Nova (Hillsboro, USA) using mixed secondary electron/back-scattered electron in-lens detection. All samples were observed after sputter-coating with Au/Pd alloy in a 208 HR equipment of Cressington Scientific Instruments (Watford, UK) with high-resolution thickness control.

UV-VIS absorbance spectral measurements were carried out using a 2401PC double-beam spectrophotometer from Shimadzu (Kyoto, Japan). The infrared spectra of selected bienzyme organic-inorganic complexes were obtained in a FTIR-4600 apparatus of JASCO (Tokyo, Japan) at room temperature, with a resolution of 4 cm^{-1} accumulating up to 64 spectra for optimum signal-to-noise ratio. The pulverulent samples were studied with an attachment for attenuated total reflection (ATR).

Synchrotron wide-angle (WAXS) and small-angle X-ray scattering (SAXS) measurements were performed at the NCD-SWEET beamline of the ALBA synchrotron facility in Barcelona, Spain. Two-dimensional detectors were used, namely, LH255-HS (Rayonix, USA) and Pilatus 1M (Dectris, Switzerland) for registering the WAXS and SAXS patterns, respectively. The sample-to-detector distance was set to 150.3 mm for WAXS and 2696.5 mm for SAXS measurements. The size of the incident beam was 0.35 × 0.38 mm (h × v) with a wavelength λ = 0.1 nm. The 2D data from the two detectors were reduced to linear profiles using the pyFAI software [52]. For further processing of the WAXS and SAXS patterns, the commercial package Peakfit 4.12 by SeaSolve was implemented.

3.3. Synthesis of PA6 Microparticles

The PA6 MP were synthesized by AAROP, as previously described by Dencheva et al. [43,44]. Typically, 0.5 mol of ECL was added to 100 mL of a mixed hydrocarbon solvent (toluene/xylene 1:1 by volume) while stirring, under a nitrogen atmosphere, refluxing the reaction mixture for 10–15 min. Subsequently, 3.0 mol% of DL and 1.5 mol% of C20 were added at once. The reaction time was 1 h from the point of catalytic system addition, and the temperature was maintained in the 125–135 °C range at a constant stirring of about 800 rpm. The scheme of AAROP is presented in Figure S1 of Supplementary Materials. The PA6 MP formed as fine powder that was separated from the reaction mixture by hot vacuum filtration, washed several times with methanol, and dried for 30 min in a vacuum oven at 60 °C. Additional Soxhlet extraction for 4 h with methanol was applied to remove the low molecular fractions. The resulting neat PA6 MP were kept in a desiccator.

3.4. Preparation of Conventional Hybrid Organic/Inorganic Nanoflowers (NF)

3.4.1. Synthesis of Single NF Containing Only GOx or HRP

A modified procedure reported by Sun et al. [35] was followed. In a typical experiment, 34 µL of aqueous $CuSO_4$ solution (120 mM) was added to 5 mL of 0.02 M phosphate buffer saline (PBS) with pH 7.4 containing different concentrations of GOx or HRP in the range of 0.01–0.5 mg/mL, followed by incubation at 20 °C, for 72 h. Then, the solutions were centrifuged, and the supernatant was decanted and preserved for determination of the residual protein, while the solid NF products were subjected to activity tests. The synthesis of these monoenzyme samples was necessary to determine the optimum concentration of the two enzymes so as to use it further in the synthesis of bienzyme nanoflowers. The optimum concentration for GOx was found to be 0.5 mg/mL and for HRP—0.1 mg/mL.

3.4.2. Synthesis of Bienzyme NF Carrying GOx and HRP

34 µL of aqueous $CuSO_4$ solution (120 mM) were added to 5 mL of PBS (pH 7.4) containing 0.5 mg/mL of GOx and 0.1 mg/mL of HRP, followed by incubation at 20–23 °C for 72 h. This sample was designated as NF GH. In all of the NF preparations, molecular biology grade water was used.

3.5. Preparation of Polymer-Assisted Hybrid Organic-Inorganic Bienzyme Nanoflowers (PANF))

3.5.1. Synthesis of Bienzyme NF Planted on PA6 MP

Typically, 20 mg of Soxhlet-extracted and dried PA6 MP and 34 μL of aqueous $CuSO_4$ solution (120 mM) were added to 1.5 mL of PBS (pH 7.4), 2.5 mL of PBS (pH 7.4) containing 1 mg/mL of GOx (final concentration 0.5 mg/mL), and 1 mL of PBS (pH 7.4) containing 0.5 mg/mL of HRP (final concentration 0.1 mg/mL). The mixture was subjected to incubation at 20–23 °C for 72 h. The resulting PANF sample was designated as NF GH/PA. A similar process was applied for the preparation of the NF GH&PA sample, but there, the PA6 MP were added 24 h after the NF incipient formation at room temperature, followed by incubation for additional 48 h at 20–23 °C.

3.5.2. Synthesis of GOx NF Planted on MP with Preliminary Immobilized HRP

In the first step, non-covalent immobilization by adsorption of HRP on PA6 MP was carried out. Typically, 50 mg of Soxhlet extracted and dried PA6 MP were added to 5 mL of 0.02M PBS (pH 7.4) containing 0.5 mg/mL of HRP, followed by incubation at 37 °C for 24 h. These MP samples were designated as PAiH. In the second step, 25 mg of PAiH and 34 μL of aqueous $CuSO_4$ solution (120 mM) were added to 5 mL of PBS (pH 7.4) containing 1 mg/mL of GOx, followed by incubation for 72 h. This sample was designated as NF G/PAiH. For the other PANF sample of this series (NF G&PAiH), a similar process was applied, in which the PAiH microparticles were added 24 h after the incipient NF formation, followed by additional incubation for 48 h.

3.6. Determination of the Total Amount of Protein in NF and PANF Complexes

After completion of the NF and PANF syntheses, all samples were centrifuged, the supernatant was decanted, and subjected to UV analysis to determine the residual protein and calculate the amount of the total protein, TP, incorporated into the NF and PANF complexes, expressed as:

$$TP = C_0 - C_s,\ [mg] \tag{1}$$

where C_0 was the starting protein content and C_s was the protein content in the resultant supernatant after the NF synthesis. To estimate C_s, direct UV quantification of the absorption peak at $\lambda \approx 277$ nm was applied. This peak (often referred to as A280) was ascribed to the tryptophan and tyrosine units of the GOx and HRP protein matrices, comprising substituted benzene rings. This direct method provided reliable data with standard deviation between the individual measurements in the range of 3–5%. The indirect Bradford or bicinchoninic acid colorimetric assays for C_s determination produced data dispersions of up to 15% and 30%, respectively. Thus, for all PANF samples, the absorbance at $\lambda = 277$ nm was measured and C_s was calculated using a standard calibration plot. For the samples containing pre-immobilized on PA6 MP enzyme, the amount of the adsorbed HRP was first determined by direct UV measurement of the respective supernatant and taken into account in the estimation of the TP content in the final NF G/PAiH and NF G@PAiH samples.

3.7. Activity Studies

The overall catalytic activity of the GOx/HRP bienzyme system was determined using colorimetric assay. Initially, glucose was oxidized by O_2 in the presence of GOx to gluconolactone and H_2O_2. Then, the HRP used this H_2O_2 to oxidize a chromophore compound (TMB) to a blue-colored product with absorbance at $\lambda_{max} = 652$ nm. The scheme of this cascade reaction was discussed in Section 2.5. and presented schematically in Figure 8.

3.7.1. Benchmark Activity Studies

The following protocol was applied for the activity test of the free GOx/HRP enzyme dyad. First, 0.2 mL of 5 mM glucose solution in PBS (pH 7), were added to 1.98 mL of

GOx solution in PBS (0.5 mg/mL). After incubation at 37 °C for 30 min, 0.160 mL from this reaction mixture were introduced into the final volume of a 2 mL UV/VIS cuvette containing also 1.72 mL acetate buffer (0.2 M, pH 4), 0.02 mL HRP solution (1 mg/mL in PBS), and 0.1 mL of 4.2 mM TMB solution in ethanol. The absorbance increase at 652 nm was measured as a function of time at 20–23 °C for a period of 2 min. Then, the slope of the linear segment of this curve between 0–30 s was used to determine the initial rate of the process.

For the activity studies of the NF and PANF complexes, 3–5 mg wet samples from each system (the amount depended on the total protein content),) were added to a cuvette containing 1.884 mL of 0.2 M acetate buffer (pH 4). The volumes of glucose and TMB were the same as in the previous assay and corresponded to 40 μM glucose and 0.2 mM TMB. The UV-VIS measurements to determine the initial rate of the cascade reaction were the same as in the COx/HRP free enzymes assay. When analyzing a single enzyme NF system produced with GOx or HRP, the respective amounts of the other component of the dyad were added as free enzyme.

All overall activities were expressed in microkatals, μkat. In this study, 1.0 μkat is the amount of enzyme(s) required to convert 1.0 μmol of TMB to the charge transfer colored product within 1 s at pH 4 and temperature of 20–23 °C. The molar extinction coefficient ε of the TMB chromophore is 39,000 M^{-1} cm^{-1}. These activity assays were performed in triplicate, the standard deviation between the parallel values being below 5%. In this study, the activities are expressed in μkat L^{-1}.

All activities were normalized by the total protein content in the respective sample and referred to as "specific activity". The specific activity of the Gox/HRP free enzyme dyad was assumed to be 100% and used as reference in the calculation of the relative activity the all NF and PANF samples.

3.7.2. Activity Studies at Different pH

The influence of the acidity of the reaction medium on the catalytic performance of all PANF was studied in the interval of pH 3–8 using the following 0.2 M buffers: citrate buffers for pH 3 and pH 5, acetate buffer for pH 4, phosphate buffer for pH 7, and potassium-phosphate buffer for pH 8. The activity measurements were performed at 20–23 °C, following the protocol described in Section 3.7.1. The specific activity of the free Gox/HRP system at pH 4 was assumed as 100% and used to calculate the relative activities of the PANF samples.

3.7.3. Activity Studies at Different Temperatures

These studies were also carried out using the protocol described in Section 3.7.1. Before the UV-VIS analysis, the samples were kept in acetate buffer (0.2 M, pH 4) for 1 h in an oven at different temperatures in the 20–60 °C interval. The specific activity of the free GOx/HRP at 20 °C was taken as the basis for determining the relative activities of the PANF samples.

3.8. Cascade Kinetics Studies

The cascade kinetic tests were carried out at 20–23 °C in 0.2 M acetate buffer, varying the glucose substrate concentration in the range of 50–175 μM, at a constant TMB concentration of 2 mM, whereas those of GOx and HRP were kept at 1–2 μkat L^{-1}. The absorbance increase at 652 nm was measured for a period of 2 min and the initial rates were determined from the slope of the linear Abs_{652}/time dependences. Then, the double reciprocal Lineweaver-Burk plots were constructed according to Equation (2):

$$\frac{1}{V_0} = \frac{K_m}{V_{max}} \cdot \frac{1}{S} + \frac{1}{V_{max}} \quad (2)$$

wherein V_0 is the initial rate of the cascade reaction, and S is the glucose substrate concentration. These linear plots were used to calculate the maximum velocity V_{max} from the x-axis intercept of the linear dependence for all bienzyme systems studied, which values were then used to determine the Michaelis constant K_m from the slope.

3.9. Reusability and Shelf Life of PANF

The reusability of the conventional NF or PANF complexes was investigated for six cycles by using 3–5 mg wet sample to catalyze the cascade oxidation reaction of 40 μM glucose in the presence of 0.2 mM TMB under standard conditions: 0.2 M acetate buffer, pH 4, at room temperature (21–23 °C). The enzyme activity was measured by UV-VIS as explained in 3.7.1. Then, each sample was centrifuged, filtered, washed with 0.02 M PBS (pH 7.4), and reused for the next cycle. The relative activity (%) was calculated considering the activity of the free GOx/HRP system during the first cycle as 100%.

The storage stability of the free GOx/HRP system and the PANF complexes was also analyzed by keeping them in 0.02 M phosphate buffer saline (PBS), pH 7.4 at 4 °C for 60 days. The enzyme activity was measured at time intervals following the protocol in 3.7.1. The relative activity was calculated by taking the activity of the respective sample on the first day as 100%.

4. Conclusions

This is the first report on bienzyme hybrid cascade complexes for glucose detection based on GOx and HRP nanoflowers planted on PA6 microparticles denominated as PANF. Altogether, four PANF samples have been prepared by a facile auto-assembly process, varying the enzyme co-localization. HRP becomes localized either in the nanoflowers comprising also GOx, or, by physical adsorption, on PA6 MP. The SEM/EDX results prove the expected morphology of all samples prepared. Evidence is also provided that the PA6 support can stimulate the formation of Cu (II) phosphate platelets in the absence of enzymes. The chemical composition of PANF is validated by FTIR, and their crystalline micro and nanostructure—by synchrotron X-ray scattering. It has been demonstrated by WAXS that the formation of the inorganic nanoflowers in the presence of PA6 MP changes some of the crystalline reflections of the $Cu_3(PO_4)_2$ triclinic unit cell, thus suggesting physical interaction between the nanoflowers and PA6. Irrespective of the PANF design method applied, partial adsorption of GOx, HRP, or both into the pores of the PA6 MP support always occurs, which is proved by SAXS. Extensive activity studies with all four PANF complexes as compared to the free GOx/HRP dyad or the GOx/HRP nanoflowers without PA6 have been performed in the 20–60 °C range and pH between 3 and 8. At the optimum conditions of 40 °C and pH = 6, the NF GH@PA complex displays relative activities being more than three times higher than the free enzyme dyad. For all other temperatures and pH values, this PANF system continues to be better than the free enzymes. The NF GH/PA complex is the next in the order of activity reaching up to 150–175% of the activity of the free enzymes. These two PANF containing the two enzymes in the NF domains maintain 20–30% of their initial catalytic activity after six operation cycles. It is comparable to the activity of the conventional GOx/HRP NF during its only catalytic cycle. Moreover, the NF GH@PA and NF GH/PA complexes possess an excellent storage stability at 4 °C, losing only 2–3% of their initial activity after 60 days. All these properties correlate perfectly with the Michaelis-Menten kinetic parameters V_{max} and K_m.

Based on this study, it can be acknowledged that the PANF complexes comprising both GOx and HRP in the nanoflowers component are significantly more active and more stable than the free enzyme GOx/HRP or the stand-alone classic bienzyme NF. This can potentially open the way of the PANF biotechnological application.

Supplementary Materials: The following supporting information can be downloaded at: https://www.mdpi.com/article/10.3390/molecules28020839/s1, Figure S1: Chemical reactions occurring during AAROP of ε-caprolactam to PA6 MP; Table S1: Elemental analysis of PANF samples by SEM/EDX with ZAF corrections; Figure S2: WAXS linear profiles of PA6 MP supports before and after immobilization of GOx, HRP or GOx + HRP; Figure S3: Deconvolution of the WAXS linear profiles of PA6 MP support before (a) and after (b) adsorption immobilization of GOx+HRP dyad; Figure S4: SEM micrographs of polymer-assisted hybrid nanoflowers (PANF) after the 6th catalytic cycle.

Author Contributions: Conceptualization, N.V.D.; methodology, N.V.D. and J.F.B.; software, M.M. and Z.Z.D.; validation, N.V.D. and M.M.; formal analysis, N.V.D. and J.F.B.; investigation, J.F.B.; resources, Z.Z.D.; data curation, J.F.B. and N.V.D.; writing—original draft preparation, J.F.B. and N.V.D.; writing—review and editing, N.V.D. and Z.Z.D.; visualization, J.F.B.; supervision, N.V.D. and Z.Z.D.; project administration, Z.Z.D.; funding acquisition, N.V.D. and Z.Z.D. All authors have read and agreed to the published version of the manuscript.

Funding: The authors gratefully acknowledge the financial support of Fundação para a Ciência e Tecnologia (FCT), project UID/CTM/50025/2019. Special thank is due to the ALBA synchrotron governance for financing our WAXS/SAXS experiments at NCD SWEET beamline in the framework of the project ID 2018/022726. N. Dencheva is also grateful for the financial support of FCT in the frames of the personal program contract CTTI-51/18-IPC. J. Braz acknowledges the support of the FCT UI/BD/150854/2021 Ph.D. grant.

Institutional Review Board Statement: Not applicable.

Informed Consent Statement: Not applicable.

Data Availability Statement: All the data in this research are presented in the manuscript and in the Supplementary Materials.

Conflicts of Interest: The authors declare no conflict of interest.

Sample Availability: Samples of the PANF complexes and their polymer precursors are available from the corresponding authors.

References

1. Ren, S.; Li, C.; Jiao, X.; Jia, S.; Jiang, Y.; Bilal, M.; Cui, J. Recent progress in multienzymes co-immobilization and multienzyme system applications. *J. Chem. Eng.* **2019**, *373*, 1254–1278. [CrossRef]
2. Liu, J.; Tian, K.; Li, Z. Chapter 20—Biocatalysis for cascade reactions to produce high-value chemicals. In *Biomass, Biofuels, Biochemicals*; Advances in Enzyme Catalysis and Technology; Singh, S., Pandey, A., Singhania, R.R., Larroche, C., Li, Z., Eds.; Elsevier: Amsterdam, The Netherlands, 2020; pp. 427–447. [CrossRef]
3. Schrittwieser, J.H.; Velikogne, S.; Hall, M.; Kroutil, W. Artificial biocatalytic linear cascades for preparation of organic molecules. *Chem. Rev.* **2018**, *118*, 270–348. [CrossRef] [PubMed]
4. Gonzalo, G.; Paul, C.E. Recent trends in synthetic enzymatic cascades promoted by alcohol dehydrogenases. *Curr. Opin. Green Sustain. Chem.* **2021**, *32*, 100548. [CrossRef]
5. Lee, D.S.; Song, J.W.; Voss, M.; Schuiten, E.; Akula, R.K.; Kwon, Y.U.; Bornscheuer, U.; Park, J.B. Enzyme cascade reactions for the biosynthesis of long chain aliphatic amines from renewable fatty acids. *Adv. Synth. Catal.* **2019**, *361*, 1359–1367. [CrossRef]
6. Wu, S.; Li, Z. Whole-cell cascade biotransformations for one-pot multistep organic synthesis. *ChemCatChem* **2018**, *10*, 2164–2178. [CrossRef]
7. Nikulin, M.; Švedas, V. Prospects of using biocatalysis for the synthesis and modification of polymers. *Molecules* **2021**, *26*, 2750. [CrossRef] [PubMed]
8. Arana-Pena, S.; Carballares, D.; Morellon-Sterlling, R.; Berenguer-Murcia, A.; Alcántara, A.R.; Rodrigues, R.C.; Fernandez-Lafuente, R. Enzyme co-immobilization: Always the biocatalyst designers' choice or not? *Biotechnol. Adv.* **2021**, *51*, 107584. [CrossRef]
9. Zdarta, J.; Meyer, A.S.; Jesionowski, T.; Pinelo, M. A general overview of support materials for enzyme immobilization: Characteristics, properties, practical utility. *Catalysts* **2018**, *8*, 92. [CrossRef]
10. Xu, K.; Chen, X.; Zheng, R.; Zheng, Y. Immobilization of multienzymes on support materials for efficient biocatalysis. *Front. Bioeng. Biotechnol.* **2020**, *8*, 660. [CrossRef]
11. Bauler, P.; Huber, G.; Leyh, T.; McCammon, J.A. Channeling by proximity: The catalytic advantages of active site colocalization using Brownian dynamics. *J. Phys. Chem. Lett.* **2010**, *1*, 1332–1335. [CrossRef]
12. Wheeldon, I.; Minteer, S.D.; Banta, S.; Barton, S.C.; Atanassov, P.; Sigman, M. Substrate channeling as an approach to cascade reactions. *Nat. Chem.* **2016**, *8*, 299–309. [CrossRef]

13. Fu, J.; Yang, Y.R.; Johnson-Buck, A.; Liu, M.; Liu, Y.; Walter, N.G.; Woodbury, N.W.; Yan, H. Multienzyme complexes on DNA scaffolds capable of substrate channeling with an artificial swinging arm. *Nat. Nanotechnol.* **2014**, *9*, 531–536. [CrossRef] [PubMed]
14. Idan, O.; Hess, H. Origins of activity enhancement in enzyme cascades on scaffolds. *ACS Nano* **2013**, *7*, 8658–8665. [CrossRef] [PubMed]
15. Zhang, Y.; Tsitkov, S.; Hess, H. Proximity does not contribute to activity enhancement in the glucose oxidase–horseradish peroxidase cascade. *Nat. Commun.* **2016**, *7*, 13982. [CrossRef] [PubMed]
16. Chado, G.R.; Stoykovich, M.P.; Kaar, J.L. Role of dimension and spatial arrangement on the activity of biocatalytic cascade reactions on scaffolds. *ACS Catal.* **2016**, *6*, 5161–5169. [CrossRef]
17. Papamichael, E.M.; Stergiou, P.Y. Chapter 12—Enzyme immobilization strategies and bioprocessing applications. In *Biomass, Biofuels, Biochemicals, Advances in Enzyme Catalysis and Technology*; Singh, S.P., Pandey, A., Singhania, R.R., Larroche, C., Li, Z., Eds.; Elsevier: Amsterdam, The Netherlands, 2020; pp. 217–241. [CrossRef]
18. Sheldon, R.A.; Pelt, S.V. Enzyme immobilization in biocatalysis: Why, what and how. *Chem. Soc. Rev.* **2013**, *42*, 6223–6235. [CrossRef]
19. Bülow, L.; Mosbach, K. Multienzyme systems obtained by gene fusion. *Trends Biotechnol.* **1991**, *9*, 226–231. [CrossRef]
20. Gad, S.; Ayakar, S. Protein scaffolds: A tool for multienzyme assembly. *Biotechnol. Rep.* **2021**, *32*, e00670. [CrossRef]
21. Hwang, E.T.; Lee, S. Multienzymatic cascade reactions via enzyme complex by immobilization. *ACS Catal.* **2019**, *9*, 4402–4425. [CrossRef]
22. Basso, A.; Serban, S. Industrial applications of immobilized enzymes—A review. *Mol. Catal.* **2019**, *479*, 110607. [CrossRef]
23. El-Zahab, B.; Jia, H.; Wang, P. Enabling multienzyme biocatalysis using nanoporous materials. *Biotechnol. Bioeng.* **2004**, *87*, 178–183. [CrossRef]
24. Wilner, O.I.; Weizmann, Y.; Gill, R.; Lioubashevski, O.; Freeman, R.; Willner, I. Enzyme cascades activated on topologically programmed DNA scaffolds. *Nat. Nanotechnol.* **2009**, *4*, 249–254. [CrossRef] [PubMed]
25. Kreft, O.; Prevot, M.; Mohwald, H.; Sukhorukov, G.B. Shell-in-shell microcapsules: A novel tool for integrated, spatially confined enzymatic reactions. *Angew. Chem. Int. Ed.* **2007**, *46*, 5605–5608. [CrossRef] [PubMed]
26. Vriezema, D.M.; Garcia, P.M.L.; Oltra, N.S.; Hatzakis, N.S.; Kuiper, S.M.; Nolte, R.J.M.; Rowan, A.E.; van Hest, J.C.M. Positional assembly of enzymes in polymersome nanoreactors for cascade reactions. *Angew. Chem. Int. Ed.* **2007**, *46*, 7378–7382. [CrossRef] [PubMed]
27. Elani, Y.; Law, R.V.; Ces, O. Vesicle-based artificial cells as chemical microreactors with spatially segregated reaction pathways. *Nat. Commun.* **2014**, *5*, 5305. [CrossRef]
28. Caruso, F.; Schüler, C. Enzyme multilayers on colloid particles: Assembly, stability, and enzymatic activity. *Langmuir* **2000**, *16*, 9595–9603. [CrossRef]
29. Ge, J.; Lei, J.; Zare, R.N. Protein–inorganic hybrid nanoflowers. *Nat. Nanotechnol.* **2012**, *7*, 428–432. [CrossRef]
30. Lee, S.W.; Cheon, S.A.; Kim, M.I.; Park, T.J. Organic–inorganic hybrid nanoflowers: Types, characteristics, and future prospects. *J. Nanobiotechnol.* **2015**, *13*, 54. [CrossRef]
31. Cui, J.; Jia, S. Organic–inorganic hybrid nanoflowers: A novel host platform for immobilizing biomolecules. *Coord. Chem. Rev.* **2017**, *352*, 249–263. [CrossRef]
32. Zhu, J.; Wen, M.; Wen, W.; Du, D.; Zhang, X.; Wang, S.; Lin, Y. Recent progress in biosensors based on organic-inorganic hybrid nanoflowers. *Biosens. Bioelectron.* **2018**, *120*, 175–187. [CrossRef]
33. Tran, T.D.; Kim, M.I. Organic-inorganic hybrid nanoflowers as potent materials for biosensing and biocatalytic applications. *BioChip J.* **2018**, *12*, 268–279. [CrossRef]
34. Subramani, I.G.; Perumal, V.; Gopinath, S.C.B.; Fhan, K.S.; Mohamed, N.M. Organic-inorganic hybrid nanoflower production and analytical utilization: Fundamental to cutting-edge technologies. *Crit. Rev. Anal. Chem.* **2021**, *52*, 1488–1510. [CrossRef] [PubMed]
35. Sun, J.; Ge, J.; Liu, W.; Lan, M.; Zhang, H.; Wang, P.; Wang, Y.; Niu, Z. Multienzyme co-embedded organic-inorganic hybrid nanoflowers: Synthesis and application in colorimetric sensor. *Nanoscale* **2014**, *6*, 255–262. [CrossRef] [PubMed]
36. Li, Z.; Zhang, Y.; Su, Y.; Ouyang, P.; Ge, J.; Liu, Z. Spatial co-localization of multienzymes by inorganic nanocrystal–protein complexes. *Chem. Commun.* **2014**, *50*, 12465–12468. [CrossRef] [PubMed]
37. Song, Y.; Ji, H.; Wang, M.; He, L.; Song, R.; Zhang, Z. Feasible synthesis of protein-templated zinc phosphate-supported Pt nanoparticle with enhanced electrocatalysis for methanol oxidation. *Appl. Surf. Sci.* **2017**, *422*, 228–238. [CrossRef]
38. Lee, H.R.; Chung, M.; Kim, M.I.; Ha, S.H. Preparation of glutaraldehyde-treated lipase-inorganic hybrid nanoflowers and their catalytic performance as immobilized enzymes. *Enzym. Microb. Technol.* **2017**, *105*, 24–29. [CrossRef] [PubMed]
39. Guimarães, J.R.; Carballares, D.; Tardioli, P.W.; Rocha-Martin, J.; Fernandez-Lafuente, R. Tunning immobilized commercial lipase preparations features by simple treatment with phosphate salts. *Molecules* **2022**, *27*, 4486. [CrossRef]
40. Patel, S.K.S.; Otari, S.V.; Li, J.; Kim, D.R.; Kim, S.C.; Cho, B.K.; Kalia, V.C.; Kang, Y.C.; Lee, J.K. Synthesis of cross-linked protein-metal hybrid nanoflowers and its application in repeated batch decolorization of synthetic dyes. *J. Hazard. Mater.* **2018**, *347*, 442–450. [CrossRef]
41. Zhao, F.; Wang, Q.; Dong, J.; Xian, M.; Yu, J.; Yin, H.; Chang, Z.; Mu, X.; Hou, T.; Wang, J. Enzyme-inorganic nanoflowers/alginate microbeads: An enzyme immobilization system and its potential application. *Process Biochem.* **2017**, *57*, 87–94. [CrossRef]
42. Luo, M.; Li, M.; Lu, Y.; Xia, M.; Zhao, Q.; Wang, D. In-situ growth of multienzyme-inorganic hybrid nanoflowers on PVA-co-PE nanofibrous strip for colorimetric biosensor. *Colloids Surf. A Physicochem. Eng. Asp.* **2022**, *640*, 128419. [CrossRef]
43. Dencheva, N.; Denchev, Z.; Lanceros-Méndez, S.; Sanz, T.E. One-step in situ synthesis of polyamide microcapsules with inorganic payload and their transformation into responsive thermoplastic composite materials. *Macromol. Mater. Eng.* **2016**, *301*, 119–124. [CrossRef]

44. Denchev, Z.; Dencheva, N. Polyamide Microcapsules and Method to Produce the Same. Patent Application Number PT107679, 3 June 2014.
45. Dencheva, N.; Oliveira, S.; Braz, J.; Getya, D.; Malfois, M.; Denchev, Z.; Gitsov, I. Magnetically responsive PA6 microparticles with immobilized laccase show high catalytic efficiency in the enzymatic treatment of catechol. *Catalysts* **2021**, *11*, 239. [CrossRef]
46. Wu, Z.F.; Wang, Z.; Zhang, Y.; Ma, Y.L.; He, C.Y.; Li, H.; Chen, L.; Huo, Q.S.; Wang, L.; Li, Z.Q. Amino acids-incorporated nanoflowers with an intrinsic peroxidase-like activity. *Sci. Rep.* **2016**, *6*, 22412. [CrossRef] [PubMed]
47. Cho, I.S.; Kim, D.W.; Lee, S.; Kwak, C.H.; Bae, S.T.; Noh, J.H.; Yoon, S.H.; Jung, H.S.; Kim, D.W.; Hong, K.S. Synthesis of Cu_2PO_4OH hierarchical superstructures with photocatalytic activity in visible light. *Adv. Funct. Mater.* **2008**, *18*, 2154–2162. [CrossRef]
48. Lin, J.L.; Palomec, L.; Wheeldon, I. Design and analysis of enhanced catalysis in scaffolded multienzyme cascade reactions. *ACS Catal.* **2014**, *4*, 505–511. [CrossRef]
49. Jo, S.M.; Kim, J.; Lee, J.E.; Wurm, F.R.; Landfester, K.; Wooh, S.S. Multimodal enzyme-carrying suprastructures for rapid and sensitive biocatalytic cascade reactions. *Adv. Sci.* **2022**, *9*, 2104884. [CrossRef]
50. Chen, S.; Wen, L.; Svec, F.; Tana, T.; Lv, Y. Magnetic metal–organic frameworks as scaffolds for spatial co-location and positional assembly of multi-enzyme systems enabling enhanced cascade biocatalysis. *RSC Adv.* **2017**, *7*, 21205–21213. [CrossRef]
51. Zore, O.V.; Pattammattel, A.; Gnanaguru, S.; Kumar, C.V.; Kasi, R.M. Bienzyme polymer−graphene oxide quaternary hybrid biocatalysts: Efficient substrate channeling under chemically and thermally denaturing conditions. *ACS Catal.* **2015**, *5*, 4979–4988. [CrossRef]
52. Ashiotis, G.; Deschildre, A.; Nawaz, Z.; Wright, J.P.; Karkoulis, D.; Picca, F.E.; Kieffer, J. The fast azimuthal integration Python library: pyFAI. *J. Appl. Crystallogr.* **2015**, *48*, 510–519. [CrossRef]

molecules

MDPI

Article

Solvent-Free Method of Polyacrylonitrile-Coated LLZTO Solid-State Electrolytes for Lithium Batteries

Xuehan Wang, Kaiqi Zhang, Huilin Shen, Hao Zhang, Hongyan Yao, Zheng Chen * and Zhenhua Jiang

Engineering Research Center of Special Engineering Plastics, Ministry of Education, National and Local Joint Engineering Laboratory for Synthetic Technology of High Performance Polymer, College of Chemistry, Jilin University, Qianjin Street 2699, Changchun 130012, China; xuehan23@mails.jlu.edu.cn (X.W.)
* Correspondence: chenzheng2013@jlu.edu.cn

Abstract: Solid-state electrolytes (SSEs), particularly garnet-type $Li_{6.4}La_3Zr_{1.4}Ta_{0.6}O_{12}$ (LLZTO), offer high stability and a wide electrochemical window. However, their grain boundaries limit ionic conductivity, necessitating high-temperature sintering for improved performance. Yet, this process results in brittle electrolytes prone to fracture during manufacturing. To address these difficulties, solvent-free solid-state electrolytes with a polyacrylonitrile (PAN) coating on LLZTO particles are reported in this work. Most notably, the PAN-coated LLZTO (PAN@LLZTO) electrolyte demonstrates self-supporting characteristics, eliminating the need for high-temperature sintering. Importantly, the homogeneous polymeric PAN coating, synthesized via the described method, facilitates efficient Li^+ transport between LLZTO particles. This electrolyte not only achieves an ionic conductivity of up to 2.11×10^{-3} S cm^{-1} but also exhibits excellent interfacial compatibility with lithium. Furthermore, a lithium metal battery incorporating 3% PAN@LLZTO-3%PTFE as the solid-state electrolyte and $LiFePO_4$ as the cathode demonstrates a remarkable specific discharge capacity of 169 mAh g^{-1} at 0.1 °C. The strategy of organic polymer-coated LLZTO provides the possibility of a green manufacturing process for preparing room-temperature sinter-free solid-state electrolytes, which shows significant cost-effectiveness.

Keywords: composite solid-state electrolytes; solvent-free method; high ionic conductivity

1. Introduction

In recent years, as a representative of the energy industry, the scale of lithium batteries [1,2] is constantly expanding. At present, the cathodes of automobile batteries, both at domestic and international levels, are mainly $LiFePO_4$ [3] and $LiCoO_2$ [4], etc. Nevertheless, most of these batteries have reached their highest theoretical energy density and suffer from safety risks. Solid-state batteries (SSBs) [5,6] are considered to be one of the most promising categories, replacing the organic electrolyte and diaphragm in existing lithium batteries with solid-state electrolytes (SSEs) [7,8]. These batteries are noted for their enhanced cycle life, compact form factor, scalability, adaptability of design, and straightforward packaging, which are attributes that align with the progressive evolution of energy storage solutions.

SSBs are differentiated based on their electrolyte composition into two primary categories: organic and inorganic. Organic solid-state electrolytes are typified by polyethylene oxide (PEO) [9,10], polyacrylonitrile (PAN) [11–14], etc. Inorganic solid electrolytes are mainly characterized by oxides and sulfides, etc. The prevalent electrolytes in SSBs are oxides, sulfides, and polymers, with oxide electrolytes exhibiting a better comprehensive performance and substantial developmental prospects. Oxide solid electrolytes can be classified into five principal types: (1) garnet (e.g., $Li_{6.4}La_3Zr_{1.4}Ta_{0.6}O_{12}$, LLZTO) [15–18], (2) Lithium-ion Conductor (LISICON, e.g., $Li_{14}Zn(GeO_4)_4$) [19], (3) Lithium ion-containing Na^+ Super Ionic Conductor (NASICON, e.g., $Li_2Ti_2(PO_4)_3$) [20], (4) perovskite (e.g., $Li_{0.33}La_{0.55}TiO_3$), and anti-perovskite (e.g., Li_2OHCl). Among these, the garnet type stands out for its higher ionic conductivity, robust lithium metal stability, and significant long-term developmental potential [21,22]. In particular, LLZTO has the advantages of Li friendliness, high electrochemical stability, and a wide electrochemical window (0–6 V vs. Li^+/Li) [23].

Current research underscores significant advancements in the mass production and market adoption of garnet electrolytes. Notably, Wachsman's group [24] focused on industrial-scale applications, endeavors to realize high-capacity, high-energy-density solid-state lithium–sulfur batteries via a cost-effective strategy. By eliminating the impurity phase from the Ta-LLZTO surface, they have created batteries with a runtime exceeding 300 days and a capacity retention rate of 80%. As another example, Guo's research group [25] has introduced LLZTO with ceramic particles coated by a polyacrylonitrile (PAN) layer, synthesized via a flow-assisted method. This innovation has resulted in the diminished interfacial impedance between garnet ceramics and lithium metal, concurrently enhancing the electrochemical stability of the developed solid-state batteries. However, LLZTO has several critical limitations in research and practical applications. (1) Environmental sensitivity: The surface of LLZTO can easily generate lithium-hating impurities, e.g., LiOH and Li_2CO_3 [26], which are harmful to the stable cycling of batteries. (2) Insufficient softness: LLZTO solid-state electrolytes do not have enough "softness" to adapt to the deformation strain of the electrode material, although the isotropic shear modulus (60 GPa) [27] is higher than the critical value required to suppress lithium dendrites (8.5 GPa) [28]. (3) High application requirements: Unlike sulfide-based solid electrolytes, LLZTO particles cannot support fast inter-particle Li^+ conduction [29] at loosely stacked grain/particle boundaries and generally need the additional work of sintering the particles or applying external pressure. (4) Difficulty in production: Due to the high-hardness behavior of LLZTO SSEs (Young's modulus ≈ 150 GPa, Bulk modulus ≈ 100 GPa), electrolyte membranes with a thickness of less than 200 μm [30] are tough to produce. The above series of problems seriously prevent the usage of LLZTO SSEs. Therefore, it is necessary to overcome the obstacle of the heterogeneous phase on the SSEs' surface, high hardness, harsh preparation conditions, and processing. Currently, novel LLZTO SSEs materials with both excellent Li^+ conductivity and simple fabrication and application urgently need to be developed.

In this study, we coated 3% homogeneous PAN on the surface of LLZTO particles. The close chemical interactions at the PAN/LLZTO interface contribute to inducing partial de-hydrocyanation of PAN and result in a localized conjugated structure, which can benefit the transport of Li^+ among 3% PAN@LLZTO particles. Nevertheless, the enhancement of high safety and energy density (>500 wh kg^{-1}) batteries [31] is difficult to achieve with only a single material. We discovered that the solvent-free method [32–35] used to prepare electrode films is also suitable for the preparation of LLZTO SSEs membranes. This simple and energy-efficient preparation method requires only a small percentage of polytetrafluoroethylene binder without a solvent. Controlled mixing, microfibrillation, and fibrillation steps could be employed to repeatedly calendar 3% PAN@LLZTO particles to obtain thin electrolyte membranes. Finally, we observed that electrolyte membranes (≈50 μm) prepared by biaxial calendaring at 80 °C using 3%PTFE can achieve an ionic conductivity of >10^{-3} S cm^{-1} without sintering or applying external pressure. Notably, we used a lithium metal battery with 3% PAN@LLZTO-3%PTFE as the SSEs and $LiFePO_4$ as the cathode exhibits an excellent discharge-specific capacity (169 mAh g^{-1}) at 0.1 °C. The investigation into solvent-free composite solid-state electrolytes (SSEs) represents a seminal advancement in the field, enabling the feasibility of sinter-free SSEs at room temperature. This sophisticated methodology is poised to redefine the frontiers of cost-effectiveness within the domain of energy storage technologies.

2. Results and Discussion

The synthesis of 3% PAN@LLZTO composite electrolyte powder was carried out by the solution precipitation method, and the specific process is shown in Figure 1a (refer to the Experimental Section). Keeping a uniform distribution of LLZTO particles in the polymer network is crucial. Scanning electron microscope (SEM) images of LLZTO and 3% PAN@LLZTO powders are shown in Figure 1b,c. The electrostatic interaction between

PAN, LLZTO, and DMSO occurs during the synthesis process, which can promote the even distribution of the polymer coating on the surface of the particles and can prevent the aggregation of LLZTO particles [25]. Therefore, PAN has significant advantages as a coating material for LLZTO. As shown in the XRD image (Figure 1d), the 3% PAN@LLZTO powder did not undergo a chemical change compared to the LLZTO. This result illustrates that the PAN surface coating cannot change the crystal structure of LLZTO.

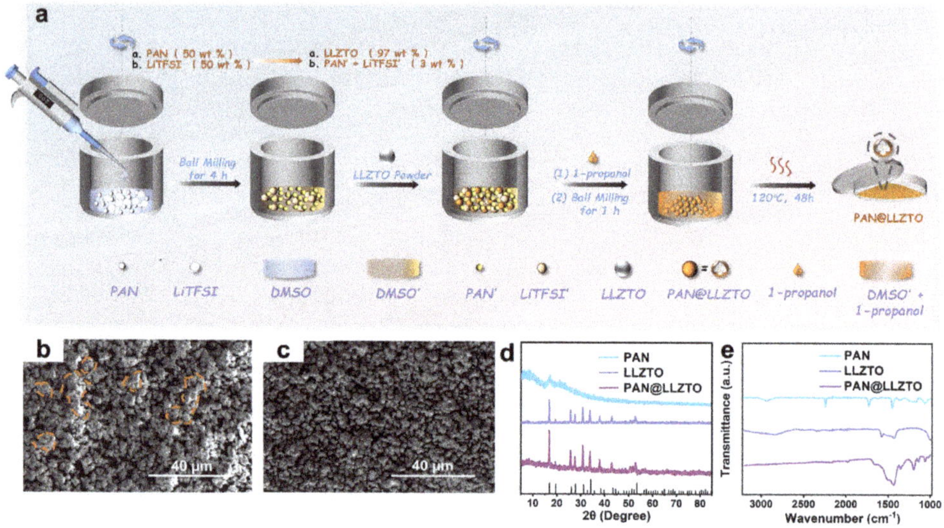

Figure 1. (**a**) Synthesis of 3% PAN@LLZTO composite electrolyte powder; SEM images: (**b**) LLZTO powder and (**c**) 3% PAN@LLZTO powder; (**d**) XRD image and (**e**) FT-IR spectra of PAN, LLZTO, and 3% PAN@LLZTO.

In the FT-IR spectra (Figure 1e), the peak intensities of 3% PAN@LLZTO at 2936 cm^{-1} (stretching vibrational mode of the C-H bond) and 2242 cm^{-1} (stretching vibrational mode of the -CN bond) are obviously lower than PAN. The observation of a peak at 2212 cm^{-1}, which we hypothesize is indicative of the formation of a novel compound LiCN, suggests that the polyacrylonitrile (PAN) may have experienced a dehydrogenation reaction. Notably, pure PAN is characterized by a white color, in contrast to the brownish-yellow appearance of PAN-coated LLZTO, as depicted in Figure S1a,b. The interaction between DMSO and LLZTO is believed to induce a redistribution of electron density within the DMSO molecule, thereby promoting the dehydrogenation of PAN on the LLZTO surface and leading to the emergence of a conjugated structure [25]. Additionally, the introduction of pure DMSO to PAN and LiTFSI results in a light-yellow coloration, as illustrated in Figure S2a,b, with the hue becoming more pronounced in the presence of both DMSO and LLZTO. It should be noted that the addition of LiTFSI is solely intended to enhance the ionic conductivity of the solid-state electrolyte, and it does not react chemically. These color changes, alongside the results from the FT-IR spectroscopy, collectively support the inference of a dehydrogenation reaction taking place. Furthermore, when exposed to air, LLZTO is prone to the formation of alkaline surface impurities, including LiOH and Li_2CO_3. Interestingly, LiOH has been identified as a catalyst for the de-hydrocyanation of PAN [36]. As a result, the modification of LLZTO with the ionic conductor PAN is posited as a multifaceted strategy. This approach not only enhances the conductivity of Li^+ at the ceramic/polymer interface but also contributes to the mitigation of surface contamination on LLZTO.

The preparation of the SSEs membrane by the 3% PAN@LLZTO powder is shown in the preparation process in Figure 2a. Pre-mixing, microfibrillating, and fibrillating LLZTO with commercial PTFE followed by calendaring can yield 3% PAN@LLZTO-PTFE membranes. The specific procedure was as follows: Firstly, the 3% PAN@LLZTO electrolyte and PTFE particles were uniformly mixed using a ball mill without zirconium balls for 15 min in Figure 2b. Subsequently, the mixed powder described above was microfibrillated in a ball mill with zirconium balls added for 5 min to obtain a flocculated PAN@LLZTO-PTFE matrix in Figure 2c. Finally, the above-obtained flocculent was calendared with a roller press to a defined thickness, which provided an enhancement of the fibrillation in Figure 2d and allowed the fibers to be clearly seen. After that, all the LLZTO particles were distributed in the tightly connected PTFE fibers to obtain a solvent-free solid electrolyte membrane of 3% PAN@LLZTO-PTFE. The top view of the PAN@LLZTO-PTFE membrane shows complete fibrosis in Figure 2e. The FT-IR spectra (Figure S3) of the PAN@LLZTO powder and PAN@LLZTO-PTFE membrane show a significant shift compared to the PTFE in the bands at 1207 and 1151 cm^{-1}. It proves that PTFE and LLZTO form interactions during the preparation process.

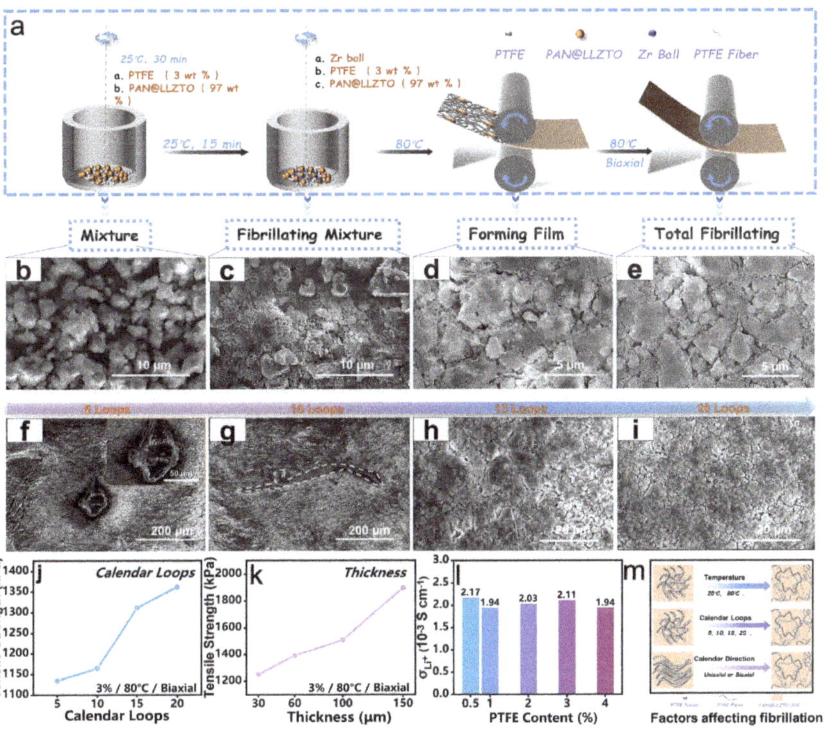

Figure 2. (**a**) Synthesis of 3% PAN@LLZTO-PTFE membranes; SEM of: (**b**) the mixture of 3% PAN@LLZTO and PTFE, (**c**) 3% PAN@LLZTO powder and micro-fibrillated PTFE flocculent, (**d**) enhanced fibrillation of the flocculent structure, (**e**) total fibrillating of the 3% PAN@LLZTO-PTFE; SEM of 3% PAN@LLZTO-PTFE calendaring after: (**f**) 5 times; (**g**) 10 times; (**h**) 15 times; (**i**) 20 times; Variation of (**j**) tensile strength with the number of calendaring cycles, (**k**) tensile strength with electrolyte film thickness, and (**l**) ionic conductivity with PTFE content; (**m**) factors influencing the fibrillation degree of solvent-free SSE films.

In this research, we have conducted an in-depth investigation into the factors influencing the fabrication of PTFE-LLZTO electrolyte films. Scanning electron microscopy (SEM) was employed to analyze the microstructural alterations of the materials, as depicted in

Figure 2f–i. SSEs membranes were fabricated using varying calendaring times (5, 10, 15, and 20), while maintaining consistent PTFE content. The PAN@LLZTO-PTFE samples subjected to 5 and 10 cycles of calendaring exhibited a fibrillated structure (Figure 2f,g), with distinct PAN@LLZTO and microfibrillated PTFE phases being discernible. SEM images (Figure 2h,i) reveal that the tensile strength of the PAN@LLZTO-PTFE composites increase with the number of calendaring cycles as the surface becomes more compact. In Figure 2j, there is a significant enhancement in the tensile strength of the SSEs film when the calendaring cycles reach 15 times (288 kPa) or more. This finding suggests a correlation between the number of calendaring cycles and the extent of fibrillation. However, the degree of fibrillation is not solely dependent on the number of calendaring cycles; it was also observed that the tensile strength of the SSEs membranes increases with thickness when thickness is the sole variable (Figure 2k). It is important to note that variations in PTFE content can fundamentally affect the degree of fibrillation, which subsequently impacts the intimate contact between the PAN@LLZTO particles, as reflected in the differences in Li^+ conductivity. Although the PTFE binder is capable of binding LLZTO particles to form membranes, the incorporation of excessive insulating PTFE may diminish the Li^+ conductivity of the resulting membranes. This underscores the delicate balance required in the formulation of PTFE-LLZTO electrolyte films to optimize both mechanical strength and ionic conductivity.

To investigate the influence of PTFE content on the electrochemical behavior of SSE membranes, SS | PAN@LLZTO-PTFE | SS were assembled with PTFE membranes of uniform thickness (60 $\pm$ 2 μm). The Li^+ conductivity of the SSE membranes was determined from the electrochemical impedance spectra (EIS). In Figure 2l, although the highest ionic conductivity (2.17×10^{-3} S cm^{-1}) was observed at a PTFE ratio of 0.5%, the resulting SSE membrane structure was too loose for practical application. Consequently, the SSE membrane with the next highest ionic conductivity was selected, which, at a PTFE content of 3%, exhibited an ionic conductivity of 2.11×10^{-3} S cm^{-1}. In contrast, an excessive PTFE content, such as 4%, resulted in a decrease in ionic conductivity. The optimal PTFE content facilitates close contact between LLZTO particles, thereby maximizing Li^+ conductivity. For all electrochemical studies in this work, PAN@LLZTO membranes prepared with 3% PTFE (3% PAN@LLZTO-3%PTFE) were utilized. Furthermore, the fibrillation of SSE membranes must account for the effects of temperature and calendaring direction. Given that the glass transition temperature (Tg) of PTFE exceeds 80 °C [37], all SSE membranes were calendared at this temperature in the current study. Unidirectional calendaring tends to create aggregated one-dimensional (X direction or Y direction) fibers, whereas biaxial-directional (X direction and Y direction) calendaring creates randomly dispersed fibers. Hence, biaxial-directional calendaring was employed throughout this work. In conclusion, various conditions, including PTFE content, calendaring frequency, film thickness, temperature, and calendaring direction, affect the fibrillation degree of solvent-free SSE membranes, as illustrated in Figure 2m. Therefore, to enhance the physical and electrochemical properties of SSE membranes, a multifaceted consideration [38] of these conditions is necessary, rather than focusing on a single factor.

The 60 $\pm$ 2 μm membranes (Figure 3a) used in the EIS tests, (Figure 3b) as well as the SS | LLZTO-3%PTFE | SS and SS | 3% PAN@LLZTO-3%PTFE | SS, were of uniform thickness. The impedance of the 3% PAN@LLZTO-3%PTFE membrane was found to be significantly lower (3.48 Ω vs. 6.04 Ω), signifying its superior Li^+ conductivity. This enhanced ionic conductivity is beneficial in mitigating concentration polarization during the battery's charge and discharge cycles. The operational voltage of a battery is a critical parameter that affects its performance. To assess the electrochemical stability of the electrolyte, linear scanning voltammetry (LSV) was employed on the cell SS | 3% PAN@LLZTO-3%PTFE | Li. The LSV curve shows a horizontal shape followed by a rising curve. The finding in Figure 3c indicates that this electrolyte can support normal battery operation up to a maximum voltage of approximately 4.9 V versus Li/Li^+. The electrolyte membrane fabricated in this study, 3% PAN@LLZTO-3%PTFE, demonstrated excellent interfacial compatibility with lithium metal.

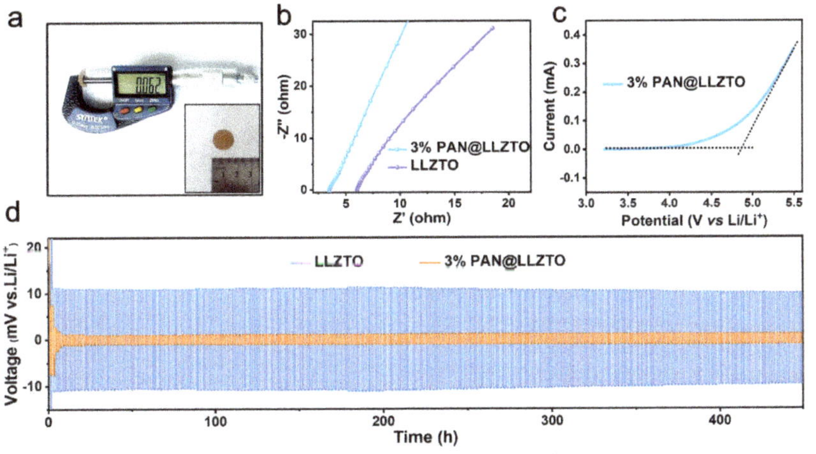

Figure 3. (**a**) Thickness measurement diagram and schematic diagram for cutting 16mm diameter of 3% PAN@LLZTO-3%PTFE; (**b**) impedance of LLZTO-3%PTFE and 3% PAN@LLZTO-3%PTFE; (**c**) linear scanning voltammetry of 3% PAN@LLZTO-3%PTFE; (**d**) Li plating/stripping curves of LLZTO-3%PTFE and 3% PAN@LLZTO-3%PTFE.

To further evaluate the dynamic stability of lithium ions at the interface between 3% PAN@LLZTO-3%PTFE and lithium metal, a Li | 3% PAN@LLZTO-3%PTFE | Li battery was subjected to the plating and stripping process at a current density of 0.1 mA cm^{-2}. The experimental outcomes, as shown in Figure 3d, reveal that these lithium cells were capable of cycling stably for over 450 h with a maintained overpotential. Additionally, the overpotential of the Li | 3% PAN@LLZTO-3%PTFE | Li was notably lower (±1.3 mV) than that of the Li | LLZTO-3%PTFE | Li (±11 mV). In addition, we summarize recent representative work on solvent-free inorganic solid-state electrolytes in Table 1. Comparisons were made in terms of thickness, PTFE content, and ionic conductivity, respectively. The comparison reveals that the electrolyte films prepared in this work have the highest conductivity and have potential for development.

Table 1. Electrochemical properties of lithium-based batteries using solvent-free electrolytes membrane.

Solvent-Free Membrane	Thickness (μm)	PTFE (wt%)	Ionic Conductivity (10^{-3} S cm^{-1}) 25 °C	Batteries	References
Li_3InCl_6/PTFE	20	0.5	1.00	Li_3InCl_6@$LiCoO_2$ | Li_3InCl_6/LPSC | Graphite/LPSC	[39]
LPSC/S/C/PTFE	93	1.0	0.63	S/C/LPSC | LPSC | Li	[32]
LPSC/PTFE	300	2.0	1.28	NCM811/LPSC | LPSC | Si	[37]
LLZTO/PTFE	20	0.5	0.52	NCM811 | LLZTO/LiTFSI in DMSO with 10% FEC | Si-C-450	[39]
LLZTO/PTFE	20	3.0	0.66	LFP | LLZTO/$LiPF_6$ in EC:PC = 1:1 vol% with 5% FEC | Li	[38]
PAN@LLZTO/PTFE	62	3.0	2.11	LFP | PAN@ LLZTO/$LiPF_6$ in EC:DEC = 1:1 vol% with 5% FEC | Li	This work

LPSC (Li_6PS_5Cl), NCM811 ($LiNi_{0.8}Mn_{0.1}Co_{0.1}O_2$), LLZTO ($Li_{6.4}La_3Zr_{1.4}Ta_{0.6}O_{12}$), LFP($LiFePO_4$).

Since 3% PAN@LLZTO has excellent Li^+ conductivity and lithium metal compatibility, it can be prepared as a solvent-free SSE (3% PAN@LLZTO-3%PTFE) and then assembled with a Li anode and $LiFePO_4$ (LFP) cathode (Figure 4a). The EIS test was performed at 25 °C for the battery before cycling and after 100 cycles. Nyquist plots captured from 1 Hz to 1 MHz at open circuit potential revealed a high-frequency semicircle, correlating to charge-transfer resistance (R_{ct}). [40,41] As depicted in Figure 4b, R_{ct} increased from 6.25×10^3 to 7.7×10^3 Ω post charge–discharge cycle. Figure 4c shows the Coulombic efficiency of the LFP | 3% PAN@LLZTO-3%PTFE | Li during the three cycles, and the values are all better than those of the LFP | LLZTO-3%PTFE | Li. In particular, the Coulombic efficiency of the 3% PAN@LLZTO-3%PTFE half-cell for the first cycle is close to 99%, indicating that its cycling reversibility is excellent. In addition, in the multiplicity test curve (Figure 4d), the discharge capacity of this cell was as high as 169 mAh g^{-1} at 0.1 °C, nearly close to its theoretical capacity of 170 mAh g^{-1}. As the current was increased from 0.1 °C to 1.0 °C and then restored to 0.1 °C, the reversible capacities of the electrodes were 70 and 159 mAh g^{-1}, respectively. Furthermore, the cyclic reversibility of the LFP | 3% PAN@LLZTO-3%PTFE | Li charge–discharge performance at 1, 2, 5, 10, 50, and 100 cycles can be seen in Figure 4e. The charge–discharge curves overlap very well, indicating the excellent reversibility, as well as the stable performance of the batteries. The capacity retention of the 3% PAN@LLZTO-3%PTFE batteries (92.8%), was significantly higher than that of the LLZTO-3%PTFE half-batteries (3.4%) for 100 cycles at 0.1 °C (Figure 4f), and this is probably attributed to the excellent Li^+ conductivity of the electrolyte. Based on the above electrochemistry of the batteries, the 3% PAN@LLZTO-3%PTFE synthesized using the solvent-free production method proposed herein holds great promise for improving the performance of lithium metal batteries.

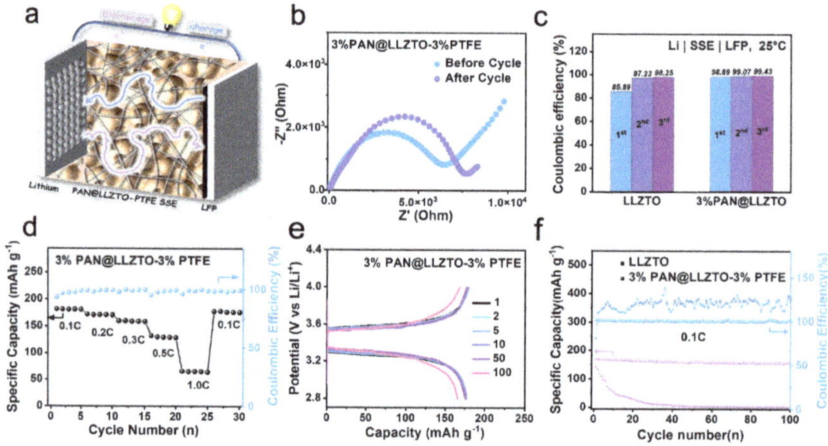

Figure 4. Electrochemical properties of LFP | 3% PAN@LLZTO-3%PTFE | Li batteries. (**a**) Battery's structure diagram; (**b**) impedance before and after cycle; (**c**) Coulombic efficiency during the three cycles; (**d**) discharge rate curves at 0.1 °C, 0.2 °C, 0.3 °C, 0.5 °C, 1.0 °C; (**e**) charge and discharge voltage profiles obtained at different cycles; (**f**) long-term cycling performance at 0.1 °C compared with LFP | LLZTO-3%PTFE | Li.

3. Materials and Methods

3.1. Materials

Polyacrylonitrile (PAN) was purchased from Shanghai Aladdin Biochemical Technology Co., Ltd. $Li_{6.4}La_3Zr_{1.4}Ta_{0.6}O_{12}$ (LLZTO) was purchased from Shenzhen Neware Co., Ltd. 1-methyl-2-pyrrolidinone (NMP) (99.7%), methyl sulfoxide (DMSO), and 1-propanol were purchased from Sigma-Aldrich (Shanghai) Trading Co., Ltd. $LiFePO_4$ (LFP), Super P, Li foil (∅ 15.6 mm, thickness: 0.45 mm), and electrolyte (1 M $LiPF_6$ in EC:DEC = 1:1 vol% with 5 vol% FEC) were purchased from Dongguan Canrud Innovative Technology Co., Ltd. PTFE binding was purchased from Daikin Fluorochemical (China) Co., Ltd.

3.2. Preparation of 3% PAN@LLZTO Powder

Initially, PAN and lithium bis(trifluoroethane)sulfonamide salt (LiTFSI) was dissolved in methyl sulfoxide (DMSO) in a mass ratio of 1:1 and ball-milled for 4 h. Subsequently, under an argon atmosphere, LLZTO was added in the mass ratio of LLZTO: PAN = 100:3 and then continued to be ball-milled for 12 h. Finally, 1-propanol was added to the above mixture and ball-milled for 1 h. The mixture was dried under a vacuum at 80 °C for 48 h to remove the solvent. The obtained 3% PAN@LLZTO powder was stored in an argon-filled glove box.

3.3. Preparation of PTFE-LLZTO Electrolyte Membrane

First, the 3% PAN@LLZTO electrolyte and PTFE particles were homogeneously mixed using a ball mill without zirconium balls for 15 min at room temperature. Subsequently, the above-mixed powder obtained was microfibrillated in a ball mill grinder with the addition of zirconium balls for 5 min to obtain a flocculent 3% PAN@LLZTO-PTFE matrix. Finally, the above matrix obtained was calendared to a specified thickness by a roller press to obtain a solvent-free solid electrolyte membrane of 3% PAN@LLZTO-PTFE.

3.4. Preparation of LFP Electrodes and Cell Assembling

The LFP cathode was prepared by mixing LFP, Super P, PVDF, LiTFSI, and PAN with a mass ratio of 70:10:10:5:5. The slurry was coated on aluminum foil and dried at 80 °C for 10 h. The electrolyte was controlled at 15 μL. The electrolyte tends to minimize the poor contact at the electrolyte/electrode interface and to avoid direct failure of the battery to operate. The mass loading of LFP was controlled within 1–1.5 mg cm^{-2}. All coin cells were assembled with designed electrodes and electrolytes in a glove box (H_2O and $O_2 \leq 0.01$ ppm).

3.5. Characterizations and Electrochemical Measurements

A Fourier transform infrared (FT-IR) spectrometer (Bruker Tensor-27, Saarbrucken, Germany) was used to examine the chemical structures. X-ray diffraction (XRD) patterns were collected on a Rigaku X-ray diffractometer (D/max 2500, Rigaku, The Woodlands, TX, USA) to identify the crystal structure. A scanning electron microscope (SEM) was used to visualize the morphology and elemental distribution (JEOL JSM-6701F, JEOL Ltd., Tokyo, Japan). Electrochemical measurement (electrochemical workstation of CHI-660E, Neware battery test system, CT-4008T-5V10mA-164, Shenzhen, China).

4. Conclusions

In summary, we report a solvent-free solid electrolyte membrane prepared from a polymer PAN-coated oxide electrolyte. The modification of LLZTO with an ionic conductor PAN not only contributes to the enhancement of Li^+ conductivity at the ceramic/polymer interface but also supports the removal of contaminants from the LLZTO surface. In addition, we tested the tensile strength of SSEs prepared with different process parameters (content of PTFE, number of calendaring, film thickness, temperature, and calendaring direction). The 3% PAN@LLZTO-3%PTFE prepared using this strategy has good compatibility with Li metal and excellent Li^+ conductivity. The electrochemical performance of LFP | Li half-cells assembled with 3% PAN@LLZTO-3%PTFE was much higher than that of LLZTO-3%PTFE without PAN modification. Therefore, the present work is feasible and tremendously cost-effective for facilitating the transition of SSB technology from the laboratory to the factory.

Supplementary Materials: The following supporting information can be downloaded at https://www.mdpi.com/article/10.3390/molecules29184452/s1. Figure S1: PAN powder and 3% PAN@LLZTO powder; Figure S2: color of DMSO/PAN and DMSO/PAN/LLZTO; Figure S3: FT-IR of PAN, PAN@LLZTO, and PAN@ LLZTO-PTFE membrane.

Author Contributions: Conceptualization, Z.C. and Z.J.; methodology, Z.C., H.Y. and Z.J.; software, X.W.; formal analysis, X.W., H.S., H.Z. and K.Z.; resources, H.S., H.Z. and K.Z.; data curation, X.W.

and H.Y.; writing—original draft preparation, X.W.; writing—review and editing, X.W. and Z.C.; supervision, H.S. and H.Z.; project administration, K.Z. and H.Z.; funding acquisition, Z.C., H.Y. and Z.J. All authors have read and agreed to the published version of the manuscript.

Funding: This work is financially supported by Jilin Province industry key core technology research project (No. 20240302101GX).

Institutional Review Board Statement: Not applicable.

Informed Consent Statement: Not applicable.

Data Availability Statement: Data are available in the source publications listed in the bibliography.

Conflicts of Interest: The authors declare no conflicts of interest.

References

1. Abdul Razzaq, A.; Yao, Y.; Shah, R.; Qi, P.; Miao, L.; Chen, M.; Zhao, X.; Peng, Y.; Deng, Z. High-performance lithium sulfur batteries enabled by a synergy between sulfur and carbon nanotubes. *Energy Storage Mater.* **2019**, *16*, 194–202. [CrossRef]
2. Yu, Z.; Wang, H.; Kong, X.; Huang, W.; Tsao, Y.; Mackanic, D.G.; Wang, K.; Wang, X.; Huang, W.; Choudhury, S.; et al. Molecular design for electrolyte solvents enabling energy-dense and long-cycling lithium metal batteries. *Nat. Energy* **2020**, *5*, 526–533. [CrossRef]
3. Li, Y.; Zhou, Q.; Weng, S.; Ding, F.; Qi, X.; Lu, J.; Li, Y.; Zhang, X.; Rong, X.; Lu, Y.; et al. Interfacial engineering to achieve an energy density of over 200 Wh kg^{-1} in sodium batteries. *Nat. Energy* **2022**, *7*, 511–519. [CrossRef]
4. Zheng, X.; Cui, P.; Qian, Y.; Zhao, G.; Zheng, X.; Xu, X.; Cheng, Z.; Liu, Y.; Dou, S.X.; Sun, W. Multifunctional Active-Center-Transferable Platinum/Lithium Cobalt Oxide Heterostructured Electrocatalysts towards Superior Water Splitting. *Angew. Chem. Int. Ed.* **2020**, *59*, 14533–14540. [CrossRef]
5. Janek, J.; Zeier, W.G. Challenges in speeding up solid-state battery development. *Nat. Energy* **2023**, *8*, 230–240. [CrossRef]
6. Liu, S.; Liu, W.; Ba, D.; Zhao, Y.; Ye, Y.; Li, Y.; Liu, J. Filler-Integrated Composite Polymer Electrolyte for Solid-State Lithium Batteries. *Adv. Mater.* **2022**, *35*, 2110423. [CrossRef] [PubMed]
7. Lewis, J.A.; Cavallaro, K.A.; Liu, Y.; McDowell, M.T. The promise of alloy anodes for solid-state batteries. *Joule* **2022**, *6*, 1418–1430. [CrossRef]
8. Wang, C.; Fu, K.; Kammampata, S.P.; McOwen, D.W.; Samson, A.J.; Zhang, L.; Hitz, G.T.; Nolan, A.M.; Wachsman, E.D.; Mo, Y.; et al. Garnet-Type Solid-State Electrolytes: Materials, Interfaces, and Batteries. *Chem. Rev.* **2020**, *120*, 4257–4300. [CrossRef]
9. Wang, C.; Yang, Y.; Liu, X.; Zhong, H.; Xu, H.; Xu, Z.; Shao, H.; Ding, F. Suppression of Lithium Dendrite Formation by Using LAGP-PEO (LiTFSI) Composite Solid Electrolyte and Lithium Metal Anode Modified by PEO (LiTFSI) in All-Solid-State Lithium Batteries. *ACS Appl. Mater. Interfaces* **2017**, *9*, 13694–13702. [CrossRef]
10. Xu, S.; Sun, Z.; Sun, C.; Li, F.; Chen, K.; Zhang, Z.; Hou, G.; Cheng, H.M.; Li, F. Homogeneous and Fast Ion Conduction of PEO-Based Solid-State Electrolyte at Low Temperature. *Adv. Funct. Mater.* **2020**, *30*, 2007172. [CrossRef]
11. Xin, S.; Chang, Z.; Zhang, X.; Guo, Y.-G. Progress of rechargeable lithium metal batteries based on conversion reactions. *Natl. Sci. Rev.* **2017**, *4*, 54–70. [CrossRef]
12. Wei, S.; Ma, L.; Hendrickson, K.E.; Tu, Z.; Archer, L.A. Metal–Sulfur Battery Cathodes Based on PAN–Sulfur Composites. *J. Am. Chem. Soc.* **2015**, *137*, 12143–12152. [CrossRef]
13. Sha, L.; Sui, B.-B.; Wang, P.-F.; Gong, Z.; Zhang, Y.-H.; Wu, Y.-H.; Zhao, L.-N.; Tang, J.-J.; Shi, F.-N. 3D network of zinc powder woven into fibre filaments for dendrite-free zinc battery anodes. *Chem. Eng. J.* **2024**, *481*, 148393. [CrossRef]
14. Liang, J.-Y.; Zeng, X.-X.; Zhang, X.-D.; Zuo, T.-T.; Yan, M.; Yin, Y.-X.; Shi, J.-L.; Wu, X.-W.; Guo, Y.-G.; Wan, L.-J. Engineering Janus Interfaces of Ceramic Electrolyte via Distinct Functional Polymers for Stable High-Voltage Li-Metal Batteries. *J. Am. Chem. Soc.* **2019**, *141*, 9165–9169. [CrossRef] [PubMed]
15. Han, F.; Westover, A.S.; Yue, J.; Fan, X.; Wang, F.; Chi, M.; Leonard, D.N.; Dudney, N.J.; Wang, H.; Wang, C. High electronic conductivity as the origin of lithium dendrite formation within solid electrolytes. *Nat. Energy* **2019**, *4*, 187–196. [CrossRef]
16. Zhang, W.; Nie, J.; Li, F.; Wang, Z.L.; Sun, C. A durable and safe solid-state lithium battery with a hybrid electrolyte membrane. *Nano Energy* **2018**, *45*, 413–419. [CrossRef]
17. Liu, Q.; Geng, Z.; Han, C.; Fu, Y.; Li, S.; He, Y.-B.; Kang, F.; Li, B. Challenges and perspectives of garnet solid electrolytes for all solid-state lithium batteries. *J. Power Sources* **2018**, *389*, 120–134. [CrossRef]
18. Tao, X.; Liu, Y.; Liu, W.; Zhou, G.; Zhao, J.; Lin, D.; Zu, C.; Sheng, O.; Zhang, W.; Lee, H.-W.; et al. Solid-State Lithium–Sulfur Batteries Operated at 37 °C with Composites of Nanostructured $Li_7La_3Zr_2O_{12}$/Carbon Foam and Polymer. *Nano Lett.* **2017**, *17*, 2967–2972. [CrossRef]
19. Okumura, T.; Takeuchi, T.; Kobayashi, H. All-Solid-State Batteries with $LiCoO_2$-Type Electrodes: Realization of an Impurity-Free Interface by Utilizing a Cosinterable $Li_{3.5}Ge_{0.5}V_{0.5}O_4$ Electrolyte. *ACS Appl. Energy Mater.* **2020**, *4*, 30–34. [CrossRef]
20. Jian, Z.; Hu, Y.S.; Ji, X.; Chen, W. NASICON-Structured Materials for Energy Storage. *Adv. Mater.* **2017**, *29*, 1601925. [CrossRef]

21. Zhang, J.; Zhao, N.; Zhang, M.; Li, Y.; Chu, P.K.; Guo, X.; Di, Z.; Wang, X.; Li, H. Flexible and ion-conducting membrane electrolytes for solid-state lithium batteries: Dispersion of garnet nanoparticles in insulating polyethylene oxide. *Nano Energy* **2016**, *28*, 447–454. [CrossRef]
22. Jin, Y.; Liu, K.; Lang, J.; Zhuo, D.; Huang, Z.; Wang, C.; Wu, H.; Cui, Y. An intermediate temperature garnet-type solid electrolyte-based molten lithium battery for grid energy storage. *Nat. Energy* **2018**, *3*, 732–738. [CrossRef]
23. Zhai, L.; Yang, K.; Jiang, F.; Liu, W.; Yan, Z.; Sun, J. High-performance solid-state lithium metal batteries achieved by interface modification. *J. Energy Chem.* **2023**, *79*, 357–364. [CrossRef]
24. Shi, C.; Hamann, T.; Takeuchi, S.; Alexander, G.V.; Nolan, A.M.; Limpert, M.; Fu, Z.; O'Neill, J.; Godbey, G.; Dura, J.A.; et al. 3D Asymmetric Bilayer Garnet-Hybridized High-Energy-Density Lithium–Sulfur Batteries. *ACS Appl. Mater. Interfaces* **2022**, *15*, 751–760. [CrossRef]
25. Chen, W.-P.; Duan, H.; Shi, J.-L.; Qian, Y.; Wan, J.; Zhang, X.-D.; Sheng, H.; Guan, B.; Wen, R.; Yin, Y.-X.; et al. Bridging Interparticle Li^+ Conduction in a Soft Ceramic Oxide Electrolyte. *J. Am. Chem. Soc.* **2021**, *143*, 5717–5726. [CrossRef] [PubMed]
26. He, K.; Cheng, S.H.S.; Hu, J.; Zhang, Y.; Yang, H.; Liu, Y.; Liao, W.; Chen, D.; Liao, C.; Cheng, X.; et al. In-Situ Intermolecular Interaction in Composite Polymer Electrolyte for Ultralong Life Quasi-Solid-State Lithium Metal Batteries. *Angew. Chem. Int. Ed.* **2021**, *60*, 12116–12123. [CrossRef]
27. Yu, S.; Schmidt, R.D.; Garcia-Mendez, R.; Herbert, E.; Dudney, N.J.; Wolfenstine, J.B.; Sakamoto, J.; Siegel, D.J. Elastic Properties of the Solid Electrolyte $Li_7La_3Zr_2O_{12}$(LLZO). *Chem. Mater.* **2015**, *28*, 197–206. [CrossRef]
28. Xu, W.; Wang, J.; Ding, F.; Chen, X.; Nasybulin, E.; Zhang, Y.; Zhang, J.-G. Lithium metal anodes for rechargeable batteries. *Energy Environ. Sci.* **2014**, *7*, 513–537. [CrossRef]
29. Zhang, Q.; Cao, D.; Ma, Y.; Natan, A.; Aurora, P.; Zhu, H. Sulfide-Based Solid-State Electrolytes: Synthesis, Stability, and Potential for All-Solid-State Batteries. *Adv. Mater.* **2019**, *31*, 1901131. [CrossRef]
30. Wu, N.; Chien, P.H.; Qian, Y.; Li, Y.; Xu, H.; Grundish, N.S.; Xu, B.; Jin, H.; Hu, Y.Y.; Yu, G.; et al. Enhanced Surface Interactions Enable Fast Li^+ Conduction in Oxide/Polymer Composite Electrolyte. *Angew. Chem. Int. Ed.* **2020**, *59*, 4131–4137. [CrossRef]
31. Bi, C.X.; Zhao, M.; Hou, L.P.; Chen, Z.X.; Zhang, X.Q.; Li, B.Q.; Yuan, H.; Huang, J.Q. Anode Material Options Toward 500 Wh kg^{-1} Lithium–Sulfur Batteries. *Adv. Sci.* **2021**, *9*, 2103910. [CrossRef] [PubMed]
32. Hu, J.-K.; Yuan, H.; Yang, S.-J.; Lu, Y.; Sun, S.; Liu, J.; Liao, Y.-L.; Li, S.; Zhao, C.-Z.; Huang, J.-Q. Dry electrode technology for scalable and flexible high-energy sulfur cathodes in all-solid-state lithium-sulfur batteries. *J. Energy Chem.* **2022**, *71*, 612–618. [CrossRef]
33. Zhang, Y.; Huld, F.; Lu, S.; Jektvik, C.; Lou, F.; Yu, Z. Revisiting Polytetrafluorethylene Binder for Solvent-Free Lithium-Ion Battery Anode Fabrication. *Batteries* **2022**, *8*, 57. [CrossRef]
34. Suh, Y.; Koo, J.K.; Im, H.-j.; Kim, Y.-J. Astonishing performance improvements of dry-film graphite anode for reliable lithium-ion batteries. *Chem. Eng. J.* **2023**, *476*, 146299. [CrossRef]
35. Tao, R.; Steinhoff, B.; Sun, X.-G.; Sardo, K.; Skelly, B.; Meyer, H.M.; Sawicki, C.; Polizos, G.; Lyu, X.; Du, Z.; et al. High-throughput and high-performance lithium-ion batteries via dry processing. *Chem. Eng. J.* **2023**, *471*, 144300. [CrossRef]
36. Duan, H.; Chen, W.P.; Fan, M.; Wang, W.P.; Yu, L.; Tan, S.J.; Chen, X.; Zhang, Q.; Xin, S.; Wan, L.J.; et al. Building an Air Stable and Lithium Deposition Regulable Garnet Interface from Moderate-Temperature Conversion Chemistry. *Angew. Chem. Int. Ed.* **2020**, *59*, 12069–12075. [CrossRef]
37. Lee, D.J.; Jang, J.; Lee, J.P.; Wu, J.; Chen, Y.T.; Holoubek, J.; Yu, K.; Ham, S.Y.; Jeon, Y.; Kim, T.H.; et al. Physio-Electrochemically Durable Dry-Processed Solid-State Electrolyte Films for All-Solid-State Batteries. *Adv. Funct. Mater.* **2023**, 33, 2301341. [CrossRef]
38. Li, Q.; Li, Y.; Chen, H.; Liu, H.; Li, L.; Song, J.; Xu, M.; Bao, S.-J. Dry approach production of a garnet solid electrolyte membrane for lithium batteries. *Inorg. Chem. Front.* **2023**, *10*, 6023–6031. [CrossRef]
39. Wang, C.; Yu, R.; Duan, H.; Lu, Q.; Li, Q.; Adair, K.R.; Bao, D.; Liu, Y.; Yang, R.; Wang, J.; et al. Solvent-Free Approach for Interweaving Freestanding and Ultrathin Inorganic Solid Electrolyte Membranes. *ACS Energy Lett.* **2021**, *7*, 410–416. [CrossRef]
40. Haripriya, M.; Manimekala, T.; Dharmalingam, G.; Minakshi, M.; Sivasubramanian, R. Asymmetric Supercapacitors Based on $ZnCo_2O_4$ Nanohexagons and Orange Peel Derived Activated Carbon Electrodes. *Chem.–Asian J.* **2024**, *19*, e202400202. [CrossRef]
41. Vasudevan, S.; D, S.T.; Manickam, M.; Sivasubramanian, R. A sol–gel derived $LaCoO_3$ perovskite as an electrocatalyst for Al–air batteries. *Dalton Trans.* **2024**, *53*, 3713–3721. [CrossRef] [PubMed]

molecules

MDPI

Article

Additively Manufactured Carbon-Reinforced ABS Honeycomb Composite Structures and Property Prediction by Machine Learning

Meelad Ranaiefar [1,*], Mrityunjay Singh [2,*] and Michael C. Halbig [1]

[1] NASA Glenn Research Center, Cleveland, OH 44135, USA; michael.c.halbig@nasa.gov
[2] Ohio Aerospace Institute, Cleveland, OH 44142, USA
[*] Correspondence: meelad.ranaiefar@nasa.gov (M.R.); mrityunjay.singh-1@nasa.gov (M.S.)

Abstract: The expansive utility of polymeric 3D-printing technologies and demand for high-performance lightweight structures has prompted the emergence of various carbon-reinforced polymer composite filaments. However, detailed characterization of the processing–microstructure–property relationships of these materials is still required to realize their full potential. In this study, acrylonitrile butadiene styrene (ABS) and two carbon-reinforced ABS variants, with either carbon nanotubes (CNT) or 5 wt.% chopped carbon fiber (CF), were designed in a bio-inspired honeycomb geometry. These structures were manufactured by fused filament fabrication (FFF) and investigated across a range of layer thicknesses and hexagonal (hex) sizes. Microscopy of material cross-sections was conducted to evaluate the relationship between print parameters and porosity. Analyses determined a trend of reduced porosity with lower print-layer heights and hex sizes compared to larger print-layer heights and hex sizes. Mechanical properties were evaluated through compression testing, with ABS specimens achieving higher compressive yield strength, while CNT-ABS achieved higher ultimate compressive strength due to the reduction in porosity and subsequent strengthening. A trend of decreasing strength with increasing hex size across all materials was supported by the negative correlation between porosity and increasing print-layer height and hex size. We elucidated the potential of honeycomb ABS, CNT-ABS, and ABS-5wt.% CF polymer composites for novel 3D-printed structures. These studies were supported by the development of a predictive classification and regression supervised machine learning model with 0.92 accuracy and a 0.96 coefficient of determination to help inform and guide design for targeted performance.

Keywords: acrylonitrile butadiene styrene (ABS); additive manufacturing; carbon-reinforced ABS; classification; fused filament fabrication; honeycomb; machine learning; mechanical properties; polymer composites; regression

1. Introduction

The continued advancement of tools and technologies to meet lightweight and high-performance needs demands the development and utilization of novel multi-functional composite structures. To this end, additive manufacturing (AM), commonly referred to as 3D-printing, represents a significant advancement in conventional manufacturing by enabling the fabrication of complex, tailored components and offering compatibility with a broad spectrum of materials. AM further deviates from traditional subtractive approaches and is instead characterized by the layer-by-layer assembly of intricate geometries directly from computer-aided design models. This has led to novel applications in the automotive, aerospace, maritime, energy, and medical sectors through the development of components, including crankshafts, heat exchangers, batteries, and dental implants [1–4].

One notable technique under the umbrella of AM is fused filament fabrication (FFF), also referred to as fused deposition modeling (FDM); it has been made popular due to its versatility, cost-effectiveness, and low barrier to entry [5]. In this process, a spool of filament material is fed through a heated nozzle and deposited onto a substrate layer-by-layer to construct bespoke three-dimensional objects. Depending on the material, such as high-performance thermoplastics and ceramics, a heated substrate and heated chamber could be necessary to address print defects, improve layer-to-layer adhesion, and enhance print quality [6]. The properties and fabrication of components from polylactic acid filament with metal-reinforcement for added multi-functionality, carbon-reinforced PEEK and PEI filament, and sustainable naturally reinforced filament with wood and hemp have also been explored [7–9]. Some FFF printers also allow the user to incorporate continuous-fiber reinforcement, such as Kevlar for added impact resistance [10]. The utilization of multiple nozzles on a single FFF printer further enables the fabrication of tailored multi-material and multi-functional components through the use of multiple filaments in a single print task [11]. A secondary nozzle can also be leveraged to incorporate sacrificial filament within a design, supporting overhangs and other typically difficult-to-print geometries; this sacrificial filament is later removed through dissolution.

Taking full advantage of the design freedom and tailorability afforded by FFF, opportunities to advance technology through topology optimization, integrated features, and part consolidation for reduced weight and improved performance will play a critical role in FFF's adoption by industry [12,13]. One specific avenue to address lightweighting through FFF of components is by incorporating lattice design, which can be achieved through the use of triangular, grid, rectilinear, and honeycomb infill patterns [14]. The honeycomb pattern, a hexagonal (hex) arrangement of cells, is a prevalent architecture in nature that is most recognized for its role in beehive structures; it provides a high strength-to-weight ratio and packing density and is currently used by the aerospace industry for the design of lightweight aluminum aircraft panels. In a study by Mansour et al. [15], an examination of the honeycomb structure and the reinforcement material determined that incorporating a hierarchical structure with chopped fiber or carbon nanotubes results in increased strength and stiffness. In this light, FFF opens the door for the development of complex multi-material and multi-functional honeycomb structures with great potential for application in the aerospace industry.

Addressing the need for both novel lightweight structures and an increased desire for sustainable closed-loop systems toward a circular economy, acrylonitrile butadiene styrene (ABS) and polylactic acid (PLA) are widely available material systems that are compatible with low-cost FFF printers and have demonstrated qualities of biodegradability and recyclability [16–18]. Relative to PLA, ABS offers better impact, heat, and electrical resistance properties and has been widely investigated in order to determine the effects of processing parameters and reinforcement materials on mechanical properties, tribological behavior, and thermal properties [19–22]. However, the numerous features involved in 3D-printing create a large and complex design space, which has not yet been fully explored and requires additional investigation. With sufficient development and understanding of the process–structure–property relationships of the ABS system along with the effect of material reinforcement, FFF components with tailored designs could be developed to meet performance needs. By leveraging machine learning and artificial intelligence approaches, these relationships and discoveries could then be utilized to accelerate, inform, and guide designs for targeted properties [23–27].

In the current work, ABS and reinforced ABS materials were printed by desktop FFF. Process parameters, including print-layer height and hex size, were modified to ascertain their effects on the resultant 3D-printed structures and properties. This was achieved through the characterization and evaluation of specimens by microscopy and compression testing. Furthermore, machine learning models trained on experimental results were developed to classify materials and predict ultimate compressive strength; these were subsequently evaluated for performance and can provide a means to guide and inform the design of these honeycomb structures.

2. Results and Discussion

2.1. Material Characterization

The structures of 3D-printed components, including defects such as porosity, are influenced by the combination of process parameters used during fabrication. Hex sizes ranging from 5.46 mm to 6.99 mm and print-layer heights of 0.2 mm and 0.3 mm were investigated for ABS, CNT-ABS, and CF-ABS materials (3DXTech, Grand Rapids, MI, USA). Figure 1 illustrates the effects of the hex size and print-layer height (LH) on FFF CNT-ABS for several process combinations. It is observed in Figure 1a that a cross-section of CNT-ABS having the smallest hex size of 5.46 mm and shortest layer height of 0.2 mm contains relatively few defects, which are highlighted in red. When increasing the hex size to the maximum-printed 6.99 mm while maintaining the layer height at 0.2 mm, as shown in Figure 1b, there is an increase in inter-layer porosity found throughout the cross-section. Maintaining the larger hex size of 6.99 mm and increasing the print-layer height to 0.3 mm is then shown to significantly increase inter-layer porosity relative to the other process parameter combinations. When maintaining all other parameters as constant, a smaller print-layer height improves packing, overlap, and adhesion [28,29]. For these reasons, relative to a 0.2 mm print-layer height, the 0.3 mm print-layer height had increased air gaps and porosity. The increase in porosity with increasing hex size could potentially stem from changes in the print duration. As the hex size increases, an individual layer's surface area decreases, and this corresponds to a decrease in print time. The resulting thermal gradient and its effect on layer adhesion could influence porosity formation. However, a comprehensive analysis of porosity at different hex sizes but at a constant print-layer height is necessary to further investigate this relationship. Additionally, across all process parameter combinations, there is porosity observed along the full length of the honeycomb vertices. The consistency of this porosity in each cross-section can then be explained by the printer's nozzle path during fabrication and possible lack of overlap or under-extrusion of filament to fully connect the honeycomb walls at the vertices. A modification to the print process parameters, such as by increasing the overlap percentage, would likely resolve this defect.

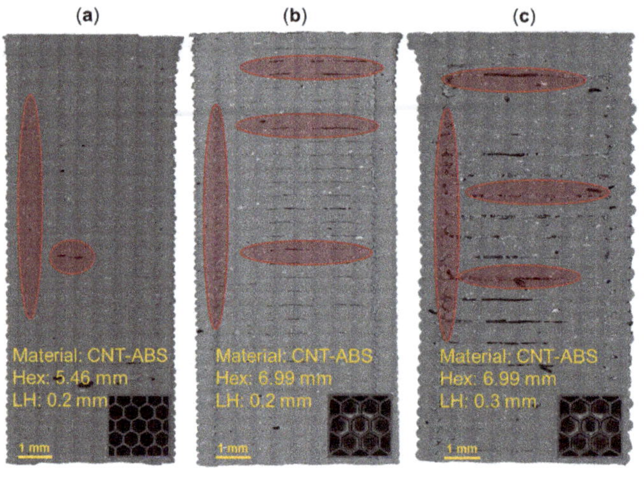

Figure 1. Optical microscopy cross-sections, in-plane with the print direction, of carbon nanotube reinforced (CNT) ABS specimens with different hex sizes and print-layer heights (LHs): (**a**) Hex: 5.46 mm | LH: 0.2 mm; (**b**) Hex: 6.99 mm | LH: 0.2 mm; (**c**) Hex: 6.99 mm | LH: 0.3 mm. A selection of porosity defects is highlighted in red. A cross-section perpendicular to the print direction corresponding to the 5.46 mm and 6.99 mm hex sizes and each having a length and width of approximately 25.4 mm × 25.4 mm is also provided for each specimen.

With this perspective, the pure ABS honeycomb cross-sections in Figure 2 illustrate the effect of layer height on porosity for the largest printed hex size. A print-layer height of 0.2 mm with a hex size of 6.99 mm (Figure 2a) demonstrates partial inter-layer porosity that is mostly contained in the upper half of the print. When the print-layer height is increased to 0.3 mm and the hex size remains constant at 6.99 mm (Figure 2b), the observed inter-layer porosity increases: extending through a larger portion of the cross-section. These ABS honeycomb specimens demonstrate an increase in porosity with increasing print-layer height, in agreement with trends founds in the CNT-ABS honeycomb specimens. However, it is also observed that the inter-layer porosity found in ABS honeycomb cross-sections is not as significant relative to CNT-ABS specimens fabricated under the same hex size and print layer height conditions. This is possibly due to a combination of process parameters and an increased propensity for micro-porosity in CNT-ABS, which are demonstrated by the speckled features throughout the cross-sections in Figure 1. The result is weaker interfaces and more extensive porosity overall compared to printing with pure ABS. In this context, a general solution to improve the print quality is to use a smaller print-layer height at the trade-off of the time savings achieved through increased print-layer heights. Additionally, similar to the CNT-ABS specimens, the ABS cross-sections demonstrate inherent porosity along the honeycomb vertices due to the nozzle path and under-extrusion. However, the increased number of discontinuous regions in Figure 2 relative to Figure 1 is a result of the chosen process and print parameters. Given constant processing conditions such as nozzle diameter, nozzle speed, and print bed temperature, the resulting print quality can vary based on adjustments to the hex size, print-layer height, and material. For the same processing parameters, the increased thermal conductivity of CNT-ABS relative to ABS could result in increased material extrusion and overlap as well as reduced discontinuities in the print. In both materials, these discontinuities could be remedied by modifying process parameters to provide additional material extrusion for connecting the honeycomb vertices. For additional microstructure analysis and porosity quantification, the reader is referred to prior work [30].

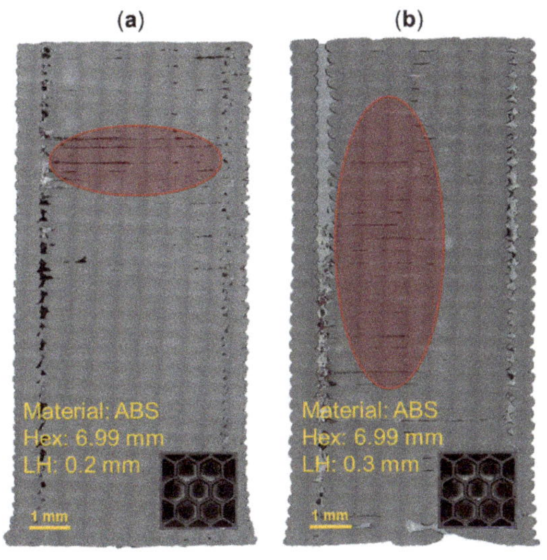

Figure 2. Optical microscopy cross-sections, in-plane with the print direction, of pure ABS specimens with a constant hex size and different layer heights (LHs): (**a**) Hex: 6.99 mm | LH: 0.2 mm; (**b**) Hex: 6.99 mm | LH: 0.3 mm. A selection of porosity defects are highlighted in red. A cross-section perpendicular to the print direction corresponding to the 6.99 mm hex size and each having a length and width of approximately 25.4 mm × 25.4 mm is also provided for each specimen.

Figure 3 provides a digital image correlation (DIC) snapshot of several honeycomb specimens during compression testing at a minimal applied load and after yielding. It is noted that the orientation of coupons within the test frame was varied such that the coupon was resting on either its first or last print layer. The 5.46 mm hex and 0.2 mm print-layer height ABS honeycomb specimen (Figure 3(1-a,1-b)) was tested while resting on its last print layer. Upon compressive yielding, the bottom of the specimen, corresponding to the last print layers, begins to demonstrate inter-layer delamination. Examination of the CNT-ABS specimen with a 6.99 mm hex and 0.3 mm print-layer height in Figure 3(2-a,2-b) revealed a similar location of failure in the region corresponding to the upper half of the print. Furthermore, failure in the upper print layers also occurred with the 6.22 mm hex and 0.2 mm print-layer-height CF-ABS honeycomb specimen in Figure 3(3-a,3-b). The proximity of the failure to the upper print layers of these honeycomb specimens was identified as a consistent trend across all tested honeycomb specimen variations, regardless of specimen orientation in the test frame. This is likely an artifact of the FFF process, processing parameters, and material. When printing the first layers of a specimen, the build plate provides an additional heat source to help reduce warping and improve adhesion between each print layer. As the print extends further away from the build plate, the effect of the heated build plate is diminished, resulting in reduced inter-layer adhesion and print quality [31,32]. This is supported by imaged cross-sections in Figures 1 and 2, where a gradual increase in porosity in the latter half of print layers can be observed. A potential solution to improve layer adhesion throughout the print is to leverage a fully enclosed chamber and a preheating period to bring the internal chamber environment to a sufficiently high and uniform temperature. As a result of increased porosity in the upper print layers, this region of the specimen is inherently weaker and susceptible to becoming the initial location of failure upon yielding. Additionally, it is observed that compressive failure in the carbon-reinforced ABS is generally more destructive than that of the pure ABS honeycomb specimens. This could, again, be explained by increased porosity in the carbon-reinforced specimen relative to pure ABS specimen, resulting in the more extreme inter-layer delamination after yielding observed in Figure 3(2-b,3-b).

2.2. Mechanical Properties

Figure 4 shows a representative subset of data obtained from compression testing of ABS, CF-ABS, and CNT-ABS honeycomb specimens with hex sizes of 5.46 mm, 5.72 mm, 6.22 mm, and 6.99 mm for print-layer heights of 0.2 mm and 0.3 mm. When maintaining a constant print-layer height while varying the material and hex size, it is observed in Figure 4a that the ABS, CF-ABS, and CNT-ABS specimens with hex sizes of 5.46 mm, 5.72 mm, and 5.46 mm, respectively, withstood larger stresses than their larger-hex-sized counterparts of 6.99 mm, 6.22 mm, and 6.99 mm, respectively. The ABS and CNT-ABS specimens with a print-layer height of 0.3 mm, shown in Figure 4b, demonstrated similar behavior, with the smaller 5.46 mm-hex-size specimens undergoing larger stresses before failure relative to the 6.99 mm hex specimens. These observations reinforce a trend of decreasing compressive strength with increasing hex size and are further corroborated by forthcoming results and analysis.

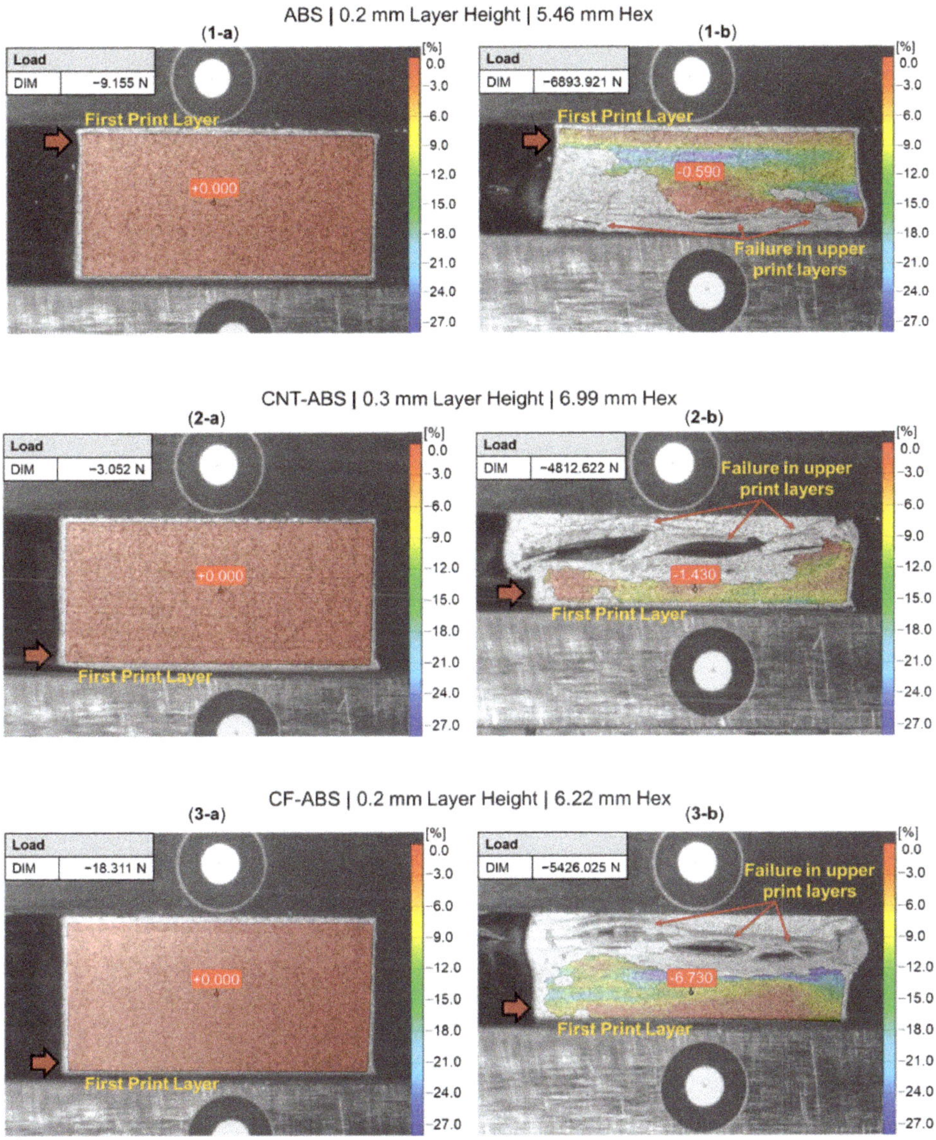

Figure 3. DIC snapshots of honeycomb specimens during compression testing: (**1**) ABS with 5.46 mm hex size and 0.2 mm layer height; (**2**) CNT-ABS with 6.99 mm hex size and 0.3 mm layer height; (**3**) CF-ABS with 6.22 mm hex size and 0.2 mm layer height; (**a**) minimal applied load; (**b**) after yielding. The first print layer and the location of failure after yielding are noted.

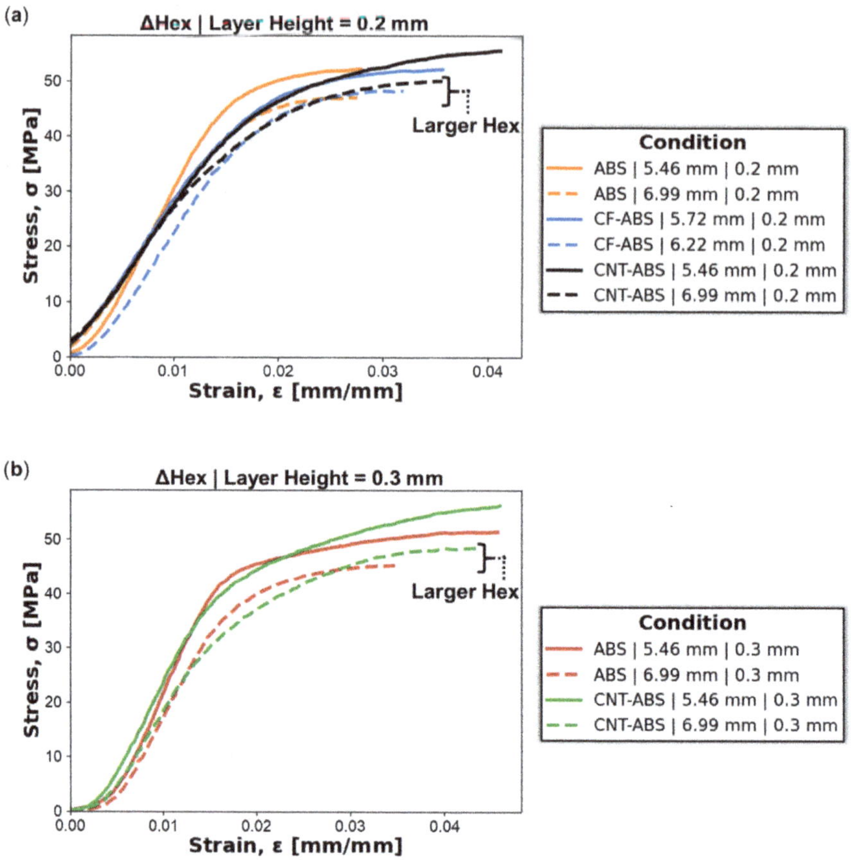

Figure 4. Stress–strain curves for the compression testing of 3D-printed honeycomb coupons: (**a**) ABS and CNT-ABS specimens with 5.46 mm and 6.99 mm hex sizes as well as CF-ABS specimens with 5.72 mm and 6.22 mm hex sizes, where all specimens have a constant print-layer height of 0.2 mm; (**b**) ABS and CNT-ABS specimens with 5.46 mm and 6.99 mm hex sizes, where all specimens have a constant print-layer height of 0.3 mm. Solid lines represent the smaller hex size (5.46 mm or 5.72 mm), and dashed lines represent the larger hex size (6.2 mm or 6.99 mm) of each material.

Figure 5 shows the average compressive yield strength, σ_{yield}, with standard deviation error bars for ABS, CNT-ABS, and CF-ABS across all hex sizes, where the compressive yield stress for each specimen is determined from the 0.2% offset plastic strain. Observing the overall trend of the data, there is a slight negative correlation of decreasing compressive yield strength with increasing hex size, in agreement with trends from Figure 4. Additionally, the yield strengths for ABS and CNT-ABS with 0.2 mm print-layer height are consistently greater than their counterparts with 0.3 mm print-layer height. This is due to increased inter-layer porosity with increased print-layer heights, as observed in Figures 1 and 2, resulting in reduced strength for the 0.3 mm print-layer-height specimens relative to the 0.2 mm print-layer-height specimens. When comparing ABS and CNT-ABS, it is shown that ABS generally achieved higher compressive yield strength across all hex sizes. For the case of CF-ABS with 0.2 mm print-layer height, the determined yield strength was lower than the values for all other feature combinations except CNT-ABS with 0.3 mm layer height. The performance of CF-ABS specimens falling between that of the CNT-ABS 0.2 mm and 0.3 mm print-layer height specimens and below that of ABS honeycombs is an expected outcome based on the increased porosity observed in the carbon-reinforced specimens. CF-ABS with 0.2 mm print-layer height likely demonstrated reduced compressive yield strength relative to CNT-ABS with 0.2 mm print-layer height due to an increased quantity of porosity within the filament and, consequently, increased porosity within the printed component, as described by Vakharia et al. [7]. If CF-ABS with 0.3 mm print-layer height had been 3D-printed, it would likely have the lowest compressive strength of all tested feature combinations.

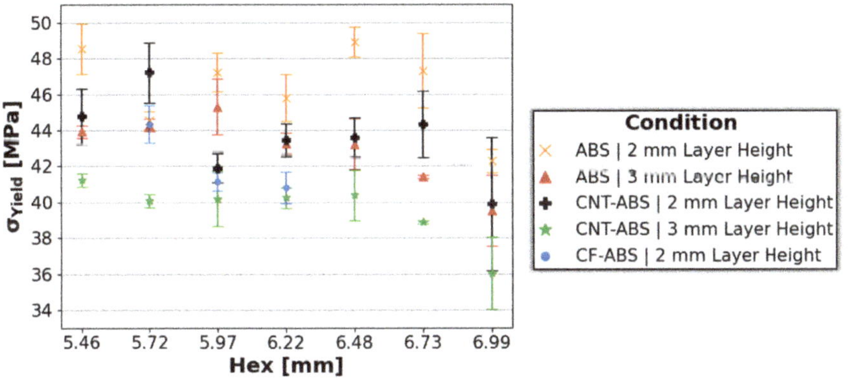

Figure 5. Average compressive yield strengths with standard deviation error bars for ABS, CNT-ABS, and CF-ABS across all hex sizes (5.46 mm, 5.72 mm, 5.97 mm, 6.22 mm, 6.48 mm, 6.73 mm, and 6.99 mm) and print-layer heights (0.2 mm and 0.3 mm).

The average ultimate compressive strength, $\sigma_{compressive}$, and standard deviation error bars, determined by the maximum stress, for ABS, CNT-ABS, and CF-ABS specimens across all hex and layer heights is captured in Figure 6. A strong negative correlation is readily observed in this instance, where the ultimate compressive strength decreases with increasing hex size, from 5.46 mm to 6.99 mm, for 0.2 mm and 0.3 mm print-layer heights. For each ABS and CNT-ABS, it is observed that the 0.2 mm print-layer height resulted in a larger ultimate strength than the 0.3 mm print-layer height, resulting from increased porosity with increasing layer height. However, unlike with compressive yield strength, the CNT-ABS specimens of both layer heights achieved larger ultimate compressive strength values than ABS honeycombs across all hex sizes. This is likely due to the initial compressive load compacting the inter-layer porosity present in coupons combined with enhanced CNT reinforcement, resulting in substantial strengthening of CNT-ABS specimens before failure. Producing the lowest ultimate strength values, CF-ABS specimens of 0.2 mm print-layer height and 5.72 mm, 5.97 mm, and 6.2 mm hex sizes likely suffered from increased porosity, contributing to an overall weaker performance relative to ABS and CNT-ABS specimens. This increased porosity is a cascade effect stemming from CF-ABS filament porosity coupled with poor fiber–matrix interfaces contributing to differing flow of the chopped fiber and polymer matrix and resulting in inner-bead porosity during filament extrusion [33]. The investigation of alternate extrusion methods, processing parameters, and optimization of the processing and treatment of the fiber-reinforced composite filament could provide a solution to mitigate porosity formation within these FFF components. Alternative designs, such as hierarchical honeycomb structures, could provide additional avenues for study. Although tested in a different orientation, Mansour et al. [15] reported ultimate strength values in the range of 32–38 MPa, 40–46 MPa, and 41–47 MPA for hierarchical honeycomb structures of ABS, CNT-ABS, and CF-ABS, respectively. These values fall below the ranges of 45–54 MPa, 46–57 MPa, and 47–53 MPa determined for ABS, CNT-ABS, and CF-ABS in the current study, and this can be explained by differences in the hexagonal structures and process parameters between the studies.

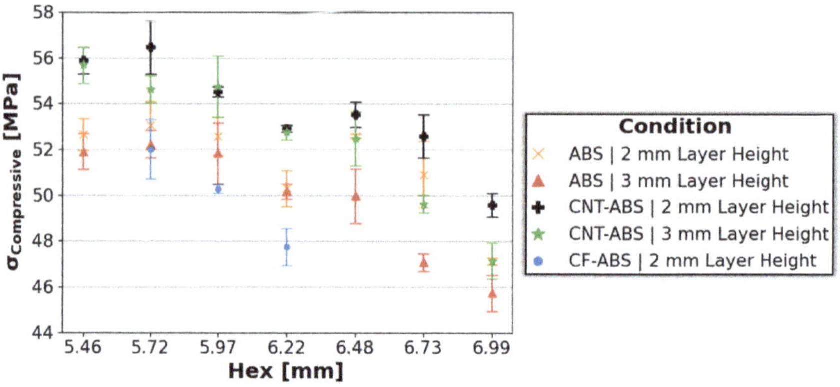

Figure 6. Average ultimate compressive strength with standard deviation error bars for ABS, CNT-ABS, and CF-ABS across all hex sizes (5.46 mm, 5.72 mm, 5.97 mm, 6.22 mm, 6.48 mm, 6.73 mm, and 6.99 mm) and print-layer heights (0.2 mm and 0.3 mm).

The average Young's modulus, E, and standard deviation error bars for the ABS, CNT-ABS, and CF-ABS honeycomb coupons, calculated from the linear elastic region of the compressive response, is shown in Figure 7a. Across all hex sizes, from 5.46 mm to 6.99 mm, the Young's modulus remained relatively constant within each material and print-layer-height combination. The values for CNT-ABS and CF-ABS in this study fall within the ranges of 1.8–4.7 GPa and 1.3–3.6 GPa, respectively, that have been reported in several studies [15,30,34]. However, the Young's modulus for ABS in the current study exceeded the range (1.6–2.2 GPa) reported by these studies [15,30,34]. This is possibly due to variation in filament properties between batches and vendors in addition to variations between printers and process parameters, yielding printed components with different mechanical properties.

The density, ρ, of each honeycomb material and print-layer-height set generally remained constant across the hex sizes as well, as captured in Figure 7b. Due to the increased propensity for porosity within CF-ABS prints, reduced density in CF-ABS relative to CNT-ABS and ABS specimens was observed. The surface area, $A_{surface}$, measurements in Figure 7c represent the honeycomb cross-sectional area perpendicular to the print direction as calculated by Keyence (Itsaca, IL, USA) analysis software version 3.0.34 for each feature combination. A negative correlation is observed between surface area and hex size for ABS, CNT-ABS, and CF-ABS at 0.2 mm and 0.3 mm print-layer heights. This is expected due to the constrained dimensions of the honeycomb coupons of approximately 25.4 mm for both the length and the width. As a result, increasing the hex size translates to fewer hexes fitting within the 25.4 mm by 25.4 mm area and a decrease in the cross-sectional surface area. This also supports the negative trends for compressive yield and ultimate strengths previously discussed. Although the same printer is used, there are still slight variations in the surface areas between specimens having the same hex size. A combination of processing parameters to include materials, print-layer heights, and fluctuations within the material filament itself impact filament extrusion and the corresponding line width of the extruded filament, providing an explanation for dimensional variance from the input honeycomb design.

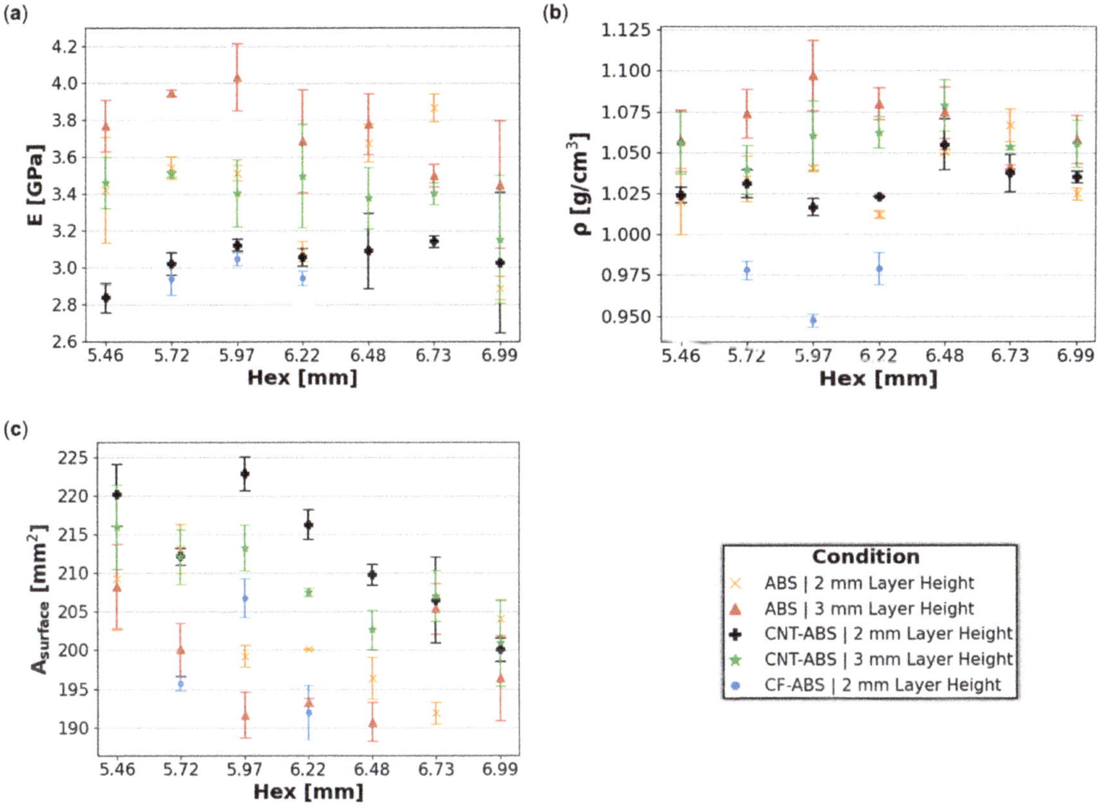

Figure 7. Average (**a**) Young's modulus (E), (**b**) density (ρ), and (**c**) surface area ($A_{surface}$) with standard deviation error bars for ABS, CNT-ABS, and CF-ABS across all hex sizes (5.46 mm, 5.72 mm, 5.97 mm, 6.22 mm, 6.48 mm, 6.73 mm, and 6.99 mm) and print-layer heights (0.2 mm and 0.3 mm).

From the 65 data points across all three materials, the Pearson correlation coefficient matrix heatmap between hex size, LH, Young's modulus, compressive yield strength, ultimate compressive strength, surface area, and density is illustrated in Figure 8. In agreement with observations from Figures 5–7, weak-to-moderate negative correlations for [hex size, compressive yield strength], [hex size, compressive ultimate strength], [hex size, surface area], [LH, Young's modulus], and [surface area, Young's modulus] are determined. Weak-to-moderate positive correlations for [LH, Young's modulus], [LH, density], [Young's modulus, compressive yield strength], [Young's modulus, density], [compressive yield strength, compressive ultimate strength], and [compressive ultimate strength, surface area] are also observed.

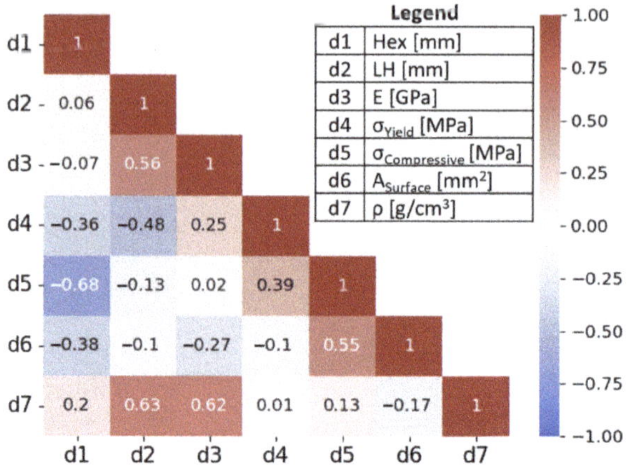

Figure 8. Pearson correlation coefficient matrix heatmap for: hex size (Hex [mm]); layer height (LH [mm]); Young's modulus (E [GPa]); compressive yield strength (σ_{Yield} [MPa]); compressive ultimate strength ($\sigma_{Compressive}$ [MPa]); surface area ($A_{surface}$ [mm^2]); density (ρ [g/cm^3]). The correlation coefficient value is provided within each grid. Color and intensity correspond to the degree of correlation: red—positive correlation; blue—negative correlation; white—no correlation.

2.3. Classification and Regression

A classification model was developed to classify materials by type, and it was trained on a selection of input features with an 80/20% train–test split on the 65 experiment data points. These features include hex size, print-layer height, surface area, mass, volume, density, maximum applied load, yield strength, Young's modulus, and specific strength. The training accuracies of multiple supervised machine learning classification algorithms following stratified eight-fold cross-validation (CV) are illustrated with the box plot in Figure 9. These algorithms include linear regression (LR), linear discriminant analysis (LDR), K-nearest neighbors (KNN), classification and regression decision tree (CART), random forest (RF), naive Bayes (NB), and support vector machine (SVM). The spread of the accuracy for each algorithm depicts the model's consistency across each cross-validation split. Here, KNN's wide spread is indicative of varied performance depending on the cross-validation set, whereas SVM might demonstrate a smaller spread but also has a consistently lower accuracy. Of these algorithms, LDA seems to perform the best, with an average accuracy of 0.98 and a standard deviation of 0.05, as summarized in Table 1. The average accuracies and standard deviations for KNN and SVM are 0.64, 0.21] and [0.48, 0.05], respectively. LR provided the second-best CV accuracy at 0.89 and a standard deviation of 0.16, followed by RF with values of 0.84 and 0.14, respectively. Although the accuracy metric alone does not comprehensively assess model behavior and could be biased toward overfitting, the stratified CV approach helps alleviate these concerns for an initial model down-selection, and these results can subsequently be supported by descriptive metrics, including recall, precision, and F1-score.

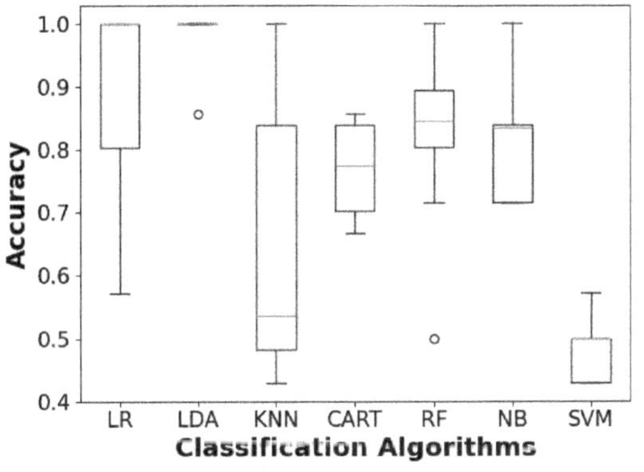

Figure 9. Spot-screening of classification algorithms and corresponding accuracies for material prediction resulting from stratified 8-fold cross-validation with training data: LR—linear regression; LDA—linear discriminant analysis; KNN—k-nearest neighbors; CART—classification and regression decision tree; RF—random forest; NB—naive Bayes; SVM—support vector machine.

Table 1. Training results from spot-screening multiple machine learning algorithms for material classification. Stratified 8-fold cross-validation (CV) average accuracies and standard deviations are provided.

Algorithm	CV—Accuracy	Std. Deviation
Linear Regression	0.89	0.16
Linear Discriminant Analysis	0.98	0.05
K-Nearest Neighbors	0.64	0.21
Decision Tree	0.81	0.10
Random Forest	0.84	0.14
Naive Bayes	0.81	0.09
Support Vector Machine	0.48	0.05

A down-selection was made based on previously discussed performance, and the LDA model was validated with the unseen test data. A confusion matrix (Figure 10) depicts the model-predicted label of ABS, CF-ABS, or CNT-ABS for a given set of honeycomb features with the true label. Entries in the top-left to bottom-right diagonal correspond to accurate predictions, while counts outside of this diagonal were inaccurately predicted. The overall LDA classification accuracy, taken as the ratio of correct predictions over the total number targets, was found to be 0.92. It was found that the model provided one incorrect prediction of ABS when the material was, in fact, CNT-ABS. This is likely due to overlap in some of the features for each material, such as compressive yield strength, Young's modulus, and density, and due to the low number of data points available for training. However, it is shown that the LDA classification model performed sufficiently well on the test data, with precision, recall, and F1-score metrics for each material class provided in Table 2. Classification of ABS had a reduced precision and F1-score due to the incorrect label prediction, CNT-ABS had a slight drop to its recall and F1-score, and the F1-score of 1 for CF-ABS demonstrates the model's well-balanced ability to recall the material class with precision. The caveat is that the small data set resulted in only a single CF-ABS data point remaining in the test set for validation, and a single error resulted in 0.75 precision for ABS classification. Although the model performed reasonably with the current data, by conducting additional experiments and gathering more data, the model could be expanded and refined with more extensive training and testing sets for added utility and improved accuracy. Also not explored is hyperparameter tuning, which could provide an alternate method for improving model performance. However, this step was not deemed necessary based on the current results.

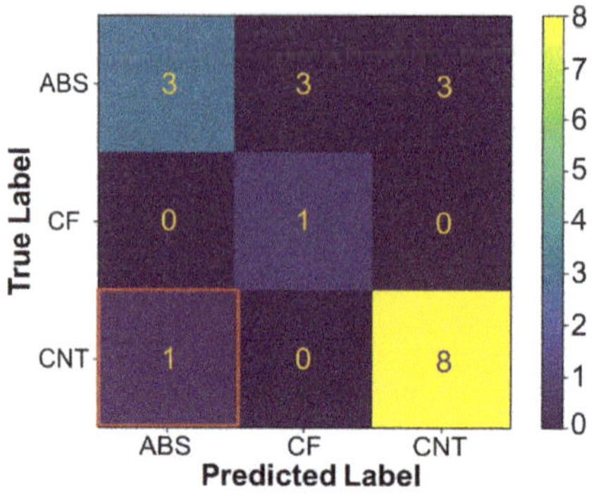

Figure 10. Confusion matrix of linear discriminant analysis predicted labels and true labels based on the test data set: ABS, CF (CF-ABS), and CNT (CNT-ABS).

Table 2. Test results for material classification with linear discriminant analysis. Precision, recall, and F1-score statistical metrics are provided.

Material	Precision	Recall	F1-Score
ABS	0.75	1.00	0.86
CF-ABS	1.00	1.00	1.00
CNT-ABS	1.00	0.89	0.94

A regression model, taking the form $y = m_i X_i + \ldots + m_n X_n + b$, was developed to predict the ultimate compressive strength of honeycomb specimens based on a selection of input features and an 80/20% train–test split of the 65 experiment data points. In this formula, X represents the features, m is the feature coefficients, and b is the intercept. These features include material, hex size, print-layer height, surface area, mass, volume, and density. Multiple regression algorithms were trained, and the coefficient of determination, R^2, to evaluate the fit of model predictions was determined through validation with the test data, with the results shown in Table 3.

Table 3. Results from screening multiple regression algorithms for the prediction of ABS, CNT-ABS, and CF-ABS compressive strength. The R^2 and mean absolute error for each algorithm is provided.

Algorithm	R^2	Mean Absolute Error [MPa]
Ridge Regressor	0.91	0.83
Linear Regression	0.96	0.63
K-Neighbors Regressor	0.70	1.70
Gradient Boosting Regressor	0.88	0.99
Random Forest Regressor	0.79	1.40
Extra Trees Regressor	0.82	1.28
Decision Tree Regressor	0.79	1.5
Lasso Regression	0.36	2.34

The trained algorithms include ridge regressor, linear regression, k-neighbors regressor, gradient boosting regressor, random forest regressor, extra trees regressor, decision tree regressor, and lasso regression. Although R^2 has limitations, such as neglecting prediction accuracy, this metric is still informative for evaluating regression performance [35]. Of these algorithms, lasso regression had the weakest performance, with an R^2 of 0.36, whereas the top-performing algorithm was linear regression, with an R^2 of 0.96. Ridge regressor and gradient boosting regressor fell short, with R^2 values of 0.91 and 0.88, respectively. The mean absolute error (MAE) values of these algorithms also provide insight to model performance and are calculated as the average absolute error between the predicted and the actual ultimate compressive strength. The linear regression MAE of 0.63 MPa corresponds to highly accurate predictions given the range of ultimate compressive strength falls between 46 MPa and 58 MPa. The MAE also tends to decrease with decreasing R^2 and has a value of 2.34 MPa for the lasso regression algorithm. Based on these findings, the linear regression model predictions evaluated against corresponding experimental measurements are further analyzed in Figure 11. The y = x line demarcates predictions that perfectly correspond to experiment measurements, with under-predicted values falling below the line and over-predicted values above. It is observed that an equal distribution of over- and under-predicted values occurs across the full range of ultimate compressive strength values within the test data set. This signals that the model is not apparently biased toward over- or under-predicting and can perform well across the design space for which it was trained. These observations are supported by the root mean square error (RMSE), 0.74 MPa, and the mean absolute percentage error (MAPE), 1.2%, for the 13 test samples, which represent a good measure of predictive accuracy for the model when compared with experiments. Table 4 summarizes the intercept, features, and corresponding coefficients for this linear regression model.

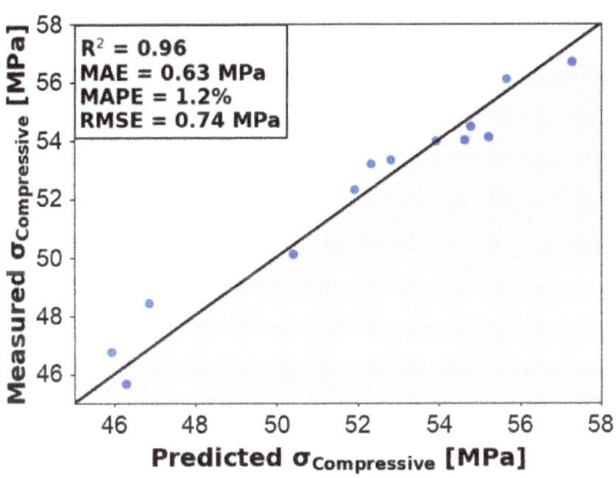

Figure 11. Predicted ultimate compressive strength from trained linear regression model evaluated against experimental measurements for ABS, CNT-ABS, and CF-ABS, resulting in an R^2 of 0.96, mean absolute error (MAE) of 0.63 MPa, mean absolute percentage error (MAPE) of 1.2%, and a root mean square error (RMSE) of 0.74 MPa. Predictions falling on the y = x correspond perfectly with experiments.

Table 4. The intercept, features, and corresponding coefficients for a linear regression model with one-hot encoding for material, hex size, and print-layer height: b—intercept; SA—surface area; V—volume; D—density; M—mass; H1 (hex 1)—5.46 mm; H2—5.71 mm; H3—5.97 mm; H4—6.22 mm; H5—6.48 mm; H6—6.73 mm; H7—6.99 mm; LH1 (layer height 1)—0.2 mm; LH2—0.3 mm.

b	SA	V	D	M	ABS	CF-ABS	CNT-ABS	H1	H2	H3	H4	H5	H6	H7	LH1	LH2
−20.0	−0.107	19.9	59.7	−7.20	−1.25	0.0150	1.23	2.06	2.21	1.40	−0.256	−0.105	−1.7	−3.62	0.963	−0.963

One-hot encoding was utilized to incorporate categorical features into the previously described linear regression model. In this method, a value of 0 or 1 is assigned to the feature depending on whether it is false or true, respectively. These categorical features included material, hex size, and layer height. An approach with hex size and print-layer height as continuous values will help reduce the number of features and result in a more generalized linear regression model. Table 5 summarizes the intercept, features, and corresponding coefficients for the generalized linear regression model. Here, mass was also dropped as a feature due its redundancy with volume and density resulting in a negligible effect on model performance. This linear regression model demonstrated good predictive accuracy and similar performance to the first model, yielding an R^2, MAE, MAPE, and RMSE of 0.95, 0.61 MPa, 1.2%, and 0.77 MPa, respectively, for the 13 test samples. It should be noted that, similar to the classification model, additional experiments would be beneficial for increasing the training data available and for expanding the utility of the model. However, the linear regression model in its current state performed well at accurately predicting the ultimate compressive strength of the honeycomb specimens in this work.

Table 5. The intercept, features, and corresponding coefficients for the generalized linear regression model with continuous values for hex size and print-layer height and one-hot encoding for material: b—intercept; SA—surface area; V—volume; D—density.

b	SA	V	D	ABS	CF-ABS	CNT-ABS	Hex Size	Layer Height
−10.0	−0.205	24.5	63.7	−1.97	1.72	0.250	−3.15	−1.69

3. Materials and Methods

The specimens in this work were 3D-printed on a Makerbot Replicator 2X (MakerBot Industries, LLC One MetroTech Center, Brooklyn, NY, USA). In addition to material type, print process parameters that were varied included the dimensions of the hexagons comprising the honeycomb structures and the print-layer height, as summarized in Table 6. The 3D-printed filament material, manufactured by 3DXTech, included pure ABS, multi-walled carbon nanotube reinforced ABS, and 5 wt.% chopped carbon fiber reinforced ABS, which will now be referred to as ABS, CNT-ABS, and CF-ABS, respectively. Hex sizes, defined as the distance between opposite walls of a hexagon, were 5.46 mm, 5.72 mm, 5.97 mm, 6.22 mm, 6.48 mm, 6.73 mm, and 6.99 mm. Specimens were printed with individual layer heights of 0.2 mm and 0.3 mm and were printed to overall length, width, and height dimensions of 25.4 mm, 25.4mm, and 12.7 mm, respectively. Each specimen had an open face on the top and bottom layers and maintained a constant wall thickness. Processing features that remained constant across all prints included nozzle diameter, 0.4 mm; filament diameter, 1.75 mm; print speed, 120 mm/s; print temperature, 230 °C; and bed temperature, 110 °C.

Table 6. Processing features for testing.

Feature	Properties
Material	ABS I CF-ABS I CNT-ABS
Hex Size [mm]	5.46 I 5.72 I 5.97 I 6.22 I 6.48 I 6.73 I 6.99
Layer Height [mm]	0.2 I 0.3

Figure 12 shows a selection of coupons across the range of hex sizes that fabricated for compression testing; they were loaded at a rate of 6 mm/min and nominally conformed to ASTM C365 [36]. Figure 12a–c show hex sizes 5.46 mm, 6.22 mm, and 6.99 mm, respectively. The overall lengths, widths, and heights of the specimens remained constant, resulting in a reduced number of hexes for coupons with increased hex sizes. Several coupons for each hex size and layer height were fabricated with ABS and CNT-ABS materials, while CF-ABS specimens were of ancillary focus and were only printed with a 0.2 mm layer height and hex sizes of 5.72 mm, 5.97 mm, and 6.22 mm.

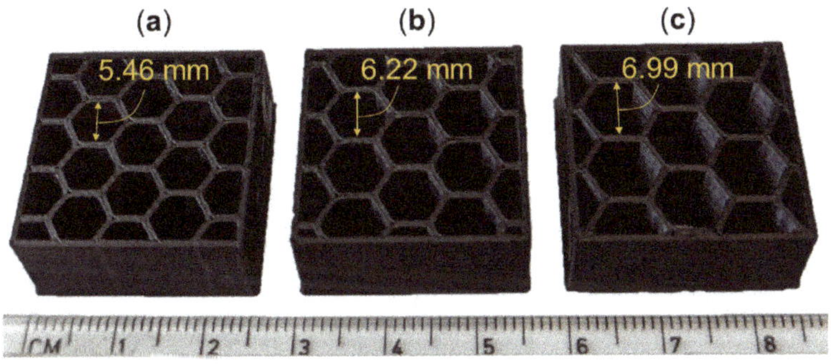

Figure 12. Honeycomb specimens with several hex sizes 3D-printed by FFF: (**a**) 5.46 mm hex; (**b**) 6.22 mm hex; (**c**) 6.99 mm hex.

A subset of specimens fabricated to the aforementioned specifications were machined to obtain sample cross-sections in-plane with the print direction, as highlighted in Figure 13. These samples were mounted in epoxy and were subsequently polished for examination by optical microscope. It should be noted that Vakharia et al. [7,30] reported pull-out of particulate reinforcement from PLA and ABS composites due to the machining and polishing steps. Additional examination of cross-sections perpendicular to the print direction and surface area measurements were conducted with a VHX-7000N (Keyence, Itsaca, IL, USA) digital microscope. Individual images were stitched together with the microscope's software (version 3.0.34) to obtain panoramas of the entire cross-section. From these panoramas, the software calculated values for cross-sectional area. Prior to testing, the mass and several measurements each for the length, width, and height by digital caliper (Mitutoyo, Aurora, IL, USA), of all coupons was recorded. A prior investigation by Vakharia and co-authors [30] conducted thermogravimetric analysis, scanning electron microscopy, energy dispersive spectroscopy, and porosity quantification of pure ABS, CNT-ABS, and CF-ABS material systems, and this is not repeated in this work.

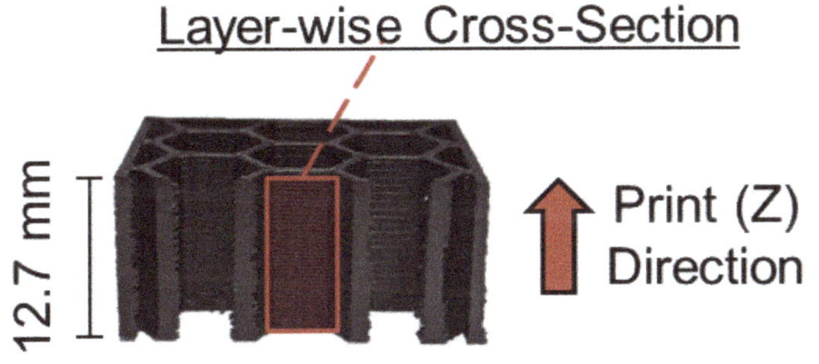

Figure 13. Cross-section of a 3D-printed honeycomb compression coupon in-plane with the print (Z) direction. The region of interest for the optical microscope is highlighted.

For compression testing, coupons were placed in an Instron 8562 test frame, with the open hexagonal cells in contact with flat steel plates on either end, as illustrated in Figure 14a. The load was applied through the honeycomb cross-sectional area (Figure 14b), which had to be determined for each honeycomb due to changing hex sizes. Digital image correlation was conducted with an ARAMIS (Trilion Quality Systems, Plymouth Meeting, PA, USA) system. Two specimens for each ABS and CNT-ABS test condition and three specimens for each CF-ABS test condition were tested. The Young's modulus, compressive yield strength, and compressive strength were determined from the test data. Although an increased sample size could be beneficial, the obtained data and confidence intervals provided in Figures 5–7 are sufficient to discern trends in mechanical properties across materials, hex sizes, and print-layer heights. Supervised machine learning models leveraging publicly available Scikit-learn [37] packages were developed to classify materials and predict the compressive strength based on process parameters and material features. An 80/20% train–test split of the processing features and experimental results was utilized, with the test data reserved for model validation. This split helps alleviate concerns related to missing features from a minimized training set and provides a sufficient test set to reduce potential performance bias. A stratified k-fold cross-validation with eight splits was implemented to ensure an equal proportion of data from each material was included in each split. Following training, the model was validated on the previously unseen test data.

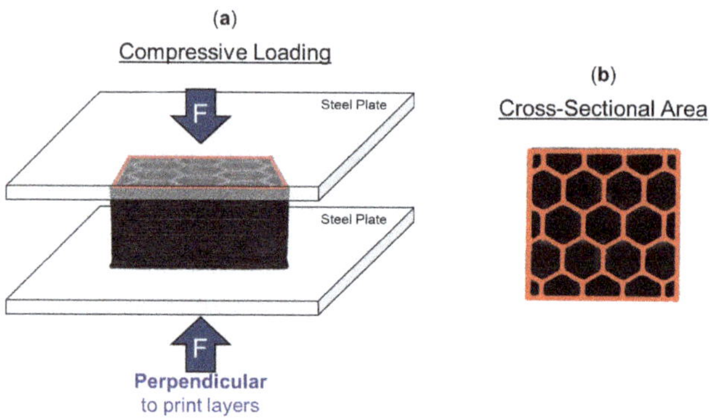

Figure 14. Schematic of honeycomb specimen compression testing: (**a**) honeycomb coupon sandwiched between two steel plates within a test frame, where force is applied perpendicular to the print layers; (**b**) cross-sectional area of honeycomb where load is applied.

4. Conclusions

Several 3D-printed honeycomb specimens of ABS, CNT-ABS, and CF-ABS demonstrated property variations stemming from porosity within the as-supplied filament and due to the influence of processing conditions during fabrication. Examination of honeycomb cross-sections across multiple hex sizes and print-layer heights revealed strong correlation with inter-layer porosity. By increasing the hex size and print-layer height for both CNT-ABS and ABS, a significant increase in porosity was realized within the FFF specimens. Furthermore, the incorporation of carbon reinforcement promoted porosity in terms of both size and quantity, where CNT-ABS specimens contained visibly increased porosity relative to pure ABS specimens. This is, in part, due to treatment with reinforcement material and the resulting adhesive strength with the matrix influencing the quality of extruded filament. These process and structure relationships demonstrated a cascading effect on properties and performance. Increased porosity and weaker inter-layer adhesion resulted in repeated failure of compression coupons in their upper print layers. Although ABS specimens generally reported higher compressive yield strength than carbon-reinforced specimens, CNT-ABS specimens reached higher ultimate compressive strength values due to the compaction of porosity upon yielding and subsequent strengthening. CF-ABS was weaker than ABS and CNT-ABS in terms of both yield and ultimate compressive strengths, having an increased magnitude of porosity, as affirmed by its reduced density.

Based on these findings, it is recommended that these materials are printed with the smallest print-layer height to achieve reduced porosity and improved performance under compression. The raster path and filament extrusion width should be monitored to ensure proper cohesion of adjacent paths, especially at vertices. By leveraging a fully sealed enclosure to attain a uniform elevated temperature prior to printing, the inter-layer adhesion throughout the print and the mechanical properties are expected to improve. Due to the increased propensity for porosity in carbon-reinforced ABS, these materials are more likely to benefit from refinement of processing procedures.

Classification and regression supervised machine learning models were developed from the experimental data by leveraging publicly available tools. Although only a relatively small data set was available for model training and testing, a stratified sampling approach and a broad sweep of available classification and regression algorithms resulted in well-performing models. The linear discriminant analysis material classification model demonstrated a classification accuracy of 0.92: only predicting one label incorrectly. The precision, recall, and F1-score metrics also supported the capability of this model to predict the ABS, CNT-ABS, and CF-ABS material classes based on process parameters and mechanical properties. For predicting the ultimate compressive strength, the linear regression model proved to be unbiased and accurate, with an R^2 of 0.96 and its mean absolute error of 0.63 falling well within an acceptable margin for values in the range of 46 MPa to 58 MPa. By conducting additional testing, the performance and potential utility of these models to assist with informing and guiding carbon-reinforced honeycomb design for novel 3D-printed structures can be expanded.

Author Contributions: Conceptualization, M.S. and M.C.H.; methodology, M.R., M.S. and M.C.H.; investigation, M.R.; data analysis, M.R., M.C.H. and M.S.; software, M.R.; writing—original draft preparation, M.R., M.C.H. and M.S.; writing—review and editing, M.R., M.C.H. and M.S.; supervision, M.C.H and M.S.; funding acquisition, M.C.H. All authors have read and agreed to the published version of the manuscript.

Funding: This research was supported by the NASA Revolutionary Vertical Lift Technology (RVLT) Project.

Institutional Review Board Statement: Not applicable.

Informed Consent Statement: Not applicable.

Data Availability Statement: The raw data supporting the conclusions of this article will be made available by the authors on request.

Acknowledgments: The authors would like to thank Andrew Ring for assistance with the digital image correlation, Jonathan Salem for assistance with mechanical testing, and the NASA Internship Project, previously known as the Lewis' Educational and Research Collaborative Internship Project (LERCIP), for the support provided to Anton Salem and Lily Kuentz and for their assistance with 3D-printing.

Conflicts of Interest: The authors declare no conflicts of interest. The funders had no role in the design of the study; in the collection, analyses, or interpretation of data; in the writing of the manuscript; or in the decision to publish the results.

Abbreviations

The following abbreviations are used in this document:

ABS	Acrylonitrile butadiene styrene
AM	Additive manufacturing
CART	Classification and regression decision tree
CNT	Carbon nanotube
CF	5 weight-percent chopped carbon fiber
CV	Cross-validation
DIC	Digital image correlation
FDM	Fused deposition modeling
FFF	Fused filament fabrication
KNN	K-nearest neighbors

LDA	Linear discriminant analysis
LH	Layer height
LR	Linear regression
MAE	Mean absolute error
MAPE	Mean absolute percentage error
NB	Naive Bayes
PLA	Polylactic acid
RF	Random forest
RMSE	Root mean square error
SEM	Scanning electron microscope
SVM	Support vector machine

References

1. Salifu, S.; Desai, D.; Ogunbiyi, O.; Mwale, K. Recent development in the additive manufacturing of polymer-based composites for automotive structures—A review. *Int. J. Adv. Manuf. Technol.* **2022**, *119*, 6877–6891. [CrossRef]
2. Careri, F.; Khan, R.H.; Todd, C.; Attallah, M.M. Additive manufacturing of heat exchangers in aerospace applications: A review. *Appl. Therm. Eng.* **2023**, *235* , 121387. [CrossRef]
3. Sun, C.; Wang, Y.; McMurtrey, M.D.; Jerred, N.D.; Liou, F.; Li, J. Additive manufacturing for energy: A review. *Appl. Energy* **2021**, *282*, 116041. [CrossRef]
4. Kumar, R.; Kumar, M.; Chohan, J.S. The role of additive manufacturing for biomedical applications: A critical review. *J. Manuf. Process.* **2021**, *64*, 828–850. [CrossRef]
5. Dey, A.; Roan Eagle, I.N.; Yodo, N. A review on filament materials for fused filament fabrication. *J. Manuf. Mater. Process.* **2021**, *5*, 69. [CrossRef]
6. Nötzel, D.; Eickhoff, R.; Pfeifer, C.; Hanemann, T. Printing of zirconia parts via fused filament fabrication. *Materials* **2021**, *14*, 5467. [CrossRef]
7. Vakharia, V.S.; Kuentz, L.; Salem, A.; Halbig, M.C.; Salem, J.A.; Singh, M. Additive manufacturing and characterization of metal particulate reinforced Polylactic Acid (PLA) polymer composites. *Polymers* **2021**, *13*, 3545. [CrossRef]
8. Diouf-Lewis, A.; Farahani, R.D.; Iervolino, F.; Pierre, J.; Abderrafai, Y.; Lévesque, M.; Piccirelli, N.; Therriault, D. Design and characterization of carbon fiber-reinforced PEEK/PEI blends for Fused Filament Fabrication additive manufacturing. *Mater. Today Commun.* **2022**, *31*, 103445. [CrossRef]
9. Fico, D.; Rizzo, D.; Casciaro, R.; Esposito Corcione, C. A review of polymer-based materials for fused filament fabrication (FFF): Focus on sustainability and recycled materials. *Polymers* **2022**, *14*, 465. [CrossRef]
10. Cersoli, T.; Yelamanchi, B.; MacDonald, E.; Carrillo, J.G.; Cortes, P. 3D printing of a continuous fiber-reinforced composite based on a coaxial Kevlar/PLA filament. *Compos. Adv. Mater.* **2021**, *30*, 26349833211000058. [CrossRef]
11. Baca, D.; Ahmad, R. The impact on the mechanical properties of multi-material polymers fabricated with a single mixing nozzle and multi-nozzle systems via fused deposition modeling. *Int. J. Adv. Manuf. Technol.* **2020**, *106*, 4509–4520. [CrossRef]
12. Romero, P.E.; Arribas-Barrios, J.; Rodriguez-Alabanda, O.; González-Merino, R.; Guerrero-Vaca, G. Manufacture of polyurethane foam parts for automotive industry using FDM 3D printed molds. *CIRP J. Manuf. Sci. Technol.* **2021**, *32*, 396–404. [CrossRef]
13. Ngo, C.V.; Nguyen, Q.A.; Le, N.; Le, N.L.L.; Nguyen, Q.H. Fabrication of Airy, Lightweight Polymer Below-Elbow Cast by a Combination of 3D Scanning and 3D Printing. In Proceedings of the Advances in Asian Mechanism and Machine Science: Proceedings of IFToMM Asian MMS, Hanoi, Vietnam, 15–18 December 2021 ; Springer: Berlin/Heidelberg, Germany, 2022; pp. 628–637.
14. Cabreira, V.; Santana, R.M.C. Effect of infill pattern in Fused Filament Fabrication (FFF) 3D Printing on materials performance. *Revista Matéria* **2020**, *25* . [CrossRef]
15. Mansour, M.T.; Tsongas, K.; Tzetzis, D. 3D printed hierarchical honeycombs with carbon fiber and carbon nanotube reinforced acrylonitrile butadiene styrene. *J. Compos. Sci.* **2021**, *5*, 62. [CrossRef]
16. Zhao, P.; Rao, C.; Gu, F.; Sharmin, N.; Fu, J. Close-looped recycling of polylactic acid used in 3D printing: An experimental investigation and life cycle assessment. *J. Clean. Prod.* **2018**, *197*, 1046–1055. [CrossRef]
17. Vidakis, N.; Petousis, M.; Maniadi, A.; Koudoumas, E.; Vairis, A.; Kechagias, J. Sustainable additive manufacturing: Mechanical response of acrylonitrile-butadiene-styrene over multiple recycling processes. *Sustainability* **2020**, *12*, 3568. [CrossRef]
18. Charles, A.; Bassan, P.M.; Mueller, T.; Elkaseer, A.; Scholz, S.G. On the assessment of thermo-mechanical degradability of multi-recycled ABS polymer for 3D printing applications. In Proceedings of the International Conference on Sustainable Design and Manufacturing, Budapest, Hungary, 4–5 July 2019; Springer: Berlin/Heidelberg, Germany, 2019; pp. 363–373.
19. Abdellah, M.Y.; Fathi, H.I.; Abdelhaleem, A.M.; Dewidar, M. Mechanical properties and wear behavior of a novel composite of acrylonitrile–butadiene–styrene strengthened by short basalt fiber. *J. Compos. Sci.* **2018**, *2*, 34. [CrossRef]
20. Sonsalla, T.; Moore, A.L.; Meng, W.; Radadia, A.D.; Weiss, L. 3-D printer settings effects on the thermal conductivity of acrylonitrile butadiene styrene (ABS). *Polym. Test.* **2018**, *70*, 389–395. [CrossRef]
21. Abbott, A.; Tandon, G.; Bradford, R.; Koerner, H.; Baur, J. Process-structure-property effects on ABS bond strength in fused filament fabrication. *Addit. Manuf.* **2018**, *19*, 29–38. [CrossRef]

22. Kaur, G.; Singari, R.M.; Kumar, H. A review of fused filament fabrication (FFF): Process parameters and their impact on the tribological behavior of polymers (ABS). *Mater. Today: Proc.* **2022**, *51*, 854–860. [CrossRef]
23. Ziadia, A.; Habibi, M.; Kelouwani, S. Machine Learning Study of the Effect of Process Parameters on Tensile Strength of FFF PLA and PLA-CF. *Eng* **2023**, *4*, 2741–2763. [CrossRef]
24. Ranaiefar, M.; Honarmandi, P.; Ye, J.; Zhang, C.; Xue, L.; Elwany, A.; Karaman, I.; Schwalbach, E.J.; Arróyave, R. Uncertainty quantification and propagation across a multi-model computational framework for the tailored design of additively manufactured shape memory alloys. *Addit. Manuf.* **2023**, *68*, 103506. [CrossRef]
25. Sarkon, G.K.; Safaei, B.; Kenevisi, M.S.; Arman, S.; Zeeshan, Q. State-of-the-art review of machine learning applications in additive manufacturing; from design to manufacturing and property control. *Arch. Comput. Methods Eng.* **2022**, *29*, 5663–5721. [CrossRef]
26. Nasiri, S.; Khosravani, M.R. Machine learning in predicting mechanical behavior of additively manufactured parts. *J. Mater. Res. Technol.* **2021**, *14*, 1137–1153. [CrossRef]
27. Veeman, D.; Sudharsan, S.; Surendhar, G.; Shanmugam, R.; Guo, L. Machine learning model for predicting the hardness of additively manufactured acrylonitrile butadiene styrene. *Mater. Today Commun.* **2023**, *35*, 106147. [CrossRef]
28. Brackett, J.; Cauthen, D.; Condon, J.; Smith, T.; Gallego, N., Kunc, V.; Duty, C. The impact of infill percentage and layer height in small-scale material extrusion on porosity and tensile properties. *Addit. Manuf.* **2022**, *58*, 103063. [CrossRef]
29. Fischer, D.; Eßbach, C.; Schönherr, R.; Dietrich, D.; Nickel, D. Improving inner structure and properties of additive manufactured amorphous plastic parts: The effects of extrusion nozzle diameter and layer height. *Addit. Manuf.* **2022**, *51*, 102596. [CrossRef]
30. Vakharia, V.S.; Singh, M.; Salem, A.; Halbig, M.C.; Salem, J.A. Effect of Reinforcements and 3-D Printing Parameters on the Microstructure and Mechanical Properties of Acrylonitrile Butadiene Styrene (ABS) Polymer Composites. *Polymers* **2022**, *14*, 2105. [CrossRef]
31. Sun, Q.; Rizvi, G.; Bellehumeur, C.; Gu, P. Effect of processing conditions on the bonding quality of FDM polymer filaments. *Rapid Prototyp. J.* **2008**, *14*, 72–80. [CrossRef]
32. Vanaei, H.R.; Raissi, K.; Deligant, M.; Shirinbayan, M.; Fitoussi, J.; Khelladi, S.; Tcharkhtchi, A. Toward the understanding of temperature effect on bonding strength, dimensions and geometry of 3D-printed parts. *J. Mater. Sci.* **2020**, *55*, 14677–14689. [CrossRef]
33. Tekinalp, H.L.; Kunc, V.; Velez-Garcia, G.M.; Duty, C.E.; Love, L.J.; Naskar, A.K.; Blue, C.A.; Ozcan, S. Highly oriented carbon fiber–polymer composites via additive manufacturing. *Compos. Sci. Technol.* **2014**, *105*, 144–150. [CrossRef]
34. Thaler, D.; Aliheidari, N.; Ameli, A. Mechanical, electrical, and piezoresistivity behaviors of additively manufactured acrylonitrile butadiene styrene/carbon nanotube nanocomposites. *Smart Mater. Struct.* **2019**, *28*, 084004. [CrossRef]
35. Chicco, D.; Warrens, M.J.; Jurman, G. The coefficient of determination R-squared is more informative than SMAPE, MAE, MAPE, MSE and RMSE in regression analysis evaluation. *PeerJ Comput. Sci.* **2021**, *7*, e623. [CrossRef]
36. *ASTM C365-22*; Standard Test Method for Flatwise Compressive Properties of Sandwich Cores. ASTM International: West Conshohocken, PA, USA, 2022.
37. Pedregosa, F.; Varoquaux, G.; Gramfort, A.; Michel, V.; Thirion, B.; Grisel, O.; Blondel, M.; Prettenhofer, P.; Weiss, R.; Dubourg, V.; et al. Scikit-learn: Machine Learning in Python. *J. Mach. Learn. Res.* **2011**, *12*, 2825–2830.

molecules

MDPI

Article

Control of Pore Sizes in Epoxy Monoliths and Applications as Sheet-Type Adhesives in Combination with Conventional Epoxy and Acrylic Adhesives

Yoshiyuki Kamo [1,2,*] and Akikazu Matsumoto [2,3]

[1] Advanced Technology R&D Center, Mitsubishi Electric Corporation, 8-1-1, Tsukaguchi-Honmachi, Amagasaki, Hyogo 661-8661, Japan
[2] Department of Applied Chemistry, Graduate School of Engineering, Osaka Prefecture University, 1-1 Gakuen-cho, Naka-ku, Sakai 599-8531, Osaka, Japan
[3] Department of Applied Chemistry, Graduate School of Engineering, Osaka Metropolitan University, 1-1 Gakuen-cho, Naka-ku, Sakai 599-8531, Osaka, Japan
* Correspondence: kamo.yoshiyuki@ct.mitsubishielectric.co.jp

Abstract: Materials with monolithic structures, such as epoxy monoliths, are used for a variety of applications, such as for column fillers in gas chromatography and HPLC, for separators in lithium-ion batteries, and for precursor polymers for monolith adhesion. In this study, we investigated the fabrication of epoxy monoliths using 1,3-bis(*N*,*N*-diglycidylaminomethyl)cyclohexane (TETRAD-C) as the tetrafunctional epoxy and 4,4′-methylenebis(cyclohexylamine) (BACM) as the amine curing agent to control pore diameters using polyethylene glycols (PEGs) of differing molecular weights as the porogenic agents. We fabricated an epoxy monolith with micron-order pores and high strength levels, and which is suitable for the precursors of composite materials in cases where smaller PEGs are used. We discussed the effects of the porous structures of monoliths on their physical properties, such as tensile strength, elongation, elastic modulus, and glass transition temperatures. For example, epoxy monoliths prepared in the presence of PEGs exhibited an elastic modulus less than 1 GPa at room temperature and Tg values of 175–187 °C, while the epoxy bulk thermoset produced without any porogenic solvent showed a high elastic modulus as 1.8 GPa, which was maintained at high temperatures, and a high Tg of 223 °C. In addition, the unique adhesion characteristics of epoxy monolith sheets are revealed as a result of the combinations made with commercial epoxy and acrylic adhesives. Epoxy monoliths that are combined with conventional adhesives can function as sheet-type adhesives purposed with avoiding problems when only liquid-type adhesives are used.

Keywords: monolith; epoxy resin; porous structure; mechanical property; solubility parameter; adhesive property

1. Introduction

A monolith is a highly porous material with a co-continuous structure that consists of three-dimensionally continuous through-holes and a network frame [1–4]. Materials with a monolithic structure are attracting attention as separation media used in chromatography, immobilized catalysts, adsorbents, and sensors [5–9]. In addition, polymer monoliths can be easily modified by functional groups or the usage of filler additives, although they are inferior to inorganic monoliths in terms of heat resistance and mechanical strength. The development and application of polymer monoliths using epoxy, acrylics, and styrene are thus underway [10–13].

Generally, organic monoliths are formed through microphase separation processes classified into the categorizations of polymerization-induced phase separation (PIPS) and nonsolvent-induced phase separation. Epoxy monoliths are one of the co-continuous porous materials formed by PIPS, and the phase separation is caused by spinodal decomposition during the heat curing of the epoxy resin and crosslinker in the presence of porogenic solvents [3,14–19]. The phase-separated structure of epoxy resin is fixed as a result of crosslinking reactions. Therefore, the rate of the crosslinking reaction greatly influences the shape and pore size of the monolith [20,21]. Efforts are being made to take advantage of continuous pores and to arrange catalysts on monoliths, imitating monoliths so as to increase their activity levels [22,23].

In recent years, epoxy monoliths have been used for the construction of composite materials. Shimizu et al. improved the tribological properties of monolithic polymer films by tuning a fine pore structure for soft and resilient porous polymer lubricant systems [24]. Tominaga et al. reported on the mechanical strength and filling conditions using X-ray CT when a second component was placed inside the porous materials [25,26]. Furthermore, Sugimoto et al. provided a report on a monolith bonding method with an advantage of a continuous porous structure [27,28]. The anchor effect is used to increase adhesive strength by creating an epoxy monolith on the adherend when bonding difficult-to-adhere objects or dissimilar materials. In these applications, monolith pores are filled with liquid or viscous materials, and the incorporated materials are then solidified or cured. Thus, controlling the pore sizes and three-dimensional structures of monoliths is important. It is also necessary to consider increasing the tensile strength of the monolith and improving its heat resistance so that it can be deployed for usage in various environments. A tetrafunctional epoxy was used to increase the crosslinking density and improve upon both the heat resistance and toughness compared to a monolith using a difunctional epoxy [29,30]. Meanwhile, previous reports have shown that monoliths using tetrafunctional compounds have small pore diameters and are not suitable for applications filling any polymer material in the monolith [31,32]. It has already been reported that it is possible to form an appropriate pore size by changing the fabrication conditions of monoliths [19,26,33].

An adhesion process often requires spacers, such as polymer or glass beads, which are used for the controlling of adhesive thickness. In addition, the viscosity or fluidity of adhesives should be finely justified since conventional adhesives are supplied in a liquid form [34]. An epoxy monolith sheet with a unique porous structure is expected to function as the gap adjuster for adhesives by the combination with commercial epoxy and acrylic adhesives. Studies already show that it is possible to form an appropriate pore size by changing the fabrication conditions of monoliths. It would be convenient if the application process could be completed by simply putting things together, such as double-sided pressure-sensitive adhesive (PSA) tape [35–37]. Here, we described the results for the controlling of the porous structure and the physical properties of the epoxy monolith for constructing a new kind of sheet-type adhesive resulting from combinations with liquid-type adhesives. It was also expected that a new type of bonding using epoxy and acrylic adhesives in combination with epoxy monolith sheets would be achieved, because there are great operational advantages for using a liquid adhesive as the material impregnated in the epoxy monolith sheet. In the case of the single-use of liquid adhesives, it is actually necessary to regulate the application procedures and carefully control the thickness of the adhesives, but these troublesome and time-consuming problems will be avoided with the new adhesion method using the monolith sheets. The validity of monolith sheets as porous molds deployed for adhesion using epoxy and acrylics as liquid adhesives has thus been demonstrated.

2. Experimental Method

2.1. Fabrication of Epoxy Monoliths

The chemical structures of the reagents used in this study are shown in Figure 1. 1,3-Bis(*N*,*N*-diglycidylaminomethyl)cyclohexane (TETRAD-C) manufactured by Mitsubishi Gas Chemical Corporation, Ltd., Tokyo, Japan, was used as the base epoxy resin. Hydrogenated bisphenol-A type epoxy resin (HBPA) manufactured by New Japan Chemical Corporation, Ltd., Osaka, Japan, was used as the additive. 4,4′-Methylenebis(cyclohexylamine) (BACM) was purchased from Tokyo Chemical Industry Corporation, Ltd., Tokyo, Japan, and was used as the crosslinking agent. Ethylene glycol (PEG50), diethylene glycol

(PEG100), triethylene glycol (PEG150), and polyethylene glycol (PEG, Mn = 200), which were used as the porogenic solvents, were purchased from Tokyo Chemical Industry Corporation, Ltd., Tokyo, Japan. Table 1 shows the solubility parameters of the chemicals [38]. All reagents and solvents were used without further purification, unless stated otherwise. Polyvinyl alcohol (PVA, Mn = 500) was purchased from Fujifilm Wako Pure Chemical Corporation, Ltd., Osaka, Japan, and was used as the sacrificial layer placed on the glass plate for easily peeling off the monolith sheet after the fabrication process.

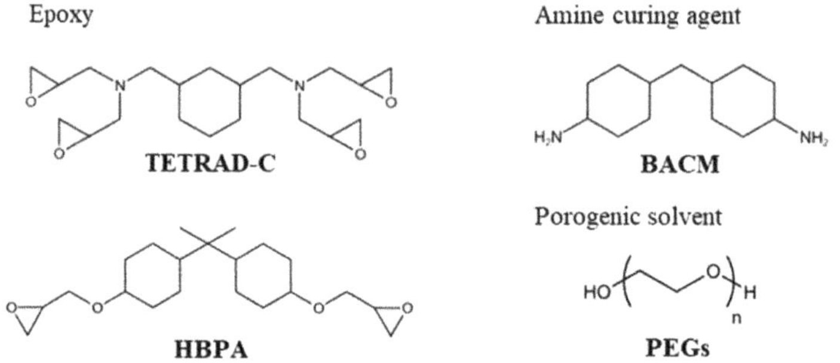

Figure 1. Chemical structures of the materials used in this study.

Table 1. Solubility parameters of the materials used for monolith fabrication.

Category	Materials	Solubility Parameter [a] (−)
Epoxy	TETRAD-C	10.1
	HBPA	9.2
Amine curing agent	BACM	9.7
Porogenic agent	PEG50	14.1
	PEG100	12.6
	PEG150	12.1
	PEG200	11.6

[a] Estimated using the Fedors method [38].

Figure S1, which is contained within the Supplementary Materials, shows the procedure for preparing the epoxy monolith sheets. The monolith sheet was prepared as follows. A solution of 1 wt% PVA in ion-exchanged water was applied onto the glass and heated at 110 °C for 1 h to create a sacrificial layer of PVA on the glass surface. TETRAD-C, HBPA, BACM, and a porogen (70 wt% for PEGs) were mixed well for 10 min using a planetary centrifugal mixer (ARE-250; Thinky Corporation, Ltd., Tokyo, Japan) at a ratio of 2[NH2]/[epoxy] (γ value) = 1.0. The mixture of the chemicals was spread onto the PVA-coated glass plate and sandwiched with another glass plate while maintaining the desired thickness. The reaction mixture was cured at 130 °C for 30 min using an oven. After curing, the samples were washed with ion-exchanged water using ultrasonics for 15 min to remove the porogenic solvent. They were then stored in water overnight. The samples were dried at 40 °C for 12 h in a vacuum oven (LCV-233; Espec Corporation, Ltd., Osaka, Japan). The thickness of the epoxy monolith was determined using a digital thickness gauge (547–401; Mitutoyo Corporation, Ltd., Kanagawa, Japan). Epoxy monolith sheets with thicknesses of 200–250 μm and 80–120 μm were used for the measurement of physical properties and for the adhesive tests, respectively.

2.2. Measurements

The differential scanning calorimetry (DSC) measurements were carried out using a DSC-X7000 (Hitachi High-Tech Corporation, Ltd., Tokyo, Japan) with a heating rate of

10 °C/min to determine the curing temperature in a sealed pan. The epoxy monolith sheet was carefully cut using a sharp blade after cooling it with liquid nitrogen. The cross-section of the epoxy monolith was observed using a TM4000 plus (Hitachi High-Tech Corporation Ltd., Tokyo, Japan) with an acceleration voltage of 15 kV and with Pt vapor deposition for pretreatment. Monolith sheets prepared under different conditions were cut into strips of 10 mm × 40 mm using a stainless-steel cutter. The tensile test was carried out at room temperature with a tensile rate of 1.0 mm/min using an Autograph AGS-10KN X (Shimadzu Corporation, Ltd., Kyoto, JAPAN). The dynamic mechanical analysis (DMA) measurement was carried out using an EXSTER6000 (Hitachi High-Tech Corporation, Ltd., Tokyo, Japan) in an atmospheric environment to examine the glass transition temperatures and storage modulus of the monolith sheets. This test was conducted in a tensile mode, at a heating rate of 2 °C/min and a frequency of 1 Hz.

2.3. Adhesion Test

Steel Plate Cold Commercial (SPCC) produced by Engineering Test Service Corporation, Ltd., Hyogo, Japan, was used as the adherend for the measurement of the lap-shear adhesion strength. The surface of the SPCC was wiped with acetone and modified with deep ultraviolet light using a low-pressure mercury lamp (11–13 mW/cm^2) for 5 min before the preparation of the test pieces. The specimens were prepared according to the ISO 4587 method [39]. The adhesive area used was 25.0 mm × 12.5 mm. The adhesive thickness was adjusted at ca. 0.1 mm using glass beads or monolith sheets. Figure 2 shows the procedure for adhesive structures consisting of commercial liquid-type adhesives and the epoxy monolith used as the porous sheet. The commercial adhesives used in this study were either heat curing-type epoxy adhesives or acrylic adhesives. Epoxy adhesive (CLS-1194) manufactured by ADEKA Corporation, Ltd., Tokyo, Japan, was used, and an acrylic adhesive (AS-6704) was purchased from Asec Corporation, Ltd., Kanagawa, Japan. The adhesives were cured under the conditions recommended by each manufacturer. CLS-1194 was cured at 180 °C for 30 min and AS-6704 was cured at 120 °C for 30 min. The tensile test was carried out at room temperature at a rate of 10 mm/min.

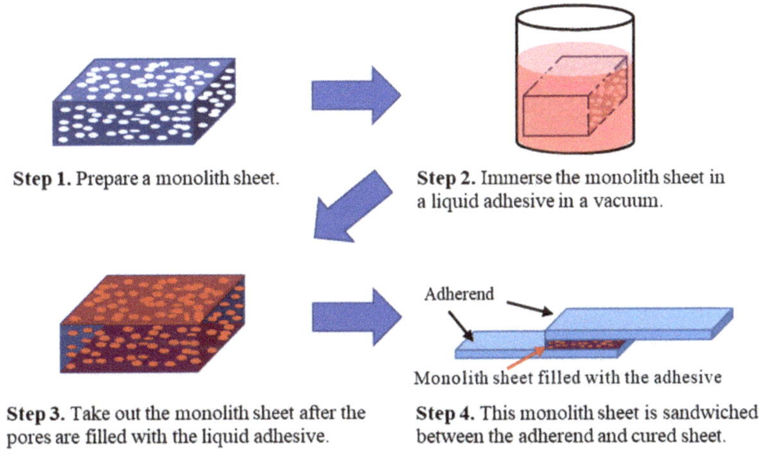

Figure 2. The procedure for a sheet-type adhesive structure using an epoxy monolith sheet in combination with liquid-type conventional adhesives.

3. Results and Discussion

3.1. DSC Measurement

Table 2 shows the results of the DSC measurement for an epoxy curing system consisting of TETRAD-C, BACM, and PEGs in the absence and presence of HBPA used as an additional epoxy. The onset and peak temperatures and the enthalpy values of the exothermic reactions observed during thermal curing increased as the molecular weight of glycol increased. This was interpreted by the acceleration of epoxy curing in a polar medium or circumstance [40–42]. For example, strong hydrogen-bonding compounds (such as alcohols) can stabilize transitions for the curing reactions and accelerate the nucleophilic attack of amines. The acceleration of curing reactions by hydroxy groups has been theoretically and experimentally demonstrated in the literature [43–48]. These results support the idea that hydrogen bonding plays a more significant role in assisting the curing reaction between epoxy and amine when glycols with a low molecular weight are used.

Table 2. Peak temperature and enthalpy for the thermal curing of the TETRAD-C/HBPA/BACM/PEG system determined by DSC.

HBPA Content (mol%)	PEG	Onset Temperature (°C)	Peak Temperature (°C)
0	PEG50	69	112
0	PEG100	74	120
0	PEG150	79	133
0	PEG200	86	139
5	PEG200	72	138
10	PEG200	72	138
15	PEG200	71	138
20	PEG200	70	136
30	PEG200	68	136

As the molecular weight of PEGs becomes lower, the number of hydroxy groups that can come into contact with an epoxy group increases. The reaction gradually started just after the reagents were mixed at room temperature when PEG50 was used. As a result, phase separation occurs more frequently because the SP values of low-molecular-weight PEGs are larger than those of the epoxies and amines, as shown in Table 1.

On the other hand, the enthalpy value decreased without changing the reaction temperature when HBPA was added. This was because the number of epoxy groups included in the curing system decreased as the HBPA content of the bifunctional epoxy increased. It is likely that the reactivities of the glycidyl groups of TETRAD-C and HBPA are similar to each other, resulting in the constant onset and peak temperatures of the DSC curves.

3.2. SEM Observations

Figure 3 shows the cross-sectional SEM images of the epoxy monoliths, which were fabricated using different PEGs with curing at 130 °C for 30 min. Based on the cross-sectional images, it is clear that the pore diameters increase as the molecular weights of the PEGs decrease. PEGs with lower molecular weights have solubility parameter values that are larger than those for TETRAD-C and BACM, as shown in Table 1. The pore and monolith skeleton sizes change depending on the rate of Ostwald ripening during the phase separation of the epoxy monolith curing system in the presence of PEGs as porogenic solvents [19,33]. When there is a high level of miscibility of the epoxy or the amine in relation to the porogenic solvents, phase separation occurs more slowly and the domain size of the phase-separated layers decreases. The phase-separated structure is fixed at any stage of the curing reaction due to the formation of the crosslinked network structure of the epoxy. When PEG50 is used, the pores are too large and the epoxy monolith is difficult to maintain in the shape of the sheet.

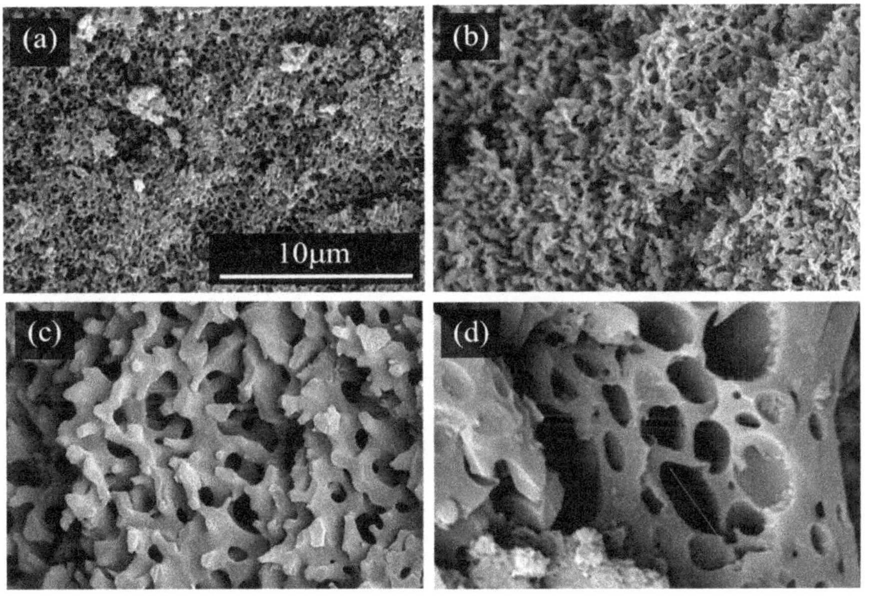

Figure 3. Cross-sectional SEM images of the epoxy monolith fabricated from the TETRAD-C/BACM/PEG system using different PEGs after curing at 130 °C for 30 min: (**a**) PEG200, (**b**) PEG150, (**c**) PEG100, and (**d**) PEG50.

We also observed the changes in pore sizes depending on the curing temperature. Figure 4 shows the cross-sectional SEM images of the epoxy monolith obtained using PEG200 at various temperatures. The lower the curing temperature, the larger the pores of the monolith. Similar observations of the temperature effects were reported in the literature [24,32,34]. The magnitude of the change occurring as a result of different curing temperatures is moderate when compared with the drastic changes in pore sizes controlled by the molecular weights of the PEGs.

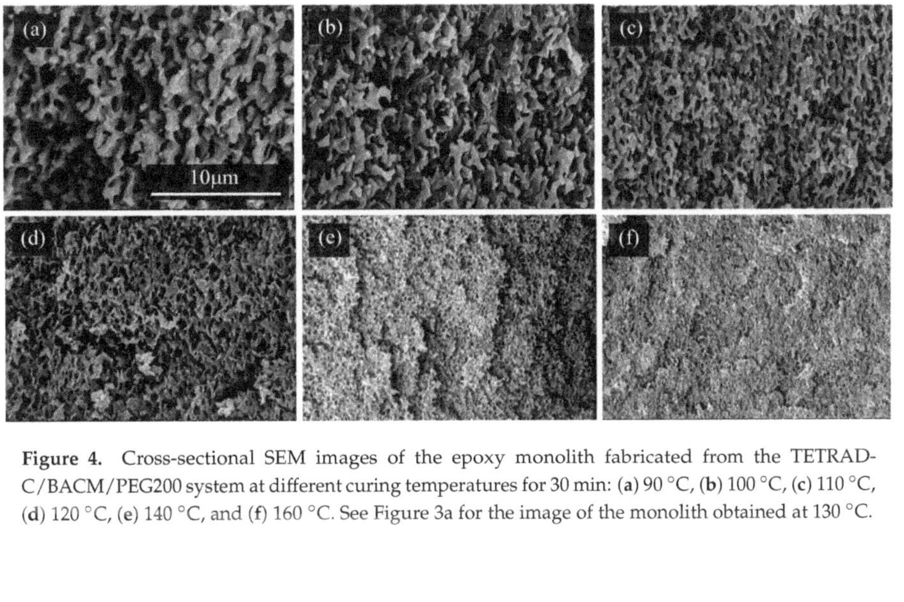

Figure 4. Cross-sectional SEM images of the epoxy monolith fabricated from the TETRAD-C/BACM/PEG200 system at different curing temperatures for 30 min: (**a**) 90 °C, (**b**) 100 °C, (**c**) 110 °C, (**d**) 120 °C, (**e**) 140 °C, and (**f**) 160 °C. See Figure 3a for the image of the monolith obtained at 130 °C.

Figure 5 shows the additional effect of HBPA on the TETRAD-C/BACM/PEG200 curing system. The pore diameter of the monoliths increased according to the content of HBPA present in the system. When 30 mol% of HBPA was added, aggregates of particles were produced and the crumbly monolith sheet could not stand by itself. In this case, the phase separation mode changed from the spinodal decomposition mechanism to the nucleation and growth mechanism [49,50] when a large amount of HBPA was added. The drastic change in the pore size is accounted for by the retardation of the fixing of the phase-separated structure by the addition of HBPA as the difunctional epoxy to the curing system of TETRAD-C as the tetrafunctional epoxy.

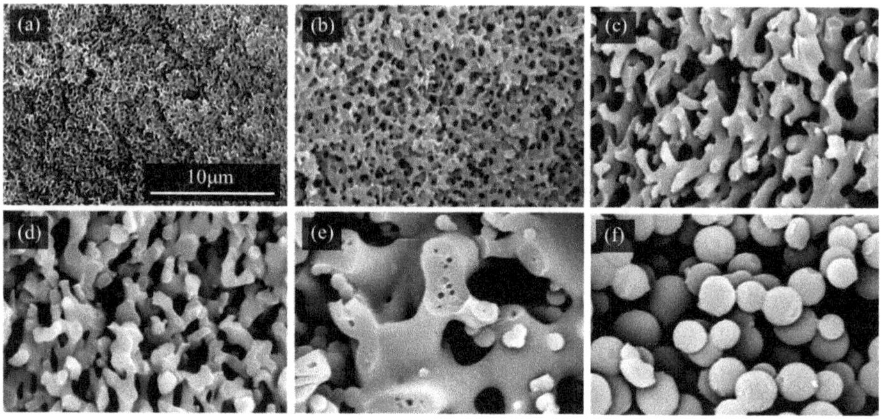

Figure 5. Cross-sectional SEM images of the epoxy monolith fabricated from the TETRAD-C/HBPA/BACM/PEG200 system cured at 130 °C for 30 min: (**a**) HBPA 0 mol% (**b**) 5 mol%, (**c**) 10 mol%, (**d**) 15 mol%, (**e**) 20 mol%, and (**f**) 30 mol%.

3.3. Mechanical and Physical Properties of the Monolith Sheets

Figure 6a shows the tensile strength and elongation at the point of breakage of the monolith sheets (thickness: 200–250 μm, N = 6). The figure includes no data for PEG50 because the epoxy monolith with the largest phase separation structure could not maintain its whole sample shape. Both the strength and elongation values for the monolith sheets become higher when PEGs with lower molecular weights are used. As shown in Figure 3, the smaller PEGs provide epoxy monoliths with larger pores. It is interesting that the presence of coarse pores exhibits favorable mechanical properties, such as high strength and high elongation. Figure 6b shows the effect of the addition of HBPA on the tensile strength and elongation. As shown in Figure 5, the pore diameter increases according to the amount of HBPA added to the curing system. The tensile strength is constant, regardless of the HBPA amount. With the addition of 10 mol% of HBPA, the elongation drastically increased. The test pieces that included more than 20 mol% of HBPA were too brittle, and thus difficult to use for determining the mechanical properties. In the systems that were controlled by the molecular weights of the PEGs and the amounts of HBPA added, the relationship between the pore sizes and the tensile strength was inconsistent. This is due to that the network density of the cured epoxy also decreases with the addition of HBPA, resulting in a decrease in the tensile strength of the monoliths. In contrast, the elongation values increase when the pore size is larger for both series, a fact that demonstrates a positive agreement with the results previously reported [26].

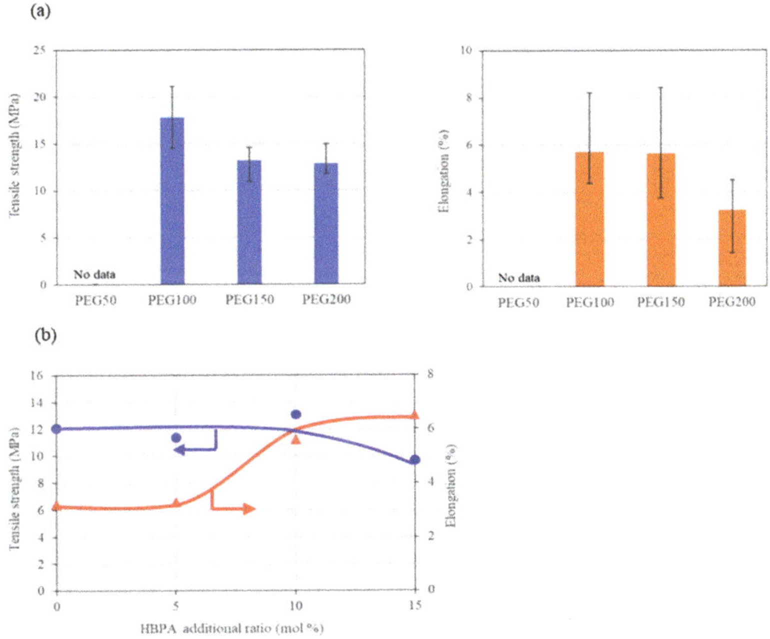

Figure 6. Tensile strength and elongation at breakage for the tensile tests of the epoxy monolith fabricated from (**a**) the TETRAD-C/BACM/PEG system and (**b**) the TETRAD-C/HBPA/BACM/PEG200 system. (•) Tensile strength and (▲) elongation.

Figure 7 shows the results for the DMA measurement of the epoxy monoliths prepared under various additive conditions. The structural, mechanical, and physical properties of the monoliths and the bulk thermoset are summarized in Table 3. The epoxy bulk thermoset obtained using TETRAD-C and BACM without any porogenic solvent was originally a tough resin with a high elastic modulus (1.8 GPa), and which maintained evenness at high temperatures, since the Tg value of the bulk thermoset reached a value as high as 223 °C, as shown in Figure 7a and Table 3. In contrast, the epoxy monoliths prepared in the presence of PEGs exhibited an elastic modulus less than 1 GPa at room temperature, and the Tg values were as low as 175–187 °C. The Tg values slightly increased with an increase in the molecular weights of the PEGs. The softening characteristics of the epoxy monolith arise due to the plasticizing effect caused by the PEGs incorporated into the epoxy monolith skeletons during the fabrication process of the monolith, as was pointed out in our previous report [33]. Figure 7b shows the effect of the addition of HBPA on the DMA properties. The properties of the cured epoxy containing HBPA reflected the softening effects caused by the presence of bifunctional epoxy units, resulting in a slight decrease in the Tg value and an increase in the elongation at breakage. Thus, it has been demonstrated that it is possible to fabricate epoxy materials with the desired mechanical and thermal properties by the introduction of a monolith structure into hard epoxy resin.

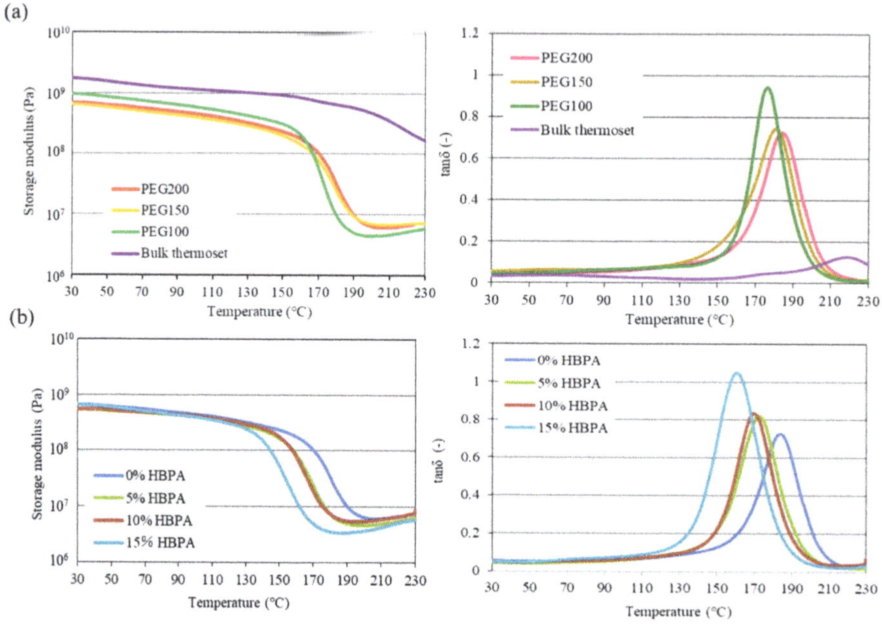

Figure 7. DMA curves for the epoxy monoliths and the bulk thermoset fabricated under various additive conditions. (**a**) Epoxy monoliths prepared in the presence of various PEGs and bulk thermoset prepared in the absence of PEG, and (**b**) epoxy monoliths prepared with the addition of various amounts of HBPA.

Table 3. Structural, mechanical, and physical properties of monolith sheets prepared in the presence or absence of PEGs and HBPA.

HBPA (mol%)	PEG	Mean Pore Diameter (μm)	Tensile Test		DMA	
			Strength (MPa)	Elongation (%)	T_g (°C)	Storage Modulus at 25 °C (GPa)
0	PEG50	3.80	-	-	-	-
0	PEG100	1.91	17.9	5.8	175	1.01
0	PEG150	0.63	13.3	5.6	177	0.66
0	PEG200	0.28	12.5	3.2	187	0.68
5	PEG200	0.87	11.4	3.3	172	0.56
10	PEG200	1.21	13.2	5.6	170	0.56
15	PEG200	3.15	9.5	6.5	161	0.59
0	-	-	-	-	223	1.81

3.4. Adhesive Property

Figure 8 shows the results for the adhesion test of SPCC using the monolith sheets fabricated in this study in combination with commercial adhesives. The test pieces used for the lap-shear tensile test were prepared using SPCC plates as adherends and the commercially available and most typical epoxy or acrylic adhesives in the presence of an epoxy monolith sheet (with a thickness of 80–120 μm) used as the adhesive layer (see Figure 2). As a result, the installation of an epoxy monolith sheet onto an adhesive layer resulted in a decrease in the lap-shear strength in both cases using the epoxy and acrylic adhesives (CLS1194 and AS6704). The adhesive strength of the epoxy adhesive varied depending on the kind of monolith sheets with different pore sizes. On the other hand, the constant strength value was obtained using the acrylic adhesive. This difference occurred due to the viscosity of the adhesive. The viscosities of CLS1194 and AS6704 are 4.0 Pa-s and 0.10 Pa-s, respectively. Thus, the penetration of the epoxy adhesive with high viscosity into the small pores of the monolith is difficult in cases involving the usage of PEG150 and PEG200. Moreover, the deterioration of adhesive strength is attributable the fact that the monolith layer was more likely to break in comparison with the higher storage modulus values of the adhesive.

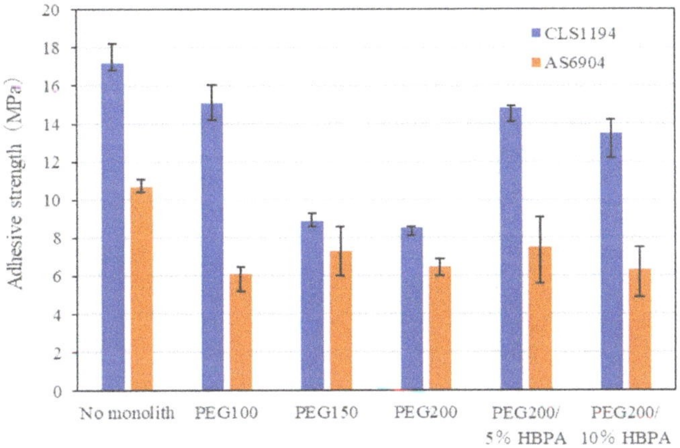

Figure 8. Comparison of adhesive strength for the lap-shear tensile test of SPCC using composite adhesive materials consisting of epoxy monolith sheets and commercial epoxy and acrylic adhesives.

Thus, we have succeeded in the demonstration of a new type of bonding using epoxy and acrylic adhesives in combination with epoxy monolith sheets. Because there are great operational advantages for using a liquid adhesive as the state impregnated in the epoxy monolith sheet, the future challenge is to achieve greater adhesive strength by optimizing the physical properties of the monolith sheet. Thus, in the future, monolith adhesive structures will be applied, regardless of the type of adhesive.

4. Conclusions

We fabricated epoxy monoliths using TETRAD-C as the tetrafunctional epoxy and BACM as the amine curing agent in the presence of PEGs with different molecular weights used as porogenic solvents. It has been demonstrated that we can control the pore diameter and mechanical properties by utilizing the differences of the molecular structures and solubility parameters of porogenic solvents. This also affects the mechanical and physical properties of epoxy monoliths, such as the tensile strength, elongation, elastic modulus, and Tg values. For example, epoxy monoliths prepared in the presence of PEGs exhibited an elastic modulus less than 1 GPa at room temperature and the Tg values of 175–187 °C, while the epoxy bulk thermoset produced without any porogenic solvent showed a high elastic modulus of 1.8 GPa, which is maintained at high temperatures, and a high Tg of 223 °C. It has been revealed that adhesive structure consisting of an epoxy monolith sheet in combination with commercial epoxy and acrylic adhesives functions as a sheet-type adhesive. The test pieces for the lap-shear tensile test were prepared using SPCC plates as adherends and commercially available epoxy and acrylic adhesives in the presence of an epoxy monolith sheet. The installation of an epoxy monolith sheet onto an adhesive layer resulted in a slight decrease in the lap-shear strength, but a constant strength value was obtained using an acrylic adhesive. The deterioration of adhesive strength is attributable the fact that the monolith layer was more likely to break in comparison with the higher storage modulus values of the adhesive. In the case of the single-use of liquid adhesives, it is necessary to regulate application procedures and carefully control the thickness of the adhesives. In contrast, these troublesome and time-consuming problems could be avoided with the new adhesion method using the monolith sheets.

Supplementary Materials: The following supporting information can be downloaded at: https://www.mdpi.com/article/10.3390/molecules29092059/s1, Figure S1: The procedure for preparing an epoxy monolith sheet; Figure S2: DSC traces for thermal curing of the epoxy systems; and Figure S3: Photographs of the fracture surfaces of the test pieces after the tensile test in Figure 8.

Author Contributions: Conceptualization, Y.K.; methodology, Y.K. and A.M.; validation, Y.K.; formal analysis, Y.K. and A.M.; investigation, Y.K.; resources, Y.K.; data curation, Y.K. and A.M.; writing—original draft preparation, Y.K.; writing—review and editing, A.M.; visualization, Y.K. and A.M.; supervision, A.M. All authors have read and agreed to the published version of the manuscript.

Funding: This research received no external funding.

Institutional Review Board Statement: Not Applicable.

Informed Consent Statement: Not applicable.

Data Availability Statement: The data presented in this study will be available on request.

Acknowledgments: The authors would like to thank Mitsubishi Gas Chemical Corporation and New Japan Chemical Corporation for providing the epoxy resin in this study. We also thank ADEKA corporation for providing the adhesives.

Conflicts of Interest: The authors declare no conflicts of interest.

References

1. Gajdosik, M.S.; Clifton, J.; Josic, D. Sample displacement chromatography as a method for purification of proteins and peptides from complex mixtures. *J. Chromatogr. A* **2012**, *1239*, 1–9. [CrossRef] [PubMed]
2. Tsujioka, N.; Hira, N.; Aoki, S.; Tanaka, N.; Hosoya, K. A new preparation method for well-controlled 3D skeletal epoxy resin-based polymer monoliths. *Macromolecules* **2005**, *38*, 9901–9903. [CrossRef]
3. Zhu, T.; Row, K.H. Preparation and Applications of Hybrid Organic-Inorganic Monoliths: A Review. *J. Sep. Sci.* **2012**, *35*, 1294–1302. [CrossRef] [PubMed]
4. Zajickova, Z. Advances in the Development and Applications of Organic-Silica Hybrid Monoliths. *J. Sep. Sci.* **2017**, *40*, 25–48. [CrossRef] [PubMed]
5. Nguyen, A.M.; Irgum, K. Epoxy-based Monoliths. A Novel Hydrophilic Separation Material for Liquid Chromatography of Biomolecules. *Chem. Mater.* **2006**, *18*, 6308–6315. [CrossRef]
6. Hasegawa, G.; Kanamori, K.; Ishizuka, N.; Nakanishi, K. New Monolithic Capillary Columns with Well-Defined Macropores Based on Poly(styrene-co-divinylbenzene). *ACS Appl. Mater. Interfaces* **2012**, *4*, 2343–2347. [CrossRef] [PubMed]
7. Liu, K.; Aggarwal, P.; Lawson, J.S.; Tolley, H.D.; Lee, M.L. Organic monoliths for highperformance reversed-phase liquid chromatography. *J. Sep. Sci.* **2013**, *36*, 2767–2781. [CrossRef] [PubMed]
8. Jandera, P. Advances in the development of organic polymer monolithic columns and their applications in food analysis—A review. *J. Chromatogr. A* **2013**, *1313*, 37–53. [CrossRef] [PubMed]
9. Svec, F.; Lv, Y. Advances and recent trends in the field of monolithic columns for chromatography. *Anal. Chem.* **2015**, *87*, 250–273. [CrossRef]
10. Zhao, J.; Wang, J.; Liu, Z.; Zhang, Y.; Mi, Y. Preparation of Porous Epoxy Resin Materials by Reaction-Induced Phase Separation Regulated Using Fumed Silica. *ACS Appl. Polym. Mater.* **2023**, *5*, 2859–2866. [CrossRef]
11. Okada, K.; Nandi, M.; Maruyama, J.; Oka, T.; Tsujimoto, T.; Kondoh, K.; Uyama, H. Fabrication of mesoporous polymer monolith: A template-free approach. *Chem. Commun.* **2011**, *47*, 7422–7424. [CrossRef] [PubMed]
12. Liu, H.; Diao, J.; Wang, Q.; Gu, S.; Chen, T.; Miao, C.; Yang, W.; Su, D. A nanodiamond/CNT–SiC monolith as a novel metal free catalyst for ethylbenzene direct dehydrogenation to styrene. *Chem. Commun.* **2014**, *50*, 7810–7812. [CrossRef] [PubMed]
13. Wang, X.; Li, X.; Jiang, X.; Dong, P.; Liu, H.; Bai, L.; Yan, H. Preparation of a poly(styrene-co-DPHA-co-EDMA) monolith and its application for the separation of small molecules and biomacromolecules by HPLC. *Talanta* **2017**, *165*, 339–345. [CrossRef] [PubMed]
14. Hara, A.; Inoue, R.; Takahashi, N.; Nishida, K.; Kanaya, T. Trajectory of Critical Point in Polymerization-Induced Phase Separation of Epoxy/Oligoethylene Glycol Solutions. *Macromolecules* **2014**, *47*, 4453–4459. [CrossRef]
15. Bonyadi, S.; Chung, T.S.; Krantz, W.B. Investigation of Corrugation Phenomenon in the Inner Contour of Hollow Fibers during the Non-Solvent Induced Phase-Separation Process. *J. Membr. Sci.* **2007**, *299*, 200–210. [CrossRef]
16. Xin, Y.; Fujimoto, T.; Uyama, H. Facile Fabrication of Polycarbonate Monolith by Non-Solvent Induced Phase Separation Method. *Polymer* **2012**, *53*, 2847–2853. [CrossRef]

17. Lloyd, D.R.; Kinzer, K.E.; Tseng, H.S. Microporous Membrane Formation via Thermally Induced Phase Separation 1. Solid Liquid-Phase Separation. *J. Membr. Sci.* **1990**, *52*, 239–261. [CrossRef]
18. Yoneda, S.; Han, W.; Hasegawa, U.; Uyama, H. Facile Fabrication of Poly(methyl methacrylate) Monolith via Thermally Induced Phase Separation by Utilizing Unique Cosolvency. *Polymer* **2014**, *55*, 3212–3216. [CrossRef]
19. Tsujioka, N.; Ishizuka, N.; Tanaka, N.; Kubo, T.; Hosoya, K. Well-Controlled 3D Skeletal Epoxy-Based Monoliths Obtained by Polymerization Induced Phase Separation. *J. Polym. Sci. Part A Polym. Chem.* **2008**, *46*, 3272–3281. [CrossRef]
20. Plieva, F.M.; Andersson, J.; Galaev, I.Y.; Mattiasson, B. Characterization of polyacrylamide based monolithic columns. *J. Sep. Sci.* **2004**, *27*, 828–836. [CrossRef]
21. Wang, J.; Zhang, C.; Du, Z.; Xiang, A.; Li, H. Formation of porous epoxy monolith via concentrated emulsion polymerization. *J. Colloid Interface Sci.* **2008**, *325*, 453–458. [CrossRef] [PubMed]
22. Sakai, M.; Kato, T.; Ishizuka, N.; Isobe, T.; Nakajima, A.; Fujishima, A. Bio-inspired highly hydrophobic surface with ecdysis behavior using an organic monolithic resin and titanium dioxide photocatalyst. *J. Sol-Gel Sci. Technol.* **2016**, *77*, 257–265. [CrossRef]
23. Kubo, T.; Tsujioka, N.; Tanaka, N.; Hosoya, K. Co-continuous monolithic titania prepared by organic polymer monolith as pore template. *Mater. Lett.* **2010**, *64*, 177–180. [CrossRef]
24. Shimizu, Y.; Sakakibara, K.; Yasuda, S.; Tsujii, Y. Lubrication Characteristics of Epoxy Resin-Based Monolithic Thin Films with a Well-Defined Porous Morphology. *Adv. Eng. Mater.* **2019**, *21*, 1900683. [CrossRef]
25. Tominaga, R.; Nishimura, Y.; Suzuki, Y.; Takeda, Y.; Kotera, M.; Matsumoto, A. Co-continuous network polymers using epoxy monolith for the design of tough materials. *Sci. Rep.* **2021**, *11*, 1431–1441. [CrossRef] [PubMed]
26. Tominaga, R.; Takeda, Y.; Kotera, M.; Suzuki, Y.; Matsumoto, A. Non-destructive observation of internal structures of epoxy monolith and co-continuous network polymer using X-ray CT imaging for elucidation of their unique mechanical features and fracture mechanism. *Polymer* **2022**, *263*, 125433. [CrossRef]
27. Sugimoto, Y.; Nishimura, Y.; Uehara, F.; Matsumoto, A. Dissimilar Materials Bonding Using Epoxy Monolith. *ACS Omega* **2018**, *3*, 7532–7541. [CrossRef] [PubMed]
28. Sakata, N.; Takeda, Y.; Kotera, M.; Suzuki, Y.; Matsumoto, A. Interfacial Structure Control and Three-Dimensional X-ray Imaging of an Epoxy Monolith Bonding System with Surface Modification. *Langmuir* **2020**, *36*, 10923–10932. [CrossRef]
29. Capricho, J.C.; Fox, B.; Hameed, N. Multifunctionality in epoxy resins. *Polym. Rev.* **2020**, *60*, 1–41. [CrossRef]
30. Mustata, F.; Bicu, I. Multifunctional Epoxy Resins: Synthesis and Characterization. *Sci. Appl. Polym. Sci.* **2000**, *77*, 2430–2436. [CrossRef]
31. Liu, X.; Xiao, Y.; Luo, X.; Liu, B.; Guo, D.; Chen, L.; Wang, Y. Flame-Retardant multifunctional epoxy resin with high performances. *Chem. Eng. J.* **2022**, *427*, 132031. [CrossRef]
32. Sakakibara, K.; Kagata, H.; Ishizuka, N.; Sato, T.; Tsujii, Y. Fabrication of surface skinless membranes of epoxy resin-based mesoporous monoliths toward advanced separators for lithium ion batteries. *J. Mater. Chem. A* **2017**, *5*, 6866–6873. [CrossRef]
33. Aragishi, K.; Takeda, Y.; Suzuki, Y.; Matsumoto, A. Study on the Deformation and Fracture of Epoxy Monoliths through Mechanical Tensile and Compressive Tests and X-ray CT Imaging. *Polym. J.* **2024**, *56*, 529–540. [CrossRef]
34. Jojibabu, P.; Zhang, Y.X.; Prusty, B.G. A review of research advances in epoxy-based nanocomposites as adhesive materials. *Int. J. Adhes. Adhes.* **2020**, *96*, 102454. [CrossRef]
35. Wagle, P.G.; Tamboli, S.S.; More, A.P. Peelable coatings: A review. *Prog. Org. Coat.* **2021**, *150*, 106005. [CrossRef]
36. Yang, S.; Liu, C.; Sun, Z.; Xu, M.; Feng, Y. Effects of resin pre-coating treatment and fibre reinforcement on adhesive bonding between CFRP sheet and concrete. *Compos. Struct.* **2022**, *292*, 115610. [CrossRef]
37. Zhu, Z.; Tu, Y.; Ye, M.; Zeng, Q.; Rao, J.; Chen, N. A formaldehyde-free bio-composite sheet used as adhesive with excellent water-wet bonding performance. *Ind. Crops Prod.* **2023**, *198*, 116680. [CrossRef]
38. Fedors, R.F. A method for estimating both the solubility parameters and molar volumes of liquids. *Polym. Eng. Sci.* **1974**, *14*, 147–154. [CrossRef]
39. *ISO 4587:2003*; Adhesives Determination of Tensile Lap-Shear Strength of Rigid-to-Rigid Bonded Assemblies. International Organization for Standardization: Geneva, Switzerland, 2003.
40. Smith, I.T. The mechanism of the crosslinking of epoxide resins by amines. *Polymer* **1961**, *2*, 95–108. [CrossRef]
41. Shundo, A.; Yamamoto, S.; Tanaka, K. Network formation and physical properties of epoxy resins for future practical applications. *JACS Au* **2022**, *2*, 1522–1542. [CrossRef]
42. Kinloch, A.J. Mechanics and mechanisms of fracture of thermosetting epoxy polymers. *Adv. Polym. Sci.* **1985**, *72*, 45–67.
43. Ratna, D.; Simon, G.P. Thermomechanical properties and morphology of blends of a hydroxy-functionalized hyperbranched polymer and epoxy resin. *Polymer* **2001**, *42*, 8833–8839. [CrossRef]
44. Xing, A.; Sun, Q.; Meng, Y.; Zhang, Y.; Li, X.; Han, B. A hydroxyl-containing hyperbranched polymer as a multi-purpose modifier for a dental epoxy. *React. Funct. Polym.* **2020**, *149*, 104505. [CrossRef]
45. Choi, S.; Janisse, A.P.; Liu, C.; Douglas, E.P. Effect of Water Addition on the Cure Kinetics of an Epoxy-Amine Thermoset. *Polym. Chem.* **2011**, *49*, 4650–4659. [CrossRef]
46. Ehlers, J.; Rondan, N.G.; Huynh, L.K.; Pham, H.; Marks, M.; Truong, T.N. Theoretical Study on Mechanisms of the Epoxy-Amine Curing Reaction. *Macromolecules* **2007**, *40*, 4370–4377. [CrossRef]
47. Wei, J.; Ma, S.; Yue, H.; Wang, S.; Zhu, J. Comparison of Hydrogenated Bisphenol A and Bisphenol A Epoxies: Curing Behavior, Thermal and Mechanical Properties, Shape Memory Properties. *Macromol. Res.* **2018**, *26*, 529–538. [CrossRef]

48. Yamamoto, S.; Kuwahara, R.; Tanaka, K. Dynamic behaviour of water molecules in heterogeneous free space formed in an epoxy resin. *Soft Matter* **2021**, *17*, 6073–6080. [CrossRef] [PubMed]
49. Nunes, S.P.; Inoue, T. Evidence for spinodal decomposition and nucleation and growth mechanisms during membrane formation. *J. Mem. Sci.* **1996**, *111*, 93–103. [CrossRef]
50. Skripov, V.P.; Skripov, A.V. Spinodal decomposition (phase transitions via unstable states). *Sov. Phys. Uspekhi* **1979**, *22*, 389–410. [CrossRef]

molecules

MDPI

Article

The New Elastomeric Compounds Made of Butyl Rubber Filled with Phyllosilicates, Characterized by Increased Barrier Properties and Hydrophobicity and Reduced Chemical Degradation

Aleksandra Smejda-Krzewicka [1,*], Emilia Irzmańska [2], Konrad Mrozowski [1], Agnieszka Adamus-Włodarczyk [2], Natalia Litwicka [2], Krzysztof Strzelec [1] and Małgorzata I. Szynkowska-Jóźwik [3]

1 Institute of Polymer and Dye Technology, Lodz University of Technology, Stefanowskiego 16, 90-537 Lodz, Poland; konrad.mrozowski@dokt.p.lodz.pl (K.M.); krzysztof.strzelec@p.lodz.pl (K.S.)
2 Central Institute for Labour Protection—National Research Institute, Department of Personal Protective Equipment, Czerniakowska 16, 00-701 Warsaw, Poland; emirz@ciop.lodz.pl (E.I.); agada@ciop.lodz.pl (A.A.-W.); nalit@ciop.lodz.pl (N.L.)
3 Institute of General and Ecological Chemistry, Lodz University of Technology, Zeromskiego 116, 90-924 Lodz, Poland; malgorzata.szynkowska@p.lodz.pl
* Correspondence: aleksandra.smejda-krzewicka@p.lodz.pl

Abstract: The aim of the study was to produce new elastomeric materials containing butyl rubber (IIR) filled with silica and phyllosilicates (vermiculite, montmorillonite, perlite or halloysite tubes) with enhanced hydrophobicity and barrier properties and reduced chemical degradation. It was found that the filler type had a significant impact on the degree of cross-linking of butyl rubber and the properties of its vulcanizates. The highest degree of cross-linking and the highest mechanical strength were achieved for IIR composites filled with Arsil with perlite or halloysite tubes. The highest surface hydrophobicity (119°) was confirmed for the IIR vulcanizates with Arsil and montmorillonite. All tested samples showed high barrier properties because both the gas diffusion rate coefficient and the permeability coefficient reached low values. Both unfilled and filled IIR vulcanizates retained chemical resistance in contact with methanol for 480 min. Hour-long contact of a polar solvent (methanol) with each of the vulcanizates did not cause material degradation, while the presence of a non-polar solvent (n-heptane) worsened the mechanical parameters by up to 80%. However, the presence of fillers reduced the chemical degradation of vulcanizates (in the case of cured IIR filled with Arsil and halloysite tubes by 40% compared to the composite without fillers).

Keywords: butyl rubber; phyllosilicate; barrier properties; hydrophobicity; chemical degradation

1. Introduction

Rubbers have become essential materials in the industry with the advancement of plastic technology. The popularity of rubbers is due to good performance properties and the possibility of functionalization [1]. The areas of particular interest in elastomer research include such features as self-healing, shape memory effect [2,3], superhydrophobicity [4], reinforced mechanical properties, improved barrier properties [5], and enhanced damping properties [6,7].

The opportunities provided by rubbers make it possible to create both elastomeric materials that perfectly dampen vibrations and those that can be repaired after breaking. For this reason, elastomers are widely used in such fields as automotive, aerospace, industrial equipment, and household appliances [2–4,7–9]. Noteworthy among rubbers is butyl rubber (IIR). It is a synthetic elastomer that is formed by the polymerization of isobutylene and isoprene. IIR vulcanizates exhibit very high anti-aging resistance and limited permeability to gases and liquids. In addition, they are resistant to weathering, ozone, hot air, acids, and bases. Butyl rubber is mainly used for products with reduced permeability to gases and liquids such as tire inner tubes, hoses, gaskets, and membranes [10–12]. However, the latest research in the area of butyl rubber application shows that its use could be much wider.

El-Sabbagh S.H. et al. [13] investigated the compatibility of natural rubber (NR) with butyl rubber to achieve elastomeric blends featuring both IIR and NR properties. The authors showed that macrophase separations form in the IIR/NR mix, which is disadvantageous for the properties of such blends. Moreover, a compatibilizer was used, which decreased the extent of phase separation and increased interfacial adhesion between phases, consequently reducing the size of dispersed phase domains, which led to better mechanical properties including tensile strength and a higher value of Young's modulus. Sukharev N. et al. [14] studied elastomeric blends containing butyl rubber and ethylene-propylene-diene rubber (EPDM). The study determined the influence of the proportion of elastomers in uncured blends and their effect on phase structure formation patterns in multicomponent polymer systems, changes in molecular mobility, and ozone resistance. The research has shown that there is such a ratio of EPDM to IIR to obtain the highest possible resistance to ozone. These materials can be successfully considered for use in applications where ozone concentrations are higher, i.e., at heights above 15 km or in the aerospace industry. A novel strategy to prepare interpenetrating polymer networks (IPNs) based on butyl rubber and poly(n-octadecyl acrylate) (PC18A) was developed by Tavsanli B. et al. [15]. Solvent-free UV polymerization of the n-octadecyl acrylate monomer in the IIR melt at ambient temperature resulted in IPNs with self-healing and shape memory functions.

According to Guo X. et al. [16], butyl rubber can be applied to materials with high damping electromagnetic interference or multi-absorbing materials, including absorbing electromagnetic waves. The study showed that the composites of butyl rubber with single-walled carbon nanotubes (IIR/SWCNT) achieved good mechanical performance (tensile strength reached 15 MPa), and the total electromagnetic shielding efficiency of the material increased to 23.8 dB. In addition, the authors showed that water-induced modification of the composite achieved good dispersion of SWCNTs to enhance electromagnetic shielding while maintaining a wide damping temperature range from −55 °C to 40 °C with a damping factor above 0.2. Chameswary J. et al. [17] studied butyl rubber filled with barium titanate ($BaTiO_3$) with micro- or nanometric sizes. The research showed that such composites exhibited good mechanical properties, and they are flexible and absorb radio frequency vibrations. These results prove that these composites are proper candidates for the core of flexible dielectric waveguides and applications in flexible microwave substrates. Hao, S., et al. [18] tested butyl rubber filled with graphene. Studies have shown that the proper formulation of such a composite makes it possible to achieve excellent mechanical properties, high conductivity, and high barrier to water vapor. The conductivity of the IIR/graphene nanocompound at a graphene content of 3.76% prepared by Liquid Phase Redispersion reached more than seven orders of magnitude higher than the conventional twin-roll mixing method. Due to the existence of homogeneously distributed networks of segregated graphene, the tensile strength and elongation at break for the IIR/graphene nanocomposites increased by 410 and 126%, respectively, at a graphene content of 3.76%. The IIR/graphene vulcanizate exhibited such electrical properties that it can be used as a wearable sensor and physiological signal detection.

The presented examples of filled IIR compositions show the advantages of using innovative substances such as elastomer fillers. However, for decades, the rubber industry, especially the tire industry, has mainly used carbon black as a reinforcing filler. Since its structure and surface can be changed over a wide range, carbon black can meet a wide variety of requirements. However, with the passing of time and the necessity to invent alternatives to carbon black, various types of silicas have been developed. In the 1940s and 1950s, the specific surface area and structure of silica were constantly adjusted to satisfy new requirements in the field of rubber materials [19–21]. As the development proceeded, the advantages and disadvantages of silica compared to carbon black began to be perceived. Therefore, the research began for new and more easily available fillers for rubber, which will not cause as much carbon dioxide emissions as carbon black. The first and most obvious reason for using fillers is to improve mechanical properties, that is, to enhance the elastomer [19,22]. Fillers such as carbon black or silica can reinforce cured rubber, improving its tensile strength, durability, and wear resistance. This is particularly important in applications where the rubber products will be subjected to significant stress or abrasion. In addition to conventional fillers, various types of synthetic or natural substances are used, which also reinforce the composite by forming elastomer–filler and filler–filler interactions, thus creating a spatial network in the structure of the elastomer, which is responsible for carrying stresses [19,21,22]. Another reason for using fillers is to reduce the cost of the rubber product. Fillers then act as diluents and reduce the amount of expensive rubber in the product. This is important to keep the right properties reducing production costs. Another reason is to control physical properties; fillers are used to adjust hardness, elasticity, and electrical and thermal conductivity. This allows manufacturers to customize rubber to suit specific requirements. In addition, the use of fillers allows the dimensional stability of rubber composites to be maintained. The incorporation of a filler into an elastomer matrix reduces shrinkage and increases the ability to retain shape over time. This is important in terms of providing the accuracy of manufactured parts, the efficiency of seals and gaskets, and the overall performance and reliability of rubber components in various applications [20–22]. Additionally, some fillers improve the processability of rubber compounds by facilitating molding, extrusion, or other processing during production. The above reasons have contributed to the search for various fillers with unique properties, and the specific choice of filler will depend on the requirements of the product and the manufacturing process. Therefore, in recent years, more and more research has been conducted on alternatives to silica and carbon black [19–23]. Substitution of standard reinforcing fillers with montmorillonite (MMT) has improved some of the desired properties of elastomeric compounds in addition to some reductions of the final weight and price of the final products [24]. Montmorillonite organoclay has also been used as a filler in the IIR composites to improve their physical and mechanical properties [25]. In this study, melt mixing in an internal mixer was selected as the method of dispersing organoclay in the elastomeric matrix, and the intercalation of rubber chains into the clay gallery was deduced from the increase in basal spacing of the silicate layers as was measured by XRD. The highest basal spacing was detected for the amount of MMT equal to 3 phr. Dispersion and distribution of the organoclay were observed by SEM. The authors proved that organoclay content and structure had a large impact on the mechanical and rheological properties of nanocomposites as well as the permeability of carbon dioxide gas through their films.

This study has focused on layered silicates and their effects on butyl rubber vulcanizates because it is a very promising group of compounds, that indicate suitable properties in the case of novel elastomeric materials. In this paper, the main focus was on the hydrophobic and barrier properties of butyl rubber vulcanizates filled with various phyllosilicates. A material whose surface exhibits hydrophobic or superhydrophobic properties is the target of much scientific research [4]. The term "superhydrophobic" describes surfaces with a Young's angle of more than 150° and indicates very low wettability. This angle is defined as the mechanical equilibrium of the drop under the action of three interfacial tensions: solid–vapor, γ_{sv}; solid–liquid, γ_{sl}; and liquid–vapor, γ_{lv} [26–28]. This equilibrium relation is known as Young's Equation (1):

$$\gamma_{lv}\cos\theta_Y = \gamma_{sv} - \gamma_{sl} \tag{1}$$

where γ_{lv}, γ_{sv}, and γ_{sl} represent the liquid–vapor, solid–vapor, and solid–liquid interfacial tensions, respectively, and θ_Y is the Young's contact angle [28].

The popularity of such surfaces was started by the "lotus effect" described by Barthlott and Neinhus [29,30]. The lotus leaves owe their unusually high contact angle to a characteristic structure that scientists want to replicate for superhydrophobic materials. Examples of applications for such products could include self-cleaning materials [31], anti-corrosion [32], anti-icing [32,33], protective coatings, and impermeability to liquids.

Hydrophobicity microscopically is the tendency of chemical molecules to repel water molecules from each other, while hydrophobicity macroscopically is the property of a material's surface to be non-wettable by water. Associated with this phenomenon in polymeric materials are such terms as surface tension, surface free energy, and contact angle [34]. The wettability of polymer surfaces, including elastomers, depends on the contact angle, e.g., the angle formed by the tangent to the surface of a droplet on the surface of a solid, and it is highly dependent on the surface tension [35–37].

The various values of Young's angle lead to different phenomena at the solid–liquid interface. Therefore, several values of θ are distinguished as follows:

- $\theta = 0$, when the liquid wets the surface of the solid and tends to spontaneously spread on it; this state is defined as the critical value of the surface tension of the liquid;
- $0 < \theta < \pi/2$, when the liquid spills on the surface and tends to shrink on the surface within a limited range;
- $\theta > \pi/2$, when the liquid does not spread on the surface and tends to shrink on the surface of the solid, forming droplets [35–37].

In the context of the water–polymer interface, it must be considered that the contact angle is a measure of the hydrophilicity of the surface of a polymer material. Because water exhibits the highest surface tension compared to other liquids, the character of the polymer surface can be determined by the value of the contact angle. A hydrophilic surface is distinguished by the fact that the applied droplet spreads, thereby wetting the surface (the angle θ is small), for example, it is worth mentioning the surfaces of non-oxidized metal or glass ($\theta < 20°$). On the opposite side, according to the literature, a hydrophobic surface is characterized by the fact that the applied water droplet does not spread on the polymer surface (the angle θ is large), such as the surface of polypropylene ($\theta = 110°$) or polystyrene ($\theta > 97°$) [34].

Other important parameters describing elastomeric materials are their barrier properties. These are the features that determine whether a type of polymer can be used as, for example, a film or protective coating against a toxic substance. Three coefficients are used to describe them: permeability coefficient, diffusion coefficient, and solubility coefficient [5,38,39]. To describe the permeation of gases through an elastomer, a diffusion mechanism is used, which occurs across the material due to a pressure gradient. In this case, the constant volume and variable pressure method is used to measure this coefficient. A vacuum was applied to both sides of the polymeric material, and the permeability coefficient P is described by the following Formula (2):

$$P = \frac{V \cdot t_f}{A \cdot R \cdot T \cdot \Delta p} \cdot \left(\frac{dp}{dt}\right) \quad (2)$$

where V is the total amount of gas permeation through the sample into a cell, t_f is the sample thickness, A is the sample area, R is the gas constant, T is the absolute temperature, Δp is the pressure gradient across the sample, dp/dt is the transmission rate [5,38,39].

In addition, the permeability coefficient (P) combines the effects of both the diffusion coefficient (D) and the solubility coefficient (S) and can be explained as Formula (3):

$$P = D \cdot S \quad (3)$$

where D is the diffusion coefficient, which describes the kinetic aspect of transport, S is the solubility coefficient, which is related to the affinity of the penetrating substance (gas or liquid).

The above relation holds true when the value of D is independent of concentration and the value of S follows Henry's law. It is used to describe gas transport in polymeric composites reinforced with impermeable nanofillers. In the above diffusion–solubility model, penetrant molecules initially dissolve into the high-pressure face of a film then diffuse across it through thickness and finally desorb at the low-pressure face. Thus, the permeability of a penetrant depends on both its diffusivity and solubility. These properties can be changed by the molecular structure of the polymer and environmental factors [38,40].

To increase the barrier properties of elastomers, layered fillers such as montmorillonite, graphene, vermiculite, or halloysite are introduced [39–45]. The layers of this filler must "split" and form a difficult path for the substance to penetrate the material under investigation. This process is called exfoliation and is not simple to achieve when the filler is incorporated in situ into the polymer. Studies show that the most effective way to stop gas diffusion is to position the filler layers in an orientation perpendicular to the movement of the penetrating gas and increase the interactions at the polymer–filler interface [41–43,46].

It is common knowledge that the morphology and dispersion of layered aluminosilicates are key factors affecting the barrier properties of the composite. Therefore, proper dispersion and achieving a high degree of exfoliation of the layered nanofiller in the polymer matrix are the most important challenges in producing nanocomposites with improved barrier properties. According to the literature, there are three possible morphologies in nanocomposites (Figure 1) [38,47,48]:

- Tactoid (conventional composite), when there is no separation of layered filler packages, micrometer-sized structures are present in the polymer medium;
- Intercalated nanocomposite, when the polymer is located between parallel filler galleries, separation of filler layers occurs;
- Exfoliated nanocomposite, when the structure with the highest degree of dispersion of the filler in the polymer matrix, the polymer chains cause separation of the filler layers [49–51].

The study aimed to produce new elastomeric materials containing butyl rubber and characterized them by enhanced hydrophobicity and barrier properties and reduced chemical degradation. To test the research, six various rubber compositions were prepared, which differed in the kind of filler. Phyllosilicates were chosen as fillers in the tested IIR compositions because they have the ability to interact with organic matter [49–55], as well as the tendency to separate the packages that constitute them into nanometer-thick layers, which theoretically allows the obtaining of materials classified as nanocomposites [56–61].

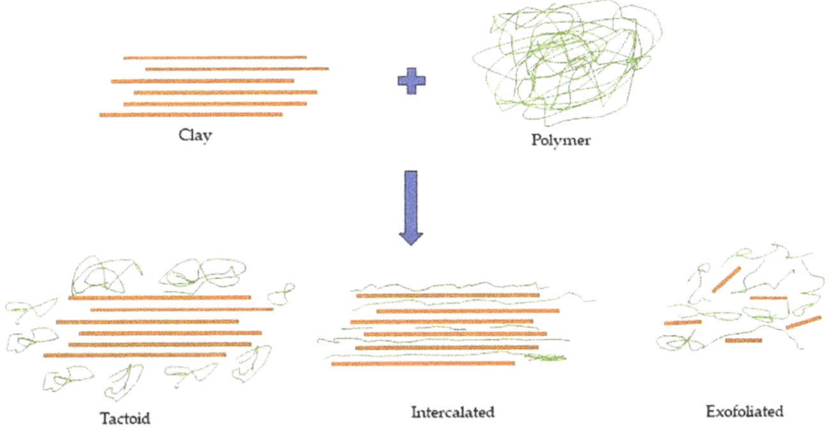

Figure 1. The possible morphologies of polymer/clay nanocomposites [38].

2. Results and Discussion

2.1. Characteristics of Phyllosilicates

Silicates, i.e., natural minerals, are most often formed as a result of the transformation of dust or volcanic rocks. Phyllosilicates have different structures. Two-layer silicates with a 1:1 structure are important for the rubber industry. This group mainly includes kaolinite, halloysite, and perlite. These compounds have unique properties and after adding them to a polymer matrix, they can form nanocomposites by intercalating the packages [62,63]. Halloysite (HNT) used in the tested IIR compositions is a mineral characterized by high porosity and specific surface (which is very desirable for elastomer fillers), high ion exchange (ability to absorb heavy metals), and easy chemical and mechanical processing [64]. It is an aluminosilicate containing approx. 45% silica and approx. 40% aluminum oxide. The remaining ingredients are water and trace amounts of metal oxides, i.e., TiO_2, Al_2O_3, FeO, MgO, CaO, Na_2O, and K_2O. Halloysite usually forms very small, tubular crystals visible under very high magnification. The mechanical, thermal, rheological, and barrier properties of HNT-modified polymers depend on the two most important factors: the degree of particle dispersion and the compatibility of the polymer with the nanofiller. Pearlite (PER) is a two-phased, lamellar compound composed of alternating layers of ferrite (87.5%) and cementite (12.5%). Its chemical formula depends on the rock compositions, but perlite consists of the following metal oxides: SiO_2 (71–75%), Al_2O_3 (12–18%), Na_2O (3–4%), K_2O (4–5%), Fe_2O_3 (0.5–1%), and MgO (0.1–1.5%). It is most often used as a filler for polyethylene, polypropylene, or poly(vinyl chloride) [65].

Three-layer silicates include compounds such as montmorillonite, mica, talc, and vermiculite. Silicates with three-layer packages with a 2:1 structure type are distinguished by the fact that the octahedral layer is located between two tetrahedral layers with their vertices facing each other [62]. As the most famous of this group of silicates, montmorillonite (MMT) is a nanofiller. It is a mineral from the group of dioctahedral smectites with the chemical formula $(Al_{4-x}Mg_x)[Si_8O_{20}](OH)_4$, where x = 0.67 and the Si:Al ratio is approximately 5:2. Another silicate with a three-layer structure is vermiculite (VER) [66]. It is a phyllosilicate that has a layered structure composed of an octahedral layer and two tetrahedral layers. The chemical composition of this compound is $(Mg,Fe,Al)_3[(Al,Si)_4O_{10}](OH)_2{\cdot}4H_2O$ [62]. During the heating, the layers expand, increasing the volume of the mineral. Vermiculite is light, inexpensive, non-toxic, chemically inert, and resistant to thermal decomposition. These properties make this material ideal as an insulator or filler in insulating applications [62,63,66–70].

The filler structure was examined based on images acquired with a scanning electron microscope (SEM), with sample pictures presented in Figure 2.

The Arsil (ARS) particles are finely aggregated and agglomerated (Figure 2a). The structure of montmorillonite (Figure 2c) and halloysite tubes (Figure 2e) looks very similar. The vermiculite particle grains are completely different from the other fillers used and they form large and flat grains, and there are many empty areas between them (Figure 2b). The perlite particles have strongly jagged shapes and are loosely bound together (Figure 2d).

2.2. The Influence of Phyllosilicates on the Course of IIR Cross-Linking

The recipe for the compounding of the IIR composites filled with phyllosilicates is given in Table 1.

The vulcanization parameters of rubber mixes depend on several factors, for example the type of rubber, the cross-linking agent, and the filler type. The minimal rheometric torque (T_{min}) is responsible for the viscosity of the mix, while the rheometric torque increment (ΔT) corresponds to the degree of cross-linking of butyl rubber. The next parameters are the scorch time (t_{02}) and optimal vulcanization time (t_{90}), which determine how long the mix will be vulcanized. The results in Table 2 clearly show that the addition of montmorillonite increased the vulcanization time (16.0 min), while the sample filled with vermiculite achieved the shortest cross-linking time (9.4 min). The other filled composites achieved similar t_{90} values (from 10.3 to 12.5 min) compared to the unfilled IIR. The addition of phyllosilicates resulted in shorter scorch times, as each filled sample reached a lower t_{02} value than the reference sample, which proves that the filling of butyl rubber is related to a reduced time for safe processing.

Figure 2. Scanning electron microscope (SEM) images of the fillers used in IIR composites: Arsil (**a**), vermiculite (**b**), montmorillonite (**c**), perlite (**d**), halloysite tubes (**e**).

Table 1. Compositions of the IIR mixes and vulcanizates.

Ingredient	Amount (phr)					
IIR	100	100	100	100	100	100
S	2	2	2	2	2	2
ZnO	5	5	5	5	5	5
TMTD	2	2	2	2	2	2
SA	2	2	2	2	2	2
ARS	-	10	10	10	10	10
VER	-	-	20	-	-	-
MMT	-	-	-	20	-	-
PER	-	-	-	-	20	-
HNT	-	-	-	-	-	20
Sample designation	0	ARS	VER	MMT	PER	HNT

IIR—butyl rubber, S—sulfur, ZnO—zinc oxide, TMTD—tetramethylthiuram disulfide, SA—stearic acid, ARS—silica, VER—vermiculite, MMT—montmorillonite, PER—perlite, HNT—halloysite tubes, phr—parts per hundred of rubber.

Table 2. Vulcametric parameters of IIR composites filled with silica and phyllosilicates.

Properties	Sample Designation					
	0	ARS	VER	MMT	PER	HNT
T_{min} (dNm)	0.66	1.03	1.06	1.34	1.05	1.07
ΔT_{max} (dNm)	3.27	3.44	2.75	2.37	3.87	3.93
α	0.95	0.96	0.97	0.84	0.96	0.94
t_{02} (min)	4.0	2.0	1.2	2.7	1.9	0.9
t_{90} (min)	12.0	10.4	9.2	16.0	10.3	12.3
CRI (1/min)	12.5	11.9	12.5	7.5	11.9	8.8
Q_v (mL/mL)	3.54 ± 0.06	2.58 ± 0.04	3.11 ± 0.06	3.79 ± 0.21	2.62 ± 0.05	2.58 ± 0.03
$-Q_w$ (mg/mg)	0.27 ± 0.01	0.14 ± 0.01	0.27 ± 0.01	0.27 ± 0.01	0.26 ± 0.01	0.26 ± 0.01
V_R	0.212 ± 0.002	0.280 ± 0.003	0.243 ± 0.003	0.209 ± 0.007	0.275 ± 0.003	0.279 ± 0.002
α_c	0.28 ± 0.02	0.39 ± 0.03	0.32 ± 0.03	0.26 ± 0.02	0.38 ± 0.04	0.39 ± 0.03

T_{min}—minimal rheometric torque; ΔT_{max}—maximum rheometric torque increment; α—degree of conversion (vulcanization); t_{02}—scorch time; t_{90}—optimal vulcanization time; CRI—cure rate index; Q_v—equilibrium volume swelling in toluene; $-Q_w$—content of the eluted fraction in toluene; V_R—volume fraction of rubber in swollen material in toluene; α_c—degree of cross-linking.

The filled samples achieved a higher viscosity than the unfilled sample, which is shown in Table 2 by comparing the T_{min} values (IIR: 0.66 dNm; IIR/VER: 1.06 dNm). Moreover, each of the filled samples achieved a rheometric torque increment equal to or higher than the unfilled compound. This indicates that added phyllosilicates cause an increase in the viscosity of the elastomer and an increase in the degree of cross-linking. The IIR/PER and IIR/HNT samples recorded a higher increase in rheometric torque (3.87 dNm and 3.93 dNm, respectively). The MMT-filled IIR sample reached the lowest ΔT_{max} (2.37 dNm), while its minimal rheometric torque was the largest (1.34 dNm). The degree of conversion (e.g., degree of transformation of plastic material into elastic material) calculated on the basis of vulcametric results also achieved the lowest result (0.84) for the sample containing MMT. The α_c value for the remaining IIR compositions was comparable and higher by 14% (α = 0.95). These results indicate a lower degree of cross-linking of butyl rubber if it was filled with montmorillonite. Table 2 also summarizes the CRI values for all composites, so it is observed that the addition of MMT or HNT significantly lowered the CRI values (7.5 and 8.8 min^{-1}, respectively) relative to the refereed compound, whereas for the other samples, this parameter was like the IIR sample.

Figure 3 shows the time relationship of the torque for all samples. It is noted that the course of the curves for all filled samples (except MMT) differed at the beginning of the cross-linking process in relation to the unfilled IIR. On the curves of the IIR and IIR/MMT samples, it is not noticed that the torque decrease during the scorch time was so pronounced; moreover, it was longer than for the other mixes. In addition, these two compositions were characterized by the least torque increment. The other curves were very similar; only curves 3 and 6 could be differentiated due to their scorch times, which were the shortest among the tested compositions. According to the data in Table 2, Figure 3 also shows that IIR/PER and IIR/HNT composites achieved the highest values of rheometric torque increment. All curves regardless of whether the sample was filled with phyllosilicates showed no reversion ability.

Besides the vulcametric parameters, the degree of cross-linking of vulcanizates could be determined by the equilibrium swelling. For this purpose, several parameters were determined. The first is the volumetric equilibrium swelling (Q_v), which accounts for how much a given vulcanizate is cross-linked. The next is the content of the eluted fraction during the swelling ($-Q_w$), which determines how much substance has been eluted after contact with a selected solvent. The volume fraction of rubber in swollen material (V_R) determines the degree of cross-linking of the vulcanizate; the higher the value of this parameter indicates the greater the degree of cross-linking (α_c). Table 2 shows the results of equilibrium swelling and states that the addition of Arsil or Arsil with halloysite tubes or perlite improves the degree of cross-linking of the IIR vulcanizates. The values of α_c for these samples are 0.38–0.39, while this parameter for the reference sample was 0.28. Thus, it was observed that there was a significant increase in the degree of cross-linking by filling the vulcanizates. The Q_v values obtained for all filled samples (without IIR/MMT) prove that the addition of phyllosilicates increased the degree of cross-linking. The IIR/ARS and IIR/HNT samples obtained the smallest Q_v value (2.58 mL/mL), which indicates the highest degree of cross-linking. The contents of the eluted fraction, except for the IIR/ARS sample ($-Q_w$ = 0.14 mg/mg), were at a similar level of values to the reference sample, i.e., about 0.27 mg/mg. The volume fraction of rubber in the swollen material also confirmed that the samples containing ARS, PER, and HNT had the highest degree of cross-linking, as the V_R values were the largest among the samples tested (IIR/ARS: 0.280). It is worth emphasizing that the equilibrium swelling results of the produced IIR vulcanizates presented in this part correlate well with the vulcametric results.

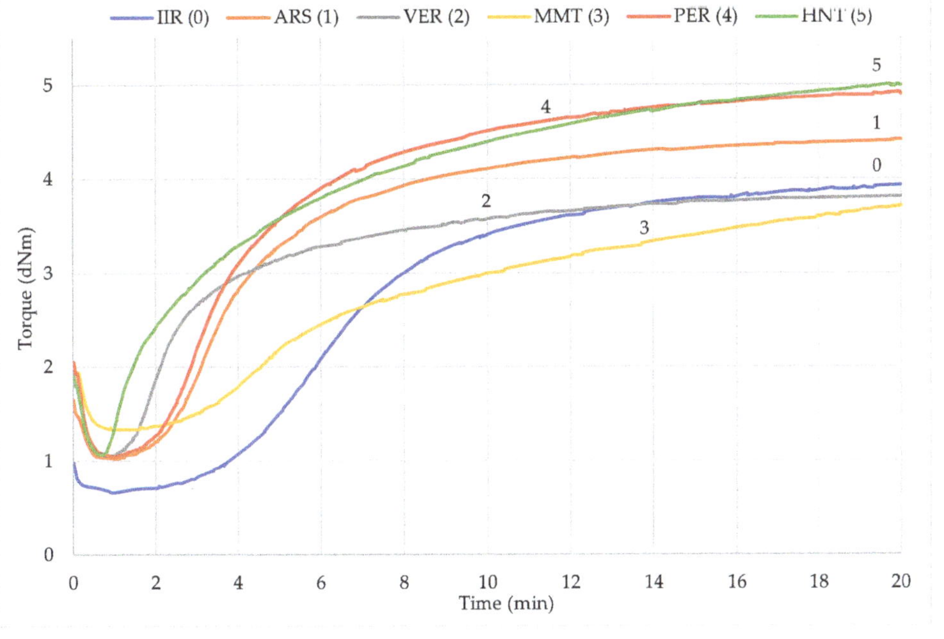

Figure 3. Vulcametric kinetics of IIR composites filled with phyllosilicates.

2.3. The Influence of Phyllosilicates on the Mechanical Properties of IIR Vulcanizates

The results of the tensile strength test before and after thermo-oxidative aging are summarized in Table 3. The obtained results before aging show that the addition of Arsil or Arsil with perlite or halloysite tubes slightly increased the stiffness of the samples. The composite samples filled with vermiculite and montmorillonite reached comparable S_{e100} values as the reference sample. In Table 3, TS_b results confirmed the fact that the addition of phyllosilicates to IIR mix increased the mechanical strength of the vulcanizate. Each of the tested composites increased its tensile strength value by a minimum of 50%. Such a correlation might be due to the formation of a spatial network of bonds between the filler in the elastomer matrix, which results in the rigidity and enhancement of the elastomeric composite. The greatest TS_b values were obtained by the IIR/HNT (13.4 MPa), IIR/ARS (11.1 MPa), and IIR/PER (11.0 MPa) vulcanizates, which indicates the best mechanical strength of these two compositions. The elongation at break for all filled samples is more than twice that of the unfilled IIR vulcanizate (IIR: 590%, IIR/PER: 1289%). However, in addition to the strength parameters before aging, Table 3 summarizes the results after aging. This is important because as time passes, the mechanical properties of rubber products undergo various changes, because of which the products may become useless due to increased stiffness and cracking. Among the factors influencing the degradation of elastomer properties, oxygen and increased temperature are the most important. The results in Table 3 show that after thermo-oxidative aging, the tested composite became more rigid, which could be seen to a greater degree for the IIR/HNT, IIR/PER, and IIR/VER samples. The values of S_{e100} and S'_{e100} for the IIR/VER vulcanizate were 0.48 MPa and 0.66 MPa, respectively, and this was since under the influence of increased temperature and the presence of oxygen the vulcanizate begins to cross-link, making it more rigid and resistant to deformation. This relationship was visible for silicate-filled composites, while for the unfilled vulcanizates, no major changes in strain stress can be observed. The tensile strength after aging increased for the IIR/MMT and IIR/VER samples (5.7 MPa and 8.7 MPa, respectively), while it decreased for the other compositions. The IIR/PER vulcanizate recorded the greatest decrease in tensile strength among those tested (from 11.0 MPa to 6.6 MPa), while the others decreased by a few tenths of an MPa. After aging, a decrease in elongation values at break was also observed for all vulcanizates. The filled samples, however, continued to reach values greater than the referenced composition. The filled vulcanizate with the highest resistance to thermo-oxidative aging was the IIR/MMT composition, as it achieved an aging factor of AF = 0.67, closely behind was the IIR/VER vulcanizate with a value of AF = 0.63. These values can be explained by the fact that the two samples have the lowest degree of cross-linking and after thermo-oxidative aging, the vulcanizates undergo a process of cross-linking, while the other vulcanizates undergo a process of degradation due to the higher degree of cross-linking.

Table 3. Strength properties of IIR vulcanizates filled with phyllosilicates before and after the thermo-oxidative aging process.

Properties	Sample Designation					
	0	ARS	VER	MMT	PER	HNT
S_{e100} (MPa)	0.56 ± 0.03	0.62 ± 0.03	0.48 ± 0.03	0.56 ± 0.01	0.66 ± 0.02	0.68 ± 0.01
S_{e200} (MPa)	0.83 ± 0.02	0.88 ± 0.05	0.68 ± 0.04	0.76 ± 0.02	0.90 ± 0.02	0.96 ± 0.02
S_{e300} (MPa)	1.14 ± 0.02	1.18 ± 0.08	0.88 ± 0.04	0.96 ± 0.03	1.18 ± 0.03	1.27 ± 0.03
TS_b (MPa)	2.5 ± 0.0	11.1 ± 1.1	7.9 ± 0.6	5.3 ± 0.4	11.0 ± 0.4	13.4 ± 1.0
E_b (%)	590 ± 57	1260 ± 56	1277 ± 19	1286 ± 1	1289 ± 2	1287 ± 2
S'_{e100} (MPa)	0.58 ± 0.08	0.62 ± 0.06	0.66 ± 0.02	0.67 ± 0.04	0.86 ± 0.04	0.82 ± 0.05
S'_{e200} (MPa)	0.87 ± 0.03	0.88 ± 0.06	0.90 ± 0.02	0.89 ± 0.05	1.12 ± 0.05	1.20 ± 0.09
S'_{e300} (MPa)	1.22 ± 0.09	1.16 ± 0.05	1.11 ± 0.02	1.12 ± 0.07	1.40 ± 0.08	1.56 ± 0.14
TS'_b (MPa)	2.3 ± 0.1	10.5 ± 1.4	8.7 ± 0.4	5.7 ± 0.4	6.6 ± 0.6	13.0 ± 0.7
E'_b (%)	558 ± 27	634 ± 34	732 ± 2	802 ± 24	673 ± 10	733 ± 9
AF (-)	0.85	0.48	0.63	0.67	0.31	0.55

S_{e100}, S_{e200}, and S_{e300}—stress at elongation of 100%, 200%, and 300%, respectively; TS_b—tensile strength; E_b—elongation at break; S'_{e100}, S'_{e200}, and S'_{e300}—stress at elongation of 100%, 200%, and 300%, respectively, after thermo-oxidative aging; TS'_b—tensile strength after thermo-oxidative aging; E'_b—elongation at break after thermo-oxidative aging; AF—aging factor.

Other mechanical properties are summarized in Table 4. The results of the hardness test of the vulcanizates showed that the addition of phyllosilicates increased hardness. The IIR/PER and IIR/HNT samples achieved the highest Shore A hardness values (57.6 °ShA and 56.2 °ShA, respectively). The IIR/MMT vulcanizate, despite the weakest mechanical strength among the filled compositions, showed the highest tear resistance (T_s = 5.46 N/mm); moreover, the addition of silicates made this parameter improved for each sample relative to the unfilled sample. To determine the damping properties of the rubber, hysteresis loops were determined on a graph at five tensile cycles and the Mullins effect was calculated. This test was performed only for filled composites, due to the study of filler–elastomer interactions.

In Table 4, the results of hysteresis losses during the first and five stretching cycles show that the IIR/HNT sample recorded the highest losses (ΔW_1 = 51.2 N·mm and ΔW_5 = 29.9 N·mm). The composite that achieved the fewest losses is the IIR/ARS vulcanizate (ΔW_1 = 39.9 N·mm and ΔW_5 = 27.4 N·mm), so it is considered the best damping composite among those tested. The value of the Mullins effect (E_M) shows that the smaller filler dispersion was better in the elastomeric matrix or indicates small agglomerates. The IIR/ARS composite had the smallest value of E_M (13.5%), which determines that the filler in this sample was well dispersed.

Table 4. Other mechanical properties of IIR vulcanizates filled with phyllosilicates.

Properties	Sample Designation					
	0	ARS	VER	MMT	PER	HNT
HA (°ShA)	42.8 ± 1.6	49.6 ± 2.4	50.8 ± 2.3	53.7 ± 1.5	57.6 ± 1.2	56.2 ± 2.4
T_s (N/mm)	2.29 ± 0.06	3.81 ± 0.53	4.45 ± 0.39	5.46 ± 0.25	3.62 ± 0.33	3.87 ± 0.32
ΔW_1 (N·mm)	-	39.9	43.2	50.1	51.1	51.2
ΔW_5 (N·mm)	-	27.4	25.1	24.1	26.4	29.9
E_M (%)	-	13.5	23.3	32.9	26.7	20.7

HA—hardness; T_s—tear resistance; ΔW_1, ΔW_5—hysteresis losses during the first and fifth sample stretching cycles, E_M—Mullins effect.

2.4. The Influence of Phyllosilicates on the Dynamic Properties of IIR Vulcanizates

The dynamic properties of rubber materials depend mainly on the quantity and activity of the filler used. The distribution of filler particles in the elastomeric matrix is also an important factor. Table 5 compares the results of the dynamic properties of the filled vulcanizates tested. The unfilled IIR composite was not considered in this study. The highest value of the storage modulus G' was achieved by the IIR/PER composite (0.278 MPa), indicating the greatest resistance to deformation of this sample and the greatest filler–filler interactions, which cause reinforcement of the composite. The other vulcanizates obtained smaller values of the storage modulus. The loss modulus represents the viscous part, or the amount of energy dissipated in the sample, and is related to the material's ability to dissipate stress through heat. The IIR/ARS sample (G''_{max} = 0.012 MPa) dissipates energies through the heating and shows the best damping properties due to the smallest value of G'', which agrees with the study of damping properties by hysteresis loops and the Mullins effect.

Table 5. Dynamic properties of IIR vulcanizates filled with phyllosilicates.

Properties	Sample Designation				
	ARS	VER	MMT	PER	HNT
G'_{max} (MPa)	0.066	0.122	0.183	0.278	0.172
G''_{max} (MPa)	0.012	0.022	0.033	0.067	0.022
$\Delta G'$ (MPa)	0.040	0.116	0.181	0.256	0.144

G'_{max}—maximum storage modulus; G''_{max}—maximum loss modulus; $\Delta G'$—Payne effect.

The largest Payne effect values were achieved by the IIR/PER and IIR/MMT samples (0.256 MPa and 0.181 MPa, respectively). Therefore, these vulcanizates probably had larger aggregates and the spaces between them were smaller than the others, which was also confirmed by the SEM analysis (Figure 2). The IIR/ARS composite probably had the smallest aggregates due to the smallest value of $\Delta G'$ = 0.040 MPa. In addition, each sample had a storage modulus greater than the loss modulus, which was why these materials were mainly elastic.

The variation of the storage modulus from the increased oscillation strain is shown in Figure 4. The IIR/PER vulcanizate (curve 4) had the highest stiffness, and therefore the greatest value of the storage modulus. The addition of perlite caused the formation of filler–filler bonds in the elastomer matrix, which made the G′ value increase. The strong Payne effect in this case was probably caused by the structure of pearlite, and its particles are highly corrugated (Figure 2d), increasing its specific surface area and interactions with butyl rubber macromolecules. The IIR/MMT sample (curve 3) also showed a high value of G′ (0.183 MPa), indicating the appearance of a large amount of filler–filler bonds. As the oscillation strain increased, the storage modulus for each of the tested compositions decreased. After crossing 10%, a noticeable decrease in the value of the storage modulus was observed for each curve in Figure 4. In addition, it was observed that as the deformation increased, the elastic properties of the tested vulcanizates decreased, which was due to the deterioration of the spatial network structures of the filler in the elastomer matrix.

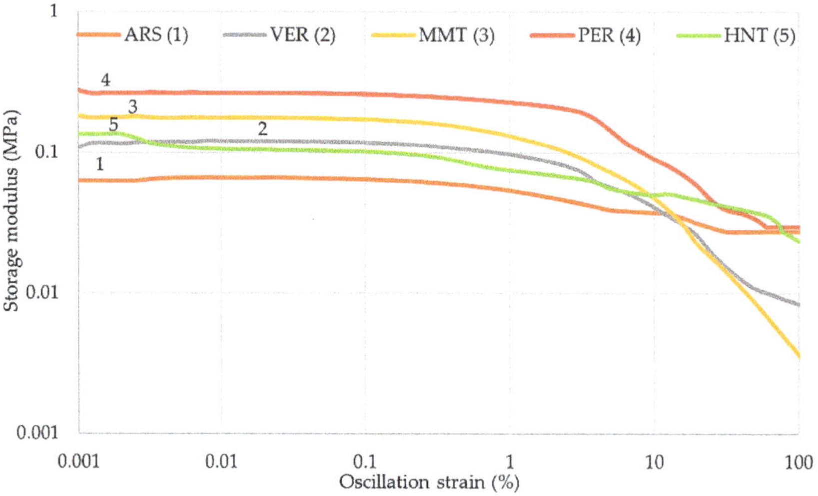

Figure 4. Relationship of storage modulus to oscillation strain of IIR composites filled with phyllosilicates.

Figure 5 shows the variation of loss modulus with increasing oscillation strain. Loss modulus measures the energy dissipated or lost as heat per cycle of sinusoidal deformation. The loss value was the largest for the IIR/PER sample (curve 4), the other compositions achieved a lower sinusoidal peak value. In addition, a peak shift was observed with the IIR/HNT vulcanizate (curve 5). This sample had its peak at about a 60% oscillation strain, while the other compositions had this peak at about a 5% oscillation strain. The IIR/ARS vulcanizate (curve 1) in the graph shown did not have a peak at all, indicating little energy loss through heat. The shift of the peak perhaps indicated specific interactions between the filler and the elastomeric matrix in the IIR/HNT sample.

2.5. The Influence of Phyllosilicates on the Hydrophobicity of IIR Vulcanizates

The contact angle is aimed at determining the nature of the surface of the polymer under test. The surface can be hydrophobic or hydrophilic. For polymers, the addition of selected fillers is often important, as they allow the surface to remain hydrophobic. In Figure 6 the results of the contact angle are summarized. It is noted that filled samples obtain greater hydrophobicity than the unfilled IIR vulcanizate. Also, the addition of phyllosilicates improved the hydrophobicity of butyl rubber materials. The largest contact angle was obtained by the IIR/MMT composition (Φ = 118.9°). The other filled samples also obtained such a value of Φ that they are considered materials with improved hydrophobicity. Among the filled products, the IIR/VER vulcanizate had the lowest contact angle value, which was related to the poor dispersion of vermiculite in the elastomeric matrix. When the fillers are unequally distributed in the matrix, the agglomerates and aggregates are formed, which causes disturbances in the character of the composite surface. The presented results clearly show that IIR/HNT (Φ = 116.2°) and IIR/PER (Φ = 110.4°) vulcanizates had better dispersion filler than the IIR/VER composition. In addition, an important aspect influencing the hydrophobicity of the rubber material is the shape and size of the filler particles. According to Figure 2, it is important to note that vermiculite has the largest particles with regular shapes among the tested fillers. According to [71–75], fillers, which have a smaller particle size (nano), cause an increase in the roughness of the material's surface and this results in an increase in the hydrophobicity of the composite according to the Wenzel model and Cassie–Baxter model [72,76,77]. Thus, the IIR/MMT, IIR/PER, and IIR/HNT samples achieved higher contact angle values than the other samples, although all fillers belong to the hydrophilic. In summary, the degree of dispersion, the degree of surface roughness, and the choice of filler and its size also affect the hydrophobicity of the composite [71–73]. Nanofillers cause greater surface roughness, but it is more difficult to disperse them evenly in the elastomeric matrix; therefore, the selection of appropriate components when creating new rubber materials is very important [74,75].

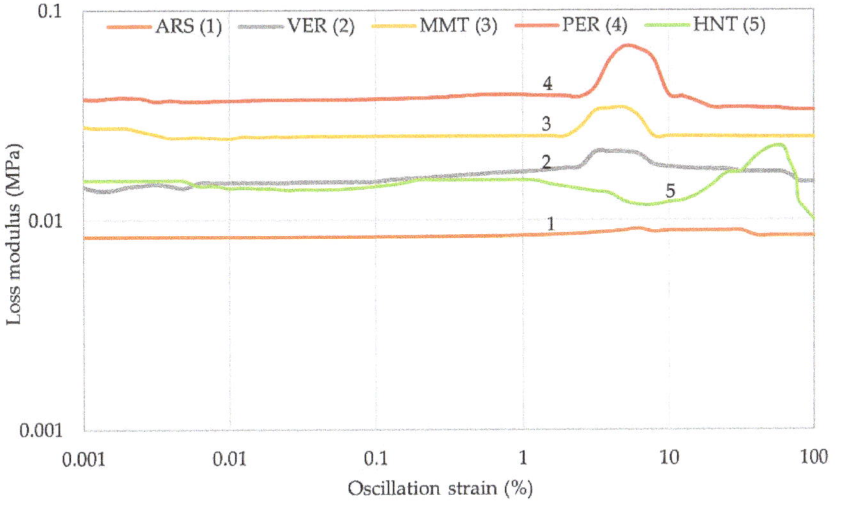

Figure 5. Relationship of loss modulus to oscillation strain of IIR composites filled with phyllosilicates.

2.6. The Influence of Phyllosilicates on the Barrier Properties of IIR Vulcanizates

The barrier test was conducted to determine the effect of added fillers on air permeability. The calculated gas transmission rate and permeability coefficients are summarized in Table 6. Analyzing the results, it was noted that the addition of fillers increased both GTR and P values, making the filled composites more permeable than the reference sample. The IIR/PER and IIR/HNT vulcanizates were the least permeable, as they achieved the lowest GTR and P values. Filled vulcanizates should show lower permeability than the unfilled sample, but the problem could be the lack of modification of the fillers or the process of incorporating them into the elastomeric matrix. An important issue in achieving increased barrier properties is to provide the right morphology of the composite (torturous diffusion path). Aiming for exfoliation of filler layers is essential to raise the barrier properties of the product. In this case, it is possible to achieve a very low degree of exfoliation of layered filler packages, which translated into an unsatisfactory result.

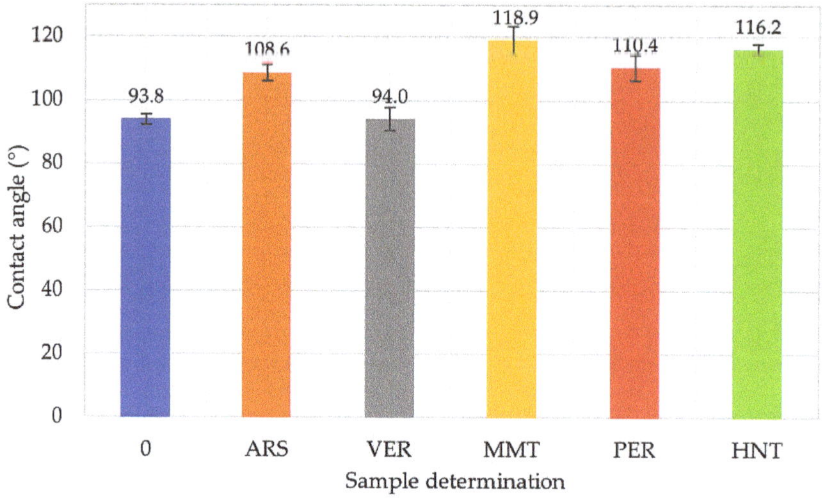

Figure 6. Contact angle values of IIR vulcanizates filled with phyllosilicates.

Table 6. Barrier properties of IIR vulcanizates filled with phyllosilicates.

Properties	Sample Designation					
	0	ARS	VER	MMT	PER	HNT
GTR ($\frac{mol}{m^2 \cdot s \cdot Pa}$)	2.27×10^{-9}	1.10×10^{-8}	1.12×10^{-8}	1.10×10^{-8}	9.59×10^{-9}	1.06×10^{-8}
P ($\frac{mol}{m \cdot s \cdot Pa}$)	2.62×10^{-12}	1.10×10^{-8}	1.21×10^{-11}	1.54×10^{-11}	1.09×10^{-11}	1.09×10^{-11}

GTR—gas transmission rate; P—coefficient gas permeability.

2.7. Resistance to Permeation by Liquid Chemical Substances

Resistance to permeation by the test chemical was determined by measuring the breakthrough detection time. Breakthrough detection time is the elapsed time from the start of the test to the time the test chemical first breaks through the barrier material and can be measured in the collecting medium. Resistance to permeation by liquid chemicals was conducted according to the procedures described in the Materials and Methods section, with the results given in Table 7. The tests involved two substances: methanol, which is a polar solvent, and n-heptane, which is non-polar.

Table 7. Test results for permeation by liquid chemicals.

Properties	Test Substance	Sample Designation					
		0	ARS	VER	MMT	PER	HNT
Breakthrough time (min)	methanol	>480					
	n-heptane	249	280	210	380	289	295

Previous research [78] indicates interest in the effects of fillers on the gas permeability of polymeric materials. In their work, Takahashi et al. observed that the addition of 30 wt% vermiculite as a nanofiller to butyl rubber caused a more than 20-fold decrease in the permeability of helium, hydrogen, oxygen, nitrogen, methane, and carbon dioxide through the resulting vulcanizates as well as a higher diffusion coefficient for those gases (by two orders of magnitude) as compared to IIR without the filler. While the cited study revealed a beneficial (reducing) effect of nanofillers on the gas permeation rate through elastomeric materials, these findings cannot be directly extrapolated to liquid permeation due to differences in the size and shape of gas and liquid molecules. In the present study, the barrier properties of non-polar IIR were found to depend on solvent polarity as well as on the surface properties of the nanofillers added to the system.

In the case of a polar solvent, the addition of fillers does not change the resistance to penetration of liquid chemicals. The sample retained chemical resistance in contact with methanol for 480 min, and in all cases, the highest performance level of 6 against the permeation of liquid chemicals has been achieved by the EN ISO 374-1:2016 standard [79]. The presence of fillers did not reduce the protective parameter.

Fillers had no effect on the permeation of a polar solvent (methanol) through the non-polar material (IIR), which exhibits a greater thermodynamic affinity to IIR than n-heptane. In the case of n-heptane, the vulcanizates showed lower resistance than for contact with the polar solvent.

Differences in the time taken for the liquid chemical substance to penetrate and change were observed depending on the filler addition and times in the range from 210 to 380 min were recorded. The lowest value of chemical resistance was observed for the IIR/VER vulcanizate, in which the tested material achieved a performance level of 4. In other cases, an effectiveness level of 5 was achieved by the EN ISO 374-1:2016 standard, and the longest permeation resistance time of 380 min was obtained for the IIR/MMT sample.

While the n-heptane breakthrough time for unfilled IIR was 249 min, it increased by approx. 30–45 min (20%) for the IIR/ARS, IIR/PER, and IIR/HNT composites (without significant differences between nanofillers used). However, the breakthrough time for the IIR/MMT sample was as long as 380 min, which means a 50% increase as compared to the unfilled composite. The differential effects of the studied nanofillers on breakthrough time for IIR with respect to liquids of varying degrees of thermodynamic affinity are probably related both to differences in liquid sorbability and to the heterogeneous distribution of surface energy in fillers with dissimilar chemical structures. Some influence on the barrier properties of IIR composites may also be exerted by differences (if any) in the degree of dispersion and aggregation of nanofiller particles [80].

Differences in breakthrough time for composites with nanofillers may result from different amounts of n-heptane retained in the nano- and mesopores of aggregates of filler particles (with the fillers differing in the surface polarity of their solid state particles). Differences in polar liquid breakthrough times for composites also depend on the liquid desorption rate, which is associated with the immobilization of its particles in aggregates of nano- and mesopores and with different forces bonding the nonpolar liquid with the surfaces of the applied fillers of different polarities. Among the studied fillers, the highest polarity was found for MMT, which also exhibited the highest chemical resistance when challenged with a non-polar solvent [81].

2.8. Chemical Degradation

Chemical degradation due to the action of liquid substances is defined as the percentage change in mechanical properties, i.e., piercing force, as a result of contact with chemical agents. Chemical degradation tests for the studied vulcanizates were conducted according to the procedures described in the Materials and Methods section, with the results given in Table 8. As a result of the analysis, it is concluded that the polar solvent does not degrade any of the tested vulcanizates. The opposite effect is present when vulcanizates come into contact with a non-polar solvent, i.e., n-heptane. It is observed that the addition of nano-fillers weakens the chemical degradation process, compared to the pure vulcanizate, by up to 40%. The IIR/ARS and IIR/MMT as well as IIR/VER and IIR/PER composites achieved similar chemical degradation values.

According to the literature data, cured butyl rubber is characterized by a similar resistance to chemicals, including solvents and mineral oils, as silicone rubber. However, as compared to butyl rubber, the sorptive capacity of silicone is almost 40% greater for non-polar solvents (e.g., benzene) and much higher for polar chemicals, such as ethanol (leading to a mass increase by 15 units after 168 h of immersion in the solvent). Consequently, the current findings concerning chemical degradation and resistance to liquid permeation indicate an improvement in the barrier properties of filled vulcanizates (e.g., IIR/ARS) as compared to non-polar ones (IIR). Szadkowki et al. also reported an improvement in barrier properties by the addition of materials such as perlite and vermiculite to the polymeric matrix, especially at high concentrations, when mineral particles may form an impermeable layer due to their structure [82]. Vulcanizates filled with VER and PER, with a finer structure, exhibit a higher mechanical strength following a controlled process of immersing a vulcanizate surface in a non-polar solvent (i.e., n-heptane). The other fillers with a looser and finer structure, as well as those with similar structures, are characterized by even higher barrier properties, with permeation resistance results indicating longer n-heptane breakthrough times for the IIR/MMT and IIR/HNT vulcanizates as compared to composites containing perlite and vermiculite. Halloysite nanotubes led to the best results, that, is, the lowest degree of chemical degradation of vulcanizates (approx. 38%), which is consistent with the literature. Other authors tentatively attributed this phenomenon to the infiltration of macromolecular polymeric chains through pores in nanotubes [83].

The surface of vulcanizates following exposure to n-heptane was examined using a stereoscopic microscope under 7× magnification (Opta-Tech series SK, Warsaw, Poland), with sample images presented in Figure 7.

Table 8. Test results for methanol-induced chemical degradation.

Properties	Test Substance	Sample Designation					
		0	ARS	VER	MMT	PER	HNT
Results for methanol-induced chemical degradation							
Mean puncture force (N)	Pre-exposure	51.51	58.65	32.80	27.92	37.68	41.68
	Post-exposure	50.96	48.84	35.23	28.06	39.00	41.39
Sample degradation (%)		1.07	16.74	−7.39	−0.49	−3.49	0.70
Results for n-heptane-induced chemical degradation							
Mean puncture force (N)	Pre-exposure	51.51	58.65	32.80	27.92	37.68	41.68
	Post-exposure	10.60	21.79	14.43	10.53	17.24	25.48
Sample degradation (%)		79.42	62.84	56.01	62.30	54.24	38.88

2.9. Abrasion Resistance

The abrasion resistance of the fabricated vulcanizates was tested according to the procedures described in Section 3.10. In the case of this test, a deviation from the test in accordance with the standard EN 388:2016+A1:2018 [84] was used. Due to the high friction of the samples, the test was stopped after 500 rub cycles. The surface of the fabricated vulcanizates before tests and after 500 rub cycles was examined based on the acquired SEM images, with sample pictures presented in Figure 8.

The literature shows that the vulcanizate hardness increases with silicate concentration and with the diameter of filler particles due to higher cross-linking density resulting from polymer–filler interactions. Mostafa et al. observed greater abrasion-related mass loss for vulcanizates containing fillers (vs. unfilled ones) and for vulcanizates with higher filler concentrations [85].

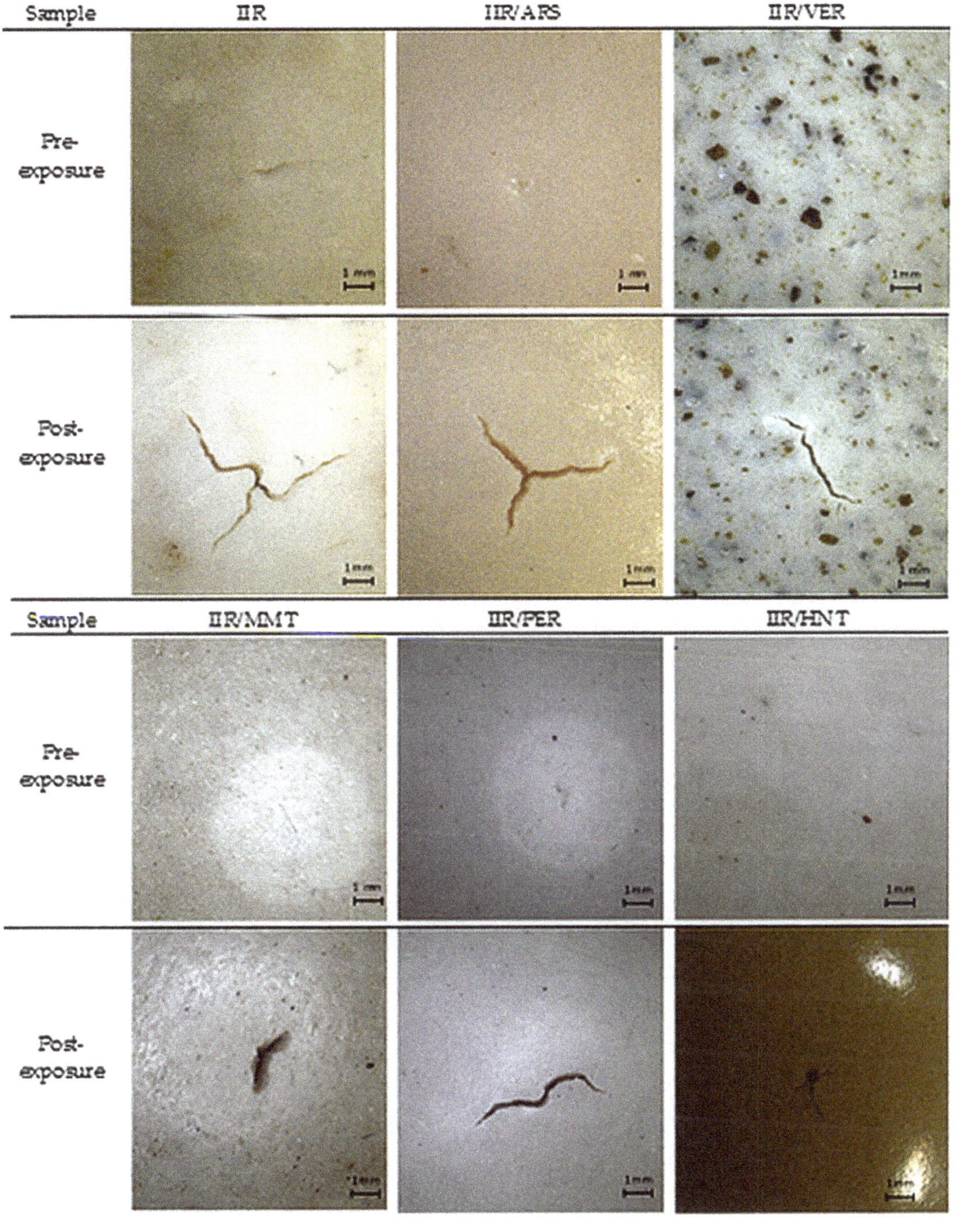

Figure 7. Stereoscopic microscope images of the surface of vulcanizates before and after n-heptane-induced chemical degradation.

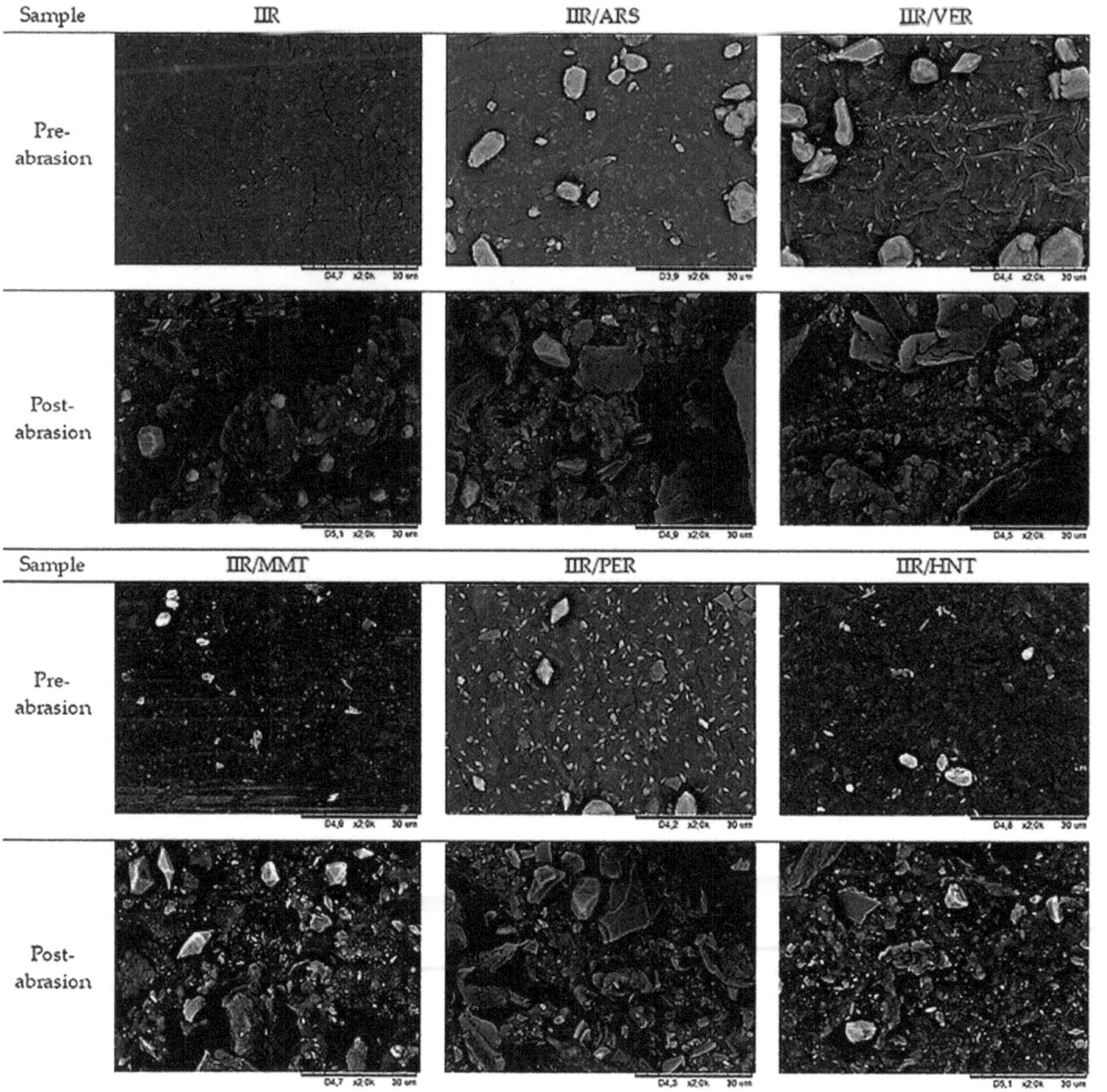

Figure 8. SEM images of the surface of fabricated vulcanizates before and after abrasion tests.

3. Materials and Methods

3.1. Materials

In this work, butyl rubber, IIR (type: Butyl 206) with 2.3% mol. unsaturated bonds, a density of 0.91 g/mL, and Mooney viscosity of ML 1 + 8 (125 °C): 51 (delivered by ExxonMobil Chemicals&Specialties, Irving, TX, USA).

The curing system consisted of the following substances:

- Sulfur (S) as a cross-linking agent, with a density of 1.8–2.36 g/mL (delivered by Chempur, Piekary Śląskie, Poland);
- Zinc oxide (ZnO) as a cross-linking activator, with a density of 5.6 g/mL (delivered by Chempur, Piekary Śląskie, Poland);

- Tetramethylthiuram disulfide (TMTD) as a cross-linking accelerator, with a density of 1.29 g/mL (delivered by Brenntag Polska Sp. z o. o., Kędzierzyn-Koźle, Polska);
- Stearic acid (SA) as a cross-linking activator and dispersing agent, with a density of 0.94 g/mL (delivery by Chempur, Piekary Śląskie, Poland).

The following fillers were used:

- Silica (ARS), type: Arsil, with a density of 2.20 g/mL (delivered by Zakłady Chemiczne "Rudniki" S.A., Rudniki, Poland);
- Vermiculite (VER), FlameHunter VE MIC, with a density of 0.85–1.00 g/mL, and an average particle size of 250–710 μm (>80%) (delivered by NYSA Chem® Sp. z o. o., Wrocław, Poland);
- Montmorillonite (MMT, NanoBent ZR), modified with a quaternary ammonium salt with two short- and two long-chain alkyl substituents, with an average particle size of 20–60 μm (56%), ≤20 μm (44%) and layer separation of 2.0–2.4 nm (delivered by ZGM "Zębiec", Zębiec, Poland);
- Perlite (PER), type: EP100F, with a density of 0.06–0.14 g/mL (delivered by Perlipol, Bełchatów, Poland);
- Halloysite tubes (HNT), with a density of 2.53 g/mL (delivered by Sigma-Aldrich Chemie, Steinheim am Albuch, Germany).

3.2. Compounding and Vulcanization

The IIR composites were prepared using a two-roll mill (type: Laborwalzwerk, Krupp-Gruson, Magdeburg-Buckau, Germany) with a roll diameter of 200 mm and a roll length of 450 mm, at a roll temperature of 30–35 °C. The total time to create the composition was 5–8 min. First, the rubber was plasticized and the ingredients were incorporated in the following order: stearic acid, ZnO, filler, accelerator, and sulfur. The obtained rubber composites were stored separately in tightly closed foils at room temperature.

The produced mixes were vulcanized in hydraulic presses in appropriate metal molds. The vulcanization parameters were a temperature of 160 °C, a pressure of 150–180 bar, and a cure time of 30 min.

3.3. Characteristics of the Cross-Linking Process

The vulcanization process is characterized by determining the cure kinetics, and the equilibrium swelling. The cure kinetics of the IIR composites were determined using the Alpha Technologies (MDR 2000) oscillating disk rheometer (Alpha Technologies, Hudson, OH, USA) at 160 °C (ASTM D5289-17 standard [86]), which was employed to determine the following parameters: scorch time (t_{02}); vulcanization time (t_{90}); and minimal torque (T_{min}); maximum torque increment (ΔT_{max}), which is the difference between the torque after heating and minimal torque values. The degree of vulcanization (conversion) in the vulcametric studies was defined according to Formula (4) [87].

$$\alpha = \frac{T_t - T_0}{T_h - T_0} \quad (4)$$

where T_t is the torque value at a given time during the vulcanization (in this case: T_{15}), T_0 is the torque value at time zero, T_h is the torque value at the end of vulcanization (in this case: T_{30}).

The cure rate index (CRI) was designated according to Formula (5):

$$CRI = \frac{100}{t_{90} - t_{02}} \quad (5)$$

Swelling behavior was assessed using toluene (according to ASTM D471 [88]). From each vulcanizate, four test pieces of 25–60 mg of different shapes were cut out, weighed using an electrical balance, and swollen in toluene until equilibrium was reached (for 72 h). After this time, the swollen samples were removed from toluene and washed with diethyl ether, and their weights were determined again. The samples were dried to a constant weight at a temperature of 50 °C and then reweighed.

Equilibrium volume swelling (Q_v) was calculated using Formula (6):

$$Q_v = Q_w \times \frac{d_v}{d_s} \tag{6}$$

where Q_w is the value of the equilibrium mass swelling (mg/mg), d_v is the vulcanizate density (g/mL), and d_s is the solvent density (g/mL).

Equilibrium weight swelling was calculated from Formula (7):

$$Q_w = \frac{m_s - m_d}{m_d^*} \tag{7}$$

where m_s is the swollen sample weight (mg), m_d is the dry sample weight (mg), and m_d^* is the reduced sample weight (mg). The reduced sample weight was calculated from Formula (8):

$$m_d^* = m_d - m_0 \cdot \frac{m_m}{m_t} \tag{8}$$

where m_0 is the initial sample weight (mg), m_m is the mineral content in the blend (mg), and m_t is the total weight of the blend (mg).

Negative equilibrium weight swelling ($-Q_w$), interpreted as the amount of leaching substances, was calculated from Formula (9):

$$-Q_w = \frac{m_0 - m_d^*}{m_0} \tag{9}$$

The rubber volume fraction (V_R) was calculated from Formula (10):

$$V_R = \frac{1}{1 + Q_v} \tag{10}$$

The degree of cross-linking (α_c) was determined using Formula (11):

$$\alpha_c = \frac{1}{Q_v} \tag{11}$$

3.4. Determination of Surface Morphology

The morphology of the vulcanizates was assessed using a scanning electron microscope (SEM). This was a Hitachi Tabletop Microscope TM-1000 (Tokyo, Japan) product. The preparation of the samples for measurement consisted of placing a double-sided self-adhesive foil onto a special table and gluing the testing sample to it. Then, a gold layer was applied to the prepared sample using the Cressington Sputter Coater 108 auto vacuum sputtering machine (Redding, CA, USA) at a pressure greater than 40 mbar for 60 s. The samples prepared in this way were placed into the scanning electron microscope chamber, and the measurement was performed.

3.5. Determination of Dynamic and Mechanical Properties

For the vulcanizates, the following properties were determined: strength properties, hysteresis losses and Mullins effect, tear resistance, hardness, elasticity, loss modules, and Payne effect.

Measurements of the tensile properties were carried out using a testing machine (Zwick1435/Roell GmbH & Co. KG, Ulm, Germany). The parameters determined from this test were stress at elongation of 100, 200, and 300% (S_{e100}, S_{e200}, S_{e300}); tensile strength (TS_b); and relative elongation at break (E_b). Each property was determined for five samples. The test was conducted at a constant speed of 500 mm/min.

The hysteresis losses were determined using a testing machine (Zwick1435/Roell GmbH & Co. KG, Ulm, Germany). Each test was conducted for three samples, which were stretched five times to 200% elongation at a stretching speed of 500 mm/min, and the initial force was 0.1 N. The Mullins effect was determined according to Formula (12):

$$E_M = \frac{W_1 - W_5}{W_1} \times 100\% \quad (12)$$

where W_1 is the hysteresis loss at the first extension of the sample (N·mm) and W_5 is the hysteresis loss at the fifth extension of the sample (N·mm).

The tear strength (T_s) was tested in accordance with method A of the standard ISO 34-1:2022 [89] using a testing machine (Zwick1435/Roell GmbH & Co. KG, Ulm, Germany). Rectangular specimens with dimensions of 100 mm × 15 mm and a cut of 40 mm were used for the tests.

Hardness (HA) was tested on the Shore A scale using a Zwick/Roell hardness tester according to ISO-48-4:2018 [90]. Each test was performed ten times. The samples were in the shapes of cylinders, with diameters of 80 mm and heights of 6 mm.

The dynamic properties of the vulcanizates were determined by the minimum and maximum storage modulus (G'_{min}, G'_{max}), the maximum loss modulus (G'_{max}), and the Payne effect ($\Delta G'$, Formula (13)) at room temperature. The test was performed using the Ares G2 rotational rheometer (New Castle, UK) according to ISO 4664-1:2022 [91]. The tested samples, in the form of discs with dimensions of 25 mm × 2 mm, were placed between special measuring plates of the apparatus. The parameters that were used were as follows: a soak time of 10 s, an angular frequency of 10 rad/s, a logarithmic sweep with strain from 0.005 to 70% s, 20 points per decade, and an initial force of 5 N.

$$\Delta G' = G'_{max} - G'_{min} \quad (13)$$

where G'_{max} is the maximum storage modulus (MPa) and G'_{min} is the minimum storage modulus (MPa).

3.6. Resistance to Thermo-Oxidative Aging

The thermal aging of the IIR vulcanizates was carried out in a forced circulating aging oven at 70 °C for 7 days. After conditioning at room temperature for 24 h, the changes of mechanical properties (stress at 100%, 200%, or 300% strain, tensile strength, elongation at break) were evaluated based on the aging factor (AF) according to Formula (14):

$$AF = \frac{TS'_b \cdot E'_b}{TS_b \cdot E_b} \quad (14)$$

where TS'_b is the tensile strength after thermo-oxidative aging (MPa), TS_b is the tensile strength before thermo-oxidative aging (MPa), E'_b is the elongation at break after thermo-oxidative aging (%), and E_b is the elongation at break before thermo-oxidative aging (%).

3.7. Determination of Hydrophobicity

The contact angle of the vulcanizates surface was determined using a goniometer from DataPhysics Instruments GmbH OCA 15EC (Filderstadt, Germany). The embedded drop method was used. At the beginning of the measurement, a drop of water with a volume of ~5 μL was placed on the surface of the vulcanizate using a Hamilton microsyringe. Then, using a special program, a photo of the drop was taken within 10 s so that the boundary between the surface and the drop was visible. The contact angle was measured by analysis in a computer program adapted for this study. A minimum of 5 drops were deposited into each sample and the average value of the contact angle was calculated.

3.8. Determination of Barrier Properties

The barrier properties were measured using a device that tests gas permeability using the manometric method. The measurement of barrier properties was based on a method that used pressure differences in measurement chambers on both sides of the tested sample. The apparatus consisted of a measuring cell in which the test sample was placed. The measuring cell was divided into two parts, i.e., an atmospheric pressure chamber and a low-pressure chamber. The test gas (air) was supplied to the chamber at atmospheric pressure. The low-pressure chamber contained a high-sensitivity sensor that measured pressure changes occurring in the measuring chamber as a result of gas permeating through the partition between both chambers containing the tested sample. A vacuum pump was connected to the low-pressure chamber, generating low pressure in the chamber (<10 Pa). From the results obtained, the gas transmission rate (GTR) was calculated according to Formula (15):

$$\mathrm{GTR} = \frac{\mathrm{V_c}}{\mathrm{R \cdot T \cdot P_u \cdot A}} \cdot (\mathrm{dp/dt}) \tag{15}$$

where $\mathrm{V_c}$ is the volume of the low-pressure chamber (l), R is the gas constant (8.31×10^3) [(l·Pa)/(K·mol)], T is the measurement temperature (K), $\mathrm{P_u}$ is the gas pressure in the high-pressure chamber (Pa), A is the area of gas permeation through the sample ($\mathrm{m^2}$), dp/dt is the pressure changes per unit of time (Pa/s).

The coefficient gas permeability (P) was determined according to Formula (16):

$$\mathrm{P} = \mathrm{GTR \cdot d} \tag{16}$$

where d is the sample thickness (m).

3.9. Chemical Degradation

Chemical degradation studies for the obtained vulcanizates were carried out in accordance with the standard EN ISO 374-4:2019 [92]. Samples with a diameter of 20 mm were acclimatized at (23 ± 2 °C) for 24 h as per EN ISO 2231:1995 (PN-EN ISO 2231:1999) [93]. Then they were secured in glass vials containing 2 mL of heptane or methanol sealed with septa having a center hole 12 mm in diameter. The samples were placed under the septa. Vials prepared in this way were inverted to make sure that the test chemical was in direct contact with the vulcanizate surface for one hour.

Subsequently, the mechanical parameter of puncture force was determined in a comparative system, namely, for vulcanizates exposed and not exposed to the test chemicals (Figure 9); the latter served as reference values. Puncture force was measured at a probe advance rate of 100 mm/min (the initial distance between the puncture probe and the sample surface was 100 mm). The mean difference (n = 3) of the puncture force for samples exposed and not exposed to the test chemicals was expressed as a percentage.

3.10. Abrasion Resistance

Abrasion resistance was tested according to the standard PN-EN 388:2017-02 [94] with a modified maximum number of rubs. Samples with a diameter of 38 mm were acclimatized at a temperature of (23 ± 2) °C and a relative humidity of (50 ± 5)% for 24 h according to EN ISO 2231:1995 (PN-EN ISO 2231:1999). The mechanical parameter of resistance to cyclical abrasion was determined after 500 rubs executed at a force of (9 ± 0.2) kPa using a Martindale apparatus (James Heal, Sterling, VA, USA).

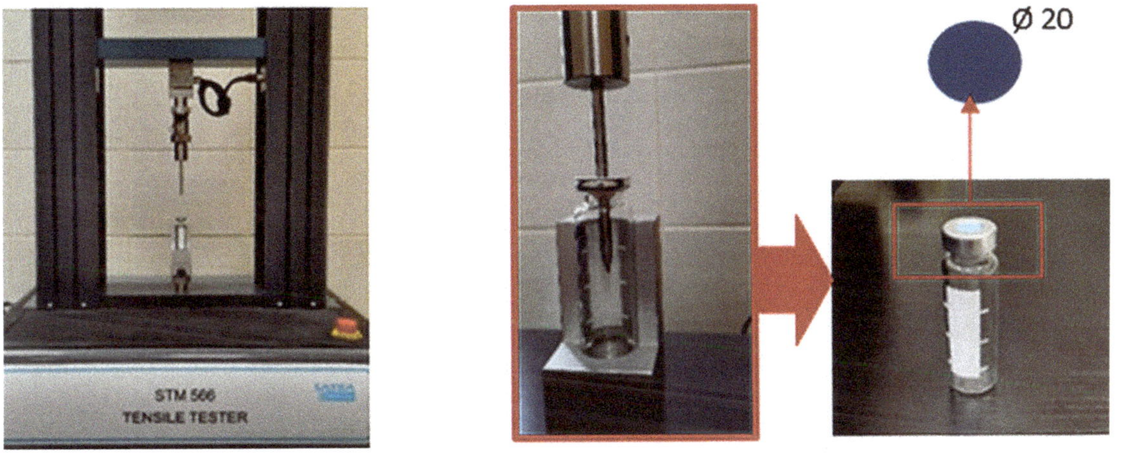

Figure 9. Measurement stand for determining chemical degradation according to EN ISO 374-4:2019 (Satra Technology, Kettering, United Kingdom).

4. Conclusions

The effect of applied fillers on the IIR composite performance illustrates a radar chart (Figure 10).

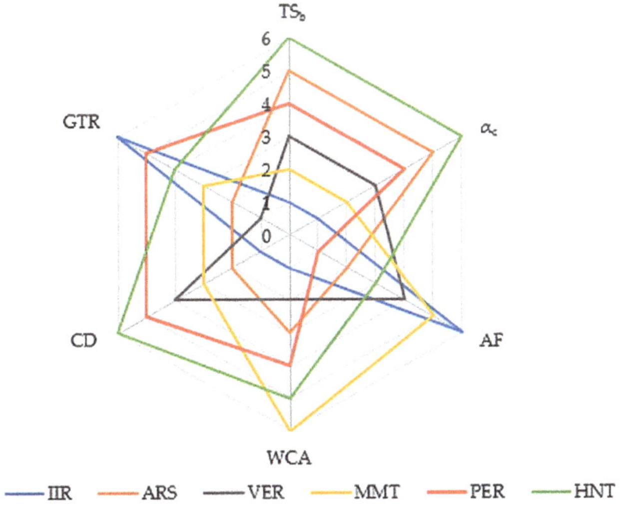

Figure 10. Comparison of properties of tested IIR vulcanizates filled with phyllosilicates (1—the weakest property; 6—the best property); TS_b—tensile strength, α_c—cross-linking degree, AF—aging factor, WCA—water contact angle, CD—chemical degradation, GTR—gas transmission rate.

Based on the research conducted, the following conclusions were drawn:

1. The IIR composite filled with Arsil and vermiculite was characterized by the shortest cross-linking time (9 min) and the highest cure rate index (12.5 min^{-1}).
2. The highest degree of cross-linking (0.39) was achieved for IIR composites filled with Arsil or Arsil with perlite or halloysite tubes. The samples with the highest degree of cross-linking also obtained the highest mechanical strength.

3. The tested vulcanizates did not show high tear resistance; however, the vulcanizates filled with Arsil and montmorillonite had the lowest tear resistance.
4. The hardness of the tested compositions did not exceed 60 °ShA. The vulcanizates filled with Arsil and perlite were characterized by the highest hardness.
5. Most of the tested compositions achieved an aging factor of approximately 0.6, which indicates good resistance to thermo-oxidative aging. The samples filled with Arsil and vermiculite or montmorillonite were characterized by the highest resistance to thermo-oxidative aging.
6. Vulcanizates filled with Arsil achieved the lowest values of the Mullins effect, which may indicate the best degree of filler dispersion and the best vibration damping effect.
7. The highest Payne effect was achieved for the IIR vulcanizate containing Arsil and perlite.
8. The highest surface hydrophobicity (119°) was confirmed for the IIR vulcanizates with Arsil and montmorillonite.
9. All tested samples showed high barrier properties because both the gas diffusion rate coefficient and the permeability coefficient reached low values.
10. No changes were observed on the surface of the tested vulcanizates, such as shrinkage, chipping, peeling, or hardening, after contact with organic solvents of different polarity (methanol and n-heptane). Permanent deformation of the sample was visible after contact of all variants of vulcanizates with heptane.
11. The sample retained chemical resistance in contact with methanol for 480 min. The presence of fillers did not reduce the protective parameter.
12. In the case of contact of vulcanizates with n-heptane, the time taken for the liquid chemical substance to penetrate changed in the range from 210 to 380 min.
13. Hour-long contact of a polar solvent (methanol) with each of the vulcanizates did not cause material degradation, while the presence of a non-polar solvent (heptane) worsened the mechanical parameters by up to 80%. However, the presence of fillers reduced the chemical degradation of vulcanizates (in the case of vulcanizates filled with Arsil and halloysite tubes by 40% compared to the vulcanizate without fillers).

Author Contributions: Conceptualization, A.S.-K.; methodology, A.S.-K. and E.I.; investigation, A.S.-K., K.M., A.A.-W., and N.L.; formal analysis, A.S.-K., K.M., and E.I.; writing—original draft preparation, A.S.-K.; writing—review and editing, A.S.-K., K.M., K.S., M.I.S.-J., and E.I. All authors have read and agreed to the published version of the manuscript.

Funding: This paper is published and based on the results of a research task on testing the protective parameters of vulcanizates carried out within the scope of the sixth stage of the National Programme "Governmental Programme for Improvement of Safety and Working Conditions" supported within the scope of state services by the Ministry of Family and Social Policy. Task no. 1.ZS.06 entitled "Method for determining the chemical degradation resistance of polymeric protective gloves, including dual-use medical gloves taking into account the requirements of the European standard PN-EN ISO 374-4:2020-03" [95]. The Central Institute for Labour Protection—National Research Institute is the programme's main coordinator.

Institutional Review Board Statement: Not applicable.

Informed Consent Statement: Not applicable.

Data Availability Statement: Data are contained within the article.

Conflicts of Interest: The authors declare no conflicts of interest.

References

1. Liu, C.; Fan, J.; Chen, Y. Design of regulable chlorobutyl rubber damping materials with high-damping value for a wide temperature range. *Polym. Test.* **2019**, *79*, 106003. [CrossRef]
2. Huang, J.; Cao, L.; Yuan, D.; Chen, Y. Design of novel self-healing thermoplastic vulcanizates utilizing thermal/magnetic/light-triggered shape memory effects. *ACS Appl. Mater. Interfaces* **2018**, *10*, 40996–41002. [CrossRef]

3. Cao, L.; Fan, J.; Huang, J.; Chen, Y. A robust and stretchable cross-linked rubber network with recyclable and self-healable capabilities based on dynamic covalent bonds. *J. Mater. Chem. A Mater.* **2019**, *7*, 4922–4933. [CrossRef]
4. Irzmańska, E.; Korzeniewska, E.; Pawlak, R.; Tomczyk, M.; Smejda-Krzewicka, A.; Adamus-Włodarczyk, A. Enhanced Hydrofobicity of Polymers for Personal Protective Equipment Achieved by Chemical and Physical Modification. *Materials* **2021**, *15*, 106. [CrossRef]
5. Koros, W.J. *Barrier Polymers and Structures*; American Chemical Society: Washington, DC, USA, 1990.
6. Kishi, H.; Nagao, A.; Kobayashi, Y.; Matsuda, S.; Asami, T.; Murakami, A. Carboxyl-terminated butadiene acrylonitrile rubber/epoxy polymer alloys as damping adhesives and energy absorbable resins. *J. Appl. Polym. Sci.* **2007**, *105*, 1817–1824. [CrossRef]
7. Zhou, X.Q.; Wang, L.; Yu, D.Y.; Zhang, C.Y. Dynamic effective equivalent stiffness analysis on the periodical honeycomb reinforced composite laminated structure filled with viscoelastic damping material. *Compos. Struct.* **2018**, *193*, 306–320. [CrossRef]
8. Chung, D.D.L. Processing-structure-property relationships of continuous carbon fiber polymer-matrix composites. *Mater. Sci. Eng. R* **2017**, *113*, 1–29. [CrossRef]
9. Amin, M.; Akbar, M.; Amin, S. Hydrophobicity of silicone rubber used for outdoor insulation. *Rev. Adv. Mater. Sci.* **2007**, *16*, 10–26.
10. Sreehari, H. Evaluation of solvent transport and cure characteristics of chlorobutyl rubber graphene oxide nanocomposites. *Mater. Today Proc.* **2022**, *49*, 1431–1435. [CrossRef]
11. Liang, Y.; Wang, Y.; Wu, Y.; Lu, Y.; Zhang, H.; Zhang, L. Preparation and properties of isobutylene–isoprene rubber (IIR)/clay nanocomposites. *Polym. Test.* **2005**, *24*, 12–17. [CrossRef]
12. Krzemińska, S.; Rzymski, W.M. The effect of conditions of solvents interactions on barier properties of butyl rubber vulcanizates. *Polimery* **2008**, *53*, 311–316. [CrossRef]
13. El-Sabbagh, S.; Ismail, M.; Mahmoud, D.; Yehia, A. Investigation of natural rubber compatibility with different types of butyl rubber. *Egypt. J. Chem.* **2023**, *66*, 309–321. [CrossRef]
14. Sukharev, N.; Sukhareva, K.; Karpova, S.; Mamin, E.; Popov, A.; Niu, Y.; Ai, T. Effect of blend ratio on the phase structure and ozone aging resistance of ethylene-propylene-diene, butyl, and chlorobutyl rubber blends. *Polym.-Plast. Tech. Mat.* **2023**, *63*, 432–446. [CrossRef]
15. Tavsanli, B.; Bilici, C.; Sungur, P.; Ide, S.; Okay, O. Butyl rubber as a macro-cross-linker in the preparation of a shape-memory and self-healing polymer. *J. Rheol.* **2022**, *66*, 1367–1378. [CrossRef]
16. Guo, X.; Liu, G. Electromagnetic Shielding Enhancement of Butyl Rubber/Single-Walled Carbon Nanotube Composites via Water-Induced Modification. *Polymers* **2023**, *15*, 2101. [CrossRef]
17. Chameswary, J.; Sebastian, M. Preparation and properties of $BaTiO_3$ filled butyl rubber composites for flexible electronic circuit applications. *J. Mater. Sci. Mater. Electron.* **2015**, *26*, 4629–4637. [CrossRef]
18. Hao, S.; Wang, J.; Lavorgna, M.; Fei, G.; Wang, Z.; Xia, H. Constructing 3D graphene network in rubber nanocomposite via liquid-phase redispersion and self-assembly. *ACS Appl. Mater. Inter.* **2020**, *12*, 9682–9692. [CrossRef]
19. Wolff, S. Chemical aspects of rubber reinforcement by fillers. *Rubber Chem. Technol.* **1996**, *69*, 325–346. [CrossRef]
20. Katz, H.; Milewski, J. *Handbook of Fillers for Plastics*, 2nd ed.; Van Nostrand: New York, NY, USA, 1987.
21. Rothon, R. *Particulate-Filled Polymer Composites*; Longman: Harlow, UK, 1995.
22. Mittal, V. Polymer layered silicate nanocomposites: A review. *Materials* **2009**, *2*, 992–1057. [CrossRef]
23. Brechet, Y.; Cavaille, J.; Chabert, E.; Chazeau, L.; Dendievel, R.; Flandin, L.; Gauthier, C. Polymer based nanocomposites: Effect of filler-filler and filler-matrix interactions. *Adv. Eng. Mater.* **2001**, *3*, 571–577. [CrossRef]
24. Azizli, M.J.; Naderi, G.; Bakhshandeh, G.R.; Soltani, S.; Askari, F.; Esmizadeh, E. Improvement in physical and mechanical properties of IIR/CR rubber blend organoclay nanocomposites. *Rubber Chem. Technol.* **2014**, *87*, 10–20. [CrossRef]
25. Razzaghi-Kashani, M.; Hasankhani, H.; Kokabi, M. Improvement in physical and mechanical properties of butyl rubber with montmorillonite organo-clay. *Iranian Polym. J.* **2007**, *16*, 671–679.
26. Duncan, D.; Li, D.; Gaydos, J.; Neumann, A. Correlation of line tension and solid–liquid interfacial tension from the measurement of drop size dependence of contact angles. *J. Colloid Interface Sci.* **1995**, *169*, 256–261. [CrossRef]
27. Pal, S.; Jana, U.; Manna, P.; Mohanta, G.; Manavalan, R. Nanoparticle: An overview of preparation and characterization. *J. Appl. Pharm. Sci.* **2011**, *1*, 228–234.
28. Sonia, T.; Sharma, C. 4—Experimental techniques involved in the development of oral insulin carriers. *Oral Deliv. Insul.* **2014**, 169–217. [CrossRef]
29. Barthlott, W.; Neinhuis, C. Purity of the sacred lotus, or escape from contamination in biological surfaces. *Planta* **2018**, *202*, 1–8. [CrossRef]
30. Si, Y.; Dong, Z.; Jiang, L. Bioinspired Designs of Superhydrophobic and Superhydrophilic Materials. *ACS Cent. Sci.* **2018**, *4*, 1102–1112. [CrossRef] [PubMed]
31. Wu, X.; Yang, F.; Gan, J.; Zhao, W.; Wu, Y. A flower-like waterborne coating with self-cleaning, self-repairing properties for superhydrophobic applications. *J. Mater. Res. Tech.* **2021**, *14*, 1820–1829. [CrossRef]
32. Wang, N.; Xiong, D.; Deng, Y.; Shi, Y.; Wang, K. Mechanically Robust Superhydrophobic Steel Surface with Anti-Icing, UV-Durability, and Corrosion Resistance Properties. *ACS Appl. Mater. Interfaces* **2015**, *7*, 6260–6272. [CrossRef]

33. Fu, K. Mechanically robust, self-healing superhydrophobic anti-icing coatings based on a novel fluorinated polyurethane synthesized by a two-step thiol click reaction. *Chem. Eng. J.* **2021**, *404*, 127110. [CrossRef]
34. Rabek, J.F. *Polimery: Otrzymywanie, Metody Badawcze, Zastosowanie*; Wydawnictwo Naukowe PWN: Warszawa, Poland, 2013.
35. Yuan, Y.; Lee, T.R. Contact angle and wetting properties. *Surf. Sci. Tech.* **2013**, *1*, 3–34.
36. Chibowski, E.; Perea-Carpio, R. Problems of contact angle and solid surface free energy determination. *Adv. Colloid. Interface Sci.* **2002**, *98*, 245–264. [CrossRef] [PubMed]
37. Liber-Kneć, A.; Łagan, S. The use of contact angle and the structure free energy as the surface characteristics of the polymers used in medicine. *Polim. Med.* **2014**, *44*, 29–372. [PubMed]
38. Cui, Y.; Kumar, S.; Kona, B.R.; van Houcke, D. Gas barrier properties of polymer/clay nanocomposites. *RSC Adv.* **2015**, *5*, 63669–63690. [CrossRef]
39. Sorrentino, A.; Gorrasi, G.; Tortora, M.; Vittoria, V. Barrier properties of polymer/clay nanocomposites. *Polym. Nanocomp.* **2006**, *17*, 273–292.
40. Wu, J.; Huang, G.; Li, H.; Wu, S.; Liu, Y.; Zheng, J. Enhanced mechanical and gas barrier properties of rubber nanocomposites with surface functionalized graphene oxide at low content. *Polymer* **2013**, *54*, 1930–1937. [CrossRef]
41. Picard, E.; Vermogen, A.; Gérard, J.-F.; Espuche, E. Barrier properties of nylon 6-montmorillonite nanocomposite membranes prepared by melt blending: Influence of the clay content and dispersion state: Consequences on modelling. *J. Memb. Sci.* **2007**, *292*, 133–144. [CrossRef]
42. Joshi, M.; Banerjee, K.; Prasanth, R.; Thakare, V. Polymer/clay nanocomposite based coatings for enhanced gas barrier property. *Indian J. Fibre Text. Res.* **2006**, *31*, 202–214.
43. Krzemińska, S.; Rzymski, W.M. Effect of layered aluminosilicates on the barier properties of butyl rubber. *Polimery* **2012**, *57*, 551–557. [CrossRef]
44. Królikowski, W.; Rosłaniec, Z. Polymer nanocomposites. *Composites* **2004**, *4*, 3–15.
45. Xu, B.; Zheng, Q.; Song, Y.; Shangguan, Y. Calculating barrier properties of polymer/clay nanocomposites: Effects of clay layers. *Polymer* **2006**, *47*, 2904–2910. [CrossRef]
46. Wong, A.; Wijnands, S.F.L.; Kuboki, T.; Park, C.B. Mechanisms of nanoclay-enhanced plastic foaming processes: Effects of nanoclay intercalation and exfoliation. *J. Nanopart. Res.* **2013**, *15*, 1–15. [CrossRef]
47. Gołębiewski, J.; Różański, A.; Gałęski, A. Study on the process of preparation of polypropylene nanocomposite with montmorillonite. *Polimery* **2006**, *51*, 374–381. [CrossRef]
48. Spasówka, E.; Rudnik, E.; Kijeński, J. Biodegradable nanocomposites. Part I. Methods of preparation. *Polimery* **2006**, *51*, 617–626. [CrossRef]
49. Alexandre, M.; Dubois, P. Polymer layered silicate nanocomposites: Preparation, properties and uses of a new class of materials. *Mater. Sci. Eng. R Rep.* **2000**, *28*, 1–63. [CrossRef]
50. Osman, M.; Mittal, V.; Suter, U. Poly(propylene)-layered silicate nanocomposites: Gas permeation properties and clay exfoliation. *Macromol. Chem. Phys.* **2007**, *208*, 68–75. [CrossRef]
51. Salahuddin, N.; Hapeez, M.; Abidin, A.; Hamzah, N.; Mustam, M. Investigation of effectiveness of phyllosilicate into polymeric materials. In Proceedings of the 2014 IEEE Student Conference on Research and Development, Penang, Malaysia, 16–17 December 2014.
52. Gopakumar, T.; Lee, J.; Kontopoulou, M.; Parent, J. Influence of clay exfoliation on the physical properties of montmorillonite/polyethylene composites. *Polymer* **2002**, *43*, 5483–5491. [CrossRef]
53. Thellen, C.; Orroth, C.; Froio, D.; Ziegler, D.; Lucciarini, J.; Farrell, R.; D'Souza, N.; Ratto, J. Influence of montmorillonite layered silicate on plasticized poly(l-lactide) blown films. *Polymer* **2005**, *46*, 11716–11727. [CrossRef]
54. Di, Y.; Iannace, S.; Maio, E.; Nicolais, L. Nanocomposites by melt intercalation based on polycaprolactone and organoclay. *J. Polym. Sci. B Polym. Phys.* **2003**, *41*, 670–678. [CrossRef]
55. Kato, M.; Usuki, A.; Okada, A. Synthesis of polypropylene oligomer-clay intercalation compounds. *J. Appl. Polym. Sci.* **1997**, *66*, 1781–1785. [CrossRef]
56. Kawasumi, M.; Hasegawa, N.; Kato, M.; Usuki, A.; Okada, A. Preparation and mechanical properties of polypropylene-clay hybrids. *Macromolecules* **1997**, *30*, 6333–6338. [CrossRef]
57. Osman, M.; Rupp, J.; Suter, U. Gas permeation properties of polyethylene-layered silicate nanocomposites. *J. Mater. Chem.* **2005**, *15*, 1298–1304. [CrossRef]
58. Mittal, V. *Novel Polymers and Nanoscience*; Adeli, M., Ed.; Research Signpost Publishers: Trivandrum, India, 2008; pp. 73–92.
59. Arias, J.; Rosa, J.; Marques, M. Influence of phyllosilicate structure on performance of polypropylene nanocomposites prepared via in-situ polymerization. *J. Nanosci. Nanotechnol.* **2019**, *19*, 1908–1922. [CrossRef]
60. Mittal, V. Polypropylene-layered silicate nanocomposites: Filler matrix interactions and mechanical properties. *J. Thermoplast. Compos. Mater.* **2007**, *20*, 575–599. [CrossRef]
61. Manias, E. A direct blending approach for polypropylene/clay nanocomposites enhances properties. *MRS Bull.* **2001**, *26*, 862–863.
62. Szczygielska, A.; Kijeński, J. The application of halloysite as a filler for modification of polypropylene. Part I. The characteristics of halloysite as a filler. *Composites* **2010**, *10*, 186–191.
63. Kunert, A.; Zaborski, M. The structure, properties and uses of layered minerals. *Przem. Chem.* **2006**, *85*, 1510–1517.

64. Zhang, Q.; Liu, Q.; Zhang, Y.; Cheng, H.; Lu, Y. Silane-grafted silica-covered kaolinite as filler of styrene butadiene rubber. *Appl. Clay Sci.* **2012**, *65–66*, 134–138. [CrossRef]
65. Wypych, G. *Handbook of Fillers*; ChemTec Publishing: Toronto, ON, Canada, 1999; p. 120.
66. Kubicz, A. Vermiculite, its properties, occurrence and use. *Przegląd Geol.* **1960**, *8*, 376–380.
67. Célini, N.; Bergaya, F.; Poncin-Epaillard, F. Grafting of hydrocarbon moieties on smectites by cold acetylene plasma and characterization of plasma-treated clay mineral polyethylene nanocomposites. *Polymer* **2007**, *48*, 58–67. [CrossRef]
68. Cetin, M.S.; Toprakci, O.; Taskin, O.S.; Aksu, A.; Toprakci, H.A.K. Expanded vermiculite-filled flexible polymer composites. *J. Elastom. Plast.* **2022**, *54*, 145–168. [CrossRef]
69. Vasilev, A.P. Mechanical and tribological properties of polytetrafluoroethylene composites with carbon fiber and layered silicate fillers. *Molecules* **2019**, *24*, 224. [CrossRef]
70. Khammassi, S.; Tarfaoui, M.; Škrlová, K.; Měřínská, D.; Plachá, D.; Erchiqui, F. Poly(Lactic Acid) (PLA)-Based Nanocomposites: Impact of Vermiculite, Silver, and Graphene Oxide on Thermal Stability, Isothermal Crystallization, and Local Mechanical Behavior. *J. Comp. Sci.* **2022**, *6*, 112. [CrossRef]
71. Namen, F.; Galan, J., Jr.; Oliveira, J.; Cabreira, R.; Costa e Silva Filho, F.; Souza, A.; Deus, G. Surface properties of dental polymers: Measurements of contact angles, roughness and fluoride release. *Mater. Res.* **2008**, *11*, 239–243. [CrossRef]
72. Mendoza, A.; Moriana, R.; Hillborg, H.; Strömberg, E. Super-hydrophobic zinc oxide/silicone rubber nanocomposite surfaces. *Surf. Interfaces* **2019**, *14*, 146–157. [CrossRef]
73. Yüce, M.; Demirel, A.; Menzel, F. Tuning the surface hydrophobicity of polymer/nanoparticle composite films in the wenzel regime by composition. *Langmuir* **2005**, *21*, 5073–5078. [CrossRef]
74. Karthika, M.; Chi, H.; Li, T.; Wang, H.; Thomas, S. Super-hydrophobic graphene oxide-azobenzene hybrids for improved hydrophobicity of polyurethane. *Compos. Part B Eng.* **2019**, *173*, 106978.
75. Mohamad Shahimin, M.; Mohd Rodzi, N.; Omar, M.; Poopalan, P.; Man, B.; Md Nor, M. Hybrid elastomer-nanotube matrix for hydrophobic surface functionalization. *J. Adhes. Sci. Technol.* **2015**, *29*, 532–542. [CrossRef]
76. He, P.; Zhang, Z.; Xia, W.; Shu, L.; Ma, X.; Gou, F.; Zhang, K. Compatibility between high-flux helium plasma irradiated molybdenum and liquid lithium. *J. Nucl. Mater.* **2018**, *509*, 736–741. [CrossRef]
77. Hanaei, H.; Assadi, M.; Saidur, R. Highly efficient antireflective and self-cleaning coatings that incorporate carbon nanotubes (CNTs) into solar cells: A review. *Renew. Sustain. Energy Rev.* **2016**, *59*, 620–635. [CrossRef]
78. Takahashi, S.; Goldberg, H.; Feeney, C.A.; Karim, D.P.; Farrell, M.; O'Leary, K.; Paul, D.R. Gas barrier properties of butyl rubber/vermiculite nanocomposite coatings. *Polymer* **2006**, *47*, 3083–3093. [CrossRef]
79. *EN ISO 374-1:2016*; Protective Gloves against Dangerous Chemicals and Micro-Organisms—Part 1: Terminology and Performance Requirements for Chemical Risks. International Organization for Standardization: Geneva, Switzerland, 2016.
80. Pinnavaia, T.J.; Beall, G.W. *Polymer-Clay Nanocomposites*; Wiley: Chichester, UK, 1997.
81. Wolf, C.; Angellier-Coussy, H.; Gontard, N.; Doghieri, F.; Guillard, V. How the shape of fillers affects the barrier properties of polymer/non-porous particles nanocomposites: A review. *J. Membr. Sci.* **2018**, *556*, 393–418. [CrossRef]
82. Szadkowski, B.; Marzec, A.; Rybiński, P.; Żukowski, W.; Zaborski, M. Characterization of ethylene-propylene composites filled with perlite and vermiculite minerals: Mechanical, barrier, and flammability properties. *Materials* **2020**, *13*, 585. [CrossRef] [PubMed]
83. Rooj, S.; Das, A.; Thakur, V.; Mahaling, R.N.; Bhowmick, A.K.; Heinrich, G. Preparation and properties of natural nanocomposites based on natural rubber and naturally occurring halloysite nanotubes. *Mater. Des.* **2010**, *31*, 2151–2156. [CrossRef]
84. *EN 388:2016+A1:2018*; Protective Gloves against Mechanical Risks. European Standard: Plzen, Czech Republic, 2018.
85. Mostafa, A.; Abouel-Kasem, A.; Bayoumi, M.R.; El-Sabaie, M.G. Rubber-Filler Interactions and Its effect in rheological and mechanical properties of filled compounds. *J. Test. Eval.* **2010**, *38*, 347–359. [CrossRef]
86. *ASTM D5289-17*; Standard Test Method for Rubber Property—Vulcanization Using Rotorless Cure Meters. ASTM International: West Conshohocken, PA, USA, 2017.
87. Sepehri, A.; Razzaghi-Kashani, M.; Ghoreishy, M.H.R. Vulcanization Kinetics of Butyl Rubber-Clay Nano Composites and Its Dependence on Clay Microstructure. *J. Appl. Polym. Sci.* **2012**, *125*, E204–E213. [CrossRef]
88. *ASTM D471*; Standard Test Method for Rubber Property—Effect of Liquids. ASTM International: West Conshohocken, PA, USA, 2016.
89. *ISO 34-1:2022*; Rubber, Vulcanized or Thermoplastic Determination of Tear Strength—Part 1: Trouser, Angle and Crescent Test Pieces. International Organization for Standardization: Geneva, Switzerland, 2022.
90. *ISO 48-4:2018*; Rubber, Vulcanized or Thermoplastic Determination of Hardness—Part 4: Indentation Hardness by Durometer Method (Shore Hardness). International Organization for Standardization: Geneva, Switzerland, 2018.
91. *ISO 4664-1:2022*; Rubber, Vulcanized or Thermoplastic Determination of Dynamic Properties—Part 1: General Guidance. International Organization for Standardization: Geneva, Switzerland, 2022.
92. *EN ISO 374-4:2019*; Protective Gloves against Dangerous Chemicals and Micro-Organisms—Part 4: Determination of Resistance to Degradation by Chemicals. International Organization for Standardization: Geneva, Switzerland, 2019.
93. *EN ISO 2231:1995/PN-EN ISO 2231:1999*; Rubber- or Plastics-Coated Fabrics. Standard Atmospheres for Conditioning and Testing. International Organization for Standardization: Geneva, Switzerland, 1999.

94. *PN-EN 388:2017-02*; Rękawice chroniące przed zagrożeniami mechanicznymi. Polish Committee for Standardization: Warszawa, Poland, 2017.
95. *PN-EN ISO 374-4:2020-03*; Rękawice chroniące przed niebezpiecznymi substancjami chemicznymi i mikroorganizmami—Część 4: Wyznaczanie odporności na degradację w wyniku działania substancji chemicznych. Polish Committee for Standardization: Warszawa, Poland, 2020.

molecules

MDPI

Article

Fluorine-Containing Flow Modifier for BN/PPS Composites Enabled by Low Surface Energy

Bo Cao [1,2,*], Xiaodan Huang [1], Wenxiang Zhang [1] and Peng Wu [1]

1 School of Materials Science and Engineering, South China University of Technology, Guangzhou 510640, China
2 Key Lab Guangdong High Property & Functional Polymer Materials, and Key Laboratory of Polymer Processing Engineering, Ministry of Education, Guangzhou 510640, China
* Correspondence: shawn78521@sina.com

Abstract: In this study, a fluorine-containing flow modifier (Si-DF) with low surface energy is successfully synthesized, which is applied to fabricate ideal electronic packaging materials (BN/PPS composites) with high thermal conductivity, excellent dielectric properties, processability, and toughness by conventional melt blending. Si-DPF is located at the interface between the BN fillers and the PPS matrix, which not only improves the dispersion of BN fillers but also strengthens the interaction. With the help of 5 wt% Si-DF, **BN/PPS/Si-DF (70/25/5)** still exhibits the high thermally conductive coefficient (3.985 W/m·K) and low dielectric constant (3.76 at 100 MHz) although BN fillers are loaded as high as 70 wt%. Moreover, the sample processes a lower stable torque value (2.5 N·m), and the area under the stress–strain curves is also increased. This work provides an efficient way to develop high-performance polymer-based composites with high thermally conductive coefficients and low dielectric constants for electronic packaging applications.

Keywords: BN/PPS composites; fluorine-containing flow modifier; dispersion; thermal conductivity; dielectric property

1. Introduction

Polyphenylene sulfide (PPS) has been widely used as electronic packaging material in the field of electronic communications due to its outstanding thermal dimensional stability, superior chemical resistance, and high flame retardance. With continuous miniaturization and increased power in electronic devices, the comprehensive performance of electronic packaging materials makes more stringent requirements [1–4]. An ideal electronic packaging material should demonstrate high thermal conductivity, excellent dielectric properties, processability, and toughness. Unfortunately, the PPS matrix has a relatively low thermally conductive coefficient (<0.4 W/m·K), which needs to be enhanced by incorporating thermally conductive particles [5–8]. Some studies have shown that introducing a large number of thermally conductive particles leads to a dramatic increase in the thermal conductivity of the PPS-based composites, such as boron nitride (BN), expanded graphite (EG), aluminum nitride (AlN), and carbon nanotubes (CNT) [9–12].

However, a high loading of particles easily forms a large number of agglomerates, deteriorating the processability and toughness of the composites [13,14]. Moreover, the serious agglomeration of particles also leads to more interfacial thermal barriers, which is not conducive to the fabrication of composites with high thermally conductive coefficients and low dielectric constants [15]. Thus, a major challenge in the fabrication of ideal electronic packaging materials (PPS-based composites) is to achieve the uniform dispersion of thermally conductive particles in the PPS matrix.

Usually, the PPS-based composites with high thermal conductivity are prepared by a hot-compression technique in which the surface-modified particles are premixed with the PPS matrix, which greatly improves the dispersion of the particles [16–19].

Yang et al. [16] modified BN fillers using a silanization reaction with γ-aminopropyl triethoxy silane/aminopropyllsobutyl polyhedraloligomeric silsesquioxane (f-BN) and found that f-BN fillers had better dispersion in the PPS matrix. The thermally conductive coefficient (λ) value of the f-BN/PPS composite with 60 wt% f-BN was improved to 1.122 W/m·K, four times higher than that of the PPS matrix (0.286 W/m·K). Ryu et al. [19] reported that the grafting of (3-Aminopropyl) triethoxysilane onto the surface of BN fillers reduced agglomeration. The λ value of 3.09 W/m·K was exhibited by the modified BN containing 60 wt%. However, due to the complexity of this method, it cannot be extended to large-scale industrial production. Therefore, the uniformly dispersed particles in the PPS matrix could be easily realized by melt blending technology, which is of prime importance for the development of ideal electronic packaging materials in industry.

In this work, a fluorine-containing flow modifier (Si-DF) is designed and synthesized (Figure 1), which is introduced into BN/PPS composites (the weight ratio of BN to PPS is 60:40 and 70:30) by a conventional melt processing technique. The -CF_3 terminal groups endow Si-DF with low surface energy, which enables Si-DF to locate at a two-phase interface [14]. It is expected that the BN fillers will be dispersed in the PPS matrix, and the BN/PPS composites will present high thermal conductivity, excellent processability, and toughness. The composite will also maintain a low dielectric constant with the help of Si-DF. In addition, the dispersion mechanism is explored.

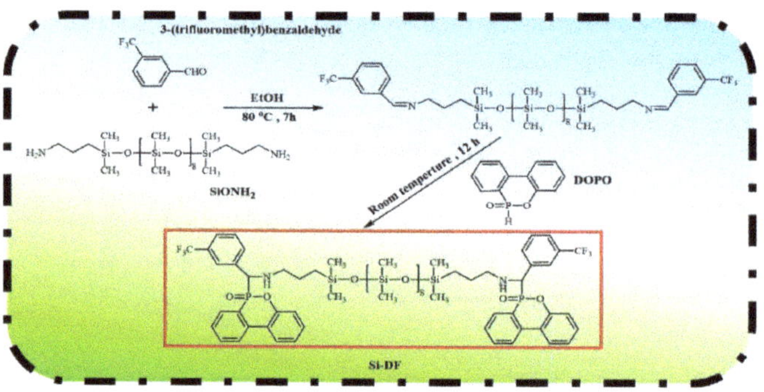

Figure 1. Synthesis route of Si-DF.

2. Experimental

2.1. Main Materials

Polyphenylene sulfide (PPS) was purchased from Nanjing Deyuan Science and Technology Co. Ltd. (Nanjing, China) Boron nitride (BN, mean diameter of 20–30 μm, thermally conductive coefficient of 33 W/m·K) was received from Qinhuangdao ENO High-Tech Materials Development Co., Ltd. (Qinhuangdao, China) Amino silicone oil was purchased from Shin-Etsu Chemical Industrial Co. Ltd. (Tokyo, Japan) 9,10-dihydro-9-oxa-10-phosphaphenanthrene 10-oxide (DOPO) and 3-(Trifluoromethyl) benzaldehyde were purchased from Shanghai Aladdin Biochemical Co. Ltd. (Shanghai, China).

2.2. Synthesis of Si-DF

Amino silicone oil (17.2 g, 0.02 mol), 3-(Trifluoromethyl) benzaldehyde (6.96 g, 0.04 mol), and 100 mL of ethanol were added sequentially to a 250 mL round-bottom glass flask equipped with a nitrogen inlet, a condensing device, and a magnetic stirrer. The mixture was stirred at 80 °C for 8 h until the complete conversion of amino silicone oil and 3-(Trifluoromethyl) benzaldehyde, as confirmed by FTIR, and then the intermediate (Si-CF) was obtained. DOPO (8.64 g, 0.04 mol) was added into the system and then further reacted for 12 h at room temperature. After the complete conversion of DOPO was tested via FTIR, the ethanol was removed using a rotary evaporator, and the residual solid was recrystallized from ethyl acetate to yield a yellowish powder named Si-DF (30.76 g, yield: 93.8 %). GPC data: M_n = 1398 g/mol, PDI = 1.22.

2.3. Preparation of BN/PPS Composites

BN and PPS were dried in a vacuum oven at 90 °C for 12 h. The BN/PPS composites were prepared by melt blending in a twin-screw extruder, and the temperatures of the extruder were set in the range from 265 °C to 295 °C at a screw speed of 120 rpm. After being granulated and dried, the pellets were injection-molded on an injection-molding machine to obtain different specimens. The molding was conducted at zone temperature profiles of 275-280-290-290-295 °C. The composite preparation process and composition are simply illustrated in Scheme 1, and the compositions and code names of the samples are listed in Table 1.

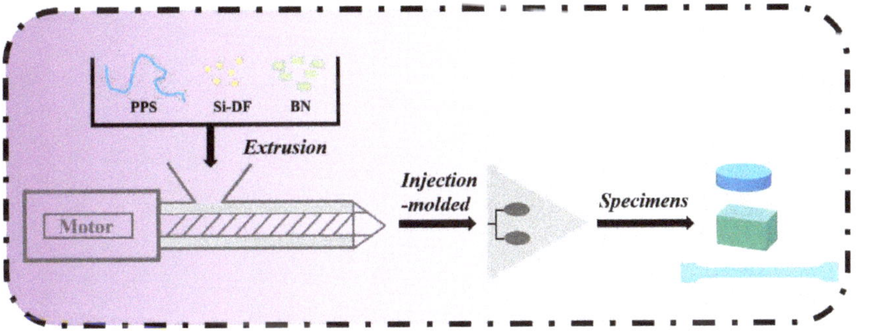

Scheme 1. Schematic illustration of PPS composites preparation.

Table 1. Compositions and code names of the samples.

Samples	BN (g)	PPS (g)	Si-DF (g)
BN/PPS (60/40)	60	40	-
BN/PPS/Si-DF (60/35/5)	60	35	5
BN/PPS (70/30)	70	30	-
BN/PPS/Si-DF (70/25/5)	70	25	5

2.4. Characterization

The FTIR spectra (scanned between 400 and 4000 cm^{-1}) were obtained on a Nicolet 50XC spectrometer (Nicolet, Glendale, WI, USA). The ^{1}H NMR and ^{31}P NMR spectra were recorded on an AV400 unity spectrometer (Bruker, Billerica, MA, USA) operated at 400 MHz with DMSO-d_6 as a solvent. The number-averaged molar mass (M_n) and polydispersity (M_w/M_n) of Si-DF were measured by gel permeation chromatography (Waters, Milford, MA, USA). The contact angles were determined on a DSA100 contact angle instrument (KRUSS, Hamburg, Germany) with water and diiodomethane at 25 °C, respectively. Scanning electron microscopy (SEM) images were obtained on a Nova NanoSEM430 scanning electron microscope (FEI, Hillsboro, OR, USA). The torque values of the samples were measured on a Haake torque XSS-300 rheometer (Haake, Vreden, Germany) at a rotation of 50 rpm at 285 °C for 10 min. The thermally conductive coefficient values of the samples were measured on a TPS2200 Hot Disk instrument (AB Corporation, Västerås, Sweden). Dielectric constant values of the samples were measured on a high-frequency Q instrument of QBG-3D (Shanghai Aiyi Electronic Equipment, Shanghai, China). According to ISO 527-2 standard, the tensile strengths of the samples were tested on an AGS-10KNI universal material testing machine (Shimadzu, Kyoto, Japan) at the speed of 10 mm/min.

3. Results and Discussion

3.1. Characterization of Si-DF

During a typical synthetic process, the intermediate Si-CF is first synthesized via a Schiff base reaction between amino silicone oil and 3-(Trifluoromethyl) benzaldehyde in an EtOH solution. The flow modifier Si-DF is then performed through the addition reaction of DOPO and Si-CF. Figure 2 shows the infrared spectra of Si-CF and Si-DF. The synthesis of the Si-CF intermediate is performed by the condensation of amino silicone oil and 3-(Trifluoromethyl) benzaldehyde. Obviously, the strong peak of -CH=N- (Schiff base) appears at 1650 cm^{-1}, and the peak of -CF_3 appears at 1650 cm^{-1}, which indicates the triggered aldimine polymerization between amino silicone oil and 3-(Trifluoromethyl) benzaldehyde. The Si-DF is performed through the addition reaction of DOPO and Si-CF. Additionally, the peak of -CH=N- is absent at the same wavenumber for Si-DF, indicating that DOPO has successfully carried out the addition reaction with the intermediate containing the -CH=N- group [14,20–22].

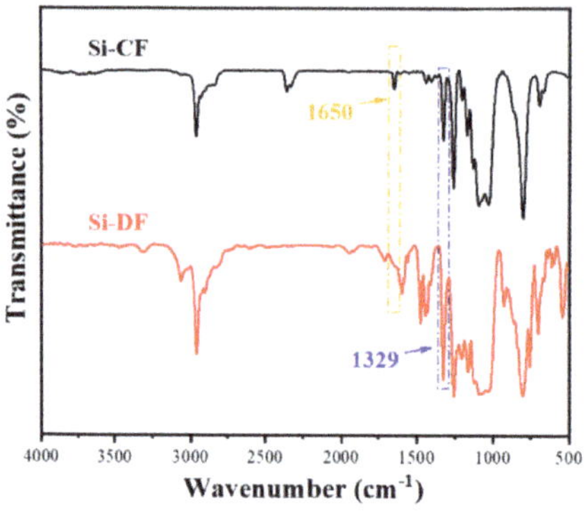

Figure 2. FTIR spectra of Si-CF and Si-DF.

The ^{1}H NMR spectrum of Si-DF is displayed in Figure 3a. The peaks at −0.12–0.16 ppm, 0.32–0.46 ppm, and 1.16–1.42 ppm are attributed to the protons in -Si-CH_3, Si-CH_2-, and -CH_2-CH_2-, respectively. The peaks at 5.29–5.32 ppm and 5.07–5.11 ppm are attributed to the protons in -NH- and P-CH. The peaks at 6.5–8.2 ppm are attributed to the protons in the benzene ring. The ^{31}P NMR spectra of DOPO and Si-DF are shown in Figure 3b. Based on the ^{31}P NMR spectrum of DOPO, the peaks at 14.40–15.75 ppm are observed, while the peaks are absent at the same wavenumber for Si-DF, and new peaks are observed at 31.50–34.85 ppm. The results from FTIR, ^{1}H NMR, and ^{31}P NMR show that the flow modifier Si-DF has been successfully synthesized.

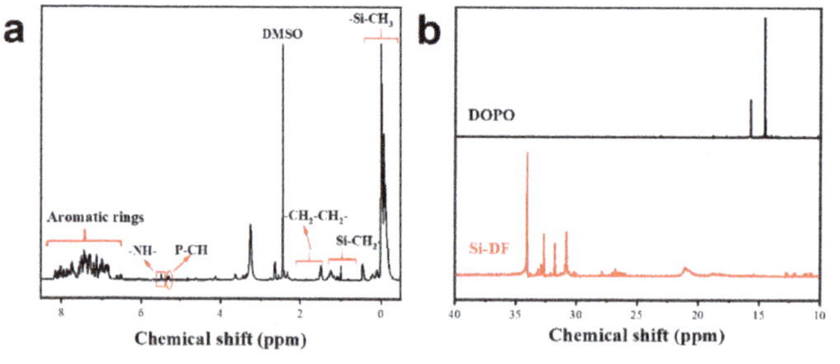

Figure 3. (**a**) ^{1}H NMR spectrum of Si-DF. (**b**) ^{31}P NMR spectra of Si-DF and DOPO.

3.2. Phase Morphology

For highly filled polymer-based composites, how to disperse the fillers in the matrix is a key issue presenting a challenge in the fabrication of high-performance composites. To display the effect of Si-DF on the dispersion of BN fillers, the morphology of the BN fillers in the samples is observed by SEM (Figure 4).

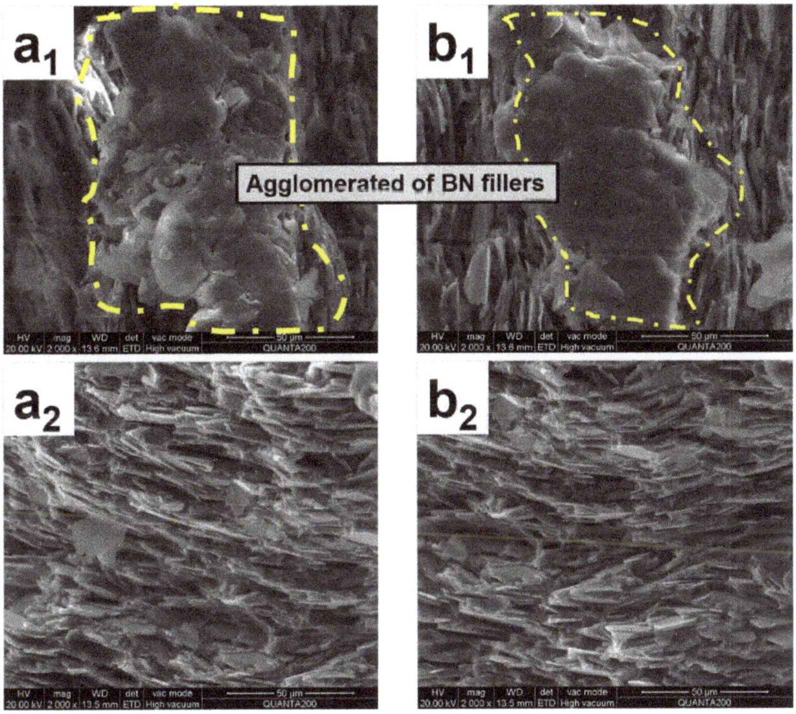

Figure 4. SEM images of BN/PPS (60/40) (**a_1**), BN/PPS/Si-DF (60/35/5) (**a_2**), BN/PPS (70/30) (**b_1**) and BN/PPS/Si-DF (70/25/5) (**b_2**).

There exists serious agglomeration of BN fillers in the sample **BN/PPS (60/40)** (Figure 4a_1) and **BN/PPS (70/30)** (Figure 4b_1). With the addition of Si-DF, the morphology of the BN fillers presents a striking contrast. It is obviously found that BN fillers exhibit a homogeneous dispersion in the samples **BN/PPS/Si-DF (60/35/5)** (Figure 4a_2) and **BN/PPS/Si-DF (70/25/5)** (Figure 4b_2).

In our previous work, we demonstrated that the fluorine-containing flow modifiers (PMFs) with low surface energy were located at the MH/LLDPE interface in the highly filled MH/LLDPE composites (80:20 by weight), which improved the dispersion of MH particles [23]. Meanwhile, the distribution of flow modifiers can be estimated by evaluating the wetting coefficient (ω_a) [24]. Herein, Young's model is used to calculate the ω_a of Si-DF. Additionally, when $-1 < \omega_a < 1$, Si-DF is preferred to locate at the interface. The surface energy of the individual components (PPS, BN, and Si-DF) is examined by measuring the contact angles, and the interfacial energies are calculated using the surface energy of the individual components using the geometric-mean equation or harmonic-mean equation [25–27]. All of the data are listed in Tables 2 and 3, respectively.

Table 2. Contact angles and surface energies of PPS, BN and Si-DF.

Samples	Contact Angle (°)		γ (mJ/m^2)	γ^d mJ/m^2)	γ^p (mJ/m^2)
	Water	Diiodomethane			
PPS	90.37	24.76	46.99	46.63	0.36
BN	89.12	64.89	26.8	22.5	4.3
Si-DF	98.45	51.20	34.08	33.78	0.3

Table 3. Preferential distribution of Si-DF in the BN/PPS composites according to interfacial energies γ and wetting coefficients ω_a.

Interfacial Energy	Based on Geometric-Mean Equation (mJ/m^2)	Based on Harmonic-Mean Equation (mJ/m^2)	ω_G	ω_H	Flow Modifiers Distribution
$\gamma_{BN/PPS}$	6.52	11.75			
$\gamma_{Si\text{-}DF/PPS}$	1.04	2.06	−0.37	−0.31	BN/PPS interface
$\gamma_{Si\text{-}DF/BN}$	3.47	5.74			

As seen, the values of ω_a for Si-DF are −0.37 and −0.31, respectively, as calculated from the geometric-mean equation, the harmonic-mean equation, and Young's model, indicating that Si-DF tends to locate at the interface between the BN fillers and PPS matrix. Combining with the phase morphology and theoretical calculations, it is concluded that the distribution of Si-DF at the two-phase interface is effectively able to reduce the formation of agglomeration in the matrix.

3.3. Thermal Conductivity

Figure 5 presents the thermally conductive coefficient (λ) values of the PPS matrix and BN/PPS composites. The λ value of the PPS matrix is 0.372 W/m·K. The composites reveal a dramatic enhancement in thermal conductivity in comparison with the PPS matrix. The λ values of **BN/PPS (60/40)** and **BN/PPS (70/30)** are 3.379 W/m·K and 3.569 W/m·K, which are approximately 9 and 9.6 times higher than that of the PPS matrix, respectively. With the addition of Si-DF, the samples present higher thermal conductivity, and the corresponding λ values are increased to 3.873 W/m·K of **BN/PPS/Si-DF (60/35/5)** and 3.985 W/m·K of **BN/PPS/Si-DF (70/25/5)**, respectively. This implies that efficient thermal transfer pathways could be formed at a high loading of BN fillers in the PPS matrix by the introduction of Si-DF [16,28]. Additionally, it is believed that the homogeneous dispersion of BN fillers in the PPS matrix is the key to the formation of thermal transfer pathways. Furthermore, Si-DF forms a protective layer on the BN surfaces, which shows relatively lower interfacial thermal barriers with the PPS matrix compared to the samples without Si-DF, resulting in higher λ values.

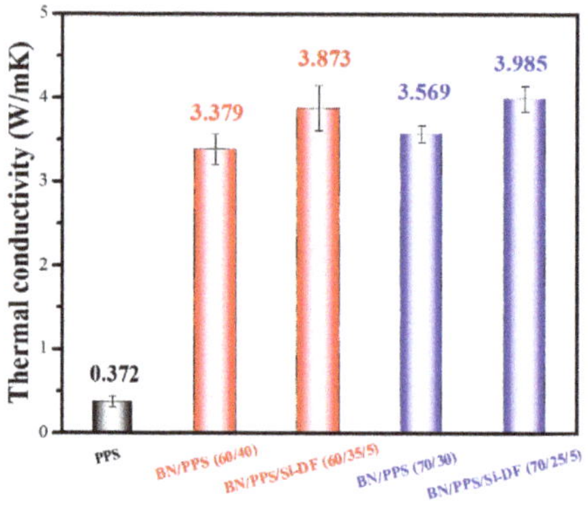

Figure 5. Thermal conductivity of PPS matrix and BN/PPS composites.

Table 4 summarizes the reported thermally conductive coefficient and thermal conductivity enhancement for the BN/PPS composites. It is noted that the BN/PPS composites with higher BN filler contents in this work show a high thermally conductive coefficient and the highest thermal conductivity enhancement. This work provides a relatively more efficient and facile method to improve the thermal conductivity of BN/PPS composites.

Table 4. Comparison of thermally conductive coefficient and thermal conductivity enhancement in BN/PPS composites.

Samples	Filler Loading [wt%]	Thermally Conductive Coefficient [W/m·K]	Thermal Conductivity Enhancement [%]	Strategy	Year [Ref]
BN/PPS	60	2.638	822.3	Hot-compression	2017 [15]
BN/PPS	60	1.122	292.3	Surface-modification, Hot-compression	2017 [16]
BN/PPS	40 (vol)	2.45	880	Hot-compression	2017 [18]
BN/PPS	60	3.1	785.8	Surface-modification	2017 [19]
BN/PPS	60	2.7	610.5	Melt blending	2017 [19]
BN/PPS	60	3.873	941.1	Melt blending	This work
BN/PPS	70	3.985	971.2	Melt blending	This work

3.4. Dielectric Property

As a key parameter of electronic packaging materials, a low dielectric constant is conducive to reducing the signal propagation time in electronic components, which is very important in practical applications. The dielectric constant-frequency curves of the PPS matrix and BN/PPS composites are displayed in Figure 6. The PPS matrix possesses a dielectric constant value of 3.32 at 100 MHz. It can be seen that the dielectric constant value of the samples is increased as the loading of BN fillers, and the corresponding values are increased to 3.75 for **BN/PPS (60/40)** and 3.98 for **BN/PPS (70/30)**. It is worth noting that the introduction of Si-DF is beneficial in decreasing the dialectic constant of the samples. For samples **BN/PPS/Si-DF (60/35/5)** and **BN/PPS/Si-DF (70/25/5)**, the dielectric constant values decreased to 3.55 and 3.76, respectively. The lower dielectric constant should be attributed to the weakening of interface polarization [29,30]. Si-DF not only improves the dispersion of BN fillers but also strengthens the interaction between the BN fillers and PPS matrix, weakening the polarization between the BN fillers and PPS matrix, resulting in relatively lower dielectric constant values of the BN/PPS composites.

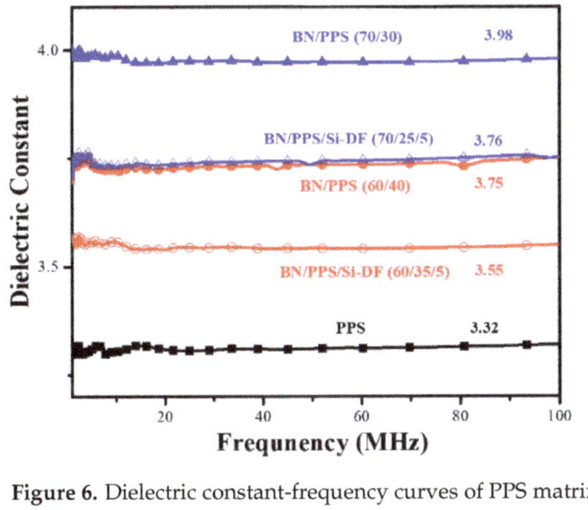

Figure 6. Dielectric constant-frequency curves of PPS matrix and BN/PPS composites.

3.5. Processability

The processability of BN/PPS composites is evaluated by the torque rheology test [31]. Figure 7 shows the torque vs. time curves for the BN/PPS composites. As expected, the stable torque values of the samples are increased with the loading of BN fillers, and a pronounced decrease in the torque values of the BN/PPS composites with the addition of Si-DF is observed. For samples **BN/PPS (60/40)** and **BN/PPS (70/30)**, the stable torque values are 3.0 N·m and 11.7 N·m. When 5 wt% of Si-DF is added, the corresponding values are decreased to 2.1 N·m of **BN/PPS/Si-DF (60/35/5)** and 2.5 N·m of **BN/PPS/Si-DF (70/25/5)**, respectively, indicating that Si-DF has an obvious advantage in improving the processability of the BN/PPS composites. Si-DF tends to be located at the interface and effectively reduces the melt viscosity, which prevents BN fillers from aggregation and decreases the friction between the BN fillers and between the BN fillers and the PPS matrix.

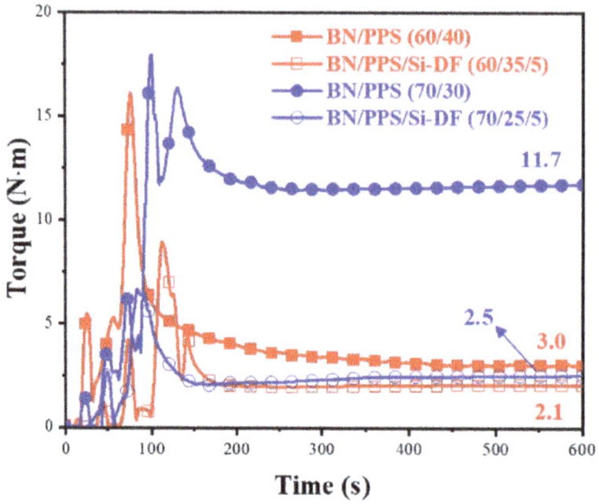

Figure 7. Torque versus time of BN/PPS composites.

3.6. Mechanical Properties

Mechanical properties are critical for composites, especially tensile strength and toughness. In this work, tensile testing of the PPS matrix and BN/PPS composites is conducted, and the results are shown in Figure 8. The tensile strength of the PPS matrix is 38 MPa. With the addition of BN fillers, the tensile strength of the samples increased, and the corresponding strength increased to 48 MPa for **BN/PPS (60/40)** and 43 MPa for **BN/PPS (70/30)**. The tensile strength is slightly decreased for the samples with Si-DF. The toughness of the samples is evaluated by calculating the fracture work derived from the area under the stress–strain curve [32,33]. The work of the fracture of the PPS matrix is 13.66 MJ/m^3, and the work of the fracture is decreased with the loading of BN fillers. For samples **BN/PPS (60/40)** and **BN/PPS (70/30)**, the work of the fracture decreased to 7.46 MJ/m^3 and 5.53 MJ/m^3, respectively. This is understandable because BN fillers aggregate easily and form agglomerations in the PPS matrix. As expected, after adding 5 wt% of Si-DF, the work of the fracture of the samples increased to 11.05 MJ/m^3 of **BN/PPS/Si-DF (60/35/5)** and 8.97 MJ/m^3 of **BN/PPS/Si-DF (70/25/5)**, indicating that Si-DPF has an obvious advantage in improving the toughness of the highly filled BN/PPS composites. This result is mainly ascribed to the distribution of Si-DF; that is, Si-DPF tends to be located at the interface and forms a "barrier layer" for BN fillers. Si-DF not only reduces the agglomeration but also transfers the external force received by the samples to the small aggregate of BN filler, which is able to act as the stress concentration points to dissipate the external energy, and the toughness of the sample is achieved.

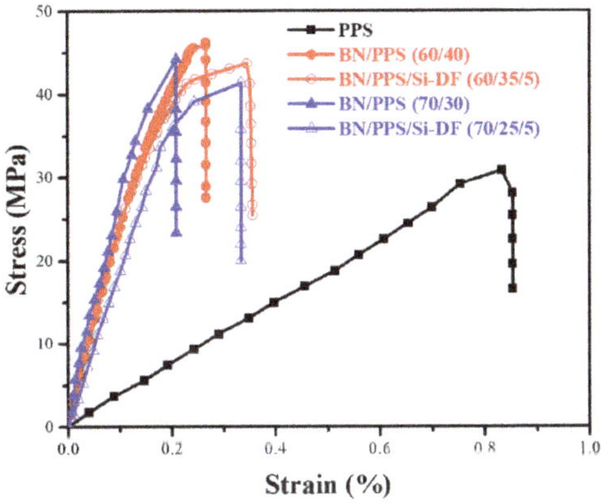

Figure 8. Strain–stress curves of PPS matrix and BN/PPS composites.

4. Conclusions

A fluorine-containing flow modifier Si-DF was successfully synthesized and applied to the fabrication of ideal electronic packaging materials (BN/PPS composites) by conventional melt blending techniques. Si-DF with low surface energy was dominantly located at the BN/PPS interface, leading to the relatively homogeneous dispersion of the BN fillers, thus successfully achieving simultaneously high thermal conductivity, excellent dielectric properties, processability, and toughness in BN/PPS composites, solving the well-known problem that polymer-based composites with high loading thermally conductive fillers usually present high thermal conductivity but poor processability and toughness due to the unavoidable aggregation of the fillers. The Si-DF located at the interface effectively constructed thermal conductive paths and reduced the interfacial heat resistance between the BN fillers and the PPS matrix as well. Meanwhile, the carefully developed additive weakened the polarization between the fillers and the matrix, which reduced the dielectric constant of the composites. In addition, the BN/PPS exhibited excellent processability and toughness. Most importantly, this work paved the way for constructing ideal electronic packaging materials through the rational design of flow modifiers.

Author Contributions: Conceptualization, B.C.; methodology, B.C. and X.H.; software, W.Z.; validation, X.H. and P.W.; investigation, B.C. and X.H.; data curation, X.H.; writing—original draft preparation, B.C.; writing—review and editing, B.C.; visualization, P.W.; supervision, W.Z. All authors have read and agreed to the published version of the manuscript.

Funding: This research was funded by Guangdong Basic and Applied Basic Research Foundation (No. 2021A1515110134) and the Opening Project of Key Laboratory of Polymer Processing Engineering (South China University of Technology), Ministry of Education (No. KFKT2102).

Institutional Review Board Statement: Not applicable.

Informed Consent Statement: Not applicable.

Data Availability Statement: The data that support the findings of this study are available from the authors upon reasonable request.

Conflicts of Interest: The authors declare no conflict of interest.

Sample Availability: Samples of the compounds are available from the authors.

References

1. Shahil, K.M.; Balandin, A.A. Graphene-Multilayer Graphene Nanocomposites as Highly Efficient Thermal Interface Materials. *Nano Lett.* **2012**, *12*, 861–867. [CrossRef] [PubMed]
2. Han, Z.; Fina, A. Thermal Conductivity of Carbon Nanotubes and Their Polymer Nanocomposites: A Review. *Prog. Polym. Sci.* **2011**, *36*, 914–944. [CrossRef]

3. Lee, G.W.; Park, M.; Kim, J.; Lee, J.I.; Yoon, H.G. Enhanced Thermal Conductivity of Polymer Composites Filled with Hybrid Filler. *Compos. Part A Appl. Sci. Manuf.* **2006**, *37*, 727–734. [CrossRef]
4. Song, S.H.; Park, K.H.; Kim, B.H.; Choi, Y.W.; Jun, G.H.; Lee, D.J.; Kong, B.S.; Paik, K.W.; Jeon, S. Enhanced Thermal Conductivity of Epoxy-Graphene Composites by Using Non-Oxidized Graphene Flakes with Non-Covalent Functionalization. *Adv. Mater.* **2013**, *25*, 732–737. [CrossRef] [PubMed]
5. Xu, L.; Chen, G.; Wang, W.; Li, L.; Fang, X. A Facile Assembly of Polyimide/Graphene Core–Shell Structured Nanocomposites with Both High Electrical and Thermal Conductivities. *Compos. Part A Appl. Sci. Manuf.* **2016**, *84*, 472–481. [CrossRef]
6. Kim, H.S.; Jang, J.U.; Yu, J.; Kim, S.Y. Thermal Conductivity of Polymer Composites Based on the Length of Multi-Walled Carbon Nanotubes. *Compos. Part B Eng.* **2015**, *79*, 505–512. [CrossRef]
7. Gu, J.; Yang, X.; Lv, Z.; Li, N.; Liang, C.; Zhang, Q. Functionalized Graphite Nanoplatelets/Epoxy Resin Nanocomposites with High Thermal Conductivity. *Int. J. Heat Mass Transf.* **2016**, *92*, 15–22. [CrossRef]
8. Huang, X.; Zhi, C.; Jiang, P.; Golberg, D.; Bando, Y.; Tanaka, T. Polyhedral Oligosilsesquioxane-Modified Boron Nitride Nanotube Based Epoxy Nanocomposites: An Ideal Dielectric Material with High Thermal Conductivity. *Adv. Funct. Mater.* **2013**, *23*, 1824–1831. [CrossRef]
9. Han, Y.; Shi, X.; Yang, X.; Guo, Y.; Zhang, J.; Kong, J.; Gu, J. Enhanced Thermal Conductivities of Epoxy Nanocomposites Via Incorporating in-Situ Fabricated Hetero-Structured SiC-BNNS Fillers. *Compos. Sci. Technol.* **2020**, *187*, 10944. [CrossRef]
10. Guo, Y.; Xu, G.; Yang, X.; Ruan, K.; Ma, T.; Zhang, Q.; Gu, J.; Wu, Y.; Liu, H.; Guo, Z. Significantly Enhanced and Precisely Modeled Thermal Conductivity in Polyimide Nanocomposites with Chemically Modified Graphene Via in Situ Polymerization and Electrospinning-Hot Press Technology. *J. Mater. Chem. C* **2018**, *6*, 3004–3015. [CrossRef]
11. Xie, S.H.; Zhu, B.K.; Li, J.B.; Wei, X.Z.; Xu, Z.K. Preparation and Properties of Polyimide/Aluminum Nitride Composites. *Polym. Test.* **2004**, *23*, 797–801. [CrossRef]
12. Jiang, Q.; Wang, X.; Zhu, Y.; Hui, D.; Qiu, Y. Mechanical, Electrical and Thermal Properties of Aligned Carbon Nanotube/Polyimide Composites. *Compos. Part B Eng.* **2014**, *56*, 408–412. [CrossRef]
13. Guan, X.; Cao, B.; Cai, J.; Ye, Z.; Lu, X.; Huang, H.; Liu, S.; Zhao, J. Design and Synthesis of Polysiloxane Based Side Chain Liquid Crystal Polymer for Improving the Processability and Toughness of Magnesium Hydrate/Linear Low-Density Polyethylene Composites. *Polymers* **2020**, *12*, 911. [CrossRef] [PubMed]
14. Cao, B.; Wang, O.; Cai, J.; Xie, W.; Liu, S.; Zhao, J. Silicone/Fluorine-Functionalized Flow Modifier with Low Surface Energy for Improving Interfaces in Highly Filled Composites. *Compos. Sci. Technol.* **2021**, *214*, 108984. [CrossRef]
15. Gu, J.; Guo, Y.; Yang, X.; Liang, C.; Geng, W.; Tang, L.; Li, N.; Zhang, Q. Synergistic Improvement of Thermal Conductivities of Polyphenylene Sulfide Composites Filled with Boron Nitride Hybrid Fillers. *Compos. Part A Appl. Sci. Manuf.* **2017**, *95*, 267–273. [CrossRef]
16. Yang, X.; Tang, L.; Guo, Y.; Liang, C.; Zhang, Q.; Kou, K.; Gu, J. Improvement of Thermal Conductivities for Pps Dielectric Nanocomposites Via Incorporating NH_2-POSS Functionalized nBN Fillers. *Compos. Part A Appl. Sci. Manuf.* **2017**, *101*, 237–242. [CrossRef]
17. Pan, C.; Kou, K.; Jia, Q.; Zhang, Y.; Wu, G.; Ji, T. Improved Thermal Conductivity and Dielectric Properties of hBN/PTFE Composites Via Surface Treatment by Silane Coupling Agent. *Compos. Part B Eng.* **2017**, *111*, 83–90. [CrossRef]
18. Jiang, Y.; Liu, Y.; Min, P.; Sui, G. BN-MWCNT/PPS core-shell Structure Particles and Their 3D Segregated Architecture Composites with High Thermal Conductivities. *Compos. Sci. Technol.* **2017**, *144*, 63–69. [CrossRef]
19. Ryu, S.; Kim, K.; Kim, J. Silane Surface Modification of Boron Nitride for High Thermal Conductivity with Polyphenylene Sulfide Via Melt Mixing Method. *Polym. Adv. Technol.* **2017**, *28*, 1489–1494. [CrossRef]
20. Chao, P.; Li, Y.; Gu, X.; Han, D.; Jia, X.; Wang, M.; Zhou, T.; Wang, T. Novel Phosphorus–Nitrogen–Silicon Flame Retardants and Their Application in Cycloaliphatic Epoxy Systems. *Polym. Chem.* **2015**, *6*, 2977–2985. [CrossRef]
21. Ding, C.; Hu, D.; He, X.; Lai, Y.; Shao, G. Fabrication and microstructure evolution of monolithic bridged polysilsesquioxane-derived SiC ceramic aerogels. *Ceram. Int.* **2022**, *48*, 25833–25839. [CrossRef]
22. Shao, G.; Shen, X.; Huang, X. Multilevel Structural Design and Heterointerface Engineering of a Host–Guest Binary Aerogel toward Multifunctional Broadband Microwave Absorption. *ACS Mater. Lett.* **2022**, *4*, 1787–1797. [CrossRef]
23. Cao, B.; Zhou, Y.; Wu, Y.; Cai, J.; Guan, X.; Liu, S.; Zhao, J.; Zhang, M. Simultaneous Improvement of Processability and Toughness of Highly Filled MH/LLDPE Composites by Using Fluorine-Containing Flow Modifiers. *Compos. Part A Appl. Sci. Manuf.* **2020**, *134*, 105900. [CrossRef]
24. Zhang, W.; Li, X.; Shan, Z.; Wang, S.; Xiao, Y. Surface Modification of Magnesium Hydroxide by Wet Process and Effect on the Thermal Stability of Silicone Rubber. *Appl. Surf. Sci.* **2019**, *465*, 740–746. [CrossRef]
25. Jalali Dil, E.; Favis, B.D. Localization of Micro- and Nano-Silica Particles in Heterophase Poly(Lactic Acid)/Poly(Butylene Adipate-Co-Terephthalate) Blends. *Polymer* **2015**, *76*, 295–306. [CrossRef]
26. Zha, X.J.; Pu, J.H.; Ma, L.F.; Li, T.; Bao, R.Y.; Bai, L.; Liu, Z.Y.; Yang, M.B.; Yang, W. A Particular Interfacial Strategy in PVDF/OBC/MWCNT Nanocomposites for High Dielectric Performance and Electromagnetic Interference Shielding. *Compos. Part A Appl. Sci. Manuf.* **2018**, *105*, 118–125. [CrossRef]
27. Wang, X.; Peng, S.; Chen, H.; Yu, X.; Zhao, X. Mechanical Properties, Rheological Behaviors, and Phase Morphologies of High-Toughness PLA/PBAT Blends by in-Situ Reactive Compatibilization. *Compos. Part B Eng.* **2019**, *173*, 107028. [CrossRef]

28. Yuan, C.; Duan, B.; Li, L.; Xie, B.; Huang, M.; Luo, X. Thermal Conductivity of Polymer-Based Composites with Magnetic Aligned Hexagonal Boron Nitride Platelets. *ACS Appl. Mater. Interfaces* **2015**, *7*, 13000–13006. [CrossRef]
29. Guo, Y.; Ruan, K.; Shi, X.; Yang, X.; Gu, J. Factors Affecting Thermal Conductivities of the Polymers and Polymer Composites: A Review. *Compos. Sci. Technol.* **2020**, *193*, 108134. [CrossRef]
30. Kim, K.; Kim, M.; Kim, J. Thermal and Mechanical Properties of Epoxy Composites with a Binary Particle Filler System Consisting of Aggregated and Whisker Type Boron Nitride Particles. *Compos. Sci. Technol.* **2014**, *103*, 72–77. [CrossRef]
31. Hidalgo, J.; Jiménez-Morales, A.; Torralba, J.M. Torque Rheology of Zircon Feedstocks for Powder Injection Moulding. *J. Eur. Ceram. Soc.* **2012**, *32*, 4063–4072. [CrossRef]
32. Zhou, Y.; Chen, C.; Zhu, S.; Sui, C.; Wang, C.; Kuang, Y.; Ray, U.; Liu, D.; Brozena, A.; Leiste, U.H.; et al. A Printed, Recyclable, Ultra-Strong, and Ultra-Tough Graphite Structural Material. *Mater. Today* **2019**, *30*, 17–25. [CrossRef]
33. Shah, D.; Maiti, P.; Jiang, D.D.; Batt, C.A.; Giannelis, E.P. Effect of Nanoparticle Mobility on Toughness of Polymer Nanocomposites. *Adv. Mater.* **2005**, *17*, 525–528. [CrossRef]

molecules

MDPI

Article

Effectiveness Evaluation of Silicone Oil Emulsion In Situ Polymerization for Dehydration of Waterlogged Wooden Artifacts

Mengruo Wu [1], Xiangna Han [1,*], Zhiguo Zhang [2,*] and Jiajun Wang [2]

[1] Key Laboratory of Archaeomaterials and Conservation, Ministry of Education, Institute for Cultural Heritage and History of Science & Technology, University of Science and Technology Beijing, Beijing 100083, China; wumengruo@163.com

[2] National Center of Archaeology, Beijing 100020, China; wangjiajun@caf.ac.cn

* Correspondence: jayna422@ustb.edu.cn (X.H.); zzgwys@126.com (Z.Z.)

Abstract: Organosilicon materials have shown potential as dehydration agents for waterlogged wooden artifacts. These materials can polymerize under normal conditions to form polymers with favorable mechanical strength, antibacterial properties, and aging resistance. However, the insolubility of most organosilicon hindered their penetration into waterlogged wood, which may lead to an unwanted cracking. This study aimed to evaluate the effectiveness of polydimethylsiloxane (PDMS) and hydroxy-terminated polydimethylsiloxane (PDMS-OH) with low viscosity and moderate reactivity for dehydrating waterlogged wooden artifacts from the *Nanhai No.1* shipwreck. Four surfactants ((3–aminopropyl) triethoxysilane (APTES), alkyl polyoxyethylene ether (APEO), trimethylstearylammonium chloride (STAC), and fatty alcohol polyoxyethylene ether (AEO)) and cosurfactant were employed to transform the two kinds of water-repellent silicone oils into eight groups of highly permeable oil-in-water (O/W) emulsions. Under the catalysis of a neutral catalyst, in situ polymerization occurred within the wood cells. Group P2-2 formulated with PDMS-OH and APEO showed the best efficiency in maintaining the dimensions of the wood during dehydration. The dehydrated wood exhibited a natural color and texture with a minimal volume shrinkage rate of 1.89%. The resulting polymer adhered uniformly to the cell walls, effectively reinforcing the wood cell structure. The weight percent gain of the wood was only 218%, and the pores of the cell lumen were well maintained for future retreatment. This method effectively controlled the sol–gel reaction process of the organosilicon and prevented damage to the wooden artifact during the dehydration process. Moreover, the dehydrated wood samples only experienced a low weight gain of 17% at 95% relative humidity (RH), indicating their great environmental stability.

Keywords: silicone oil; emulsion; in situ polymerization; waterlogged wooden artifacts; dehydration; conservation materials

1. Introduction

Archaeological wooden artifacts are rich in historical information and provide invaluable evidence for the study of the technological, artistic, and cultural aspects of ancient human societies [1,2]. However, due to degradation from combined chemical and biological factors, waterlogged archaeological wood usually exhibits high decay levels and complex degradation patterns [3]. The structural integrity of the wood cell walls made the physical and mechanical properties of the wood severely compromised [4–6]. The breaking of hydrogen bonds within the polysaccharide framework and cellulose network leads to the degradation of cellulose and hemicellulose [7]. Additionally, the substantial degradation of hemicellulose acetyl side chains, β–O–4 linkages, and the oxidation and demethylation of lignin also contribute to the deterioration process. All these degradations resulted in the loss of the structural integrity of wood cells, creating numerous pores occupied by free water within the wood [8–10]. Consequently, waterlogged archaeological wood typically exhibits increased porosity and a loose, sponge-like structure, which is easy to break down once it has gone through the dehydration process [11].

Therefore, dehydration is a critical step in the conservation of waterlogged wooden artifacts [12]. Researchers continuously seek dehydration materials that offer superior dimensional stability, high permeability, efficient dehydration, low hygroscopicity, and enhanced durability [13,14]. Organosilicon materials, known for their high strength, water resistance, antibacterial properties, and aging resistance [15], have found applications in wood modification and waterproofing [16]. These materials are also highly modifiable [17], with reports of silicone oil being used to prepare wood-based biomimetic materials [18,19]. In recent years, they have also been found to hold significant potential in cultural heritage conservation [20,21].

The application of organosilicon materials to waterlogged wooden artifacts involves in situ polymerization to support and reinforce the wood matrix. Currently, the reported organosilicon materials for dehydrating archaeological wood primarily include siloxane monomers, silane coupling agents, and silicone oils. Studies have shown that in waterlogged wood, organosilicon materials exhibit efficient sol–gel transition [22,23]. This property is advantageous for the dehydration and reinforcement of severely degraded waterlogged archaeological wood but also means the polymerization is more difficult to control.

Among the earliest attempts was the use of tetraethoxysilane (TEOS), which was commonly employed in stone artifact conservation. Although TEOS monomers provided good dimensional stability and reinforcement for wood, the stress generated during the sol–gel process could lead to wood cracking [24]. Consequently, organic segments need to be introduced to enhance the toughness of the gel network structure. In the early 1990s, researchers used ethanol solutions of methyltrimethoxysilane (MTMOS/MTMS) and 2–(heptadecafluorooctyl) ethyltrimethoxysilane (HFOETMOS) to impregnate wood under vacuum conditions, achieving improved waterproofing, dimensional stability, and UV resistance [25]. Additionally, resistance to decay fungi was also enhanced [26,27]. Recently, Broda et al. evaluated the dehydration and reinforcement effects of a dozen organosiloxanes containing various functional groups on waterlogged wooden artifacts. Among these, 3-mercaptopropyltrimethoxysilane (MPTES) demonstrated the best results for dimensional stability [28,29]. Hamilton's conservation manual introduces a method for protecting archaeological wood using a mixture of silicone oil and isobutyltrimethoxysilane. This method also requires ethanol and acetone to remove water from the wood, and due to the weaker reactivity of silicone oil compared to silane, the polymerization of silicone oil is catalyzed at 52 °C [14].

Existing studies indicate that the volumetric shrinkage rate of waterlogged archaeological wood treated with organosilicon materials can be reduced to approximately 2% to 5%. Post treatment, organosilicon primarily undergoes polymerization and solidification within the cell walls of the wood, preserving the natural porous structure of the cell cavities [30–32]. However, these methods require substantial use of organic solvents, posing significant limitations for practical applications. To eliminate the need for organic solvent replacement, the insolubility of organosilicon in water presents challenges in penetration and rapid surface polymerization. An effective way to solve this problem is to prepare aqueous dispersions or emulsions from insoluble substances [33,34].

In this study, two low-molecular-weight silicone oils were selected as the primary agents. Four commonly used surfactants were utilized to encapsulate the organosilicon within microspheres formed by the surfactants through an emulsification process, creating a water-in-oil (O/W)-type silicone oil emulsion that is easily soluble in water. This approach aimed to achieve the direct penetration and water displacement of organosilicon in waterlogged wooden artifacts. Subsequently, with the catalysis of dibutyltin dilaurate (DBTL), in situ polymerization of the silicon oil on the wood cell walls was expected to occur, to achieve dehydration and stabilization by supporting the cell wall structure and sealing the hydroxyl groups.

2. Results

2.1. Silicone Oil Emulsions

The PDMS emulsions were marked as group P1, and the PDMS-OH emulsions were marked as P2. Four different commonly used surfactants were introduced to the two kinds of silicon oils (Figure 1a) and prepared into eight groups of emulsions numbered as P1-1 to P1-4 and P2-1 to P2-4, respectively. The resulting emulsions exhibited a milky white or semi-transparent appearance, as illustrated in Figure 1b. We found that the viscosity of the emulsions was predominantly influenced by the kind of surfactant we used but not the silicone oil (Table 1). The viscosity of the emulsion was minimal in the case of the STAC (P1-4 and P2-4), imparting a high degree of flowability. However, emulsions formulated with AEO (P1-3 and P2-3) exhibited significantly high viscosity up to 105 mPa·s, which even resembled a gel-like consistency. The viscosity of the emulsions in the remaining groups fell within a moderate range, closely approximating the viscosity of pure silicon oils (~40 mPa·s). The overall variation was minor while PDMS emulsions had slightly higher viscosity than HPDMS-OH emulsions due to their molecular weight differences.

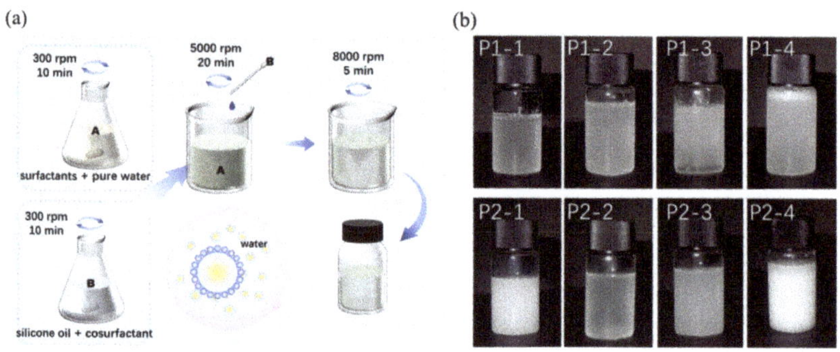

Figure 1. (**a**) Schematic diagram of the silicone oil emulsion preparation process. (**b**) The 8 groups of silicone oil emulsions.

Table 1. Viscosity of the silicone oil emulsions.

Group	P1-1	P1-2	P1-3	P1-4	P2-1	P2-2	P2-3	P2-4
Viscosity/mPa·s	54.0	76.5	105.0	21.5	49.5	68.0	89.5	26.5

2.2. Physical Properties and Appearance

The waterlogged wood was cut into sample cubes of 1 cm^3 and separately immersed in the eight groups of silicone oil emulsions for 3 days. Then, dibutyltin dilaurate (DBTL) was added to the emulsions for catalytic polymerization. After 3 days of reaction, the wood samples were cleaned and thoroughly dried together with the untreated sample in a normal condition. The dehydration process was as shown in Figure 2.

Figure 3 illustrates photos of wood samples before and after being treated with the silicone oil emulsions, with the volume shrink rate (S_V) and weight percent gain (WPG) of each group presented in Table 2. The top three groups exhibiting the best dimensional stability were P2-2, P2-4, and P1-1. The group P2-2 resulted in a minimal S_V of 4.23%, maintaining its shape well with only slight wrinkling. This formulation demonstrated moderate weight gain on the wood, with a WPG of 218.42%. The group P2-4 showed comparable dimensional stability to P2-2, with a S_V of 4.93% after dehydration. However, its WPG after treatment was even lower, at 182.37%. The P1-1 also showed an acceptable

S_v of only 4.60%, but, unfortunately, with significant shape distortion. This formulation also exhibited a low WPG at 180.61%.

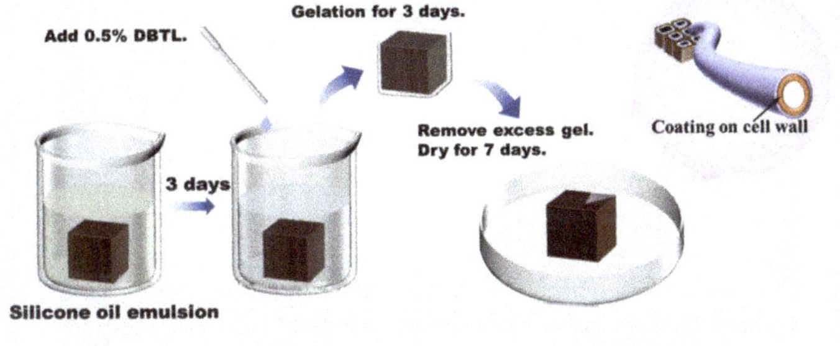

Figure 2. Schematic diagram of the dehydration process of waterlogged archaeological wood.

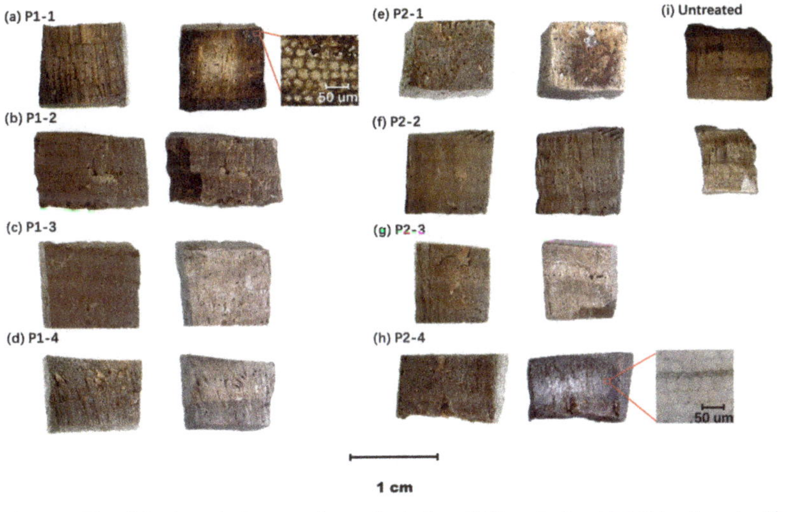

Figure 3. The dehydrated pine wood samples before (**left**) and after (**right**) treatment with silicone oil emulsions. (**a**) P1-1; (**b**) P1-2; (**c**) P1-3; (**d**) P1-4; (**e**) P2-1; (**f**) P2-2; (**g**) P2-3; (**h**) P2-4; (**i**) Untreated wood.

Table 2. The weight percent gain (WPG) and volume shrink rate (Sv) of pine wood samples treated with silicone oil emulsions.

Group	Sv	WPG
P1-1	4.60%	180.61%
P1-2	N/A *	213.16%
P1-3	14.09%	126.93%
P1-4	13.34%	110.74%
P2-1	16.16%	179.49%
P2-2	4.23%	218.42%
P2-3	N/A *	155.59%
P2-4	4.93%	182.37%

* Accurate data were not obtained due to the destruction of severely cracked samples.

The samples in groups P2-2 and P1-2 appeared to exhibit the most natural texture, just like untreated wood. Polymer accumulation on the wood surface was barely visible even on cross-sections and concave areas. In contrast, varying degrees of polymer accumulation were observed in the remaining groups. The excessive polymers mainly caused uneven whitening on the wood surfaces, while the P1-1 group exhibited an unnatural resin-like texture due to the polymer filling inside the cell lumen.

The simulated color according to the CIELab color model of each group is depicted in Figure 4. All of the dried wood showed a lighter shade compared to the color of waterlogged wood. Compared to untreated air-dried wood, P1-1, P1-2, and P2-4 colors significantly darkened, while P1-4 and P2-3 exhibited pronounced whitening due to a layer of excessive polymers on the surface. The color of P2-2 closely matched that of the untreated air-dried wood, presenting a natural pine wood hue. The colors of the remaining two groups notably shifted towards grey tones.

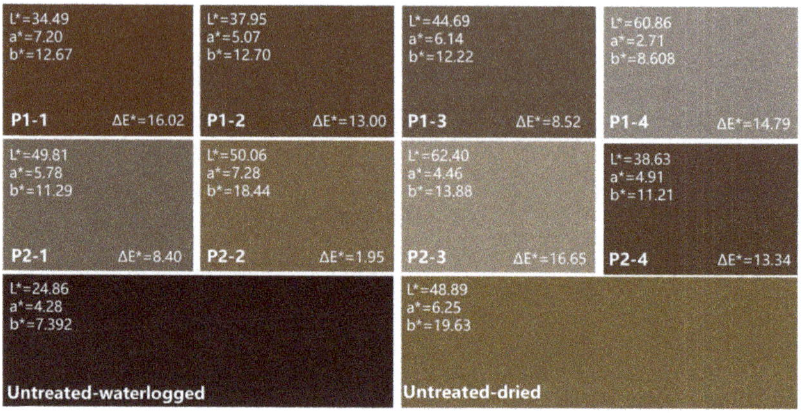

Figure 4. Simulated color according to the CIELab color model of the wood samples.

2.3. Vapor Adsorption–Desorption Test

The vapor adsorption–desorption curves at 0% to 95% RH of the wood samples are shown in Figure 5. The untreated wood samples did not exhibit a distinct moisture adsorption–desorption critical point, with a moisture absorption rate of 21% at 95% RH, displaying desorption hysteresis. After desorption, the untreated samples returned to their original weight, indicating reversible moisture adsorption.

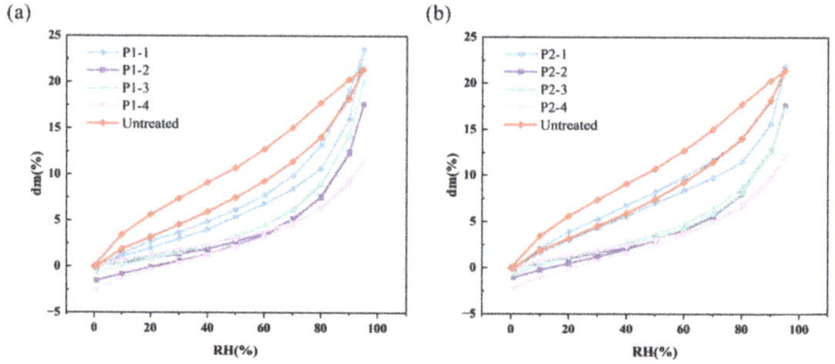

Figure 5. The vapor adsorption–desorption curves at 0% to 95% RH of the wood samples treated with (**a**) P1 and (**b**) P2.

We found that the hygroscopicity of the dehydrated wood primarily depended on the surfactants used in the emulsion, but not on the type of silicone oil. Compared to untreated wood samples, the wood treated with the emulsion using STAC (P1-4 and P2-4) had noticeably reduced hygroscopicity, with a maximum moisture absorption rate of 10%.

Group P1-1 and P2-1 showed moisture absorption rates at around 21–22.5% at 95% RH, slightly higher than untreated samples. The other two surfactants had no significant impact on the hygroscopicity of the wood, with moisture absorption rates slightly lower than untreated samples at around 17–20% at 95% RH, indicating better environmental stability.

It is also noteworthy that only the groups P1-1 and P2-1 still retained the desorption hysteresis, while all other groups were different. Same as the untreated wood, these treated wood samples were also able to desorb immediately with decreasing humidity, but differed in that their mass was slightly lower than the initial value rather than the same when RH dropped back to 0%. This mass loss of the treated wood was less than 2.5% in all groups.

2.4. Microstructural Analysis

For the microstructural analysis of wood samples, scanning electron microscopy (SEM) was utilized. The result showed that the silicone oil emulsion, after polymerizing in the wood matrix, displayed a diverse range of microstructural characteristics. This variation led us to identify three typical patterns of polymerized silicone within the wood cells, depicted in Figure 6.

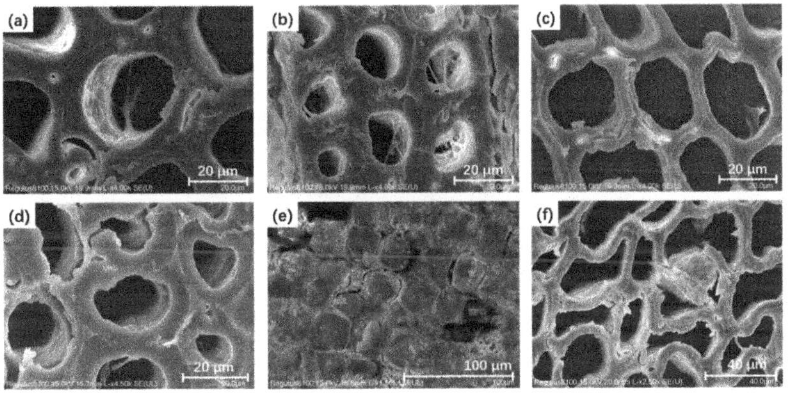

Figure 6. Typical distribution of organic silicon polymerization products in archaeological wood cells (**a**,**b**). Pattern 1: organosilicon polymers formed in the cell lumens in P1-2 and P2-2; (**c**,**d**) pattern 2: organosilicon polymers only adhere to the cell walls of earlywood and latewood cells in P2-2 samples; (**e**) pattern 3: wood cells in P2-1 samples filled with organosilicon polymers; (**f**) the control group samples without silicone oil treatment showed severe deformation of wood cells after being dried in normal conditions.

In the majority of the groups, the silicone oil emulsions predominantly entered the wood cell wall and spared the cell lumens from significant filling. Their wood cell walls became denser, which enhanced their strain resistance. Notably, both latewood and earlywood cells maintained their original shapes well with minimal distortion or collapse. Dendritic polymers were found in only a very small percentage of cells, as shown in Figure 6a,b. However, in three groups (P1-3, P1-4, and P2-1), we found the polymers accumulated and filled the cell lumens (Figure 6e).

2.5. Chemical Structure

To investigate the hydrolysis and condensation reactions within processed wood samples, infrared spectroscopy analysis was conducted. Figure 7 enumerates the primary characteristic peak observed in the infrared spectra pertinent to organosilicon compounds.

Notably, characteristic absorption peaks of Si–O–Si were detected across all sample groups, including antisymmetric and symmetric stretching vibrations at 1080 cm^{-1} and 795 cm^{-1}. These peaks confirm the extensive formation of Si–O–Si networks within the wood matrix post hydrolysis and condensation reactions. An additional Si–C stretching vibration absorption peak was observed at 744 cm^{-1}. The antisymmetric and symmetric stretching vibration absorption peaks of CH_2 within long chains were identified at 2915 cm^{-1} and 2849 cm^{-1}, respectively, while the terminal CH_3 stretching vibrations appeared at 2961 cm^{-1} and 2867 cm^{-1}. These peaks became pronounced in the spectra of the treated wood samples. Furthermore, a notable decrease in the O–H absorption peak intensity at 3400 cm^{-1}.

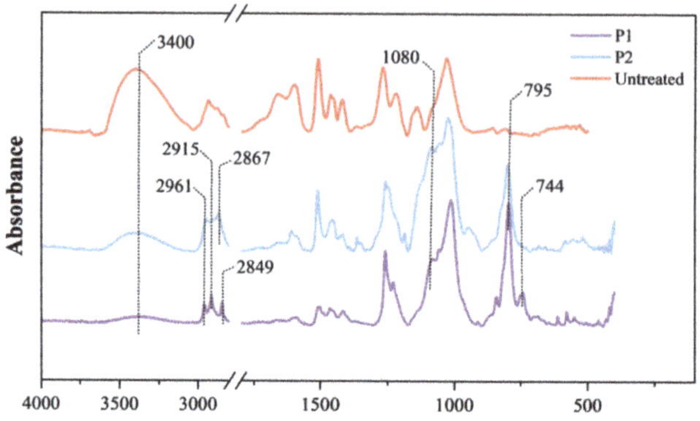

Figure 7. FT-IR spectrum of wood before and after silicone oil emulsions' dehydration treatment.

3. Discussion

By comparing the effects of eight groups of silicone oil emulsions on the dimensional stability, moisture resistance, and color of waterlogged archaeological wood samples, this study preliminarily verified the effectiveness of silicone emulsions in the dehydration of waterlogged wooden artifacts. We found that samples with higher WPG generally display lower volume shrinkage, although the S_v and WPG did not exhibit an absolute negative correlation. This observation suggested that an adequate material loading capacity may contribute to enhancing the dimensional stability effect. However, the impact of color and appearance also needs to be carefully considered. Overall, considering both physical properties and appearance, P2-2 exhibited the best dehydration effect.

To achieve higher dehydration effectiveness, the selection of surfactants was particularly crucial. In this study, the compatibility of APEO and HPDMS was the best (P2-2), while PDMS did better with APTES or APEO (P1-1 and P1-2). The applicable surfactant can disperse silicone oil droplets more evenly and reduce the viscosity of emulsion, thus making it penetrate wood more efficiently. However, the high-viscosity emulsions using AEO (P1-3 and P2-3) might not play an optimal structural support role due to uneven penetration in wood cells. In addition, we found that using STAC as a surfactant carried a risk of producing white polymers on the surface of the wood, which has a negative impact on the color of the wood after drying (P1-4 and P2-4).

We also tried to reveal the mechanism of silicone oil emulsion changing wood stability from the perspective of chemical structure and microstructure. The SEM analysis results showed that the distribution of organosilicon polymers in wood cells can be divided into three patterns: (i) the polymer appears as a branched network, with a small amount present in the cell lumens; (ii) the polymer only appeared on the cell wall; (iii) the polymer completely filled the cell lumens. An interesting observation was that all groups showing the third pattern exhibited significant shrinkage and polymer enrichment on the wood surface. In contrast, samples presenting patterns 1 and 2 generally maintain their shapes well, such as P2-2 and P1-2. This suggested that excessive silicone polymer filling did not equate to better stabilization. Premature polymerization of silicone oil before its entry into the wood substrate might lead to an uneven distribution of the resulting polymers, affecting its ability to stabilize the cellular structures. The volumetric shrinkage of silicone polymer caused by its excessive polymerization might also paradoxically increase the shrinkage. According to our experimental results, the polymer generated on the cell wall was sufficient to support the cell wall, avoid collapse and deformation during dehydration, and provide a certain degree of moisture resistance. Moreover, it is advantageous to preserve the cell lumen as much as possible for future conservation and restoration works, facilitating the infiltration of reprocessed materials without the need to remove the original dehydration agent.

Absorption peaks of silicone oil polymers were found in the infrared spectra of all samples treated with silicone oil emulsions. Variations in the peak shapes and relative intensities of these absorption features indicate the influence of functional groups from organosilicon molecular chains on the chemical structure of the wood cell walls. Simultaneously, a decrease in the intensity of the hydroxyl absorption peak was observed, which suggested potential chemical cross-linking between the organosilicon compounds and the wood cell walls, leading to a significant reduction in hydroxyl content and absorption water. This finding highlights the transformative impact of silicone oil treatment on the wood cell wall chemical composition. In DVS testing, the disappearance of the desorption hysteresis phenomenon also implied that the generated polymer effectively sealed off the hygroscopic groups on the cell walls and prevented their binding to water, thus improving the stability of the wood artifacts under changing humidity. The mass loss that occurred after desorption was presumably attributed to the incomplete hydrolysis polycondensation reaction of the silicone oil, suggesting that the encapsulation of the silicone by the surfactant limited the degree of polymerization of the silicone oil.

The low cost and the easy availability of silicone oil emulsions afford this dehydration method potential application. However, before applying it to large-sized wooden artifacts in practice, specific application techniques need to be considered in future research. The durability and microbial resistance of reinforced wood require rigorous evaluation. It is also necessary to involve archaeological wood of different tree species and decay degrees.

4. Materials and Methods

4.1. Archaeological Wood

Archaeological pine wood (*Pinus* sp.) was sourced from the 800-year-old *Nanhai No. 1* shipwreck in the Guangdong Maritime Silk Road Museum (Yangjiang, Guangdong, China) and has been immersed in pure water for four years. The long-time degradation has led to a heavy softening of the wood. The average maximum moisture content (MWC) was 488% and the average basis density was 0.19 g/cm^3, indicating a severely degraded state distinguished by the international standard [7,35]. Before dehydration, these wood samples were cut into 1 cm^3 segments and cleaned in pure water.

4.2. Emulsion Preparation

This study utilized polydimethylsiloxane (PDMS) and hydroxy-terminated polydimethylsiloxane (PDMS–OH) as the base of the emulsion agents. Four kinds of surfactants were selected, including (3–aminopropyl) triethoxysilane (APTES), alkyl polyoxyethylene ether (APEO), trimethylstearylammonium chloride (STAC), and fatty alcohol polyoxyethylene ether (AEO). A 5 wt% of 2–(2–butoxyethoxy) ethanol was used as a cosurfactant, and a 0.5 wt% of dibutyltin dilaurate (DBTL) was introduced as the catalyst. Comprehensive details on the application of these chemical reactants are outlined in Table 3.

Emulsion agents were formulated with 30 wt% of silicon oil, 10–15 wt% of surfactants, and 5 wt% of cosurfactant. The formulations are exhibited in Table 4. The preparation steps of the emulsion were as follows (shown in Figure 7): (i) mix surfactants and pure water uniformly into the solution A; (ii) combine the silicone oil with the cosurfactant to form the solution B; (iii) gradually drip the solution B into A, while stirring at 5000 rpm for 20 min; (iv) increase the stirring speed to 8000 rpm for 5 min, until the formation of a clear or milky white water-in-oil emulsion. The resultant emulsion should readily dissolve in water and remain stable upon standing for at least 24 h. The viscosity of the emulsion is tested to prevent excessive pre-polymerization during preparation.

Table 3. Materials used for the silicone oil emulsion.

Materials	Mw (g/mol)	Molecular Formula	Manufacturer/Purity
Polydimethylsiloxane (PDMS)	~770	$[-Si\ (CH_3)_2O-]_n$	Sigma-Aldrich/99%
Hydroxy terminated polydimethylsiloxane (PDMS-OH)	400~700	$HO-(C_2H_6OSi)_n-H$	Gelest/99%
2–(2–Butoxyethoxy) ethanol	162.23	$C_8H_{18}O_3$	Aladdin/99%
Dibutyltin dilaurate (DBTL)	631.56	$C_{32}H_{64}O_4Sn$	Aladdin/95%
(3–Aminopropyl)triethoxysilane (APTES)	221.37	$C_9H_{23}NO_3Si$	Aladdin/99%
Alkyl polyoxyethylene ether (APEO)	N/A	$HO(CH_2CH_2O)_n(CH_2)_{11}CH_3$	Aladdin/99.5%
Trimethylstearylammonium chloride (STAC)	348.05	$C_{21}H_{46}ClN$	Aladdin/98%
Fatty alcohol polyoxyethylene ether (AEO)	315/590	RO–(CH2CH2O) $_n$–H	Aladdin/98%

Table 4. Formulations of the silicone oil emulsion.

Group	Silicone Oil	Surfactant
P1-1	30% PDMS	10% APTES
P1-2		15% APEO
P1-3		15% AEO
P1-4		10% STAC
P2-1	30% PDMS-OH	10% APTES
P2-2		15% APEO
P2-3		15% AEO
P2-4		10% STAC

4.3. *Dehydration of Wood*

Each group of wood samples of 1 cm^3 was separately immersed in the prepared silicone oil emulsions for three days. Subsequently, a 0.5 wt% DBTL was introduced into the emulsion, and the immersing process was sustained for an additional two days. After that, gels on the wood surface were cleaned in water, and samples were then left to dry under normal conditions (at 25 °C with a relative humidity of 34.5 ± 1.5%) until the weight remained constant.

4.4. *Testing and Characterization Methods*

Photos of the wood samples were taken at 20× magnification to compare their overall appearance before and after treatment using a three-dimensional video microscope (VHX-6000, KEYENCE, Osaka, Japan).

The Basic Density (BD) and Maximum Water Content (MWC) of wood were calculated according to Equation (1) and Equation (2), respectively. The mass of the waterlogged sample with the size of about 10 mm × 10 mm × 10 mm was measured with an analytical balance (JA2003, SUNNY HENGPING Instrument, Shanghai, China) and recorded as M_1. After being dried in an oven at 103 ± 2 °C, the mass of the sample was recorded as M_0. The drainage method was used to measure the waterlogged volume V_1 and dry volume V_0 under normal conditions. On the analytical balance, the sample was completely immersed in pure water without touching the bottom and wall of the beaker. According to Archimedes' principle, the volume of the wood sample (cm^3) equals the increased weight (g) in value of the pure water.

The weight percent gain (WPG) was calculated using Equation (3), and the anti-shrinkage effect of the treatment was evaluated using the volume shrink rate (S_v) calculated with Equation (4). In Equation (3), m_1 is the dry mass of the wood after treatment and reaching equilibrium in a normal environment, and m_0 is the theoretical dry mass of the wood sample, which is estimated according to the M_1 and MWC.

$$BD = \frac{M_0}{V_1} \tag{1}$$

$$MWC = \frac{M_1 - M_0}{M_0} \quad (2)$$

$$WPG = (m_1 - m_O) \times 100\% \quad (3)$$

$$S_V = \frac{V_1 - V_0}{V_1} \times 100\% \quad (4)$$

The color difference (ΔE*) between the dried samples before and after treatment with the emulsions was measured with a spectrophotometer (CM-26D, Konica Minolta, Tokyo, Japan) under a D65 light source. The following equation was used to calculate the color difference ΔE*:

$$\Delta E^* = [(\Delta L^*)^2 + (\Delta a^*)^2 + (\Delta b^*)^2]\frac{1}{2}, \quad (5)$$

where L* ranges from 0 to 100 and indicates the change in color from black (dark) to white (light), a* varies from negative to positive and represents the change in color from green to red, and b* ranges from negative to positive and indicates the change in color from blue to yellow.

In the scanning electron microscope (SEM) analysis, an ultra-high-resolution field-emission scanning electron microscope (Regulus 8100, HITACHI, Tokyo, Japan) was used. Gold was sputtered to the surfaces of samples as a conductive coating. The microscope was operated with an accelerating voltage of 15 kV, a resolution of 0.7 nm, a working distance of 15 mm, and a secondary electron imaging (SE) working mode.

The maximal moisture content under different relative humidity (RH) conditions was measured using a high-throughput dynamic moisture-adsorption tester (SPSx-1μ, ProUmid, Ulm, Germany). Preceding the test, the samples were air-dried at 103 °C for more than 24 h until their weights no longer decreased. The humidity scope of this test ranged from 0 to 95% RH, at a gradient of 10% RH and a stable temperature of 25 °C. The default measurement frequency was every 10 min. The time taken for every gradient fell between a minimum of 60 min and a maximum of 600 min. The default weight limit had an upper cap of 1000%. The equilibrium condition was defined as $dm/dt \leq 0.01$.

In chemical structure analysis, wood samples were tested using a Fourier transform infrared spectrometer (Nicolet™ iS™5, Thermo Scientific, Waltham, MA, USA) with 32 scanning times and a resolution of 4.000 cm^{-1}. The absolute dry mass ratio of the wood sample to potassium bromide powder was 1:100.

5. Conclusions

The biggest challenge in achieving controllable and in situ polymerization of organosilicon is how to regulate the contradiction between organosilicon and water. Water-repellent organosilicon materials cannot enter the interior of the wood through the waterlogged pores and will initiate polymerization reactions once they encounter moisture, which makes the dehydration and shaping process uncontrollable.

In this study, low-molecular-weight silicone oil was prepared into oil in water emulsions, which realized the direct replacement of water in wood without organic solvents. After entering the wood cells, silicone oil tended to form on the cell wall and undergo slow polymerization. The generated organic silicon polymer effectively supported and strengthened the wood cell wall during the dehydration process.

Among the eight groups of silicone oil emulsions, P2-2 showed the best efficiency, with a shrinkage rate of only 1.63%, and the appearance of the wood is also very natural. The weight gain rate was only 218%, which was conducive to reducing the burden of external consolidants on the wooden artifacts. Overall, the filling degree of the cell lumen via organosilicon polymers was limited. Under SEM, it can be seen that the silicone oil only polymerized on the wood cell wall, with almost no filling in the cell lumen. The wood cell wall was well shaped, without severe deformation or collapse like the untreated sample.

This study indicated that this in situ polymerization reinforcement method has a good stabilizing effect on the shape of waterlogged archaeological wood. In addition, it is significant for the restoration of cultural relics that this method can effectively preserve the pore structures of wood cells so that retreatment materials can still penetrate the wooden artifacts in the future.

Author Contributions: Conceptualization, X.H. and Z.Z.; methodology, X.H.; software, M.W.; validation, X.H., Z.Z. and J.W.; formal analysis, M.W.; investigation, X.H.; resources, J.W.; data curation, Z.Z.; writing—original draft preparation, M.W.; writing—review and editing, M.W. and J.W.; visualization, M.W.; supervision, X.H. and Z.Z.; project administration, X.H., Z.Z. and J.W.; funding acquisition, Z.Z. All authors have read and agreed to the published version of the manuscript.

Funding: This research was funded by the National Key R&D Program of China, grant number 2020YFC1521803, and Shanghai Pujiang Program, grant number 23PJ1403600.

Institutional Review Board Statement: Not applicable.

Informed Consent Statement: Not applicable.

Data Availability Statement: The data presented in this study are available on request from the corresponding author.

Acknowledgments: The authors would like to thank Yafang Yin, Xiaomei Jiang, and Yonggang Zhang from the Research Institute of Wood Industry, Chinese Academy of Forestry for the wood identification support.

Conflicts of Interest: The authors declare no conflicts of interest.

References

1. Hunt, D. Properties of wood in the conservation of historical wooden artifacts. *J. Cult. Herit.* **2012**, *13*, 10–15. [CrossRef]
2. Unger, A.; Schniewind, A.P.; Unger, W. *Conservation of Wood Artifacts: A Handbook*, 4th ed.; Springer: Berlin/Heidelberg, Germany, 2001; pp. 43–49. [CrossRef]
3. Walsh, Z.; Averous, L. Recent developments in the conservation of materials properties of historical wood. *Prog. Mater. Sci.* **2019**, *102*, 167–221. [CrossRef]
4. Han, L.; Tian, X.; Keplinger, T.; Zhou, H.; Li, R.; Svedström, K.; Burgert, I.; Yin, Y.; Guo, J. Even Visually Intact Cell Walls in Waterlogged Archaeological Wood Are Chemically Deteriorated and Mechanically Fragile: A Case of a 170-Year-Old Shipwreck. *Molecules* **2020**, *25*, 1113. [CrossRef] [PubMed]
5. Han, L.; Wang, K.; Wang, W.; Guo, J.; Zhou, H. Nanomechanical and Topochemical Changes in Elm Wood from Ancient Timber Constructions in Relation to Natural Aging. *Materials* **2019**, *12*, 786. [CrossRef]
6. Blanchette, R.A. A review of microbial deterioration found in archaeological wood from different environments. *Int. Biodeterior. Biodegrad.* **2000**, *46*, 189–204. [CrossRef]
7. Christensen, B. *The Conservation of Waterlogged Wood in the National Museum of Denmark: With a Report on the Methods Chosen for the Stabilization of the Timbers of the Viking Ships from Roskilde Fjord, and a Report on Experiments Carried Out in Order to Improve upon These Methods;* Studies in Museum Technology, 1; National Museum of Denmark: Copenhagen, Denmark, 1970.
8. Guo J, Xiao L, Han L; et al. Deterioration of the cell wall in waterlogged wooden archeological artifacts, 2400 years old. *IAWA J. /Int. Assoc. Wood Anat.* **2019**, *40*, 820–845. [CrossRef]
9. Almkvist, G.; Persson, I. Distribution of iron and sulfur and their speciation in relation to degradation processes in wood from the Swedish warship Vasa. *New J. Chem.* **2011**, *35*, 1491–1502. [CrossRef]
10. Almkvist, G.; Persson, I. Degradation of polyethylene glycol and hemicellulose in the Vasa. *Holzforschung* **2008**, *62*, 64–70. [CrossRef]
11. Almkvist, G.; Persson, I. Analysis of acids and degradation products related to iron and sulfur in the Swedish warship Vasa. *Holzforschung* **2008**, *62*, 694–703. [CrossRef]
12. McCawley, J.C. Waterlogged artifacts: The challenge to conservation. *J. Can. Conserv. Inst.* **1977**, *2*, 17–26.
13. Walsh, Z.; Janeček, E.R.; Jones, M.; Scherman, A.O. Natural polymers as alternative consolidants for the preservation of waterlogged archaeological wood. *Stud. Conserv.* **2017**, *62*, 173–183. [CrossRef]
14. Hamilton, D.L. Methods of Conserving Archaeological Material from Underwater Sites. Ph.D. Thesis, Texas A&M University, College Station, TX, USA. 1999. Available online: https://liberalarts.tamu.edu/nautarch/wp-content/uploads/sites/32/2021/10/ConservationManual.pdf (accessed on 1 January 1999).
15. Mandraveli, E.; Mitani, A.; Terzopoulou, P.; Koutsianitis, D. Oil Heat Treatment of Wood—A Comprehensive Analysis of Physical, Chemical, and Mechanical Modifications. *Materials* **2024**, *17*, 2394. [CrossRef] [PubMed]

16. Zhang, Z.; Guo, Q.; Huang, X.; Zhang, Q.; Fan, J.; Huang, J. Research Progress of Reinforced Modification of Fast-Growing Wood. *Coatings* **2023**, *14*, 53. [CrossRef]
17. He, Z.; Qu, L.; Wang, Z.; Qian, J.; Yi, S. Effects of zinc chloride–silicone oil treatment on wood dimensional stability, chemical components, thermal decomposition and its mechanism. *Sci. Rep.* **2019**, *9*, 1601. [CrossRef]
18. Han, X.; Wu, J.; Zhang, X.; Shi, J.; Wei, J.; Yang, Y.; Wu, B.; Feng, Y. Special issue on advanced corrosion-resistance materials and emerging applications. The progress on antifouling organic coating: From biocide to biomimetic surface. *J. Mater. Sci. Technol.* **2021**, *61*, 46–62. [CrossRef]
19. Guan, H.; Cheng, Z.; Wang, X. Highly compressible wood sponges with a spring-like lamellar structure as effective and reusable oil absorbents. *ACS Nano* **2018**, *12*, 10365–10373. [CrossRef]
20. Salanti, A.; Zoia, L.; Zanini, S.; Orlandi, M. Synthesis and characterization of lignin–silicone hybrid polymers as possible consolidants for decayed wood. *Wood Sci. Technol.* **2016**, *50*, 117–134. [CrossRef]
21. Broda, M.; Kryg, P.; Ormondroyd, G.A. Gap-fillers for wooden artefacts exposed outdoors—A review. *Forests* **2021**, *12*, 606. [CrossRef]
22. Jespersen, K.; Fang, B. The application of TEOS in dehydration of waterlogged wood. *Sci. Conserv. Archaeol.* **1995**, *1*, 61–63. (In Chinese)
23. Ghavidel, A.; Bak, M.; Hofmann, T.; Vasilache, V.; Sandu, I. Evaluation of some wood-water relations and chemometric characteristics of recent oak and archaeological oak wood (Quercus robur) with archaeometric value. *J. Cult. Herit.* **2021**, *51*, 21–28. [CrossRef]
24. Tong, H.; Li, M.; Zhen, G.; Zhao, Y. Study on dehydration and reinforcement of waterlogged wood cultural relics by organic silicon sol-gel method. *Jianghan Archaeol.* **2014**, *1*, 60–68. (In Chinese)
25. Hisashi, M.; Shiro, S. Topochemistry of SiO_2 wood-inorganic composits for enhancing water-repellency. *J. Jpn. Soc. Test. Mater.* **1999**, *5*, 270–275. [CrossRef]
26. Saka, S.; Sasaki, M.; Tanahashi, M. Wood-inorganic composites prepared by sol-gel processing I: Wood-inorganic composites with porous structure. *J. Jpn. Wood Res. Soc.* **1992**, *38*, 1043–1049. [CrossRef]
27. Mahltig, B.; Swaboda, C.; Roessler, A.; Boettcher, H. Functionalising wood by nanosol application. *J. Mater. Chem.* **2008**, *18*, 3180–3192. [CrossRef]
28. Broda, M.; Mazela, B. Application of methyltrimethoxysilane to increase dimensional stability of waterlogged wood. *J. Cult. Herit.* **2017**, *25*, 149–156. [CrossRef]
29. Broda, M.; Mazela, B.; Dutkiewicz, A. Organosilicon compounds with various active groups as consolidants for the preservation of waterlogged archaeological wood. *J. Cult. Herit.* **2019**, *35*, 123–128. [CrossRef]
30. Qin, Z. Study on Dehydration and Dimensional stabilisation of waterlogged archaeological wood by Organosiloxane. Master's Thesis, University of Science and Technology Beijing, Beijing, China, 2022.
31. Hou, Y.; Chen, X.; Wang, K.; Hu, D. Study on the dehydration and shaping mechanism of waterlogged wooden cultural relics by γ-mercaptopropyl triethoxysilane. *Sci. Conserv. Archaeol.* **2021**, *33*, 1–11. (In Chinese) [CrossRef]
32. Broda, M.; Yelle, D.J. Reactivity of Waterlogged Archeological Elm Wood with Organosilicon Compounds Applied as Wood Consolidants: 2D 1H–13C Solution-State NMR Studies. *Molecules* **2022**, *27*, 3407. [CrossRef]
33. Lisuzzo, L.; Hueckel, T.; Cavallaro, G.; Sacanna, S.; Lazzara, G. Pickering Emulsions Based on Wax and Halloysite Nanotubes: An Ecofriendly Protocol for the Treatment of Archeological Woods. *ACS Appl. Mater. Interfaces* **2021**, *13*, 1651–1661. [CrossRef]
34. Lisuzzo, L.; Cavallaro, G.; Milioto, S.; Lazzara, G. Halloysite nanotubes filled with MgO for paper reinforcement and deacidification. *Appl. Clay Sci.* **2021**, *213*, 106231. [CrossRef]
35. Macchioni, N.; Capretti, C.; Sozzi, L.; Pizzo, B. Grading the decay of waterlogged archaeological wood according to anatomical characterisation. The case of the Fiavé site (N-E Italy). *Int. Biodeterior. Biodegrad.* **2013**, *84*, 54–64. [CrossRef]

molecules

MDPI

Review

Polymer Composites in 3D/4D Printing: Materials, Advances, and Prospects

Ayyaz Mahmood [1,2,3,4], Fouzia Perveen [5], Shenggui Chen [1,3,4,*], Tayyaba Akram [6] and Ahmad Irfan [7]

[1] School of Mechanical Engineering, Dongguan University of Technology, Dongguan 523808, China; ayyazcmc@gmail.com
[2] School of Life Science and Technology, University of Electronic Science and Technology, Chengdu 610054, China
[3] School of Art and Design, Guangzhou Panyu Polytechnic, Guangzhou 511483, China
[4] Dongguan Institute of Science and Technology Innovation, Dongguan University of Technology, Dongguan 523808, China
[5] School of Interdisciplinary Engineering & Sciences (SINES), National University of Sciences and Technology (NUST), Sector H-12, Islamabad 44000, Pakistan
[6] Department of Physics, COMSATS Institute of Information Technology, Lahore 54000, Pakistan
[7] Department of Chemistry, College of Science, King Khalid University, P.O. Box 9004, Abha 61413, Saudi Arabia
* Correspondence: dgutchensg@163.com; Tel.: +86-136-9198-4451; Fax: +86-756-9085-5702

Abstract: Additive manufacturing (AM), commonly referred to as 3D printing, has revolutionized the manufacturing landscape by enabling the intricate layer-by-layer construction of three-dimensional objects. In contrast to traditional methods relying on molds and tools, AM provides the flexibility to fabricate diverse components directly from digital models without the need for physical alterations to machinery. Four-dimensional printing is a revolutionary extension of 3D printing that introduces the dimension of time, enabling dynamic transformations in printed structures over predetermined periods. This comprehensive review focuses on polymeric materials in 3D printing, exploring their versatile processing capabilities, environmental adaptability, and applications across thermoplastics, thermosetting materials, elastomers, polymer composites, shape memory polymers (SMPs), including liquid crystal elastomer (LCE), and self-healing polymers for 4D printing. This review also examines recent advancements in microvascular and encapsulation self-healing mechanisms, explores the potential of supramolecular polymers, and highlights the latest progress in hybrid printing using polymer–metal and polymer–ceramic composites. Finally, this paper offers insights into potential challenges faced in the additive manufacturing of polymer composites and suggests avenues for future research in this dynamic and rapidly evolving field.

Keywords: additive manufacturing; 3D printing; 4D printing; polymers; shape memory

1. Introduction

The world of manufacturing is undergoing a profound and transformative evolution, and at the heart of this revolution lies the ground-breaking technology of additive manufacturing, more commonly known as 3D/4D printing [1–5]. This revolutionary approach allows for the meticulous layer-by-layer deposition of materials to create intricate and bespoke objects, thereby altering the very foundations of how we conceive, design, and manufacture [6–12]. Since its inception and subsequent commercialization, 3D printing has witnessed widespread adoption across a spectrum of industries, notably engineering, manufacturing, healthcare, aerospace, and the automotive sector. Its applications, ranging from rapid prototyping to the fabrication of intricate lightweight structures, have brought about a transformative wave of innovation in contemporary manufacturing. The impact of 3D printing on the landscape of industrial part/component design and equipment development is profound [1–4,13,14]. This revolutionary technology has not only challenged but also transcended the limitations of conventional manufacturing methods. It has empowered manufacturers and researchers to embark on endeavors that were once deemed unattainable using traditional processes. Over the past three decades, the relentless pursuit of enhancements and innovations has fueled the remarkable evolution of 3D printing technology [5,11,15–19].

Central to this innovative transformation are polymeric materials, a remarkably diverse class of substances that offer unmatched versatility, accessibility, and opportunities in the realm of additive manufacturing. Polymeric materials, also commonly referred to as plastics, have gained substantial traction within the domain of additive manufacturing [10,20]. Their appeal stems from their exceptional adaptability and widespread availability [21,22]. These materials are the cornerstone of an ever-expanding range of applications, enabling the production of prototypes, intricate components, and customized solutions across a spectrum of sectors, including aerospace, healthcare, and consumer goods. This transformative shift in the additive manufacturing landscape is significantly facilitated by the intrinsic advantages that polymeric materials bring to the table. Schematics of the overall 3D and 4D printing process are shown in Figure 1.

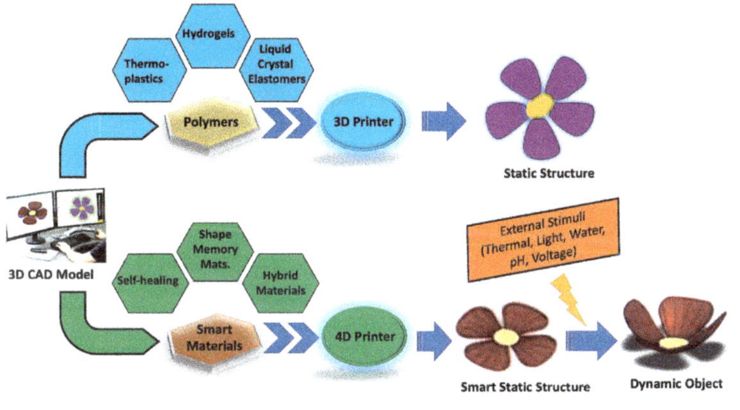

Figure 1. Schematics of the overall 3D and 4D printing process using polymeric materials.

Foremost among these advantages is the acceleration of product development cycles. Additive manufacturing, in conjunction with polymeric materials, empowers engineers and designers to seamlessly transition digital designs into tangible prototypes, facilitating rapid testing, validation, and redesign processes. This, in turn, results in remarkable time and resource savings, bolstering the speed and efficiency of product development [5,11–13,15–19]. Moreover, additive manufacturing offers a fresh perspective on the design and geometry of creations. It allows the realization of complex and intricately structured components that were previously unattainable using traditional manufacturing techniques. This capability to reimagine and innovate geometrically enables designs that are not only aesthetically appealing but also functionally superior. In addition to these transformative advantages, additive manufacturing contributes significantly to sustainability initiatives [7,18,23]. It aids in minimizing material wastage, reducing energy consumption, and minimizing the carbon footprint of various industries. By its very nature, additive manufacturing is a process that generates minimal waste and, through localized production, reduces the need for long-distance transportation. This is a crucial step forward in reducing carbon emissions and aligning the manufacturing industry with the noble objectives of a more sustainable future.

This comprehensive review navigates the intricate landscape of additive manufacturing with a focus on polymeric materials, providing an in-depth exploration of recent advancements, challenges, and future trajectories in this transformative technology. Our goal is to offer a holistic perspective on how polymeric materials within the realm of additive manufacturing are reshaping the realms of creativity, efficiency, and sustainability in diverse domains.

The Evolution of Polymer 3D Printing: A Journey through Time

The history of polymer 3D printing is a fascinating tale of innovation and technological progress. This narrative unfolds alongside the timeline of 3D printing, with polymer materials playing a pivotal role. The development of feedstock materials, especially polymers, has been instrumental in shaping the world of 3D printing as we know it today. The chronological history of 3D printing, along with the evolution of polymer materials, is depicted in Figure 2. It is important to note that the foundations for many of the polymers used in 3D printing today, such as polyamides, polylactic acid (PLA), and epoxy, were laid during the 1920s and 1940s [20]. These early discoveries set the stage for the polymer-based 3D printing revolution. The conceptual groundwork for additive manufacturing (AM) was laid over 50 years ago, highlighting the longstanding fascination with the idea of creating objects layer by layer. However, it was not until the 1980s and 1990s that we witnessed the birth and rapid growth of 3D printing as a practical technology. The watershed moment came in 1987 when stereolithography, one of the first 3D printing methods, was commercialized. This was a significant leap forward, and it marked the beginning of a new era.

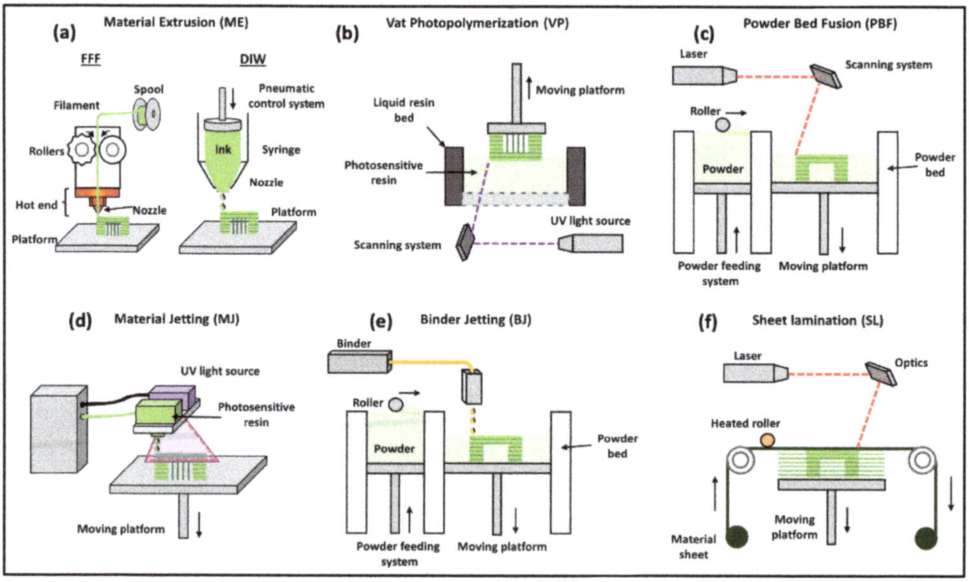

Figure 2. Illustration of the historical progression of 3D printing and the advancements in polymer materials for 3D printing, marking a shift from rapid prototyping to scalable and tailor-made production. Figure reproduced with permission from [21].

The 1980s and 1990s also saw the invention of other ground-breaking 3D printing techniques, including fused filament fabrication (FFF) and selective laser sintering (SLS) [24]. These methods, which are still widely used today, further expanded the horizons of 3D printing technology. The late 1990s brought a surge of innovation as various versions and iterations of 3D printing techniques emerged, thanks to the concurrent advancements in computer technologies. This period of rapid development paved the way for the widespread exploration of 3D printing applications in various industries. In the early 2000s, 3D printing started to find its footing in industries like medicine and aerospace, where its potential for creating complex and customized parts was particularly valuable.

Simultaneously, many early patents for 3D printing techniques began to expire, allowing numerous companies to enter the 3D printing market. This accessibility brought 3D printers within reach of the public, democratizing a technology that was once confined to the realm of research and development. The journey of polymer 3D printing is a testament to human ingenuity and the relentless pursuit of innovation. As technology continues to advance, the future of polymer 3D printing holds the promise of even more remarkable developments and widespread applications.

In the last decade, 3D printing technology has undergone a transformative journey, marked by significant advancements that have greatly elevated its precision, accuracy, and speed. The initial stages of additive manufacturing predominantly involved the adaptation of existing polymer materials to suit the demands of 3D printing. However, the landscape has rapidly evolved, and today, there is a growing emphasis on the development of polymers uniquely tailored for 3D printing. This progress has been accelerated by the rapid emergence of high-performance 3D printing materials. These cutting-edge materials encompass a broad spectrum, including intelligent polymers and advanced engineering plastics. The synergistic integration of these materials with additive manufacturing has catalyzed a new industrial revolution across various sectors. A notable example of these pioneering materials is the category of shape memory polymers [25–28], which exhibit the remarkable ability to "remember" and return to their original shape under certain conditions. Smart hydrogels [29–31], another breakthrough, demonstrate sensitivity to environmental stimuli, enabling them to change volume or shape in response to factors like temperature, moisture, or pH levels. The advent of liquid crystal polymers (LCPs) and liquid crystal elastomers (LCEs) [32–35] further exemplifies this paradigm shift. These materials, characterized by their anisotropic properties and molecular orientation control in response to temperature fluctuations, have ushered in a new era typified by 4D printing. The concept of 4D printing encompasses the creation of objects that can undergo dynamic transformations over time in response to external triggers, representing a paradigm shift in the realm of manufacturing. This innovative approach has unlocked new avenues for the production of smart devices, advanced robotics, and cutting-edge biomedical products [18,19,36–38]. The fusion of novel materials with 3D printing technology has ushered in a new era of limitless possibilities, giving industries the tools to fabricate smart, adaptive, and innovative solutions that were once considered beyond reach.

2. Polymer Materials and Design for 3D Printing

Three-dimensional printing technology is fundamentally driven by digital models, where physical objects are created by adding successive layers of printing materials. Various methods are used for the preparation of polymer composites, each with its unique advantages and disadvantages. These methods include deposition molding, selective laser sintering, ink-jet 3D printing, stereolithography, and 3D drawing, among others. While some technologies are well-established and widely adopted, others are still in the research and development stage or are used by a limited number of scientists. The choice of a specific 3D printing technology depends on several factors, such as the raw material requirements, processing speed and precision, cost considerations, and the final performance characteristics of the products to be manufactured. Each method has its own set of strengths and limitations, making it crucial to match the chosen technology with the specific requirements of the intended application [22]. These methods are briefly discussed as follows. Deposition molding: This method involves depositing successive layers of materials, often in filament or granular form, to build the final object. It is widely used and suitable for a variety of applications due to its relatively simple and cost-effective nature. Selective laser sintering (SLS): In this approach, a laser is used to selectively sinter powdered material, typically thermoplastic polymers or metals, to create the desired object. It is renowned for producing high-quality, functional parts with good mechanical properties. Ink-jet 3D printing: This method operates on the principle of depositing liquid material (often polymers) in a layer-by-layer fashion using an inkjet-style print head. It is valued for its ability to create full-color and multi-material prints, making it popular in applications requiring vibrant aesthetics. Stereolithography (SLA): SLA is a photopolymerization process that utilizes a UV laser to solidify liquid resin layer by layer. Additionally, some SLA systems use a DLP projector to display the UV image of each layer, allowing simultaneous exposure of the entire layer. The advantage of using a DLP projector lies in its ability to expose an entire layer simultaneously, potentially speeding up the printing process compared with systems that use a laser to trace each layer point by point. SLA is renowned for its high resolution and precision, making it suitable for creating detailed and intricate objects. Three-dimensional drawing: This method typically involves manually controlling a 3D printing pen, which extrudes a thermoplastic filament to create freeform structures. It is a versatile and accessible option for artists, designers, and hobbyists. Further extensive details of these printing methods can be found in the literature [4,6,12,39–43].

The choice of polymer 3D printing technology should be made after a thorough assessment of the unique requirements and constraints of the intended application. While some technologies may excel in precision and detail, others may prioritize speed and cost-effectiveness. Additionally, as the field of 3D printing continues to evolve, it is essential to stay informed about emerging technologies that may offer even more advanced capabilities in the near future. The dynamic landscape of polymer 3D printing technologies is illustrated in Figure 3, showcasing the options available and their respective applications.

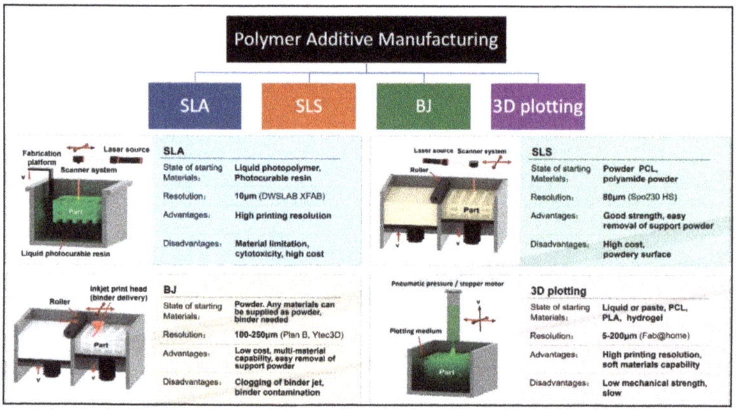

Figure 3. The evolving landscape of polymer 3D printing technologies for varied applications. Figure reproduced with permission from [12].

2.1. *Thermoplastics*

Thermoplastic polymers, commonly utilized for their ability to soften when heated and solidify upon cooling, exhibit distinctive characteristics based on their molecular structure, as depicted in Figure 4a. Amorphous thermoplastics possess a randomly ordered molecular structure without a distinct melting point, facilitating thermoforming. In contrast, semi-crystalline thermoplastics feature a highly organized molecular structure with a well-defined melting point, transitioning abruptly from solid to liquid upon heat absorption. The strong intermolecular forces in semi-crystalline polymers contribute to their favorable mechanical properties. Notably, all thermoplastic polymers offer reversible processing capabilities, rendering them well-suited for extrusion-based 3D printing.

The mechanical characteristics of pure thermoplastic materials in their pristine state might not be adequate for specific applications, necessitating modifications to bolster their properties for fused filament fabrication (FFF). Ning et al. [44] embarked on a study where they combined plastic particles, specifically acrylonitrile butadiene styrene (ABS), with carbon fiber during the FFF process. This incorporation of 5 wt% carbon fiber led to notable enhancements in bending stresses, bending moduli, and bending toughness compared with pure plastic, marking increases of 11.82%, 16.82%, and 21.86%, respectively. It is noteworthy that the samples containing 10% carbon fiber displayed the highest porosities, as indicated in Figure 5. Another investigation conducted by Tekinalp et al. [45] revolved around the utilization of short carbon fibers (ranging from 0.2 to 0.4 mm) as an additive to ABS for 3D printing using FFF technology. This approach led to remarkable increases in tensile strength and the modulus of elasticity of the 3D-printed samples, showing improvements of around 115% and 700%, respectively, in comparison with traditionally molded composites. Despite the relatively high porosities observed in 3D-printed composites, they consistently exhibited substantial tensile strengths and moduli.

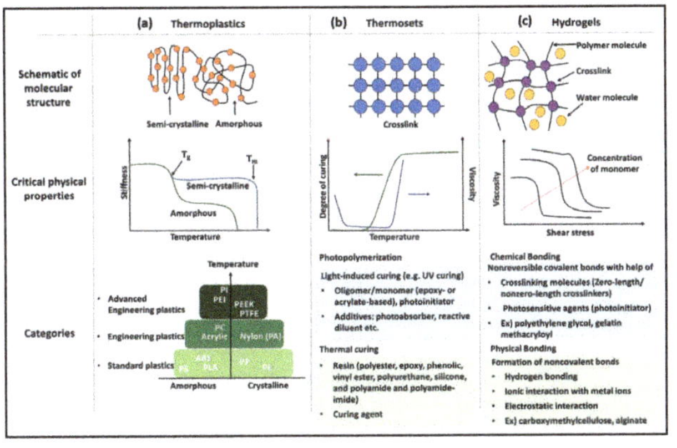

Figure 4. Frequently utilized polymers in 3D printing along with their respective properties: (**a**) thermoplastics, (**b**) thermosets, and (**c**) hydrogels.

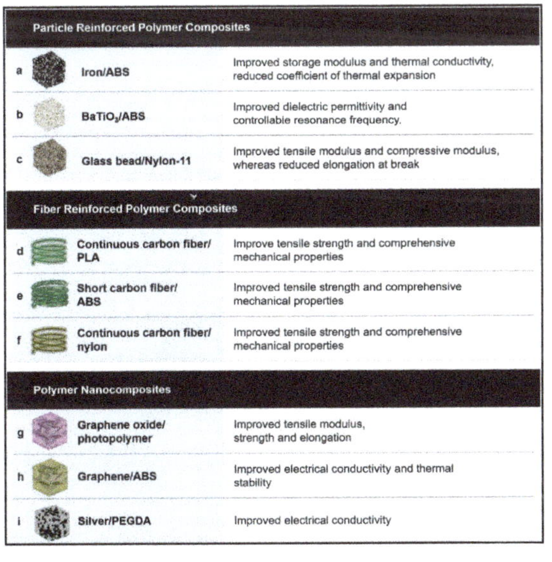

Figure 5. Polymer composite reinforcement methods and examples. Classification and instances of reinforcement techniques utilized in polymer composites: (**a**) iron/acrylonitrile butadiene styrene (ABS) composite [46], (**b**) $BaTiO_3$/ABS composite [47], (**c**) glass bead/nylon-11 composite [48], (**d**) continuous carbon fiber/polylactic acid (PLA) composite [49], (**e**) short carbon fiber/ABS composite [50], (**f**) continuous carbon fiber/nylon composite [51], (**g**) graphene oxide/photopolymer composite [52], (**h**) graphene/ABS composite [53], and (**i**) silver/poly(ethylene glycol) diacrylate (PEGDA) composite [54]. Figure reproduced with permission from [12].

Tian et al. [49] introduced long carbon fibers as a reinforcement to a polylactic acid (PLA) matrix, creating a raw silk material suitable for 3D printing using the FFF process. By fine-tuning the process parameters, they managed to achieve impressive results. The 3D-printed samples, containing 27% long carbon fibers, exhibited maximum bending tensile strengths and elastic moduli reaching 335 MPa and 30 GPa, respectively. These enhancements in mechanical properties rendered the printed samples potentially useful in aerospace applications. Furthermore, Eutionnat-Diffo et al. [55], using FFF, delved into the relationship between the tensile deformation of non-conductive and conductive PLA filaments deposited on polyethylene terephthalate (PET) fabrics, the fabric properties, and the printing platform temperature. Their research also involved the presentation of a theoretical and statistical optimization model. Notably, they observed that the non-conductive PLA printing guide demonstrated improved durability after washing or with the addition of a conductive filler. However, the washing process had an impact on the fracture stress of the woven fabric after printing.

ABS exhibits compatibility with various materials, including styrene ethylene butadiene styrene, ultrahigh-molecular-weight polyethylene [56], montmorillonite clay [57], layered silicates [58], and other petroleum-based and biobased substances. This article focuses on the recent advancements in 3D printing utilizing ABS, a versatile polymer widely applicable in automotive applications, electrical devices [59,60], custom-fit prosthetics [61], aerospace, energy storage [62,63], artifact complex designs [64], multi-purpose sensor designs [65,66], and dielectric devices [67]. ABS was reinforced with polycarbonate (PC) in different proportions (10, 20, and 30 wt%), and the resulting ABS/PC composite filaments were used in FFF-based 3D printing [68]. The mechanical properties of 3D-printed pure ABS, pure PC, and ABS/PC composite samples were assessed and compared, revealing enhanced flexural modulus, tensile strength, and flexural strength in ABS/PC composites due to modified void formation and improved interfacial bonding (Figure 6a,b). Chen et al. [69] conducted an analysis of the scratch resistance and mechanical performance of 3D-printed ABS specimens, both before and after the addition of poly(methyl methacrylate) (PMMA). The incorporation of 30% PMMA resulted in a significant improvement in scratch resistance, hardness, melt flow rate, mechanical performance, and surface glossiness of ABS. Additionally, the toughness of ABS/PMMA blends was enhanced by introducing small quantities of methacrylate–butadiene–styrene (MBS) as a toughness modifier.

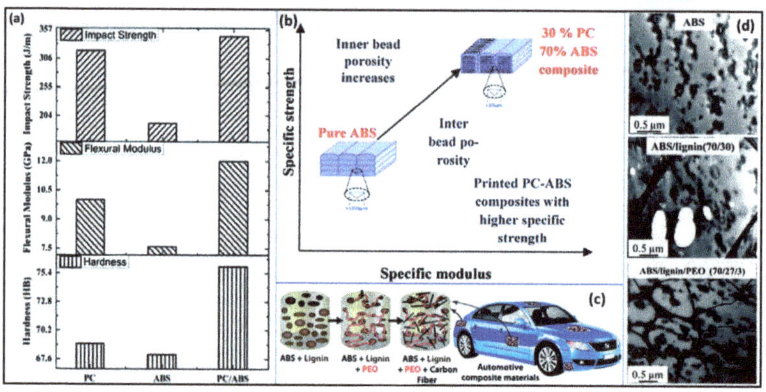

Figure 6. (**a**) A comparison of the mechanical properties of polycarbonate (PC), ABS, and PC/ABS-based specimens. (**b**) Schematic representation illustrating the difference in strength and modulus for 3D-printed pure ABS and its composites using the FFF technique. Reproduced with permission from [68]. (**c**) Schematic representation showcasing the formulation of ABS composites using lignin, poly(ethylene oxide) (PEO), and carbon fibers for automotive applications. (**d**) Transmission electron microscopy images of neat ABS, ABS/30% lignin, and ABS/27% lignin/3% PEO blends. Reproduced with permission from [70].

In a another study, the FFF-based 3D printing of ABS with biomass-derived lignin (10, 20, and 30 wt%) was investigated, following compatibilization with 10 wt% poly(ethylene oxide) (PEO) [70], as depicted in the schematic representation in Figure 6c. The findings indicated that a lower percentage of lignin in ABS led to a decrease in tensile strength, which was partially mitigated by the inclusion of PEO in ABS/lignin composites, reducing the activation energy of the blend via improved plasticization of the hard phase. The dispersion and dimensions (300–1000 nm) of lignin in ABS significantly contributed to the enhanced mechanical properties. With PEO as an interfacial adhesion promoter, the dimensions of lignin particles were further reduced to the range of 200–500 nm (as shown in Figure 6d). Additionally, the incorporation of chopped carbon fibers (20 vol%) into the ABS/lignin/PEO composite system resulted in 3D-printed samples with superior mechanical strength (77–80 MPa), demonstrating its potential application in the fabrication of automotive parts.

Another study by Shim et al. [71] investigated the influence of varying printing orientations (0°, 45°, and 90°) when using PMMA for 3D printing of denture base resin. Their comprehensive study revealed that the selected printing orientation played a pivotal role in determining key factors such as print resolution, bending strength, surface morphology, and microbial reactions. Consequently, their research highlights the significance of precise printing direction selection for achieving superior performance in denture resin 3D printing.

2.2. Thermosetting Polymers

In contrast to thermoplastic polymers that soften when heated and harden when cooled, thermosetting polymers undergo a chemical transformation during the initial heating process that sets them apart. This transformation results in the formation of cross-linked structures within the polymer matrix, making them rigid and stable upon cooling. Crucially, in contrast to thermoplastics, thermosetting polymers cannot be melted or reshaped when subjected to heat again. Some notable examples of thermosetting polymers include phenolic plastics, amino plastics, epoxy plastics, unsaturated polyester plastics, and organosilicon plastics [72]. The appeal of thermosetting polymers in 3D printing lies in their capacity to produce objects with exceptional structural stability and resilience.

Thermosetting polymers undergo irreversible hardening through the curing process of a viscous liquid prepolymer or resin. In contrast to unidirectional growth, active reaction sites within the polymer chain interact with adjacent chains, forming a closely interconnected polymer network. The mechanical strength, hardness, heat resistance, and chemical resistance of thermosets increase with higher crosslinking density. Curing can be achieved with heat, appropriate radiation, high pressure, and the inclusion of catalysts. Figure 4b illustrates the interconnected molecular structure of thermosetting polymers, delineating different components based on curing methods. Discussed below are a few further noteworthy explorations into the applications of thermosetting polymers in 3D printing:

Porous Carbon-Free Scaffolds for Biomedical Applications: Fu et al. [73] adopted a novel approach, combining 3D printing with ceramic conversion techniques, to fabricate porous carbon-free embedded scaffolds. These scaffolds, constructed from silicon resin infused with calcium carbonate filler under controlled atmospheres and high-temperature conditions, demonstrated impressive anti-tumor properties, particularly in bone tissue engineering. Additionally, these scaffolds exhibited the ability to enhance cell differentiation and foster bone tissue regeneration.

Clinical-Grade Photopolymerization Resins: Lin et al. [74] embarked on the development of photopolymerization resins ideally suited for 3D printing applications, leveraging an ultraviolet (UV) 3D printer. The results were impressive; the printed samples exhibited robust mechanical properties. Specifically, the bending strengths ranged from 60 to 90 MPa, the bending moduli fell within the range of 1.7 to 2.1 GPa, and the surface hardness values spanned from 14.5 to 24.6 HV. Such mechanical characteristics mirrored those of clinically used resin materials, pointing toward the significant clinical application potential of these 3D printable materials.

Strengthening Epoxy Resin with Nanomaterials: Hmeidat et al. [75] delved into the potential of integrating functional nano-clay into fiber-free reinforced epoxy resin, an endeavor realized using direct 3D printing technology. This pioneering approach yielded results that were nothing short of extraordinary. The samples produced demonstrated a remarkable strength range, measuring between 80 and 143 MPa. This strength exceeded any previously reported values for 3D-printed thermosetting composites, even those incorporating short fiber reinforcements. Their study underscores the promise of enhancing epoxy resin with nanomaterials to augment its viability in the context of direct 3D printing applications.

These investigations collectively emphasize the versatility and growing potential of thermosetting polymers in the exciting field of 3D printing. From innovative bone tissue scaffolds to precision denture production, these polymers are expanding the horizons of what is possible in additive manufacturing.

2.3. Hydrogels

Hydrogels are highly suitable materials for 3D printing, particularly in extrusion-based methods, due to their feasibility as ink materials [76]. The hydrogel structure consists of a 3D crosslinked hydrophilic polymer network, as depicted in Figure 4c. They can be categorized based on preparation methods into homopolymers, copolymers, and semi-interpenetrating networks (semi-IPN) [77]. Examples of homopolymers include poly(hydroxyl ethyl methacrylate) (PHEMA) and polyethylene glycol (PEG), while copolymers, such as PEG-PEGMA and carboxymethyl cellulose, incorporate two types of monomers. Semi-IPN, exemplified by acrylamide/acrylic acid copolymer, forms when a linear polymer penetrates another cross-linked network without chemical bonding. Solvents like water, ethanol, water–ethanol mixtures, and benzyl alcohol are commonly used for solution polymerization (Figure 4c) [78].

2.4. Elastomers

Within the realm of elastomeric materials for 3D printing, Polydimethylsiloxane (PDMS) stands out as a dominant choice, celebrated for its remarkable qualities. PDMS boasts outstanding mechanical flexibility, stretchability, chemical inertness, biocompatibility, and exceptional thermal stability when compared with other elastomers. Moreover, it maintains high chemical stability under demanding conditions of elevated temperatures and pressures. Wang et al. [22] used a direct ink writing 3D printer to craft specialized wetted surfaces using PDMS ink. They intricately patterned porous structures with varying geometric parameters and honed their design using parameter optimization. This meticulous approach resulted in the creation of sample films featuring a remarkable water contact angle of approximately 155 degrees. Notably, the porous PDMS films exhibited superhydrophobic properties and demonstrated exceptional durability, even when subjected to thermal aging.

Hollander et al. [79] explored the possibilities of using semi-solid extrusion 3D printing in conjunction with UV-LED cross-linking technology to produce PDMS drug delivery devices. Their work specifically focused on creating devices containing prednisolone. A standout aspect of this approach was its room temperature compatibility. By avoiding the need for elevated temperatures, it emerged as a promising method for manufacturing drug delivery devices designed for temperature-sensitive pharmaceuticals. Xiang et al. [80] ventured into the development of UV-curable silicone elastomers using thiol-ene photopolymerization. These materials exhibited excellent biocompatibility and found applications in skin wound healing as wound dressings. Their research highlighted the precision and versatility inherent in the thiol-ene photoreaction. It enabled the preparation of various soft structures with photolithography and the creation of three-dimensional elastic structures characterized by smooth surfaces and exceptional performance.

3. Smart Polymer Materials for 4D Printing

In recent years, there has been a growing interest in exploring new materials capable of exhibiting specific responses to external stimuli. This heightened interest is primarily driven by the quest to develop advanced devices, sensors, and actuators with applications spanning biomedical and electronic fields. Among the innovative materials gaining attention are smart materials, characterized by their ability to undergo changes in properties, such as shape, color, or size, in response to external stimuli like light [27], heat [25,26,34], humidity [34], or electric and magnetic fields [28,33]. This class of programmable materials introduces a unique dimension to 3D printing, referred to as 4D printing. In 4D printing, the same process as 3D printing is used, but the printed objects possess the remarkable capability to dynamically transform their shape or properties over time in response to external stimuli [19,81].

Numerous smart polymers, including SMPs, smart hydrogels, and LCPs/LCEs, have been developed; however, these materials exhibit certain limitations. Some smart polymers exhibit minimal responses over extended periods or possess limited reversibility in their transformations. Despite the need for further advancement in achieving swift and precise transformations in 4D-printed objects, the realm of 4D printing presents novel opportunities across diverse applications such as textiles [82], aerospace [18,19,83], medical industries, electronics, and robotics [18,19,84–86]. This section focuses on exploring smart polymers for 4D printing, particularly SMPs, smart hydrogels, and liquid crystals.

3.1. Shape Memory Polymers

Shape memory polymers (SMPs) have emerged as remarkable candidates for 4D printing applications due to their versatile processing capabilities, aligning perfectly with the requirements of various additive manufacturing (AM) technologies. These polymers exhibit elastic behaviors, making them well-suited for the intricate geometric transformations involved in 4D printing [8]. What sets SMPs apart is their ability to endure substantial deformation strain across a range of temperatures, allowing for dramatic and impressive shapeshifting. It is no wonder that polymeric materials have found their way into the realm of AM due to their exceptional processing attributes, making the transition to 4D printing a seamless one.

An illustrative example is a shape-changing structure under a magnetic field, depicted in Figure 7a. This structure, composed of poly(lactic acid) and magnetic iron oxide (Fe_3O_4) nanoparticles, undergoes remote heating under magnetic fields through hysteresis [28]. Utilizing direct ink writing (DIW) printing with UV curing facilitates the production of these shape-changing structures. While the transformation mechanisms align with other examples using thermal response, this structure exhibits fast, remotely actuated behaviors and magnetically guidable properties.

Another approach to activate SMPs involves leveraging a heat-shrinkable property, obviating the need for a shape-programming step. Figure 7b illustrates a 3D configuration transforming from a planar sheet to a final flower structure in response to temperature changes [25]. The release of internal strain in the polymer, generated during FFF, maintains the printed composite in a flat state when heated. Upon cooling to room temperature, the structure transforms into a flower configuration due to the mismatch in the coefficient of thermal expansion (CTE) between different materials. Additionally, Figure 7c presents another example using time-lapse to showcase the evolution of shape with temperature change [26]. In this instance, both geometry and printing patterns, including dimensions, the number of grooves, and active elements, were controlled during printing to induce the transformation of floral leaves into different shapes at specific times.

SMPs can be readily tailored to specific environments and applications by manipulating the crystalline content of the polymer. This fine-tuning enables the programming of critical transition temperatures, such as the glass transition temperature (T_g) and the melting temperature (T_m) [87,88]. A notable example of utilizing SMPs for 4D printing is the work of Ge et al., who pioneered a multi-material system for printing shape memory polymers using micro-stereolithography. Their ingenious ink formulation incorporated methacrylate-based monomers, along with essential components like the photoinitiator Phenylbis(2,4,6-trimethyl benzoyl)phosphine oxide (BAPO), Sudan I, and Rhodamine B. By adjusting the proportions of PEGDMA, BPA, DEGMA, and BMA, they could precisely tailor the T_g to achieve fixity control at distinct temperatures, such as 43 °C and 56 °C (Figure 7d). This precision allowed the bloom motion of their printed flower to unfold in two distinct phases [29].

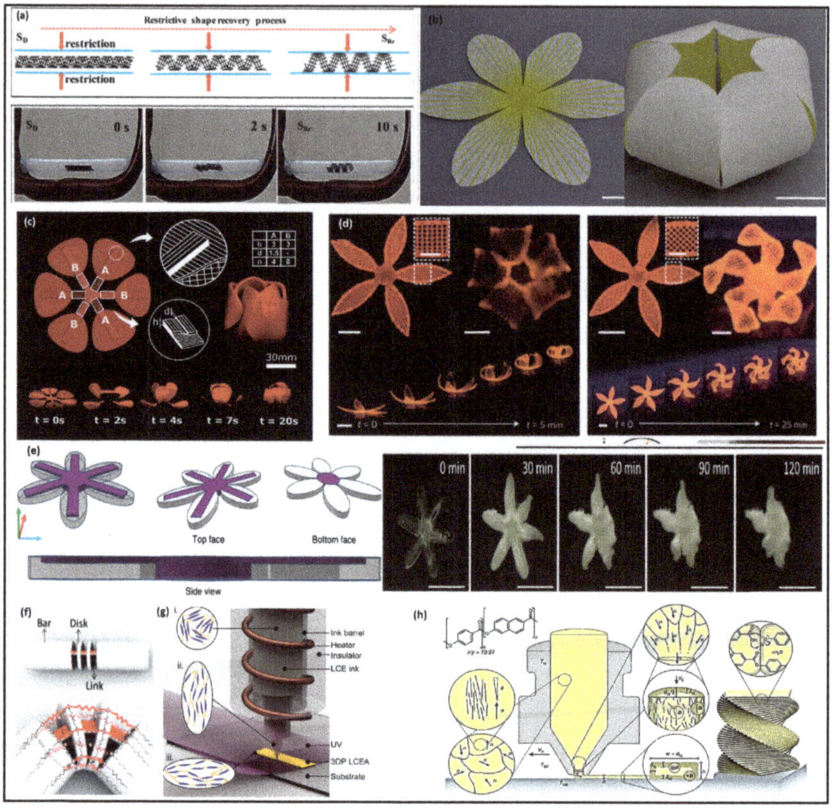

Figure 7. (**a**) A collection of polymer composites designed for 4D printing, featuring various innovative applications. Schematic and demonstration of a restrictive shape recovery process triggered by an alternating magnetic field [28]. (**b**) A flower-like 4D structure transforming from a flat sheet to a final flower structure [25]. (**c**) A time-lapse illustration showcasing the folding sequence of a tulip [26]. (**d**) Examples of 4D-printed flower morphologies with different bilayer directions (90°/0° and 45°/45°) [29]. (**e**) CAD models of multi-material sea stars (**left**), and demonstration of swelling over time in water (**right**) [30]. (**f**) An illustration of an initial joint and folding of bars with spring–mass systems [31]. (**g**) Illustration of the Liquid Crystal Elastomer (LCE) ink states during hot-DIW: disordered LCE ink within the barrel (**i**), aligned LCE ink as it moves through the nozzle (**ii**), and the resulting crosslinked LCE filaments post-printing (**iii**). Additionally, a visual representation of the printed LCE actuator with a meander-line print path showcases distinct elongation states corresponding to different temperatures [34]. (**h**) A schematic illustrating the printing of a core–shell microstructure with a highly aligned shell using fused filament fabrication (FFF) [35]. Figures reproduced with permission from corresponding references.

Smart hydrogel composites, with their ability to swell upon water immersion while maintaining structural integrity, hold significant promise as smart materials. Their applications span diverse fields, from biomedicine including drug delivery devices [89], artificial organs, and tissue engineering [19,90], to agriculture [91,92]. In 4D printing, these hydrogel composites present an innovative method for various applications such as custom-designed sensors and robotics [19,85,86]. For example, Figure 7d illustrates biomimetic hydrogel composites in the form of functional folding flowers capable of being folded and twisted [29]. The shape transformation is governed by the anisotropic swelling behavior of the hydrogel composite, which is controlled by the alignment of cellulose fibrils along the printing directions. Another approach involves incorporating different components with distinct swelling properties within a single hydrogel structure. Figure 7e showcases sea stars that actuate via spatially controlled swelling produced with Vat Polymerization (VP) [30]. The sea stars exhibit different swelling behaviors in their center region (strain-limiting area in purple) and arms (more swellable materials in white), causing gradual curling toward the center regions over time in water. Lastly, Figure 7f depicts evolving bars composed of rigid materials (white) for bars, disks, and bottom parts of links, with the top of the links made of hydrogel (red), enabling the folding of the structure [31].

Recent advancements in the realm of 4D printing have introduced the fascinating world of cross-linked liquid crystalline polymers (LCPs) and liquid crystal elastomers (LCEs). These materials have sparked innovation in the creation of actuators and soft robotics using the process of direct ink writing (DIW) [93,94]. LCEs, in particular, are a class of materials characterized by their anisotropic nature, where their properties are finely controlled by the orientation of their molecular structure, a property that can be tailored by temperature variations [95]. Their remarkable capacity for shape transformation allows them to craft intricate structures like cones, paraboloids, and even origami-like folds [96]. Moreover, external stimuli, including heat, light, and electrical fields, rendering them dynamic and versatile, trigger the mechanical responses of LCEs.

The fascinating behavior of LCEs hinges on the arrangement of rigid molecules known as mesogens, and this behavior can be fine-tuned by altering the temperature. LCEs shift from an ordered, anisotropic state to a disordered, isotropic state at a specific transition temperature, often referred to as the isotropic transition temperature (T_i or T_m) [97]. Lopez-Valdeolivas and colleagues embarked on an endeavor to create thermo-actuators using LCEs with an extrusion-based printer. In contrast to traditional methods that rely on heat-induced mesogen orientation, their innovative approach involved the extrusion process, which aligned polymer chains, effectively establishing orientational order within the mesogens. The result was an impressive 50% contraction of the printed parts when heated to 90 °C, followed by a rapid recovery to their original length upon cooling [98]. LCEs' distinct advantage lies in their rapid recovery rates, a characteristic not commonly observed in the swelling behavior of hydrogels.

Notably, LCEs offer distinctive attributes in the world of 4D printing. For instance, Yang et al. harnessed the power of infrared (IR) light (808 nm) to fabricate, repair, and assemble carbon nanotubes (CNTs) and xLCEs for soft robotic actuators designed for low-temperature environments [99]. xLCEs boast the remarkable ability to self-heal micro-cracks through photo-healing, ensuring a sustainable and efficient material life cycle. Furthermore, CNT-xLCEs are manufactured with transesterification processes carried out at temperatures exceeding 180 °C. While many of the mentioned shape memory polymers (SMPs) thus far have been thermoplastic in nature, designed to regain their shape through temperature-induced transitions involving the glass transition temperature (T_g) and melting temperature (T_m), there exists a realm where robustness is paramount. In scenarios where extreme and harsh environments demand chemically and thermally stable materials, thermoset plastics have often been the choice due to their strong chemical covalent cross-links, ensuring excellent chemical and thermal stability.

Figure 7g depicts variations in the morphologies of photopolymerizable liquid crystal elastomer (LCE) ink during direct ink writing (DIW) and the diverse elongation of the printed LCE based on the aligned printing path, depending on the temperature relative to the nematic isotropic transition. In another extrusion printing process using solid filaments, liquid crystal polymers (LCPs, Vectra A950) [35], as shown in Figure 7h, exhibit a notable alignment of nematic domains through the nozzle due to shear forces during extrusion. Despite the initial alignment, the printed filament tends to lose its orientation and solidifies from the surface, resulting in a core–shell microstructure with a highly aligned shell. The mechanical properties of this printed hierarchical structure can be further strengthened with thermal annealing, facilitating chemical crosslinking of chain ends between filaments.

A study conducted by Zhang et al. demonstrated a fascinating correlation between color transmittance and the nanostructure of a 4D-printed shape memory polymer (SMP) [100]. Using microstereolithography (MPL), the group 4D printed submicron-scale grids using VeroClearTM. When these grids were illuminated with white light, they selectively transmitted a limited range of wavelengths due to the differential scattering of the spectrum by the grid (see Figure 8). The dimensions of the grid determined which wavelengths could pass through. Upon programming the structure, i.e., heating it above the glass transition temperature (T_g) to 80 °C, distorting it, and cooling it, the structure became transparent to all wavelengths, achieving an "invisible" state. Reheating above T_g restored the nanostructure to its as-printed state, allowing only a specific range of wavelengths to pass through once again.

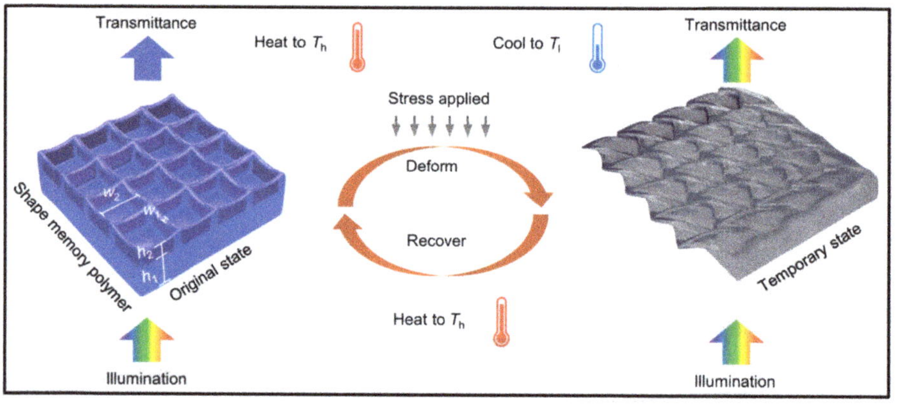

Figure 8. Illustration depicting the color and shape transformation of a constituent nanostructured element in the "invisible ink" 3D printed with a shape memory polymer. The as-printed structures feature upright grids on the left, acting as a structural color filter that selectively transmits a limited wavelength range of visible light. Deformation of the structures at elevated temperatures flattens the nanostructures on the right, resulting in a colorless state, which persists as an invisible state after cooling to room temperature. Heating restores both the original geometry and the color of the nanostructures, showcasing a submicron demonstration of 4D printing. Figure reproduced with permission from [100].

LCEs have also been successfully 3D printed in their isotropic state, providing the user with the flexibility to manually determine the nematic director of the mesogens post-printing. The resulting 4D-printed smart LCE object demonstrates the capability to switch between the high-order post-printing programmed shape and the low-order shape determined by the printing process. In the study by Barnes et al., a specially formulated LCE ink was used, comprising RM257 mesogens and chain extenders 2,2′-(ethylenedioxy)diethanethio (EDDET) and pentaerythritol tetrakis (3-mercaptopropionate) (PETMP) (Figure 9a) [101].

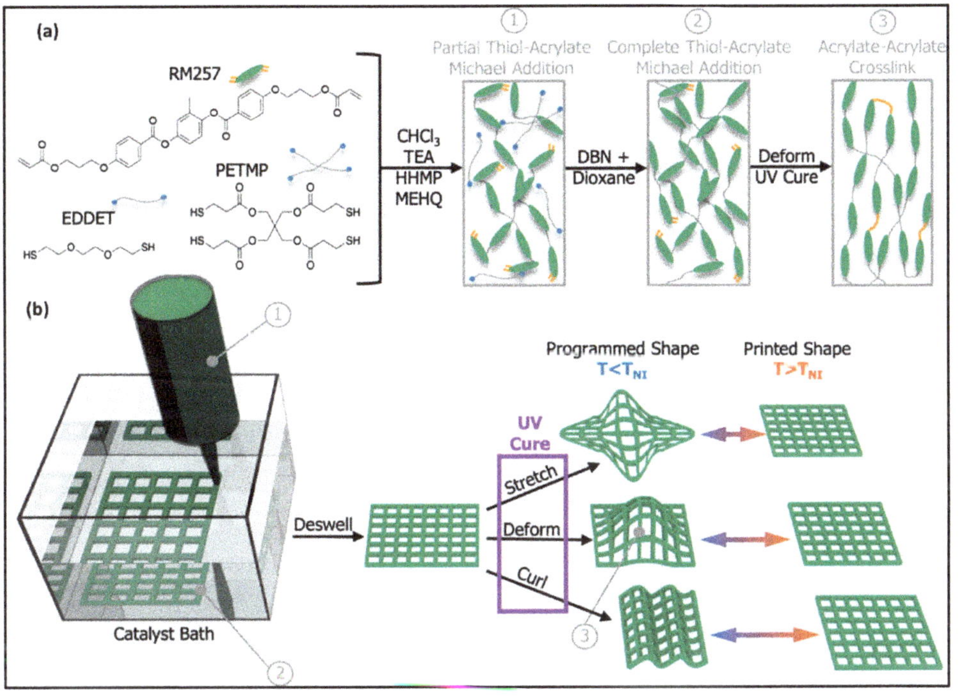

Figure 9. Schematic illustration of the reactive 4D printing process and shape programming of liquid crystal elastomers (LCEs). (**a**) Depiction of the LCE synthetic scheme, highlighting the network-forming components and three distinct reactive steps during the fabrication process. (**b**) Schematic representation of the 4D printing of LCEs, involving printing in a catalyst bath, followed by deformation and UV curing for shape programming. The resulting LCE demonstrates reversible shape changes between the printed and programmed structures when heated and cooled, respectively. Figure reproduced with permission from [101].

The ink was partially crosslinked and DIW-printed into a catalyst bath, where further crosslinking occurred, resulting in a 3D LCE structure without a defined director (Figure 9b). The printed object contained residual unreacted acrylate functional groups. By stretching the object in a specific direction, a nematic director was established parallel to that direction, and UV light was applied to crosslink the residual acrylate groups, setting the programmed shape. Upon heat treatment to the transition to nematic isotropy (TNI) at 75 °C, the mesogens relaxed, and the programmed structure reverted to the original printed structure. This shape evolution could be reversed by cooling the object.

Azobenzene moieties undergo isomerization from the trans to cis state upon UV irradiation, causing an expansion of the polymer matrix. This phenomenon has been utilized in 4D printing to create light-responsive actuating silicone bilayers [102]. In the case of 4D printable LCEs, azobenzene moieties contribute to light responsiveness at ambient temperature, inducing stress and mesogen misalignment. Ceamanos et al. demonstrated the 3D printing of an azobenzene-containing LCE strip that contracted under UV light, lifting a weight, and recovered its initial length under blue light [103]. However, spontaneous relaxation of azobenzene moieties over time posed stability challenges. Lu et al. addressed this issue by incorporating 2-ureido-4[1H]-pyrimidinone (UPy) physical crosslinkers into the LCE ink, allowing the structure to change shape via photoisomerization of azobenzene moieties under UV light. The shape was then stabilized by the reformation of UPy crosslinks when the UV light was removed [104], as shown in Figure 10.

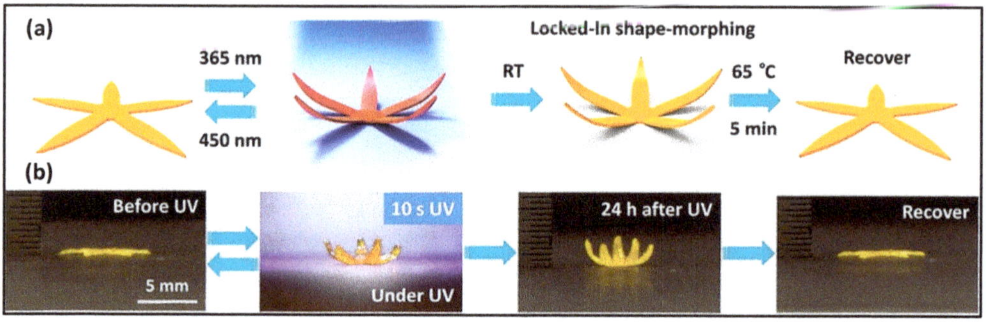

Figure 10. (**a**) Illustration and (**b**) photographs depicting the photo-switchable deformations of a flower-like actuator 4D printed with light-sensitive LCE. Under 365 nm UV light, the flower curled, and the process was reversed with 450 nm blue light. After UV irradiation, the curled shape was stabilized at room temperature via the formation of UPy crosslinks. The original flat shape was recovered by heating to 65 °C, breaking the crosslinks, and allowing the LCE to entropically switch back to the isotropic state. Figure reproduced with permission from [104].

3.2. *Self-Healing Polymer Materials*

In the context of 4D printing, self-healing materials present a remarkable capability. These materials have the unique ability to adapt and evolve over time, offering increased longevity and reliability to printed components in a variety of applications, including electronics and healthcare monitoring. Self-healing materials are a valuable addition to the toolkit of 4D printing, offering the potential for more resilient and long-lasting printed structures. By integrating self-healing properties into 4D printing materials, engineers and designers can create structures that can endure harsh environments, maintain their functionality over extended periods, and reduce the environmental impact associated with the disposal of damaged items. This remarkable feature of self-healing materials represents a significant advancement in 4D printing technology, with widespread potential across multiple industries and applications. Self-healing materials operate by autonomously initiating the repair process when defects like cracks or scratches occur. This spontaneous healing occurs as the material fills the void created by the defect with fresh material. The ultimate goal of an ideal self-healing material is to fully restore the original mechanical and chemical properties of the pristine material [105]. These self-healing materials can be broadly categorized into two main types based on their healing mechanisms. Autonomous materials are capable of immediately commencing the healing process upon sustaining damage. This spontaneous repair is facilitated through mechanisms like reversible hydrogen bonds, non-covalent bonds, or the release of self-healing substances that are embedded within the host material and triggered by damage [106]. Ton-autonomous materials, in contrast, require external stimuli to initiate the healing process. These stimuli can include factors such as exposure to UV light, application of heat, or mechanical activation. These external influences are necessary to trigger repair [106].

3.2.1. Microvascular Self-Healing Mechanisms

Self-healing mechanisms using microvascular systems incorporate a network of channels within a material, filled with a healing agent, traversing the material's matrix. Upon the occurrence of a crack or damage, capillary action initiates the release of the healing agent into the damaged area, where it solidifies to seal the crack. This concept draws inspiration from the natural arterial system [107,108]. Li et al. [109] introduced an innovative biomimetic 3D vascular design inspired by nature for self-healing cementitious systems (Figure 11a–c). The 3D printed design featured a polylactic acid (PLA) vascular network produced on an Ultimaker® 3D printer (FFF). This network incorporated the healing agent sodium silicate. In instances where a crack formed in the cement matrix, the vascular design adhered to Murray's law for circulatory blood volume transfer, thereby facilitating the healing process. Another approach was taken by Wu et al., who developed a microvascular network using a fugitive ink and hydrogel reservoir to create a self-healing printed material tailored for applications in healthcare microvascular devices [110].

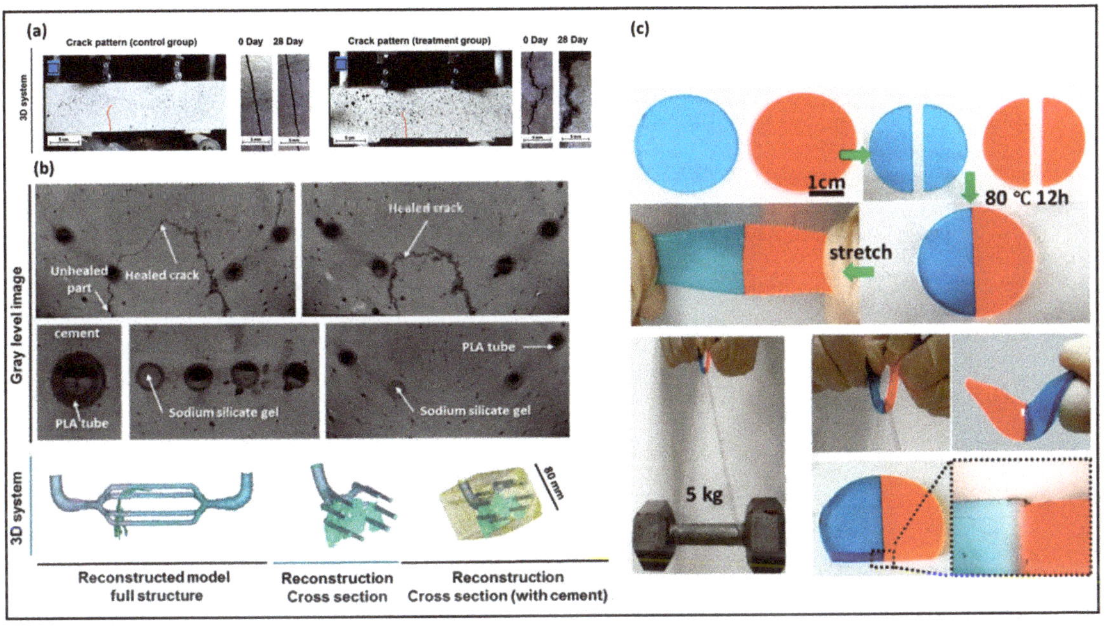

Figure 11. (**a**) Representation of a vascular beam crack pattern and microscopic imagery documenting the healing process following a 4-point bend [109]. (**b**) A CT grey level image unveiling the concrete matrix, cracks, PLA vascular tubes, and sodium silicate self-healing agent. The 3D system was reconstructed to illustrate the vascular mechanisms, with the yellow area symbolizing the cement, violet sections representing the PLA vascular structure, blue section representing sodium silicate gel, and the green area depicting the gel filling in cracks [109]. (**c**) Depictions of colored self-healing polyurethane samples that were cut, connected, and healed at 80 °C for 12 h. The healed sample is showcased while being stretched and supporting a 5 kg weight [111]. Figures reproduced with permissions.

Balancing the characteristics of printable inks with multiple functionalities is a delicate task, as it involves finding the right balance between printability properties and functional properties. In this instance, thiol groups play a role in the photocuring process, while disulfide groups contribute to self-healing properties. Li et al. also documented the printing of a photocurable elastomeric material using DLP (Figure 11c) [111].

3.2.2. Encapsulation Self-Healing Mechanisms

Encapsulation-based self-repair mechanisms utilize micro- or nano-sized capsules filled with a healing agent, such as a polymer or catalyst. Each capsule is shielded by a case or coating made from an inert material to prevent interference with the bulk material. This ensures that the healing process only commences when damage occurs, not prematurely. The capsule's shell is designed to be fragile, enabling easy rupture upon damage. These capsules are distributed throughout the material's matrix to ensure a rapid release of the healing agent upon rupture. The healing agent inside the capsules should have low viscosity to facilitate capillary action into the damaged area. For efficient healing, the healing agent should polymerize quickly at room temperature without shrinkage [112].

Davami et al. used SLA 3D printing to create structures that entrap photocurable resin within unit cells, acting as a self-healing agent. When damage occurs, the self-healing agent flows out of the cell through capillary action to the damaged site, where it is cured using UV light. Davami et al. reported a healing efficiency of 52% based on fracture toughness [113].

3.2.3. Autonomous Self-Healing Polymers: Supramolecular Polymers

Most polymers typically require the presence of a host network containing a self-healing agent, such as capsules or vascular systems, while supramolecular polymers constitute a distinct class of materials. Characterized by reversible non-covalent bonds, these polymers can reform and repair after being cleaved [114]. The inherent self-healing properties of supramolecular polymers distinguish them, allowing for repeated healing without depleting a host healing agent. Often described as "solid-liquids," these polymers consist of associated groups typically covalently bonded to chain ends or side chains of a polymer. These groups bring together liquid-like polymers into a network of non-covalently cross-linked polymers, exhibiting plastic behaviors. The reversible reforming process of supramolecular polymers involves non-covalent crosslinks like π-π stacking, hydrogen bonds, host–guest interactions, ionic interactions, and metal coordination. Evaluating the effectiveness of restoring a self-healing material involves examining various properties, such as mechanical, electrical, and thermal properties, before and after healing. This assessment, known as healing efficiency, is calculated using the ratio of these properties before and after healing [114].

4. Hybrid Materials for Printing

4.1. Polymer–Metal Composites

4.1.1. Three-Dimensional Printing of Polymer–Metal Composite Feedstocks

Polymer–metal composites involve the amalgamation of metal particles with a liquid polymer resin, typically an unsaturated polyester, in the form of water-based substances exhibiting heightened viscosities. In Figure 12, diverse methodologies are depicted for crafting resilient structures utilizing polymer–metal hybrid materials across various 3D printing techniques [115]. In specific environments, such as high-temperature settings, metal filaments and water-based metallic ink can be extruded through a nozzle, subsequently solidifying rapidly upon cooling or further sintering in a furnace (Figure 12a,b). Photo-hardening polymers, comprising liquid polymer resin, a crosslinking agent, an initiator, and a photosensitizer, can be cured with UV light, wherein the binding liquid facilitates bonding metallic particles together in powder form (Figure 12c). Objects 3D printed with metallic nano/microparticles have gained extensive application in the creation of catalytic and mechanically robust structured materials (Figure 12d).

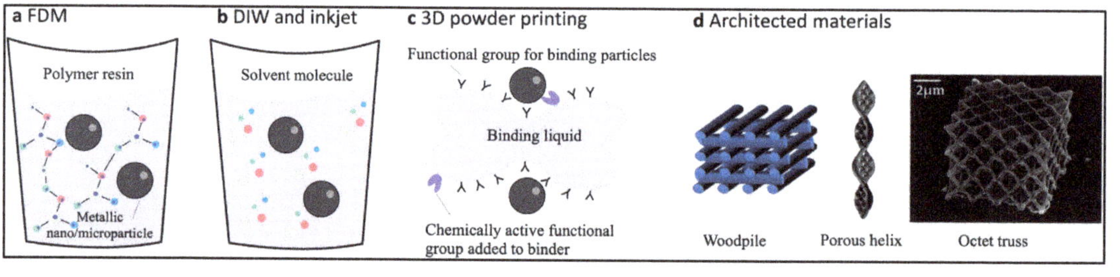

Figure 12. Deployment of metallic nano/microparticles spans various additive manufacturing techniques: (**a**) integration of polymer–metal filaments in fused filament fabrication (FFF), (**b**) utilization of water-based metal inks in direct ink writing and inkjet printing, (c) application of UV-curable metal resin in stereolithography and 3D powder printing, and (**d**) illustration of characteristic metallic structures designed for catalyst applications. On the right side: portrayal of the nickel nanolattice octet truss post-pyrolysis [115,116]. Figures reproduced with permission.

In addition to utilizing polymer–metal feedstocks, the synthesis of metal–polymer composites can be achieved with a distinct 3D printing approach—constructing a metal framework and subsequently infusing it with polymers. This method brings forth notable advantages, notably heightened conductivity in functional applications and superior mechanical properties. The synergistic effect of the composite material disperses energy efficiently, substantially enhancing fracture toughness, particularly in scenarios where traditional 3D-printed parts exhibit imperfections from the manufacturing process [117]. The stiffness balance between the soft and hard phases can be finely tuned to bolster fracture toughness in structured materials. Epoxy stands out as the preferred base material for polymers due to its commendable strength and straightforward processing. Bapari et al. [118] conducted insightful studies that illustrated the pivotal role of the soft phase in fortifying rigidity in metal–particulate polymer composites. Meanwhile, Li et al. [119] delved into the intricacies of interpenetrating phase composites (IPCs) within selectively laser-melted (SLM) metallic micro-lattices. Their experiments and simulation models unveiled that achieving uniform deformation in reinforced struts demanded more energy than the buckling of freestanding struts.

The amalgamation of polymers and metals in the fabrication of objects unveils captivating mechanical and functional properties. Despite the conventional belief that metal–polymer blends may not be ideal for swiftly manufacturing objects with high mechanical strength, significant research efforts have been devoted to exploring innovative materials. Fafenrot and colleagues [120] noted the impact of polymer–metal feedstocks containing bronze, revealing a noticeable reduction in mechanical properties. Zhu and his team presented a groundbreaking creation—hierarchical nanoporous gold with meticulously engineered nonrandom macro-architectures using DIW (direct ink writing) and dealloying techniques. The resulting nanoporous metals showcased significantly enhanced mass transport properties and reaction rates for both liquids and gases. Furthermore, Vyatskikh and collaborators pioneered a lithography-based process for producing intricate 3D nano-architected metals boasting approximately 100 nm resolution. This breakthrough transcended the limitations of existing methods, which typically offer resolutions in the range of 20 to 50 µm for 3D-printed metals.

4.1.2. Surface Coating of Metals on 3D-Printed Polymers

Nature demonstrates its mastery in developing robust and flexible materials through the creation of intricate, multi-scale hierarchical structures. This has led to the development of a unique class of biological layered lattice materials, featuring hierarchical levels ranging from the microscale to the nanoscale. These mechanical metamaterials, characterized by lightweight properties, derive their high mechanical strength from the well-established principle that "smaller is stronger", rooted in the effects observed at the nanoscale and optimized geometric arrangements. What sets these materials apart is their ability to exhibit unconventional mechanical behaviors, such as twisting or expanding in all directions under uniaxial compression—feats unattainable with traditional materials.

In the realm of architected materials, the focus of many studies is to replicate the exceptional characteristics found in natural structural materials. One approach involves the creation of core–shell composites, featuring soft polymer cores enveloped by layers of ultra-strong but brittle metallic or metallic glass (MG) coatings (Figure 13a–c). Numerous instances of 3D-printed trusses showcase feature sizes spanning various orders of magnitude, from tens of millimeters to hundreds of nanometers (Figure 14). For example, Cu octet lattices have achieved specific strengths of approximately 400 MPa, exceeding the yield strengths of bulk counterparts by an impressive 80% (Figure 14a–d) [121]. The remarkable strength of these core–shell metallic lattices is attributed to a size effect, with the material consisting of single-crystalline metal featuring submicron dimensions. These findings offer valuable insights into and a deeper understanding of the mechanical properties achieved with the utilization of hierarchical structures and innovative material design.

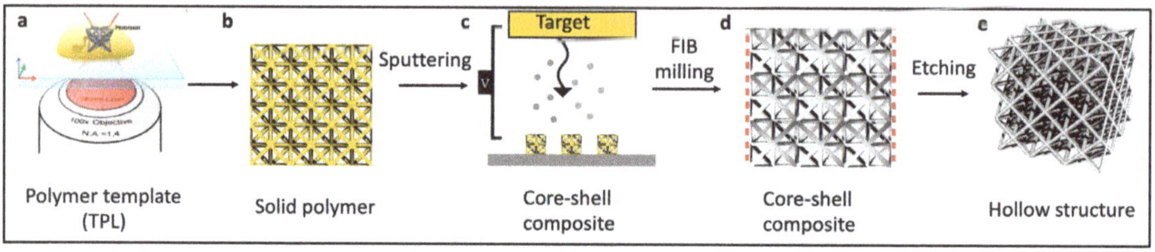

Figure 13. Illustration depicting the manufacturing procedure for creating hierarchical polymer-metal composites: (**a–c**) application of metal coatings on 3D-printed polymers, and (**d**,**e**) production of hollow-beam metallic architected materials. Figure reproduced with permission from [12].

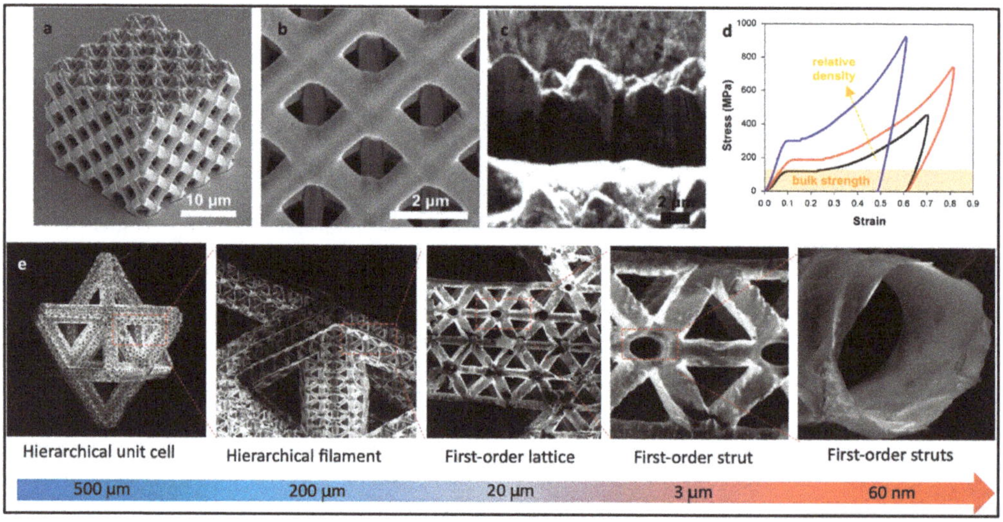

Figure 14. Hierarchical polymer–metal composites from mesoscale to microscale: (**a**,**b**) creation of core–shell Cu meso-lattices using two-photon lithography followed by electroplating, (**c**) cross-sectional milling of Cu film with a focused ion beam, (**d**) stress–strain curve of core–shell Cu lattice demonstrating higher compressive yield strengths compared with its monolithic bulk counterparts [121], and (**e**) construction of a hierarchical metamaterial composed of Ni micro-lattices with periodic hollow tubes, spanning critical feature length scales across seven orders of magnitude [122]. Figures reproduced with permission.

Thus, combining metals and polymers in 3D printing opens up opportunities to create materials with exceptional properties and characteristics. Researchers have explored various methods, such as nanoporous structures and core–shell composites, to enhance the mechanical and functional aspects of these hybrid materials. These innovations have the potential to impact fields like aerospace and medicine where advanced materials are in demand. The primary objective of the current research is to empower hierarchical polymer–metal composites to achieve remarkable levels of stiffness, strength, and toughness while maintaining very low densities. The challenge lies in finding the right balance between structure-dominated buckling failure and metal-dominated brittle failure. For instance, polymeric octet lattices with 1 μm strut diameters coated with a 10 nm NiB metallic film exhibited delayed buckling.

This elastic buckling allowed for significant rotation of the nodes, resulting in ductile-like deformation. On the other hand, thicker coatings, such as 100 nm, led to brittle failure [123]. The susceptibility to brittle failure also depends on the type of metal and the specific architecture. When utilizing High-Entropy Alloys (HEAs), the film thickness can be increased to 50 nm, overcoming the strength–recoverability trade-off [124]. While thicker coatings can better contain the polymer core, the optimal coating thickness was found to be in the range of 14–50 nm to prevent catastrophic failure. To leverage the "smaller/thinner is more ductile" size effect, it is crucial to devise innovative architectures and explore the potential of strong yet ductile metallic films.

One common approach for architectural design involves etching away the plastic core to obtain a pure, ultra-lightweight metal (Figure 13d,e). A hollow-beam nickel lattice, characterized by a nearly constant specific strength, demonstrated a high tensile elasticity of over 20%, which was unattainable for the individual metallic components [122] (Figure 14e). Future research endeavors should address both scientific and technological challenges, including the trade-off between strength and ductility [125], as well as the trade-off between speed and resolution [43].

In addition to pursuing high strength-to-density ratios and pushing the boundaries of material properties, emerging materials with innovative designs, such as chiral mechanical metamaterials and materials featuring tunable mechanical properties and programmable stimulus responses, are beginning to gain attention. To use this new category of materials in multifunctional engineering applications, it is essential to develop physical models or theories capable of capturing the architectural complexity and propagating the exceptional properties from the nano/microscale to the macroscale.

4.2. *Polymer–Ceramic Composites*

4.2.1. Three-Dimensional Printing of Polymer–Ceramic Composite Feedstocks

The primary constituents used in the 3D printing of polymer–ceramic composites include ceramics, resins, or a blend of both [126]. These materials are divided into two categories based on their form: liquid or semi-liquid inks and pastes and fine ceramic particles. Inks and pastes are a mixture of liquid or paste resin with finely dispersed ceramic particles. Their usage in 3D printing is determined by their viscosity and can include methods such as photopolymerization, inkjet printing, and extrusion [127]. Fine ceramic particles, on the other hand, are fused together either with a powder fusion and melting process, often using a heat source like a laser, or by using viscous liquid binders like non-hydroxyl resin and liquid paraffin. This kind of feedstock is suitable for 3D printing techniques such as binder jetting (BJ), selective laser sintering (SLS), or selective laser melting (SLM) [127].

4.2.2. Multi-Material Printing of Polymer–Ceramic Composite Structures

High strength and toughness are fundamental requirements for a wide range of engineering materials, but they often exhibit a mutually exclusive relationship [128]. For example, enhancing the strength of steel using cold machining typically results in a loss of toughness. Similarly, engineering ceramics surpass metals in hardness and strength but are limited by their relatively low toughness values [129]. Recent research has highlighted the remarkable combinations of strength and toughness found in natural bio-ceramics like bones or shells [130,131]. Inspired by these natural ceramics, various fabrication techniques have emerged to replicate bio-ceramics, including ice templating, layer-by-layer deposition, self-assembly, and rapid prototyping [132–134]. These studies illustrate how the trade-off between toughness and strength can be addressed by creating architected composites with well-organized microstructures.

Recently, the field of 3D printing has opened up new possibilities for fabricating microstructures with hierarchical or arbitrary geometries, utilizing a bottom-up approach similar to processes found in nature. This has led to a significant advancement in the study of bio-ceramics [125,135–137]. For example, by using multi-material 3D printers, researchers have been able to create staggered microstructures that closely resemble the architecture of bones [136]. This innovative approach resulted in a remarkable increase in fracture energy, exceeding the previous levels by tenfold. Moreover, 3D printing techniques have been used to replicate the intricate structures found in natural materials such as nacre, bone osteons, dactyl clubs, and conch shells [138–141], as illustrated in Figure 15. These endeavors aim to unravel the precise designs found in bio-ceramics. In specific cases, structures resembling dactyl clubs have been produced, exhibiting a J-shaped R-curve (crack growth resistance curve), while others have displayed Γ-shaped R-curves. A J-shaped R-curve is associated with a significantly larger critical energy release rate, making it advantageous for halting crack propagation. In contrast, a Γ-shaped R-curve is characterized by a higher critical failure stress, which effectively hinders crack initiation [142]. Additionally, researchers have explored the manipulation of magnetic fields to guide the orientation of magnetic platelets, forming intricate patterns such as concentric and layered designs [142]. As a result, 3D printing methodologies offer a versatile platform for creating complex designs to investigate the geometric influences on mechanical performance, particularly toughness, within the realm of bio-ceramics.

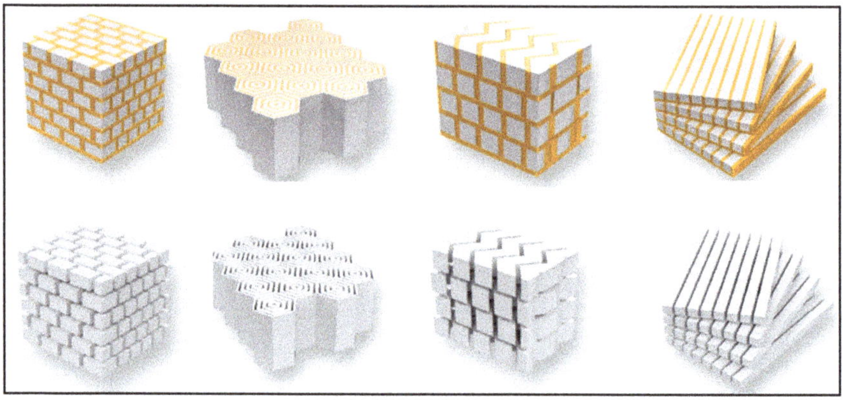

Figure 15. Emulating the structures of nacre [141], bone osteon [138], dactyl club [140], and conch shell, from left to right. The top-row representations of these structures incorporate rigid ceramic skeletons (depicted in white) and flexible polymer phases (depicted in yellow). The bottom row showcases the skeletal framework corresponding to the left. Figures reproduced with permission.

A highly promising technique in 3D printing involves the utilization of polymer-derived ceramifiable monomers, notably silicon oxycarbide. These specialized monomers are used in the 3D printing process, initially taking shape through layering and then undergoing polymerization using UV light. Following this stage, they are subjected to high-temperature sintering, a critical step that transforms them into intricate and high-quality ceramic lattices [143]. This entire process is depicted in Figure 16, which provides a visual overview of the 3D-printed polymer-derived ceramic production journey.

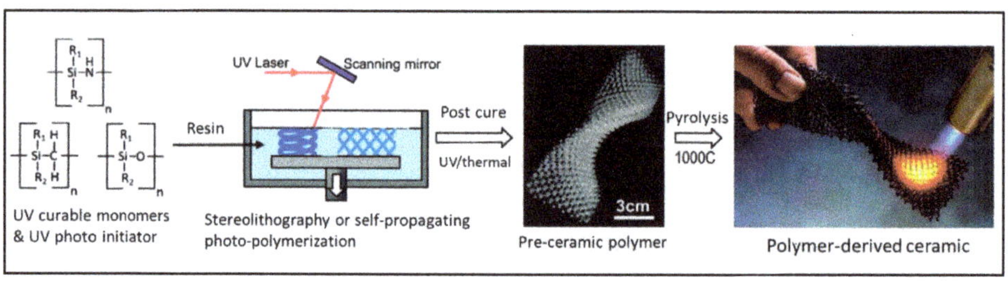

Figure 16. The process of 3D-printed polymer-derived ceramics. The depiction of the novel process where polymer-derived ceramifiable monomers, like silicon oxycarbide, are 3D printed and polymerized with UV light, followed by high-temperature sintering. Figure reproduced with permission from [143].

One notable advantage of polymer-derived ceramics is that they eliminate the need for extensive post-processing to remove the organic binder, which is often required when using ceramic fillers. This streamlined production process results in reduced manufacturing times, making it an efficient choice. The ceramic stereolithography (CSL) yields exceptionally smooth surface finishes, enhancing the overall quality of the ceramic structures it produces. However, it is essential to note that while CSL offers distinct advantages, it is considered relatively expensive due to the specialized equipment and materials required. Additionally, the selection of available materials for this technique is somewhat limited, which is viewed as one of its main drawbacks [144].

4.3. *Polymer–Metal–Ceramic Composites*

The role of printing materials is increasingly pivotal in expanding the applications of 3D printing technology. In recent years, there has been a growing trend to elevate the proportion of composite materials used to meet the intricate demands of printed products in various industries. Commonly used composite materials in 3D printing encompass combinations of metals with ceramics, polymers with ceramics, and polymers with metals. However, with the continuous evolution of "intelligent manufacturing" centered around 3D printing, the material composition of printed components often necessitates the fusion of metals, ceramics, and polymers to fulfill diverse performance criteria [145]. Although research into the domain of composite 3D printing involving metals, ceramics, and polymers is still in its nascent stages, there are emerging reports regarding its application in biomedical, electronics, surface engineering, and other domains. There are primarily two methods for printing composite materials: layered powder printing and mixed powder printing.

In layered printing, materials are progressively printed to create intricate structures. For instance, in the domain of bone tissue repair that requires load-bearing capabilities, porous scaffolds are commonly printed using metal materials, while a bio-ceramic–polymer coating is applied to the surface of the structure (Figure 17a) [146]. Such bone tissue structures composed of metals, ceramics, and polymers offer advantages such as mechanical compatibility with natural bone, corrosion resistance, and biocompatibility. In the electronics field, products like printed circuit boards (PCBs) often demand a combination of metal, ceramic, and polymer materials for their production. PCBs are initially printed with metal materials, and subsequent components like capacitors are printed directly onto PCBs using ceramics and polymer materials (Figure 17b).

In mixed powder printing, metal/ceramic/polymer powders are typically prepared with mechanical mixing. During the selective laser sintering (SLS) process, a polymer material is removed, leaving behind a porous structure (Figure 17c). Additionally, these initial porous structures formed using SLS can be infiltrated with another material to enhance their density and mechanical properties (Figure 17d). For the repair of non-load-bearing bone tissue, printing often involves the use of calcium phosphate and hydroxyapatite (HA) composite biocompatible polymers [147]. Furthermore, magnesium can be uniformly integrated into artificial bone structures during the printing process, thereby enhancing osteoinductive activity and osteoacusis performance [148].

In a multitude of industries, especially in the fields of electronics and medicine, certain critical components or products require a combination of properties associated with metals, ceramics, and polymers. The traditional methods of preparing such components are typically cumbersome and expensive. However, 3D printing technology enables the creation of complex components in a single step, eliminating the need for multiple processes. This approach offers significant advantages due to its cost-efficiency and increased speed. Nevertheless, it is important to acknowledge that the current state of 3D printing technology cannot fully support widespread applications in the production of multi-material components, and there is a relative scarcity of printable materials in this domain. Therefore, further research is required to advance this area of study.

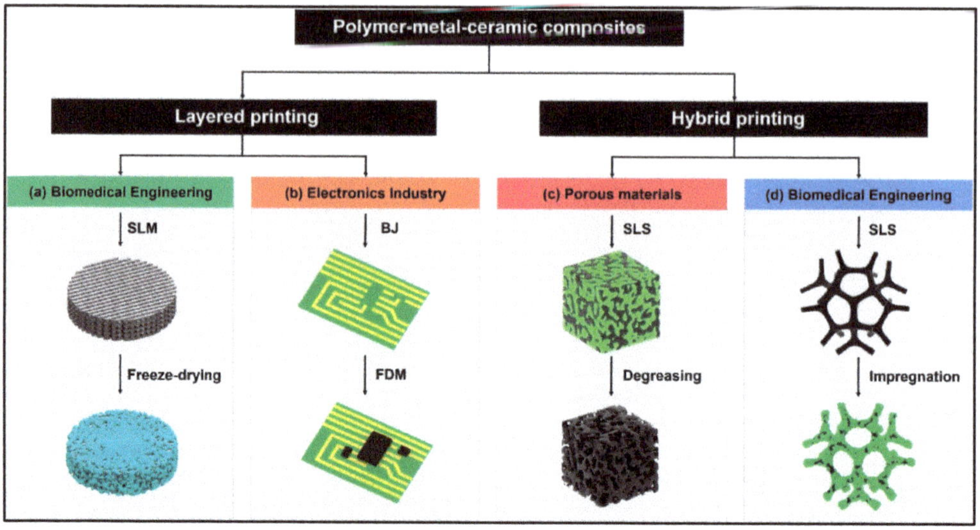

Figure 17. Potential applications and future advancements in polymer–metal–ceramic composites. Figure reproduced with permission from [12].

5. Three-Dimensional/Four-Dimensional Printing and Biomedicine

In recent years, the intersection of engineering and biomedical sciences has led to significant advancements in 3D and 4D printing technologies. These technologies have been increasingly utilized in the development of medical devices and drug delivery systems [149–151]. Bioprinting, an innovative and emerging technology of additive manufacturing, has revolutionized the biomedical sector by printing 3D cell-laden constructs in a precise and controlled manner for numerous clinical applications. This approach uses biomaterials and varying types of cells to print constructs for tissue regeneration, e.g., cardiac, bone, corneal, cartilage, neural, and skin. Furthermore, bioprinting technology helps to develop drug delivery and wound healing systems, bio-actuators, bio-robotics, and bio-sensors [149–151].

More recently, the development of 4D bioprinting technology and stimuli-responsive materials has transformed the biomedical sector with numerous innovations and revolutions [149,152]. The term 4D printing was coined to indicate the combined use of additive manufacturing, smart materials, and careful design of appropriate geometries. In this review, we report the recent progress in the design and development of smart materials that are actuated by different stimuli and their exploitation within additive manufacturing to produce biomimetic structures with important repercussions in different but interrelated biomedical areas [152]. The advent of 3D and 4D printing technologies has also led to the creation of personalized drug delivery systems [150,153,154]. These systems can be tailored to meet individual needs, addressing the challenges associated with the manufacture of pharmaceutical systems [151,153]. This has resulted in many concepts of pharmaceutical devices and formulations that can be printed and, possibly, tailored to an individual [150]. Thus, the bio-related applications of 3D and 4D printing are vast and rapidly expanding. Future research in this area promises to bring about even more innovative solutions for personalized medicine and healthcare [149–151].

In a comprehensive exploration of smart polymers, Huang et al. [155] delve into thermally responsive soft materials that adapt to environmental changes. Their review elucidates the design, characterization, and progress of these polymers, particularly in medical devices, cell therapy, and 3D printing for precision medicine. Another contribution comes from Pineda-Castillo et al. [156], who discuss state-of-the-art endovascular devices and the potential of shape memory polymers (SMPs) in addressing limitations related to complete occlusion, focusing on their application in Intracranial Aneurysm (ICA) embolization. Uboldi et al. [157] present a lab-scale film-coating process for self-expanding rod-shaped Polyvinyl Alcohol (PVA)-based devices fabricated with Hot Melt Extrusion (HME) and fused filament fabrication (FFF), offering precise control over diffusion and transport processes. In the realm of 4D bioprinting, Afzali Nanizetal [158] explores recent developments in smart materials and their applications in developing biomimetic structures with actuation capabilities, impacting pharmaceutics and biomedical research. Lastly, Sirawit et al. [159] propose a novel approach to manipulating the swelling kinetics of hydrogels with a diffusion-path architecture design, offering a unique perspective on 4D-printed hydrogel actuators.

6. Future Perspectives of Polymer AM and Challenges Ahead

In this section, we delve into the ongoing evolution of polymer 4D printing, examining its future trajectory. We identify five key areas critical for its advancement, categorized into smart materials and printing techniques. Within the smart material domain, we focus on particle–polymer composites (PPCs), "living" polymers, and polymers endowed with self-healing/welding capabilities. In the realm of printing techniques, our attention shifts to multimaterial polymer 4D printing and microscale polymer 4D printing.

In summary, this comprehensive examination reveals the significant strides made in 3D printing techniques and the pivotal role of functional polymers in reshaping the manufacturing landscape. The persistent refinement of 3D printing, with resolutions now approaching the nanoscale, paves the way for the production of high-performance products characterized by meticulously optimized structures and functions, spanning across diverse application fields [43,160]. Furthermore, by actively engineering the chemical and physical properties of polymers, we can unleash shape-changing capabilities and exercise precise control over the motion of printed products. While additive manufacturing technologies have matured and unlocked novel possibilities in a multitude of domains, they are accompanied by several challenges that merit attention in forthcoming research. These challenges serve as indispensable guiding principles for further studies, offering the potential for pioneering advancements in polymer development and applications. Despite these challenges, the future of polymer AM holds great promise. The continual development of materials, the improvement in material properties, and advancements in printer technology are expected to drive the evolution of polymer AM. This technology is on a trajectory to revolutionize manufacturing processes, expand applications, and enable the creation of innovative, high-performance components that cater to a wide range of industrial needs.

6.1. Material Diversity

The array of printable materials remains somewhat constrained due to the specific requirements of 3D printing, encompassing aspects like rheology and melting points. For example, widely used materials with low melting points and suitable viscosities, as seen in fused filament fabrication (FFF), often result in less-than-ideal mechanical and chemical properties of the final products. To meet the stringent requirements of aerospace, automotive, and structural applications, it is imperative to delve into the development of engineering-grade polymers. It is equally important to extend our perspective by considering a vast spectrum of functionalities in polymers and polymer composites for 3D printing. This expansion can take inspiration from the ingenuity of biological systems and biomaterials, spurring materials development, functional innovation, and novel structural designs.

6.2. Multi-Material and Multi-Scale AM for Various Applications

The critical drive toward multi-material and multi-scale manufacturing techniques is essential to simultaneously control material composition, ratio, functions, and internal architecture at micro- and nanoscales. Applications spanning biology, electronics, and robotics necessitate the use of multiple materials across different scales to enable complex movements or specific reactions. The evolution of multi-material and multi-scale AM represents a transformative force, reducing material waste and streamlining assembly steps across various industries. However, some 3D printing techniques encounter challenges in terms of printing with multiple materials. Innovative approaches, including the introduction of rotating vat carousels and the incorporation of microfluidic systems, have been developed to tackle these challenges [161]. Despite significant strides in composite printing, limitations persist, such as height restrictions and the difficulties encountered when fabricating products with a high loading of reinforcements, primarily due to the existing printing constraints. Looking ahead, the emergence of hybrid AM appears promising in addressing a multitude of unique customization needs, with the tantalizing prospect of 3D printing robots capable of walking off the printing plate.

6.3. Improving AM Processes for High-Quality Prints

Additive manufacturing processes, though revolutionary, introduce certain defects within printed products, impacting their mechanical properties and causing discrepancies between a digital model and a physical object. Notably, void formation remains a significant challenge in techniques such as powder bed fusion (PBF) and binder jetting (BJ). These voids materialize due to insufficient bonding between melted materials and binders. Additionally, addressing weak interlayer bonding within manufactured parts is crucial to mitigate anisotropic and subpar mechanical properties. These issues can also lead to delamination between layers. Ongoing research is dedicated to overcoming these hurdles, using tactics like introducing fillers and harnessing laser or infrared heating between printed layers. Emerging techniques, including volumetric manufacturing, hold substantial promise in surmounting these limitations [162]. Moreover, it is essential to integrate in situ monitoring and post-inspection processes to bolster quality control.

6.4. High-Throughput and Scalable Manufacturing

The speed of various 3D printing technologies during the actual printing phase remains a point of concern. The subsequent post-processing steps further diminish throughput and scalability. The industry's response involves deploying robot arms in fused filament fabrication (FFF) to circumvent the limitations posed by traditional printing plates, allowing for the production of large-scale products. Meanwhile, the application of chemical or water-soluble support materials in FFF simplifies post-processing by reducing the time spent on support cleaning. However, certain 3D printing methods are inherently limited when it comes to building size, and rapid printing of large parts can generate excess heat, leading to distortion in the final products. Recent research endeavors seek to bolster productivity by managing heat generation using mobile oil [163]. Innovative printing concepts, exemplified by volumetric AM, have begun to emerge as an enticing trend [162]. Furthermore, addressing inconsistencies in printing quality calls for the integration of feedback systems powered by machine learning, ensuring superior repeatability in scalable manufacturing [164].

6.5. Sustainability

The pursuit of a sustainable future necessitates a concerted effort to reduce waste within the realm of 3D printing. The journey toward environmental friendliness is multifaceted. It involves improving biodegradable polymers while transitioning from oil-based feedstocks to bio-based compostable plastics [23]. Additionally, managing failed prints and end-of-life products with recycling holds the promise of reducing material costs and minimizing the demand for frequent parts resupply in supply chains. The recyclability of thermoplastics stands as a notable advantage over thermosets due to the limited degradation of polymer chains. Thermoplastics are more easily recyclable compared with thermosets because they experience little to no degradation of the polymer chain when melted down. In thermoplastics, weaker interactions between polymer chains are broken, while in thermosets, the bonds between monomers deteriorate with each reuse, making it challenging to recycle thermoset polymers. An example of thermoplastic recycling is the processing of polyethylene terephthalate bottles into filaments for 3D printing [165]. These recycled filaments demonstrate comparable tensile strength and elongation, making them capable of replacing commercial filaments. Additionally, research is ongoing in other 3D printing methods, such as powder bed fusion (PBF) and Vat Polymerization (VP), to explore the reuse of unused materials, such as powder and resin, respectively.

6.6. Safety Concerns

The growing interest in and adoption of 3D printing have inevitably brought to the fore various safety concerns. Polymer-based AM processes have the potential to generate particulate matter (PM) [166,167] and emit volatile organic compounds (VOCs), including toluene, aldehydes, and ethylbenzene [167,168]. Even widely used thermoplastics like ABS and PLA have been found to release VOCs [169]. Ensuring safe printing requires the careful management of chemical reactions and the corresponding emissions throughout the manufacturing process. In-depth studies aimed at unraveling the mechanisms of PM and VOC formation across different printing methods will guide the development of more environmentally friendly 3D printers, sophisticated control strategies, and enhanced personal protection measures.

Author Contributions: Conceptualization, Writing—original draft, Data collection, A.M. and T.A.; Review and editing, F.P.; Formal analysis, Project administration, funding acquisition, S.C. and A.I. All authors have read and agreed to the published version of the manuscript.

Funding: This research is financially supported by the Key scientific research platforms and projects of Guangdong universities (2023WCXTD035 and 2022ZDZX3064 and 2023ZDZX4114 and 2023ZDZX3072 and 2023KCXTD071), the research project of Guangzhou Panyu Polytechnic (2023KJ04), the research platform of Guangzhou Panyu Polytechnic (2022KYPT03), the Guangzhou higher education teaching quality and teaching reform project (2022JSJXCXTD005 and 2022CXCYJH002). A. Irfan extends his appreciation to the Deanship of Scientific Research at King Khalid University for funding through the large research groups program under grant number R.G.P.2/276/44.

Conflicts of Interest: The authors declare that they have no known competing financial interests or personal relationships that could have appeared to influence the work reported in this paper.

References

1. Gibson, I.; Rosen, D.; Stucker, B. *Additive Manufacturing Technologies: 3D Printing, Rapid Prototyping, and Direct Digital Manufacturing*; Springer Science+Business Media: New York, NY, USA, 2015; p. 498.
2. Jacobs, P.F. *Rapid Prototyping & Manufacturing: Fundamentals of Stereolithography*; Society of Manufacturing Engineers: Southfield, MI, USA, 1993; ISBN 978-0-87263-425-1.
3. Azman, A.H.; Vignat, F.; Villeneuve, F. CAD Tools and File Format Performance Evaluation in Designing Lattice Structures for Additive Manufacturing. *J. Teknol.* **2018**, *80*, 87–95. [CrossRef]
4. Gao, W.; Zhang, Y.; Ramanujan, D.; Ramani, K.; Chen, Y.; Williams, C.B.; Wang, C.C.L.; Shin, Y.C.; Zhang, S.; Zavattieri, P.D. The Status, Challenges, and Future of Additive Manufacturing in Engineering. *Comput.-Aided Des.* **2015**, *69*, 65–89. [CrossRef]
5. Tibbits, S. 4D Printing: Multi-Material Shape Change. *Archit. Des.* **2014**, *84*, 116–121. [CrossRef]
6. Mallakpour, S.; Tabesh, F.; Hussain, C.M. 3D and 4D Printing: From Innovation to Evolution. *Adv. Colloid Interface Sci.* **2021**, *294*, 102482. [CrossRef] [PubMed]
7. Huang, S.H.; Liu, P.; Mokasdar, A.; Hou, L. Additive Manufacturing and Its Societal Impact: A Literature Review. *Int. J. Adv. Manuf. Technol.* **2013**, *67*, 1191–1203. [CrossRef]
8. Kuang, X.; Roach, D.J.; Wu, J.; Hamel, C.M.; Ding, Z.; Wang, T.; Dunn, M.L.; Qi, H.J. Advances in 4D Printing: Materials and Applications. *Adv. Funct. Mater.* **2019**, *29*, 1805290. [CrossRef]
9. Ryan, K.R.; Down, M.P.; Banks, C.E. Future of Additive Manufacturing: Overview of 4D and 3D Printed Smart and Advanced Materials and Their Applications. *Chem. Eng. J.* **2021**, *403*, 126162. [CrossRef]
10. Imrie, P.; Jin, J. Polymer 4D Printing: Advanced Shape-Change and Beyond. *J. Polym. Sci.* **2022**, *60*, 149–174. [CrossRef]
11. Mitchell, A.; Lafont, U.; Hołyńska, M.; Semprimoschnig, C. Additive Manufacturing—A Review of 4D Printing and Future Applications. *Addit. Manuf.* **2018**, *24*, 606–626. [CrossRef]
12. Bandyopadhyay, A.; Heer, B. Additive Manufacturing of Multi-Material Structures. *Mater. Sci. Eng. R Rep.* **2018**, *129*, 1–16. [CrossRef]

13. Ambrosi, A.; Pumera, M. 3D-Printing Technologies for Electrochemical Applications. *Chem. Soc. Rev.* **2016**, *45*, 2740–2755. [CrossRef] [PubMed]
14. Capel, A.J.; Rimington, R.P.; Lewis, M.P.; Christie, S.D.R. 3D Printing for Chemical, Pharmaceutical and Biological Applications. *Nat. Rev. Chem.* **2018**, *2*, 422–436. [CrossRef]
15. Bhattacharjee, N.; Urrios, A.; Kang, S.; Folch, A. The Upcoming 3D-Printing Revolution in Microfluidics. *Lab Chip* **2016**, *16*, 1720–1742. [CrossRef] [PubMed]
16. Lee, J.Y.; An, J.; Chua, C.K. Fundamentals and Applications of 3D Printing for Novel Materials. *Appl. Mater. Today* **2017**, *7*, 120–133. [CrossRef]
17. Ahmed, A.; Arya, S.; Gupta, V.; Furukawa, H.; Khosla, A. 4D Printing: Fundamentals, Materials, Applications and Challenges. *Polymer* **2021**, *228*, 123926. [CrossRef]
18. Mahmood, A.; Akram, T.; Chen, H.; Chen, S. On the Evolution of Additive Manufacturing (3D/4D Printing) Technologies: Materials, Applications, and Challenges. *Polymers* **2022**, *14*, 4698. [CrossRef] [PubMed]
19. Mahmood, A.; Akram, T.; Shenggui, C.; Chen, H. Revolutionizing Manufacturing: A Review of 4D Printing Materials, Stimuli, and Cutting-Edge Applications. *Compos. Part B Eng.* **2023**, *266*, 110952. [CrossRef]
20. Feldman, D. Polymer History. *Des. Monomers Polym.* **2008**, *11*, 13. [CrossRef]
21. Park, S.; Shou, W.; Makatura, L.; Matusik, W.; Fu, K.K. 3D Printing of Polymer Composites: Materials, Processes, and Applications. *Matter* **2022**, *5*, 43–76. [CrossRef]
22. Wang, X.; Jiang, M.; Zhou, Z.; Gou, J.; Hui, D. 3D Printing of Polymer Matrix Composites: A Review and Prospective. *Compos. Part B Eng.* **2017**, *110*, 442–458. [CrossRef]
23. Pakkanen, J.; Manfredi, D.; Minetola, P.; Iuliano, L. About the Use of Recycled or Biodegradable Filaments for Sustainability of 3D Printing. In *Sustainable Design and Manufacturing 2017*; Campana, G., Howlett, R.J., Setchi, R., Cimatti, B., Eds.; Springer International Publishing: Cham, Switzerland, 2017; pp. 776–785.
24. Wohlers, T.; Gornet, T.; Mostow, N.; Campbell, I.; Diegel, O.; Kowen, J.; Huff, R.; Stucker, B.; Fidan, I.; Doukas, A.; et al. *History of Additive Manufacturing*; Wohlers Associates, Inc.: Fort Collins, CO, USA, 2016.
25. Zhang, Q.; Zhang, K.; Hu, G. Smart Three-Dimensional Lightweight Structure Triggered from a Thin Composite Sheet via 3D Printing Technique. *Sci. Rep.* **2016**, *6*, 22431. [CrossRef] [PubMed]
26. van Manen, T.; Janbaz, S.; Zadpoor, A.A. Programming 2D/3D Shape-Shifting with Hobbyist 3D Printers. *Mater. Horiz.* **2017**, *4*, 1064–1069. [CrossRef] [PubMed]
27. Long, K.N.; Scott, T.F.; Jerry Qi, H.; Bowman, C.N.; Dunn, M.L. Photomechanics of Light-Activated Polymers. *J. Mech. Phys. Solids* **2009**, *57*, 1103–1121. [CrossRef]
28. Wei, H.; Zhang, Q.; Yao, Y.; Liu, L.; Liu, Y.; Leng, J. Direct-Write Fabrication of 4D Active Shape-Changing Structures Based on a Shape Memory Polymer and Its Nanocomposite. *ACS Appl. Mater. Interfaces* **2017**, *9*, 876–883. [CrossRef]
29. Sydney Gladman, A.; Matsumoto, E.A.; Nuzzo, R.G.; Mahadevan, L.; Lewis, J.A. Biomimetic 4D Printing. *Nat. Mater.* **2016**, *15*, 413–418. [CrossRef] [PubMed]
30. Schwartz, J.J.; Boydston, A.J. Multimaterial Actinic Spatial Control 3D and 4D Printing. *Nat. Commun.* **2019**, *10*, 791. [CrossRef]
31. Raviv, D.; Zhao, W.; McKnelly, C.; Papadopoulou, A.; Kadambi, A.; Shi, B.; Hirsch, S.; Dikovsky, D.; Zyracki, M.; Olguin, C.; et al. Active Printed Materials for Complex Self-Evolving Deformations. *Sci. Rep.* **2014**, *4*, 7422. [CrossRef]
32. van Oosten, C.L.; Bastiaansen, C.W.M.; Broer, D.J. Printed Artificial Cilia from Liquid-Crystal Network Actuators Modularly Driven by Light. *Nat. Mater.* **2009**, *8*, 677–682. [CrossRef]
33. Yuan, C.; Roach, D.J.; Dunn, C.K.; Mu, Q.; Kuang, X.; Yakacki, C.M.; Wang, T.J.; Yu, K.; Qi, H.J. 3D Printed Reversible Shape Changing Soft Actuators Assisted by Liquid Crystal Elastomers. *Soft Matter* **2017**, *13*, 5558–5568. [CrossRef]
34. Kotikian, A.; Truby, R.L.; Boley, J.W.; White, T.J.; Lewis, J.A. 3D Printing of Liquid Crystal Elastomeric Actuators with Spatially Programed Nematic Order. *Adv. Mater.* **2018**, *30*, 1706164. [CrossRef]
35. Gantenbein, S.; Masania, K.; Woigk, W.; Sesseg, J.P.W.; Tervoort, T.A.; Studart, A.R. Three-Dimensional Printing of Hierarchical Liquid-Crystal-Polymer Structures. *Nature* **2018**, *561*, 226–230. [CrossRef] [PubMed]
36. Rim, Y.S.; Bae, S.-H.; Chen, H.; De Marco, N.; Yang, Y. Recent Progress in Materials and Devices toward Printable and Flexible Sensors. *Adv. Mater.* **2016**, *28*, 4415–4440. [CrossRef] [PubMed]
37. Mao, S.; Dong, E.; Jin, H.; Xu, M.; Low, K.H. Locomotion and Gait Analysis of Multi-Limb Soft Robots Driven by Smart Actuators. In Proceedings of the 2016 IEEE/RSJ International Conference on Intelligent Robots and Systems (IROS), Daejeon, Republic of Korea, 9–14 October 2016; pp. 2438–2443.
38. Gao, B.; Yang, Q.; Zhao, X.; Jin, G.; Ma, Y.; Xu, F. 4D Bioprinting for Biomedical Applications. *Trends Biotechnol.* **2016**, *34*, 746–756. [CrossRef] [PubMed]
39. Bahnini, I.; Rivette, M.; Rechia, A.; Siadat, A.; Elmesbahi, A. Additive Manufacturing Technology: The Status, Applications, and Prospects. *Int. J. Adv. Manuf. Technol.* **2018**, *97*, 147–161. [CrossRef]
40. Mercado Rivera, F.J.; Rojas Arciniegas, A.J. Additive Manufacturing Methods: Techniques, Materials, and Closed-Loop Control Applications. *Int. J. Adv. Manuf. Technol.* **2020**, *109*, 17–31. [CrossRef]
41. Gebhardt, A.; Hötter, J.-S. (Eds.) *Additive Manufacturing*; Hanser: Munich, Germany, 2016; pp. I–XX, ISBN 978-1-56990-582-1.
42. Yuan, S.; Shen, F.; Chua, C.K.; Zhou, K. Polymeric Composites for Powder-Based Additive Manufacturing: Materials and Applications. *Prog. Polym. Sci.* **2019**, *91*, 141–168. [CrossRef]

43. Saha, S.K.; Wang, D.; Nguyen, V.H.; Chang, Y.; Oakdale, J.S.; Chen, S.C. Scalable Submicrometer Additive Manufacturing. *Science* **2019**, *366*, 105–109. [CrossRef]
44. Ning, F.; Cong, W.; Qiu, J.; Wei, J.; Wang, S. Additive Manufacturing of Carbon Fiber Reinforced Thermoplastic Composites Using Fused Deposition Modeling. *Compos. Part B Eng.* **2015**, *80*, 369–378. [CrossRef]
45. Tekinalp, H.L.; Kunc, V.; Velez-Garcia, G.M.; Duty, C.E.; Love, L.J.; Naskar, A.K.; Blue, C.A.; Ozcan, S. Highly Oriented Carbon Fiber–Polymer Composites via Additive Manufacturing. *Compos. Sci. Technol.* **2014**, *105*, 144–150. [CrossRef]
46. Nikzad, M.; Masood, S.H.; Sbarski, I. Thermo-Mechanical Properties of a Highly Filled Polymeric Composites for Fused Deposition Modeling. *Mater. Des.* **2011**, *32*, 3448–3456. [CrossRef]
47. Castles, F.; Isakov, D.; Lui, A.; Lei, Q.; Dancer, C.E.J.; Wang, Y.; Janurudin, J.M.; Speller, S.C.; Grovenor, C.R.M.; Grant, P.S. Microwave Dielectric Characterisation of 3D-Printed BaTiO3/ABS Polymer Composites. *Sci. Rep.* **2016**, *6*, 22714. [CrossRef] [PubMed]
48. Chung, H.; Das, S. Processing and Properties of Glass Bead Particulate-Filled Functionally Graded Nylon-11 Composites Produced by Selective Laser Sintering. *Mater. Sci. Eng. A* **2006**, *437*, 226–234. [CrossRef]
49. Tian, X.; Liu, T.; Yang, C.; Wang, Q.; Li, D. Interface and Performance of 3D Printed Continuous Carbon Fiber Reinforced PLA Composites. *Compos. Part A Appl. Sci. Manuf.* **2016**, *88*, 198–205. [CrossRef]
50. Compton, B.G.; Lewis, J.A. 3D-Printing of Lightweight Cellular Composites. *Adv. Mater.* **2014**, *26*, 5930–5935. [CrossRef]
51. Klift, F.V.D.; Koga, Y.; Todoroki, A.; Ueda, M.; Hirano, Y.; Matsuzaki, R. 3D Printing of Continuous Carbon Fibre Reinforced Thermo-Plastic (CFRTP) Tensile Test Specimens. *Open J. Compos. Mater.* **2015**, *6*, 18–27. [CrossRef]
52. Lin, D.; Jin, S.; Zhang, F.; Wang, C.; Wang, Y.; Zhou, C.; Cheng, G.J. 3D Stereolithography Printing of Graphene Oxide Reinforced Complex Architectures. *Nanotechnology* **2015**, *26*, 434003. [CrossRef] [PubMed]
53. Wei, X.; Li, D.; Jiang, W.; Gu, Z.; Wang, X.; Zhang, Z.; Sun, Z. 3D Printable Graphene Composite. *Sci. Rep.* **2015**, *5*, 11181. [CrossRef]
54. Fantino, E.; Chiappone, A.; Calignano, F.; Fontana, M.; Pirri, F.; Roppolo, I. In Situ Thermal Generation of Silver Nanoparticles in 3D Printed Polymeric Structures. *Materials* **2016**, *9*, 589. [CrossRef]
55. Eutionnat-Diffo, P.A.; Chen, Y.; Guan, J.; Cayla, A.; Campagne, C.; Zeng, X.; Nierstrasz, V. Stress, Strain and Deformation of Poly-Lactic Acid Filament Deposited onto Polyethylene Terephthalate Woven Fabric through 3D Printing Process. *Sci. Rep.* **2019**, *9*, 14333. [CrossRef]
56. Rocha, C.R.; Perez, A.R.T.; Roberson, D.A.; Shemelya, C.M.; MacDonald, E.; Wicker, R.B. Novel ABS-Based Binary and Ternary Polymer Blends for Material Extrusion 3D Printing. *J. Mater. Res.* **2014**, *29*, 1859–1866. [CrossRef]
57. Weng, Z.; Wang, J.; Senthil, T.; Wu, L. Mechanical and Thermal Properties of ABS/Montmorillonite Nanocomposites for Fused Deposition Modeling 3D Printing. *Mater. Des.* **2016**, *102*, 276–283. [CrossRef]
58. Alexandre, M.; Dubois, P. Polymer-Layered Silicate Nanocomposites: Preparation, Properties and Uses of a New Class of Materials. *Mater. Sci. Eng. R Rep.* **2000**, *28*, 1–63. [CrossRef]
59. Pasanen, T.P.; von Gastrow, G.; Heikkinen, I.T.S.; Vähänissi, V.; Savin, H.; Pearce, J.M. Compatibility of 3-D Printed Devices in Cleanroom Environments for Semiconductor Processing. *Mater. Sci. Semicond. Process.* **2019**, *89*, 59–67. [CrossRef]
60. Cardoso, R.M.; Mendonça, D.M.H.; Silva, W.P.; Silva, M.N.T.; Nossol, E.; da Silva, R.A.B.; Richter, E.M.; Muñoz, R.A.A. 3D Printing for Electroanalysis: From Multiuse Electrochemical Cells to Sensors. *Anal. Chim. Acta* **2018**, *1033*, 49–57. [CrossRef] [PubMed]
61. Arulmozhi, R.S.; Vaidya, M.; Poojalakshmi, M.G.; Ashok Kumar, D.; Anuraag, K. 3d Design and Printing of Custom-Fit Finger Splint. *Biomed. Eng. Appl. Basis Commun.* **2018**, *30*, 1850032. [CrossRef]
62. Singh, R.; Singh, H.; Farina, I.; Colangelo, F.; Fraternali, F. On the Additive Manufacturing of an Energy Storage Device from Recycled Material. *Compos. Part B Eng.* **2019**, *156*, 259–265. [CrossRef]
63. Bible, M.; Sefa, M.; Fedchak, J.A.; Scherschligt, J.; Natarajan, B.; Ahmed, Z.; Hartings, M.R. 3D-Printed Acrylonitrile Butadiene Styrene-Metal Organic Framework Composite Materials and Their Gas Storage Properties. *3D Print. Addit. Manuf.* **2018**, *5*, 63–72. [CrossRef]
64. Provaggi, E.; Capelli, C.; Rahmani, B.; Burriesci, G.; Kalaskar, D.M. 3D Printing Assisted Finite Element Analysis for Optimising the Manufacturing Parameters of a Lumbar Fusion Cage. *Mater. Des.* **2019**, *163*, 107540. [CrossRef]
65. Dawoud, M.; Taha, I.; Ebeid, S.J. Strain Sensing Behaviour of 3D Printed Carbon Black Filled ABS. *J. Manuf. Process.* **2018**, *35*, 337–342. [CrossRef]
66. Haque, R.I.; Chandran, O.; Lani, S.; Briand, D. Self-Powered Triboelectric Touch Sensor Made of 3D Printed Materials. *Nano Energy* **2018**, *52*, 54–62. [CrossRef]
67. Khatri, B.; Lappe, K.; Habedank, M.; Mueller, T.; Megnin, C.; Hanemann, T. Fused Deposition Modeling of ABS-Barium Titanate Composites: A Simple Route towards Tailored Dielectric Devices. *Polymers* **2018**, *10*, 666. [CrossRef] [PubMed]
68. Kumar, M.; Ramakrishnan, R.; Omarbekova, A. 3D Printed Polycarbonate Reinforced Acrylonitrile–Butadiene–Styrene Composites: Composition Effects on Mechanical Properties, Micro-Structure and Void Formation Study. *J. Mech. Sci. Technol.* **2019**, *33*, 5219–5226. [CrossRef]
69. Chen, S.; Lu, J.; Feng, J. 3D-Printable ABS Blends with Improved Scratch Resistance and Balanced Mechanical Performance. *Ind. Eng. Chem. Res.* **2018**, *57*, 3923–3931. [CrossRef]

70. Akato, K.; Tran, C.D.; Chen, J.; Naskar, A.K. Poly(Ethylene Oxide)-Assisted Macromolecular Self-Assembly of Lignin in ABS Matrix for Sustainable Composite Applications. *ACS Sustain. Chem. Eng.* **2015**, *3*, 3070–3076. [CrossRef]
71. Shim, J.S.; Kim, J.-E.; Jeong, S.H.; Choi, Y.J.; Ryu, J.J. Printing Accuracy, Mechanical Properties, Surface Characteristics, and Microbial Adhesion of 3D-Printed Resins with Various Printing Orientations. *J. Prosthet. Dent.* **2020**, *124*, 468–475. [CrossRef] [PubMed]
72. Bîrcă, A.; Gherasim, O.; Grumezescu, V.; Grumezescu, A.M. Chapter 1—Introduction in Thermoplastic and Thermosetting Polymers. In *Materials for Biomedical Engineering*; Grumezescu, V., Grumezescu, A.M., Eds.; Elsevier: Amsterdam, The Netherlands, 2019; pp. 1–28. ISBN 978-0-12-816874-5.
73. Fu, S.; Hu, H.; Chen, J.; Zhu, Y.; Zhao, S. Silicone Resin Derived Larnite/C Scaffolds via 3D Printing for Potential Tumor Therapy and Bone Regeneration. *Chem. Eng. J.* **2020**, *382*, 122928. [CrossRef]
74. Lin, C.-H.; Lin, Y.-M.; Lai, Y.-L.; Lee, S.-Y. Mechanical Properties, Accuracy, and Cytotoxicity of UV-Polymerized 3D Printing Resins Composed of Bis-EMA, UDMA, and TEGDMA. *J. Prosthet. Dent.* **2020**, *123*, 349–354. [CrossRef]
75. Hmeidat, N.S.; Kemp, J.W.; Compton, B.G. High-Strength Epoxy Nanocomposites for 3D Printing. *Compos. Sci. Technol.* **2018**, *160*, 9–20. [CrossRef]
76. Li, J.; Wu, C.; Chu, P.K.; Gelinsky, M. 3D Printing of Hydrogels: Rational Design Strategies and Emerging Biomedical Applications. *Mater. Sci. Eng. R Rep.* **2020**, *140*, 100543. [CrossRef]
77. Bahram, M.; Mohseni, N.; Moghtader, M.; Bahram, M.; Mohseni, N.; Moghtader, M. An Introduction to Hydrogels and Some Recent Applications. In *Emerging Concepts in Analysis and Applications of Hydrogels*; IntechOpen: Rijeka, Croatia, 2016; ISBN 978-953-51-2510-5.
78. Ahmed, E.M. Hydrogel: Preparation, Characterization, and Applications: A Review. *J. Adv. Res.* **2015**, *6*, 105–121. [CrossRef]
79. Holländer, J.; Hakala, R.; Suominen, J.; Moritz, N.; Yliruusi, J.; Sandler, N. 3D Printed UV Light Cured Polydimethylsiloxane Devices for Drug Delivery. *Int. J. Pharm.* **2018**, *544*, 433–442. [CrossRef] [PubMed]
80. Xiang, H.; Wang, X.; Ou, Z.; Lin, G.; Yin, J.; Liu, Z.; Zhang, L.; Liu, X. UV-Curable, 3D Printable and Biocompatible Silicone Elastomers. *Prog. Org. Coat.* **2019**, *137*, 105372. [CrossRef]
81. Khoo, Z.X.; Teoh, J.E.M.; Liu, Y.; Chua, C.K.; Yang, S.; An, J.; Leong, K.F.; Yeong, W.Y. 3D Printing of Smart Materials: A Review on Recent Progresses in 4D Printing. *Virtual Phys. Prototyp.* **2015**, *10*, 103–122. [CrossRef]
82. Mondal, S.; Hu, J.L. Temperature Stimulating Shape Memory Polyurethane for Smart Clothing. *Indian J. Fibre Text. Res. (IJFTR)* **2006**, *31*, 66–71.
83. Campbell, D.; Lake, M.; Scherbarth, M.; Nelson, E.; Six, R. Elastic Memory Composite Material: An Enabling Technology for Future Furlable Space Structures. In Proceedings of the 46th AIAA/ASME/ASCE/AHS/ASC Structures, Structural Dynamics and Materials Conference, Austin, TX, USA, 18–21 April 2005; American Institute of Aeronautics and Astronautics: Reston VA, USA, 2005.
84. Wache, H.M.; Tartakowska, D.J.; Hentrich, A.; Wagner, M.H. Development of a Polymer Stent with Shape Memory Effect as a Drug Delivery System. *J. Mater. Sci. Mater. Med.* **2003**, *14*, 109–112. [CrossRef]
85. Singh, M.; Haverinen, H.M.; Dhagat, P.; Jabbour, G.E. Inkjet Printing—Process and Its Applications. *Adv. Mater.* **2010**, *22*, 673–685. [CrossRef]
86. Cutkosky, M.R.; Kim, S. Design and Fabrication of Multi-Material Structures for Bioinspired Robots. *Philos. Trans. R. Soc. A Math. Phys. Eng. Sci.* **2009**, *367*, 1799–1813. [CrossRef]
87. Peng, W.; Zhang, G.; Liu, J.; Nie, S.; Wu, Y.; Deng, S.; Fang, G.; Zhou, J.; Song, J.; Qian, J.; et al. Light-Coded Digital Crystallinity Patterns Toward Bioinspired 4D Transformation of Shape-Memory Polymers. *Adv. Funct. Mater.* **2020**, *30*, 2000522. [CrossRef]
88. Wong, Y.S.; Venkatraman, S.S. Recovery as a Measure of Oriented Crystalline Structure in Poly(L-Lactide) Used as Shape Memory Polymer. *Acta Mater.* **2010**, *58*, 49–58. [CrossRef]
89. Vashist, A.; Vashist, A.; Gupta, Y.K.; Ahmad, S. Recent Advances in Hydrogel Based Drug Delivery Systems for the Human Body. *J. Mater. Chem. B* **2013**, *2*, 147–166. [CrossRef]
90. Bassil, M.; Davenas, J.; EL Tahchi, M. Electrochemical Properties and Actuation Mechanisms of Polyacrylamide Hydrogel for Artificial Muscle Application. *Sens. Actuators B Chem.* **2008**, *134*, 496–501. [CrossRef]
91. Azeem, M.K.; Islam, A.; Khan, R.U.; Rasool, A.; Qureshi, M.A.u.R.; Rizwan, M.; Sher, F.; Rasheed, T. Eco-Friendly Three-Dimensional Hydrogels for Sustainable Agricultural Applications: Current and Future Scenarios. *Polym. Adv. Technol.* **2023**, *34*, 3046–3062. [CrossRef]
92. Palanivelu, S.D.; Armir, N.A.Z.; Zulkifli, A.; Hair, A.H.A.; Salleh, K.M.; Lindsey, K.; Che-Othman, M.H.; Zakaria, S. Hydrogel Application in Urban Farming: Potentials and Limitations—A Review. *Polymers* **2022**, *14*, 2590. [CrossRef] [PubMed]
93. Roach, D.J.; Kuang, X.; Yuan, C.; Chen, K.; Qi, H.J. Novel Ink for Ambient Condition Printing of Liquid Crystal Elastomers for 4D Printing. *Smart Mater. Struct.* **2018**, *27*, 125011. [CrossRef]
94. Ambulo, C.P.; Burroughs, J.J.; Boothby, J.M.; Kim, H.; Shankar, M.R.; Ware, T.H. Four-Dimensional Printing of Liquid Crystal Elastomers. *ACS Appl. Mater. Interfaces* **2017**, *9*, 37332–37339. [CrossRef] [PubMed]
95. Pei, Z.; Yang, Y.; Chen, Q.; Terentjev, E.M.; Wei, Y.; Ji, Y. Mouldable Liquid-Crystalline Elastomer Actuators with Exchangeable Covalent Bonds. *Nat. Mater.* **2014**, *13*, 36–41. [CrossRef] [PubMed]
96. Guin, T.; Settle, M.J.; Kowalski, B.A.; Auguste, A.D.; Beblo, R.V.; Reich, G.W.; White, T.J. Layered Liquid Crystal Elastomer Actuators. *Nat. Commun.* **2018**, *9*, 2531. [CrossRef]

97. Kirton, J. The Physics of Liquid Crystals. *Opt. Acta Int. J. Opt.* **1975**, *22*, 158. [CrossRef]
98. López-Valdeolivas, M.; Liu, D.; Broer, D.J.; Sánchez-Somolinos, C. 4D Printed Actuators with Soft-Robotic Functions. *Macromol. Rapid Commun.* **2018**, *39*, 1700710. [CrossRef]
99. Yang, Y.; Pei, Z.; Li, Z.; Wei, Y.; Ji, Y. Making and Remaking Dynamic 3D Structures by Shining Light on Flat Liquid Crystalline Vitrimer Films without a Mold. *J. Am. Chem. Soc.* **2016**, *138*, 2118–2121. [CrossRef]
100. Zhang, W.; Wang, H.; Wang, H.; Chan, J.Y.E.; Liu, H.; Zhang, B.; Zhang, Y.-F.; Agarwal, K.; Yang, X.; Ranganath, A.S.; et al. Structural Multi-Colour Invisible Inks with Submicron 4D Printing of Shape Memory Polymers. *Nat. Commun.* **2021**, *12*, 112. [CrossRef] [PubMed]
101. Barnes, M.; Sajadi, S.M.; Parekh, S.; Rahman, M.M.; Ajayan, P.M.; Verduzco, R. Reactive 3D Printing of Shape-Programmable Liquid Crystal Elastomer Actuators. *ACS Appl. Mater. Interfaces* **2020**, *12*, 28692–28699. [CrossRef] [PubMed]
102. Hagaman, D.E.; Leist, S.; Zhou, J.; Ji, H.-F. Photoactivated Polymeric Bilayer Actuators Fabricated via 3D Printing. *ACS Appl. Mater. Interfaces* **2018**, *10*, 27308–27315. [CrossRef] [PubMed]
103. Ceamanos, L.; Kahveci, Z.; López-Valdeolivas, M.; Liu, D.; Broer, D.J.; Sánchez-Somolinos, C. Four-Dimensional Printed Liquid Crystalline Elastomer Actuators with Fast Photoinduced Mechanical Response toward Light-Driven Robotic Functions. *ACS Appl. Mater. Interfaces* **2020**, *12*, 44195–44204. [CrossRef]
104. Lu, X.; Ambulo, C.P.; Wang, S.; Rivera-Tarazona, L.K.; Kim, H.; Searles, K.; Ware, T.H. 4D-Printing of Photoswitchable Actuators. *Angew. Chem. Int. Ed.* **2021**, *60*, 5536–5543. [CrossRef] [PubMed]
105. Hager, M.D.; Greil, P.; Leyens, C.; Van Der Zwaag, S.; Schubert, U.S. Self-Healing Materials. *Adv. Mater.* **2010**, *22*, 5424–5430. [CrossRef]
106. Blaiszik, B.J.; Kramer, S.L.B.; Olugebefola, S.C.; Moore, J.S.; Sottos, N.R.; White, S.R. Self-Healing Polymers and Composites. *Annu. Rev. Mater. Res.* **2010**, *40*, 179–211. [CrossRef]
107. De Nardi, C.; Gardner, D.; Jefferson, A.D. Development of 3D Printed Networks in Self-Healing Concrete. *Materials* **2020**, *13*, 1328. [CrossRef]
108. Postiglione, G.; Alberini, M.; Leigh, S.; Levi, M.; Turri, S. Effect of 3D-Printed Microvascular Network Design on the Self-Healing Behavior of Cross-Linked Polymers. *ACS Appl. Mater. Interfaces* **2017**, *9*, 14371–14378. [CrossRef]
109. Li, Z.; de Souza, L.R.; Litina, C.; Markaki, A.E.; Al-Tabbaa, A. A Novel Biomimetic Design of a 3D Vascular Structure for Self-Healing in Cementitious Materials Using Murray's Law. *Mater. Des.* **2020**, *190*, 108572. [CrossRef]
110. Wu, W.; DeConinck, A.; Lewis, J.A. Omnidirectional Printing of 3D Microvascular Networks. *Adv. Mater.* **2011**, *23*, H178–H183. [CrossRef] [PubMed]
111. Li, X.; Yu, R.; He, Y.; Zhang, Y.; Yang, X.; Zhao, X.; Huang, W. Self-Healing Polyurethane Elastomers Based on a Disulfide Bond by Digital Light Processing 3D Printing. *ACS Macro Lett.* **2019**, *8*, 1511–1516. [CrossRef] [PubMed]
112. Ghosh, S.K. Self-Healing Materials: Fundamentals, Design Strategies, and Applications. In *Self-Healing Materials*; John Wiley & Sons Ltd.: Hoboken, NJ, USA, 2008; pp. 1–28. ISBN 978-3-527-62537-6.
113. Davami, K.; Mohsenizadeh, M.; Mitcham, M.; Damasus, P.; Williams, Q.; Munther, M. Additively Manufactured Self-Healing Structures with Embedded Healing Agent Reservoirs. *Sci. Rep.* **2019**, *9*, 7474. [CrossRef] [PubMed]
114. Yang, Y.; Ding, X.; Urban, M.W. Chemical and Physical Aspects of Self-Healing Materials. *Prog. Polym. Sci.* **2015**, *49–50*, 34–59. [CrossRef]
115. Hartings, M.R.; Ahmed, Z. Chemistry from 3D Printed Objects. *Nat. Rev. Chem.* **2019**, *3*, 305–314. [CrossRef]
116. Vyatskikh, A.; Delalande, S.; Kudo, A.; Zhang, X.; Portela, C.M.; Greer, J.R. Additive Manufacturing of 3D Nano-Architected Metals. *Nat. Commun.* **2018**, *9*, 593. [CrossRef]
117. Li, T.; Chen, Y.; Wang, L. Enhanced Fracture Toughness in Architected Interpenetrating Phase Composites by 3D Printing. *Compos. Sci. Technol.* **2018**, *167*, 251–259. [CrossRef]
118. Bapari, S.; Chokshi, A.H. The Role of the Soft Phase in Rigidity Enhancements in a Particulate Composite. *Compos. Part A Appl. Sci. Manuf.* **2019**, *123*, 114–122. [CrossRef]
119. Li, X.; Tan, Y.H.; Wang, P.; Su, X.; Willy, H.J.; Herng, T.S.; Ding, J. Metallic Microlattice and Epoxy Interpenetrating Phase Composites: Experimental and Simulation Studies on Superior Mechanical Properties and Their Mechanisms. *Compos. Part A Appl. Sci. Manuf.* **2020**, *135*, 105934. [CrossRef]
120. Fafenrot, S.; Grimmelsmann, N.; Wortmann, M.; Ehrmann, A. Three-Dimensional (3D) Printing of Polymer-Metal Hybrid Materials by Fused Deposition Modeling. *Materials* **2017**, *10*, 1199. [CrossRef]
121. Gu, X.W.; Greer, J.R. Ultra-Strong Architected Cu Meso-Lattices. *Extrem. Mech. Lett.* **2015**, *2*, 7–14. [CrossRef]
122. Zheng, X.; Smith, W.; Jackson, J.; Moran, B.; Cui, H.; Chen, D.; Ye, J.; Fang, N.; Rodriguez, N.; Weisgraber, T.; et al. Multiscale Metallic Metamaterials. *Nat. Mater.* **2016**, *15*, 1100–1106. [CrossRef] [PubMed]
123. Mieszala, M.; Hasegawa, M.; Guillonneau, G.; Bauer, J.; Raghavan, R.; Frantz, C.; Kraft, O.; Mischler, S.; Michler, J.; Philippe, L. Micromechanics of Amorphous Metal/Polymer Hybrid Structures with 3D Cellular Architectures: Size Effects, Buckling Behavior, and Energy Absorption Capability. *Small* **2017**, *13*, 1602514. [CrossRef]
124. Zhang, X.; Yao, J.; Liu, B.; Yan, J.; Lu, L.; Li, Y.; Gao, H.; Li, X. Three-Dimensional High-Entropy Alloy-Polymer Composite Nanolattices That Overcome the Strength-Recoverability Trade-Off. *Nano Lett.* **2018**, *18*, 4247–4256. [CrossRef] [PubMed]
125. Meza, L.R.; Das, S.; Greer, J.R. Strong, Lightweight, and Recoverable Three-Dimensional Ceramic Nanolattices. *Science* **2014**, *345*, 1322–1326. [CrossRef] [PubMed]

126. Wu, B.M.; Borland, S.W.; Giordano, R.A.; Cima, L.G.; Sachs, E.M.; Cima, M.J. Solid Free-Form Fabrication of Drug Delivery Devices. *J. Control. Release* **1996**, *40*, 77–87. [CrossRef]
127. Chen, Z.; Li, Z.; Li, J.; Liu, C.; Lao, C.; Fu, Y.; Liu, C.; Li, Y.; Wang, P.; He, Y. 3D Printing of Ceramics: A Review. *J. Eur. Ceram. Soc.* **2019**, *39*, 661–687. [CrossRef]
128. Ritchie, R.O. The Conflicts between Strength and Toughness. *Nat. Mater.* **2011**, *10*, 817–822. [CrossRef]
129. Mirkhalaf, M.; Dastjerdi, A.K.; Barthelat, F. Overcoming the Brittleness of Glass through Bio-Inspiration and Micro-Architecture. *Nat. Commun.* **2014**, *5*, 3166. [CrossRef]
130. Yahyazadehfar, M.; Bajaj, D.; Arola, D.D. Hidden Contributions of the Enamel Rods on the Fracture Resistance of Human Teeth. *Acta Biomater.* **2013**, *9*, 4806–4814. [CrossRef]
131. Currey, J.D.; Taylor, J.D. The Mechanical Behaviour of Some Molluscan Hard Tissues. *J. Zool.* **1974**, *173*, 395–406. [CrossRef]
132. Munch, E.; Launey, M.E.; Alsem, D.H.; Saiz, E.; Tomsia, A.P.; Ritchie, R.O. Tough, Bio-Inspired Hybrid Materials. *Science* **2008**, *322*, 1516–1520. [CrossRef] [PubMed]
133. Bonderer, L.J.; Studart, A.R.; Gauckler, L.J. Bioinspired Design and Assembly of Platelet Reinforced Polymer Films. *Science* **2008**, *319*, 1069–1073. [CrossRef] [PubMed]
134. Podsiadlo, P.; Kaushik, A.K.; Arruda, E.M.; Waas, A.M.; Shim, B.S.; Xu, J.; Nandivada, H.; Pumplin, B.G.; Lahann, J.; Ramamoorthy, A.; et al. Ultrastrong and Stiff Layered Polymer Nanocomposites. *Science* **2007**, *318*, 80–83. [CrossRef]
135. Martin, J.H.; Yahata, B.D.; Hundley, J.M.; Mayer, J.A.; Schaedler, T.A.; Pollock, T.M. 3D Printing of High-Strength Aluminium Alloys. *Nature* **2017**, *549*, 365–369. [CrossRef]
136. Dimas, L.S.; Bratzel, G.H.; Eylon, I.; Buehler, M.J. Tough Composites Inspired by Mineralized Natural Materials: Computation, 3D Printing, and Testing. *Adv. Funct. Mater.* **2013**, *23*, 4629–4638. [CrossRef]
137. Wang, L.; Lau, J.; Thomas, E.L.; Boyce, M.C. Co-Continuous Composite Materials for Stiffness, Strength, and Energy Dissipation. *Adv. Mater.* **2011**, *23*, 1524–1529. [CrossRef]
138. Wegst, U.G.K.; Bai, H.; Saiz, E.; Tomsia, A.P.; Ritchie, R.O. Bioinspired Structural Materials. *Nat. Mater.* **2014**, *14*, 23–36. [CrossRef]
139. Li, H.; Shen, J.; Wei, Q.; Li, X. Dynamic Self-Strengthening of a Bio-Nanostructured Armor—Conch Shell. *Mater. Sci. Eng. C* **2019**, *103*, 109820. [CrossRef]
140. Grunenfelder, L.K.; Suksangpanya, N.; Salinas, C.; Milliron, G.; Yaraghi, N.; Herrera, S.; Evans-Lutterodt, K.; Nutt, S.R.; Zavattieri, P.; Kisailus, D. Bio-Inspired Impact-Resistant Composites. *Acta Biomater.* **2014**, *10*, 3997–4008. [CrossRef]
141. Barthelat, F.; Tang, H.; Zavattieri, P.D.; Li, C.-M.; Espinosa, H.D. On the Mechanics of Mother-of-Pearl: A Key Feature in the Material Hierarchical Structure. *J. Mech. Phys. Solids* **2007**, *55*, 306–337. [CrossRef]
142. Jia, Z.; Wang, L. 3D Printing of Biomimetic Composites with Improved Fracture Toughness. *Acta Mater.* **2019**, *173*, 61–73. [CrossRef]
143. Eckel, Z.C.; Zhou, C.; Martin, J.H.; Jacobsen, A.J.; Carter, W.B.; Schaedler, T.A. Additive Manufacturing of Polymer-Derived Ceramics. *Science* **2016**, *351*, 58–62. [CrossRef] [PubMed]
144. Travitzky, N.; Bonet, A.; Dermeik, B.; Fey, T.; Filbert-Demut, I.; Schlier, L.; Schlordt, T.; Greil, P. Additive Manufacturing of Ceramic-Based Materials. *Adv. Eng. Mater.* **2014**, *16*, 729–754. [CrossRef]
145. Saroia, J.; Wang, Y.; Wei, Q.; Lei, M.; Li, X.; Guo, Y.; Zhang, K. A Review on 3D Printed Matrix Polymer Composites: Its Potential and Future Challenges. *Int. J. Adv. Manuf. Technol.* **2020**, *106*, 1695–1721. [CrossRef]
146. Ma, X.-Y.; Feng, Y.-F.; Ma, Z.-S.; Li, X.; Wang, J.; Wang, L.; Lei, W. The Promotion of Osteointegration under Diabetic Conditions Using Chitosan/Hydroxyapatite Composite Coating on Porous Titanium Surfaces. *Biomaterials* **2014**, *35*, 7259–7270. [CrossRef] [PubMed]
147. Backes, E.H.; Pires, L.D.N.; Beatrice, C.A.G.; Costa, L.C.; Passador, F.R.; Pessan, L.A. Fabrication of Biocompatible Composites of Poly(Lactic Acid)/Hydroxyapatite Envisioning Medical Applications. *Polym. Eng. Sci.* **2020**, *60*, 636–644. [CrossRef]
148. Li, C.; Lai, Y.; Li, L.; Cao, H.; Long, J.; Wang, X.; Qin, L. The in Vitro Biocompatibility and Osteoinductive Activity Study of Magnesium Composed PLGA/TCP Porous Scaffold for Bone Regeneration. *J. Orthop. Transl.* **2016**, *7*, 78. [CrossRef]
149. Noroozi, R.; Arif, Z.U.; Taghvaei, H.; Khalid, M.Y.; Sahbafar, H.; Hadi, A.; Sadeghianmaryan, A.; Chen, X. 3D and 4D Bioprinting Technologies: A Game Changer for the Biomedical Sector? *Ann. Biomed. Eng.* **2023**, *51*, 1683–1712. [CrossRef]
150. Willemen, N.G.A.; Morsink, M.A.J.; Veerman, D.; da Silva, C.F.; Cardoso, J.C.; Souto, E.B.; Severino, P. From Oral Formulations to Drug-Eluting Implants: Using 3D and 4D Printing to Develop Drug Delivery Systems and Personalized Medicine. *Bio-Des. Manuf.* **2022**, *5*, 85–106. [CrossRef]
151. Awad, A.; Basit, A.W. 3D and 4D Printing in Digital Healthcare. In *3D & 4D Printing Methods for Pharmaceutical Manufacturing and Personalised Drug Delivery: Opportunities and Challenges*; Lamprou, D., Ed.; AAPS Introductions in the Pharmaceutical Sciences; Springer International Publishing: Cham, Switzerland, 2023; pp. 1–23. ISBN 978-3-031-34119-9.
152. Agarwal, T.; Hann, S.Y.; Chiesa, I.; Cui, H.; Celikkin, N.; Micalizzi, S.; Barbetta, A.; Costantini, M.; Esworthy, T.; Zhang, L.G.; et al. 4D Printing in Biomedical Applications: Emerging Trends and Technologies. *J. Mater. Chem. B* **2021**, *9*, 7608–7632. [CrossRef] [PubMed]
153. Muhindo, D.; Elkanayati, R.; Srinivasan, P.; Repka, M.A.; Ashour, E.A. Recent Advances in the Applications of Additive Manufacturing (3D Printing) in Drug Delivery: A Comprehensive Review. *AAPS PharmSciTech* **2023**, *24*, 57. [CrossRef] [PubMed]
154. Tran, T.S.; Balu, R.; Mettu, S.; Roy Choudhury, N.; Dutta, N.K. 4D Printing of Hydrogels: Innovation in Material Design and Emerging Smart Systems for Drug Delivery. *Pharmaceuticals* **2022**, *15*, 1282. [CrossRef] [PubMed]

155. Huang, H.-J.; Tsai, Y.-L.; Lin, S.-H.; Hsu, S. Smart Polymers for Cell Therapy and Precision Medicine. *J. Biomed. Sci.* **2019**, *26*, 73. [CrossRef] [PubMed]
156. Pineda-Castillo, S.A.; Stiles, A.M.; Bohnstedt, B.N.; Lee, H.; Liu, Y.; Lee, C.-H. Shape Memory Polymer-Based Endovascular Devices: Design Criteria and Future Perspective. *Polymers* **2022**, *14*, 2526. [CrossRef] [PubMed]
157. Uboldi, M.; Melocchi, A.; Moutaharrik, S.; Cerea, M.; Gazzaniga, A.; Zema, L. Dataset on a Small-Scale Film-Coating Process Developed for Self-Expanding 4D Printed Drug Delivery Devices. *Coatings* **2021**, *11*, 1252. [CrossRef]
158. Afzali Naniz, M.; Askari, M.; Zolfagharian, A.; Afzali Naniz, M.; Bodaghi, M. 4D Printing: A Cutting-Edge Platform for Biomedical Applications. *Biomed. Mater.* **2022**, *17*, 062001. [CrossRef]
159. Pruksawan, S.; Lin, Z.; Lee, Y.L.; Chee, H.L.; Wang, F. 4D-Printed Hydrogel Actuators through Diffusion-Path Architecture Design. *ACS Appl. Mater. Interfaces* **2023**, *15*, 46388–46399. [CrossRef]
160. Xiong, W.; Liu, Y.; Jiang, L.J.; Zhou, Y.S.; Li, D.W.; Jiang, L.; Silvain, J.-F.; Lu, Y.F. Laser-Directed Assembly of Aligned Carbon Nanotubes in Three Dimensions for Multifunctional Device Fabrication. *Adv. Mater.* **2016**, *28*, 2002–2009. [CrossRef]
161. Choi, J.-W.; Kim, H.-C.; Wicker, R. Multi-Material Stereolithography. *J. Mater. Process. Technol.* **2011**, *211*, 318–328. [CrossRef]
162. Kelly, B.E.; Bhattacharya, I.; Heidari, H.; Shusteff, M.; Spadaccini, C.M.; Taylor, H.K. Volumetric Additive Manufacturing via Tomographic Reconstruction. *Science* **2019**, *363*, 1075–1079. [CrossRef] [PubMed]
163. Walker, D.A.; Hedrick, J.L.; Mirkin, C.A. Rapid, Large-Volume, Thermally Controlled 3D Printing Using a Mobile Liquid Interface. *Science* **2019**, *366*, 360–364. [CrossRef] [PubMed]
164. Sitthi-Amorn, P.; Ramos, J.E.; Wangy, Y.; Kwan, J.; Lan, J.; Wang, W.; Matusik, W. MultiFab: A Machine Vision Assisted Platform for Multi-Material 3D Printing. *ACM Trans. Graph.* **2015**, *34*, 129:1–129:11. [CrossRef]
165. Zander, N.E.; Gillan, M.; Lambeth, R.H. Recycled Polyethylene Terephthalate as a New FFF Feedstock Material. *Addit. Manuf.* **2018**, *21*, 174–182. [CrossRef]
166. Stephens, B.; Azimi, P.; El Orch, Z.; Ramos, T. Ultrafine Particle Emissions from Desktop 3D Printers. *Atmos. Environ.* **2013**, *79*, 334–339. [CrossRef]
167. Kim, Y.; Yoon, C.; Ham, S.; Park, J.; Kim, S.; Kwon, O.; Tsai, P.-J. Emissions of Nanoparticles and Gaseous Material from 3D Printer Operation. *Environ. Sci. Technol.* **2015**, *49*, 12044–12053. [CrossRef]
168. Yang, Y.; Li, L. Total Volatile Organic Compound Emission Evaluation and Control for Stereolithography Additive Manufacturing Process. *J. Clean. Prod.* **2018**, *170*, 1268–1278. [CrossRef]
169. Davis, A.Y.; Zhang, Q.; Wong, J.P.S.; Weber, R.J.; Black, M.S. Characterization of Volatile Organic Compound Emissions from Consumer Level Material Extrusion 3D Printers. *Build. Environ.* **2019**, *160*, 106209. [CrossRef]

molecules

MDPI

Review

The Synthesis and Polymer-Reinforced Mechanical Properties of SiO_2 Aerogels: A Review

Wang Zhan [1,*], Le Chen [2], Qinghong Kong [1,*], Lixia Li [1], Mingyi Chen [1], Juncheng Jiang [3], Weixi Li [1], Fan Shi [1] and Zhiyuan Xu [1]

[1] School of the Environment and Safety Engineering, Jiangsu University, Zhenjiang 212013, China; 1000003002@ujs.edu.cn (L.L.); 1000004815@ujs.edu.cn (M.C.); 2200904004@stmail.ujs.edu.cn (W.L.); sf_endeavor@163.com (F.S.); 2222209101@stmail.edu.cn (Z.X.)

[2] Department of Electronic Engineering, School of Electronic Science and Engineering, Nanjing University, Nanjing 210023, China; chenle26@nju.edu.cn

[3] College of Safety Science and Engineering, Nanjing Tech University, Nanjing 213000, China; jcjiang_njtech@163.com

* Correspondence: zw_freedom@126.com (W.Z.); kongqh@ujs.edu.cn (Q.K.)

Abstract: Silica aerogels are considered as the distinguished materials of the future due to their extremely low thermal conductivity, low density, and high surface area. They are widely used in construction engineering, aeronautical domains, environmental protection, heat storage, etc. However, their fragile mechanical properties are the bottleneck restricting the engineering application of silica aerogels. This review briefly introduces the synthesis of silica aerogels, including the processes of sol–gel chemistry, aging, and drying. The effects of different silicon sources on the mechanical properties of silica aerogels are summarized. Moreover, the reaction mechanism of the three stages is also described. Then, five types of polymers that are commonly used to enhance the mechanical properties of silica aerogels are listed, and the current research progress is introduced. Finally, the outlook and prospects of the silica aerogels are proposed, and this paper further summarizes the methods of different polymers to enhance silica aerogels.

Keywords: silica aerogels; polymer; mechanical properties; crosslinking

1. Introduction

Aerogels are the lightest solid materials in the world due to their high porosity and low density [1–4]. The liquid constituent of these materials are substituted with air and form intact interconnected solid structures. Relying on their unique mesoporous structure, aerogels demonstrate excellent properties and have been favored by researchers [5–8]. There are many kinds of aerogels, such as silica aerogels [9], carbon aerogels [10,11], polymer aerogels [12–15], metal oxide aerogels [16–18], metal aerogels [18], and bio-based aerogels [19,20]. These aerogels are considered as the "wonder materials" and have broad prospects of application in various fields [21–25].

Silica aerogels with excellent properties of low bulk density (0.003~0.200 g/cm^3), high porosity (80~99.8%), large specific surface area (500~1500 m^2/g), and low thermal conductivity (0.015~0.030 $W/m{\cdot}K$) are the typical representatives of aerogels [26–29]. It has been more than 90 years since silica aerogel was first invented by the American scientist Kistler in 1931 [30,31]. Although silica aerogels have a long history of development, their commercial production history only dates back to about 20 years. In 2001, Aspen realized the commercial production of silica aerogels for the first time in the United States. As Figure 1 shows, silica aerogels are widely used in construction engineering [32,33], aeronautical domains [34–36], environmental protection [37,38], flexible electronics [39], and chemical engineering [40–42]. Especially in the aerospace field, silica aerogels have made remarkable contributions to the safety of personnel and equipment [43].

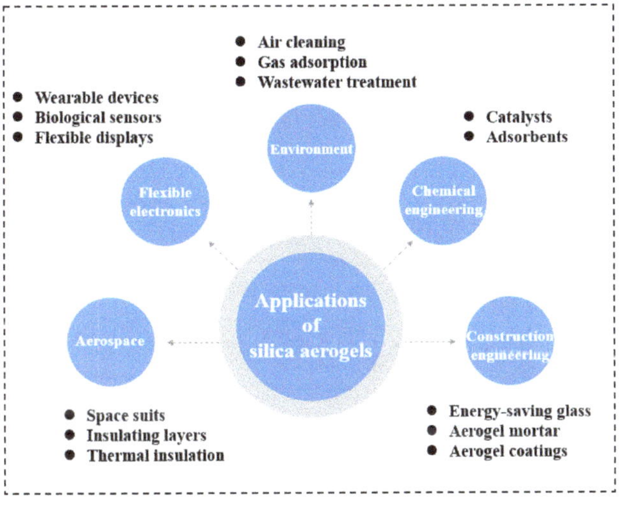

Figure 1. The applications of silica aerogels in various fields.

However, the application of silica aerogels on a large scale is still limited due to the poor mechanical properties of silica aerogels [44]. The main reason for the poor mechanical properties of silica aerogels is their pearl-necklace-like three-dimensional network structure. This network structure is connected by the interparticle necks, and the connection strength is very fragile. However, other properties of silica aerogels are excellent in the engineering field. Thus, the problem of the poor mechanical properties of silica aerogels urgently needs to be solved [45,46].

The secondary particles of silica aerogels are connected via point contact with a small contact area and weak bond force between the particles [47]. Once the aerogels have suffered from external stress, the neck region between the secondary particles break, resulting in the connection being disconnected and the gel skeleton collapsing [48]. In order to solve the above problems, scholars around the world have conducted extensive research. Focusing on the problem of the poor mechanical properties of silica aerogels, reinforcing phases have been introduced to improve the mechanical properties of silica aerogels. As shown in Figure 2, the reinforcing phases mainly include carbon, biomaterial, fiber, and polymer. The mechanical properties of silica aerogels have been enhanced to varying degrees through the experiment.

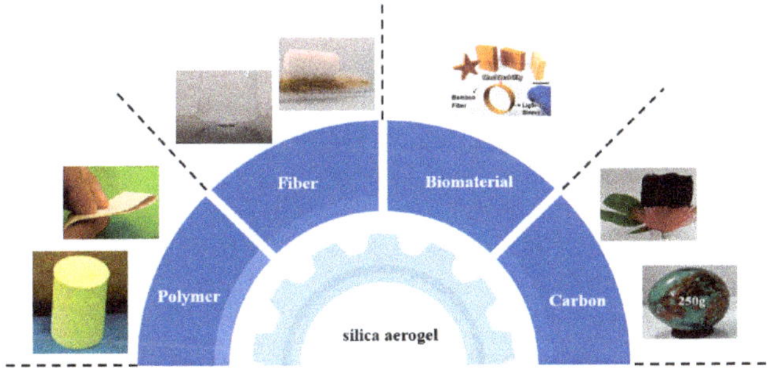

Figure 2. Materials for enhancing mechanical properties of silica aerogel. Reprinted with permission from Refs. [49–52]: Copyright 2019, Elsevier; Copyright 2022, Elsevier; Copyright 2011, American Chemical Society; Copyright 2022, American Chemical Society.

In this paper, the composites of silica aerogels with polymers are summarized, and the properties are discussed in detail.

2. Synthesis of Silica Aerogel

As shown in Figure 3, the preparation of the silica aerogels comprises (a) synthesis, (b) aging, and (c) drying [53]. The synthesis of silica aerogels mainly depends on the method of sol–gel [54]. During the sol–gel process, a three-dimensional network structure is built [55,56]. And the properties of silica aerogels are influenced by the precursors, catalysts, temperature, surface treatments, mass concentration, pH, and drying [57–60]. Then, the aging process enhances the network structure of the silica sol. Finally, the solvents are removed from the solid via the drying technology.

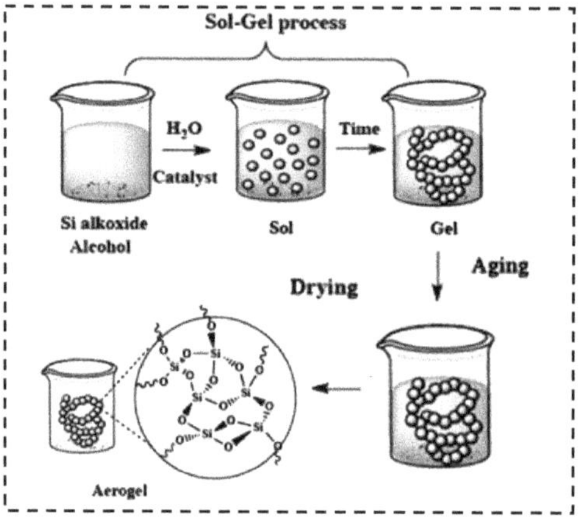

Figure 3. The preparation of aerogel using sol–gel process. Reprinted with permission from Ref. [54]: Copyright 2014, Elsevier.

2.1. Sol–Gel Chemistry

Gels are colloids that are composed of colloidal particles suspended in a solvent. The sol–gel method is a common technique used for the synthesis of wet gels, and then the silica aerogels are produced via the drying process [61,62]. During the sol–gel process, ethanol or methanol is used as a solvent. Using the solvent, the precursors form a soluble gel through hydrolysis and polycondensation reactions [63,64]. The results of the reactions are greatly influenced by the pH of the solution, the concentration of silicon precursor, the reaction time, and other factors. Moreover, the solid network of silica is formed in this process, and the particles are more closely connected [65–68]. However, the repetitive washing and the tedious water-to-alcohol solvent exchange still need more time [69]. As shown in Figure 4, the primary particles form secondary particles in the solution and then the secondary particles form a continuous solid network connected by the neck regions [70]. With the development of sol–gel chemistry, silica alkoxides are the precursors of silica aerogels, such as tetraethylorthosilicate (TEOS), tetramethylorthosilicate (TMOS), and so on [71–73]. These precursors have a great influence on the morphology and properties of silica aerogels [74]. Moreover, various precursors can also be used together to realize the self-reinforcement of silica aerogel skeletons [75,76]. Bhagat et al. [77] applied nine different co-precursors to prepare TEOS-based silica aerogels and investigated their physical properties. The pre-polymerization of silicon precursors is also a method of silica self-reinforcement that is currently being explored. Zu et al. [78] reported a new method of pre-polymerized silica-based precursors and enhanced the flexibility of aerogels. Table 1 lists some common precursors that are used for the synthesis of silica aerogels.

Table 1. Common silica precursors for silica aerogel synthesis.

Silica Precursor	Chemical Formula	Abbreviation	Physical Properties	Mechanical Properties	Thermal Properties	Ref.
Tetraethylorthosilicate	$Si(OC_2H_5)_4$	TEOS	/	G modulus: 10.7 MPa	/	[79]
Tetramethylorthosilicate	$Si(OCH_3)_4$	TMOS	Skeletal densities: 2.2 g/cm^3	/	/	[80]
Trimethylchlorosilane	$Si(CH_3)_3Cl$	TMCS	Surface area: 914.4 m^2/g; porosity: 96.16%	/	/	[81]
Methyltrimethoxysilane	$Si(OCH_3)_3CH_3$	MTMS	Shrinkage: 3.5%	/	/	[82]
Methyltriethoxysilane	$Si(OC_2H_5)_3CH_3$	MTES	Density: 0.1 g/cm^3; porosity: 95.5%	Unrecoverable strain loss: 10%	Thermal conductivity: 0.038 W/m·K	[83]
Aminopropyltrimethoxysilane	$H_2N(CH_2)_3Si(OCH_3)_3$	APTMS	/	/	Young's modulus: 14 MPa	[84]
Aminopropyltriethoxysilane	$H_2N(CH_2)_3Si(OC_2H_5)_3$	APTES	Surface area: 150.9 m^2/g	Young's modulus: 18 MPa	Thermal conductivity: 0.037 W/m·K	[85]
Propyltriethoxysilane	$C_9H_{22}O_3Si$	PTES	Density: 0.172 g/cm^3; porosity: >90%	Elastic module: 0.35 MPa	/	[86]
Vinyltrimethoxysilane	$H_2C{=}CHSi(OCH_3)_3$	VTMS	/	Elongation at break: 40~50%	Thermal conductivity: 0.06 W/m·K	[87]
Vinyltriethoxysilane	$C_8H_{18}O_3Si$	VTES	Surface area: 321 m^2/g	Compressive stress: 0.571 MPa	Thermal conductivity: 0.024 W/m·K	[88]
3-glycidoxypropyltrimethoxysilane	$C_9H_{20}O_5Si$	GPTMS	/	/	Thermal conductivity: 0.032 W/m·K	[89]
Bis [3-(triethoxysilyl)propyl]disulfide	$C_{18}H_{42}O_6S_2Si_2$	BTSPD	Density: 0.21 g/cm^3; porosity: 85.5%	Young's modulus: 2.1 MPa	/	[90]
1,6-bis(trimethoxysilyl)hexane	$C_{12}H_{30}O_6Si_2$	BTMSH	/	Strain: 50%	/	[91]
Bis(trimethoxysilylpropyl)amine	$C_{12}H_{31}NO_6Si_2$	BTMSPA	Density: 0.308 g/cm^3; porosity: 78%; surface area: 325 m^2/g	Shrink: 11%, compression Modulus: 15 MPa	/	[92]
Dimethyldiethoxysilane	$C_6H_{16}O_2Si$	DMDES	Density: 0.082 g/cm^3; surface area: 162.1 m^2/g; porosity: 94.2%	/	Maximum degradation rate: 150 °C	[93]

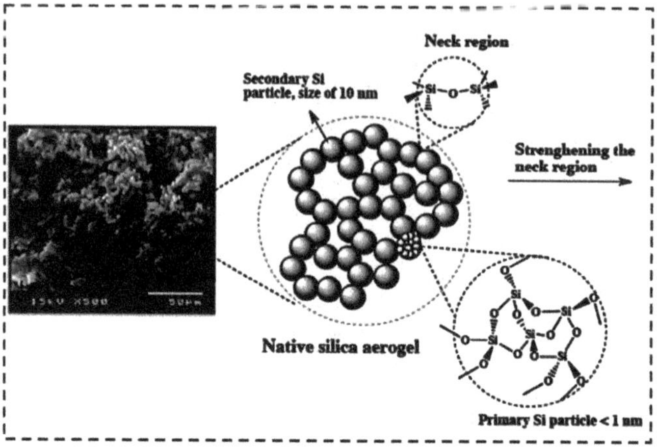

Figure 4. Primary and secondary silica particles of the silica aerogel. Reprinted with permission from Ref. [54]: Copyright 2014, Elsevier.

2.2. *Aging*

Aging is an important step in strengthening the mechanical properties of silica aerogels [94]. The aging reaction principle is ascribed to accelerating the movement of sol particles and increases the probability of collision, adding the number of siloxane connections. In the aging process, the number of siloxane linkages between particles can be increased using two different mechanisms simultaneously; thus, the mechanical properties of silica aerogels will be strengthened [95]. These two different mechanisms mainly include dissolution and re-precipitation [96]. Moreover, new monomers are transported to the neck region between the particles and form the network [97]. As an important chemical bond connecting the particles together, the O-Si bonds have a great relationship with the time and temperature of the reaction. The number of the bonds affects the mechanical properties of the silica aerogels. He et al. [98] reported that a controlled temperature and pressure can enhance the mechanical properties during the process of dissolution and reprecipitation.

However, this process requires a lot of time, and the parameters of the aging process are difficult to control. During the aging process, the silica aerogels occur different degrees of shrinkage, leading to an increase in density.

2.3. *Drying*

Drying is the procedure that transports the wet silica gel to silica aerogel [99]. In this process, gas replaces the liquid in silica wet gel, and ultimately, the solid consisting of the silica network forms the aerogel. Due to differential shrinkage, warping and cracking often occur. However, previous studies have shown that the phenomena can be prevented by controlling the drying process. At present, there are three common drying methods as follows: ambient pressure drying [100,101], freeze drying [102,103], and supercritical drying [104].

Compared with the other two drying methods, the ambient pressure drying method is widely used mainly due to its lower energy consumption and because it does not require high pressure conditions [105–107]. However, there are still drawbacks to ambient pressure drying. The most significant disadvantage is that ambient pressure drying may be affected by capillary force, and thus, still result in collapsing or cracking [108]. Freeze drying is the technology that freezes the liquid in the wet gel into a solid and then converts the solid into a gas. In order to reduce the possibility of breakage caused by ambient pressure drying, freeze drying is used to manufacture various types of aerogels. Compared with the above methods, supercritical drying is considered as the most appropriate method to prepare aerogels. It minimizes the aerogel cracking caused by shrinkage. The liquid in the pores is transformed into supercritical fluid via supercritical drying. In this state, the surface tension of the liquid disappears completely and there is no capillary force. Finally, the supercritical fluid can be separated from the solid at a temperature above the critical temperature of the liquid [109]. At present, the more mature supercritical drying technology is low temperature drying using carbon dioxide. However, this method has high requirements for synthesis equipment and the operating environment [110].

3. Polymer-Modified Silica Aerogel Composites

Based on existing reports, the polymer matrix is composed of thermoplastics and thermosetting resins [111]. Polymers demonstrate excellent mechanical properties, generally with high elastic deformation and viscoelasticity [112]. In addition to preparing polymer aerogels, polymers can also be combined with silica aerogels to prepare hybrid aerogels [113]. The process of polymerization increases the chemical bonds of O–Si to enhance their mechanical properties [114].

In the past two decades, researchers believed that combining polymer and silica is an effective way to enhance the mechanical properties of aerogels [115–118]. Meanwhile, the results indicate that the interfaces between silica gel particles and polymers also have a great influence on aerogel properties [119]. In 1994, Novak et al. [120] prepared hybrid aerogels via the method of pre-synthesized polymers or in situ polymerizing species, achieving an enhanced flexibility and compressive strength through the adsorption of energy with silica aerogels compounding the polymers [121,122]. As reported in the previous research, polymer-crosslinked aerogels were studied via the following three technical approaches: (a) modifying the surface of nanoparticles to enhance the aerogel skeleton; (b) applying different types of crosslinking agents; (c) creating the network morphology of the aerogels [123,124].

Using the method of chemical crosslinking, the polymer conformally coats the skeletal framework of the aerogels and maintains the original shape of the mesopores to reinforce their mechanical properties. As shown in Figure 5, with the addition of the polymer, the density of the aerogel will show an increasing trend. Through the growth mechanism, the number of connecting points between the secondary particles will increase, and finally resulting in the polymer-reinforcement.

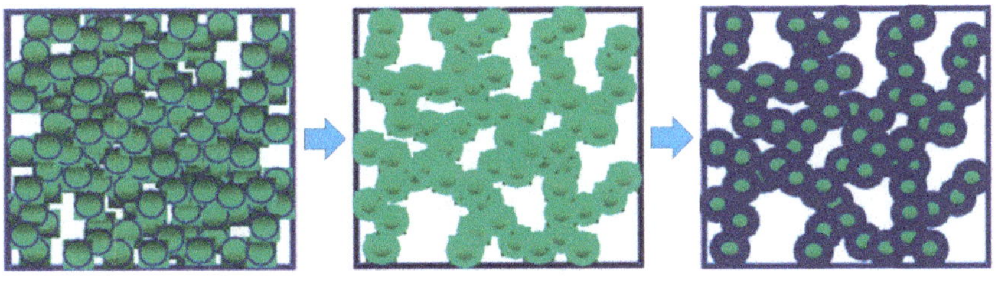

Figure 5. Technical approaches to improve the mechanical properties of aerogels. Reprinted with permission from Ref. [125]: Copyright 2011, American Chemical Society.

Although the mechanical properties of aerogels have been improved, their density and thermal conductivity have been increased as well, and the addition of the polymers even reduce their resistance to high temperatures [126]. Using well-controlled polymerization techniques, atom transfer radical polymerization can effectively enhance the performance of aerogels without significantly increasing their density. Moreover, the aggregation and poor interfacial interaction can be solved via the combination of silica aerogels and polymer [127]. The reported polymer-reinforced aerogels are listed in Table 2, and the contents are introduced in further detail.

Table 2. Overview of reported mechanical properties of polymer-reinforced aerogels.

Precursor Formulation	Polymer Matrix	Enhanced Properties	Ref
GPTMS/VTMS	Epoxide	➢ Elastic deformation: 3~5%	[128]
/	Epoxide	➢ Elastic modulus: 35%, tensile strength: 62%, toughness: 126%	[129]
/	Epoxide	➢ (Hydrophobic aerogel) contact angles: 107°, fracture toughness: improved by up to ~70%, impact strength: improved by up to ~120%	[130]
/	Epoxide	➢ Elastic modulus: 3770 ± 71 MPa, stress at yield point: 43.2 ± 1.8 MPa, strain at yield point: 1.24 ± 0.03 MPa, ultimate tensile strength: 51.0 ± 2.1 MPa, strain at break point: 3.3 ± 0.3%, toughness: 1.29 ± 0.08 J/m^3	[131]
TEOS/APTES	Epoxide	➢ Strain: 80% (18 N)	[132]
TEOS/APTES	Epoxide	➢ Tensile strength: 45.05± 4.56 MPa, modulus of elasticity: 4363.88± 209.57 MPa, break strain: 1.19 ± 0.17%	[133]
TEOS	Epoxide	➢ Density: 0.419 g/cm^3, porosity: 89%, compressive strength: 0.438 MPa, thermal conductivity: 0.0273 W/m·K	[134]
/	Epoxide	➢ (Warp direction) strength: 464.3 MPa, modulus: 1.76 GPa, (weft direction) strength: 410.2 MPa, modulus: 1.68 GPa	[135]
TMOS	Polyurea	➢ Shrinkage: 14.6 ± 0.7%, bulk density: 0.594 ± 0.026 g/cm^3, skeletal density: 1.290 ± 0.003 g/cm^3, porosity: 54%	[136]
TEOS/APTES	Polyurea	➢ Bulk density: 0.046 g/cm^3, flexural modulus: 0.14 MPa	[137]
TEOS/APTES	Polyurea	➢ Linear shrinkage: 15.73%, bulk density: 0.392 g/m^3, average elastic modulus: 14.57 MPa	[138]
TEOS /MTEOS	Polyurethane	➢ Density: 0.190 ± 0.006 g/m^3, yield strength: 2.15 ± 0.04 MPa, Young's modulus: 50 ± 0.09 MPa	[139]
/	Polyurethane	➢ Heat resistance index: 193.6%, char yield: 31.6%, bulk density: 0.580 g/mL	[140]
TEOS/APTES	Polyurethane	➢ BET surface area: 242.9 m^2/g, BJH desorption average pore diameter: 10.8 nm	[141]
TEOS	Polyurethane	➢ Density: 117.68 kg/m^3, porosity: 92.3%, linear shrinkage: −8.38%, thermal conductivity: 0.014 ± 0.00033 W/m·K	[142]
MTMS	Polyimide	➢ Compressive strain: 50%, thermal conductivity: 0.0212 W/m·K	[143]
/	Polyimide	➢ Surface area: 609 m^2/g, thermal conductivity: 0.017.5 W/m·K	[144]
TEOS/APTES	Polyimide	➢ Compressive strength: 3.82 MPa, Young's modulus: 44.16 MPa	[145]
TEOS/APTES	Polyimide	➢ Density: 0.145 g/cm^3, strain: 9%, strength: 0.29 MPa, Young's modulus: 3.22 MPa	[146]
TEOS	Polyimide	➢ Compressive modulus:1.96 MPa, thermal conductivity: 0.0311~0.0585 W/m·K	[147]
TMOS/APTES	Polystyrene	➢ Density: 0.41~0.77 g/cm^3, surface area: 213~393 m^2/g, thermal conductivity: 0.041 W/m·K, contact angles: 120°	[148]
TMOS	Polystyrene	➢ Density: 0.13~0.17 g/cm^3, surface area: 350~780 m^2/g, thermal conductivity: 0.03~0.04 W/m·K	[149]
MTMS/VTMS/TMOS	Polystyrene	➢ Bulk density: 163.1 ± 11.7 kg/cm^3, porosity: 88%, surface area: 227 m^2/g, thermal conductivity: 0.072 ± 0.001 W/m·K, Young's modulus: 91 kPa, compression strength: 68 kPa	[87]
TMOS	Polystyrene	➢ Modulus: 3 MPa	[150]

3.1. Epoxide

The mechanical properties of silica aerogels can be enhanced via epoxide. The functional groups of the epoxide can react with the amino groups on the surface of the gel skeleton. Thus, the addition of epoxide in the matrix of the silica aerogel can change its fragile properties and enhance its mechanical properties [151]. Thus, researchers have focused on applying epoxide to improve the mechanical properties of silica aerogels.

Rezaei et al. [128] have shown a new method to prepare the hybrid silica aerogel with the insulative and flexible properties. As shown in Figure 6, the researchers applied an epoxide ring containing the silica precursor and inserted flexible ether groups into the main chain using the method of ring-opening polymerization. The brittleness properties of the silica aerogels were enhanced due to the non-particulate structure. The results demonstrate that the elastic deformation of the aerogel was increased to 15%, and the mechanical properties were proportional to the density. Moreover, the aerogels have superinsulation properties with a thermal conductivity of only 0.0159 W/m K.

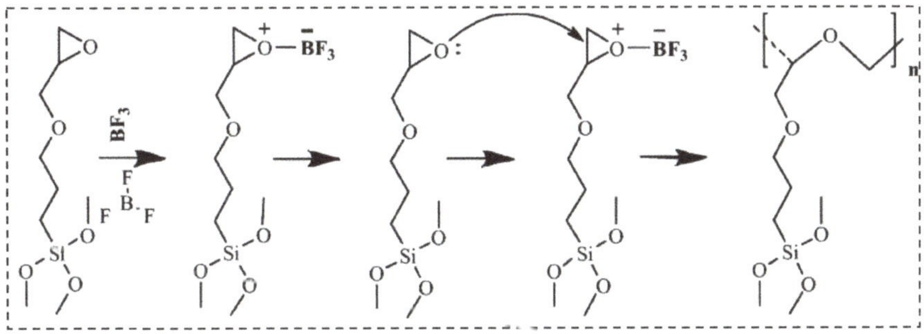

Figure 6. Polymerization reaction of GPTMS and molecular structure of PGPTMS. Reprinted with permission from Ref. [128]: Copyright 2020, Elsevier.

Salimian et al. [129] prepared the silica aerogel–epoxy nanocomposites and investigated the fracture and toughening mechanisms. By analyzing the mechanical and thermal properties, the results suggest that the viscosity of the nanocomposite suspension was increased from 1% to 6% with the silica aerogel addition. In addition, the storage modulus, *Tg*, Young's modulus, tensile strength, and toughness were increased by 11%, 5 °C, 35%, 62%, and 126%, respectively. As Figure 7 shows, the epoxy polymers are infiltrated into the mesopores of the silica aerogel. The fracture and toughening mechanisms are explained by the (a) crack pinning and deflection and the (b) plastic deformation.

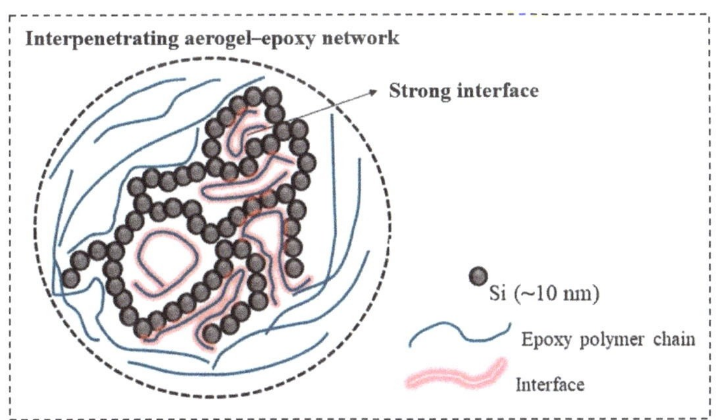

Figure 7. The structural model of the "interpenetrating organic-inorganic network". Reprinted with permission from Ref. [129]: Copyright 2018, Elsevier.

Salimian et al. [130] prepared the epoxy nanocomposites using two different types (hydrophobic and hydrophilic) of silica aerogels. As shown in Figure 8, the ≡Si−O−C≡ bonds are formed between the silica surface and the epoxy polymer network. Using this method, the storage modulus, viscosity, *Tg*, fracture toughness, and impact strength were enhanced. Moreover, the fracture toughness (K_{Ic}) and impact strength increased with the increase in the hydrophobic aerogel content.

Albooyeh et al. [131] studied the influences of silica aerogel on the mechanical, vibrational, and morphological properties of epoxy. The tensile, bending, compressive, dynamic mechanical thermal analyses, and a series of tests were conducted to verify the Euler–Bernoulli beam theory. The results indicate that silica aerogels can effectively reduce the density of materials. Meanwhile, the tensile, flexural, compressive modulus and hardness of the materials significantly increased when the addition of silica aerogel was 4%.

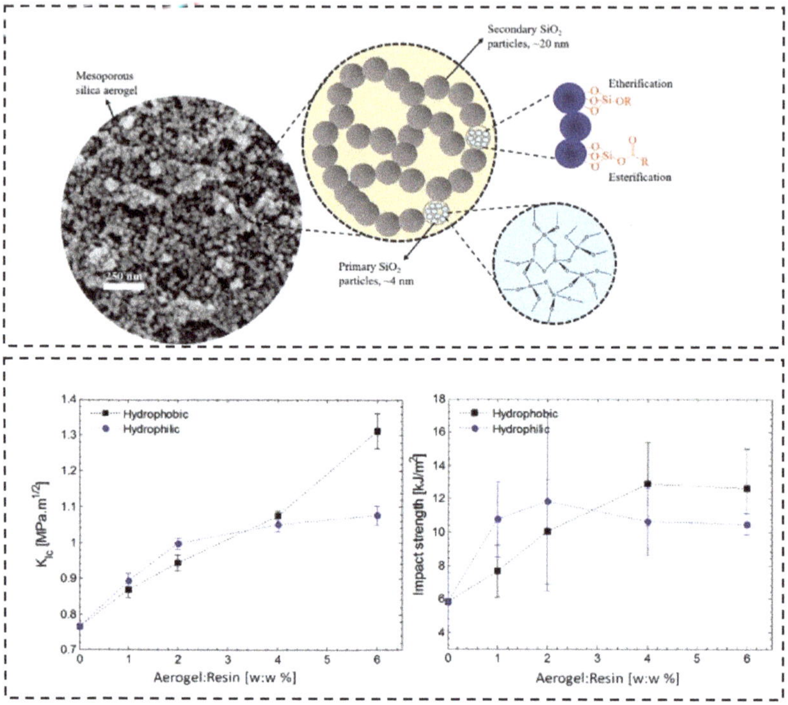

Figure 8. Schematic illustration of the silica aerogel particle network structure and possible covalent interactions between the epoxy and silica surface. Reprinted with permission from Ref. [130]: Copyright 2018, American Chemical Society.

According to the experimental data, the researchers found that the flexibility and robustness of the pore structure can be enhanced via the combination with polymers. Domènech et al. [132] synthesized a porous organic–inorganic hybrid material composed of silica and epoxy resin via a one-pot sol–gel process and subsequent supercritical drying. These results prove that controlling the bridged alkoxide proportions to enhance the mechanical properties of the silica aerogel is feasible and that the strain at 18 N can achieve 80%.

Selay et al. [133] applied silica aerogel powders with ionic liquid as a nanofiller to prepare the nanocomposite. The silica aerogels with ionic liquid possessed a lower density (0.16 g/cm^3), higher porosity (93%), and higher thermal stability (400 °C). Moreover, the composites (silica aerogel with 1wt% ionic liquid) demonstrated better mechanical properties, such as modulus of elasticity (4156.27 MPa) and tensile strength (51.96 MPa).

Epoxide is also a common gelation initiator. He et al. [134] applied epoxides as gelation accelerators to prepare the ZrO_2–SiO_2 aerogels via aging and supercritical drying. The results demonstrate that the epoxides can accelerate the gelation of sol. Considering that the decomposition of the polymer leads to a decrease in the high temperature resistance of aerogels, mullite fibers were introduced as the skeletons for the aerogels. The compressive strength of the M/ZrO_2–SiO_2 aerogel reached 0.438 MPa and the thermal conductivity was only 0.027 W/m·K. Selver et al. [135] investigated the influences of epoxy on the mechanical, nondestructive, and thermal properties of silica aerogel composites. The results show that the epoxy composites with a 1% addition of silica aerogel exhibited better flexural strength, impact, and energy absorption. However, the thermal conductivity of the 1% silica aerogel composites increased due to the void inside of the epoxy resin being filled up.

3.2. Polyurea

Polyurea consists of aromatic isocyanate segments and soft polyamine chains, which are synthesized via the reaction of the amino compound with the isocyanate component. This material has excellent anti-corrosion, waterproof, and mechanical properties. Polyurea-crosslinked silica-based aerogels have the characteristics of being nano-porous and mesoporous, so they can exhibit unique thermal management behavior.

Fu et al. [152] used the material point method (MPM) to study the mechanical behavior of the silica aerogels whose skeletal framework was coated by the polyurea at high strain rates. The researchers found that the MPM can model the compression of complex mesoporous structures and that the conformal polymer coating has a reinforcing function. The histograms of the distribution of material points versus the stress level at 19% strains is shown in Figure 9; the data show that the model with a 50% porosity has a wider range and exhibits more material points when under stress.

Churu et al. [136] also investigated the mechanical properties of the polymer-crosslinked templated silica aerogel (CTSA). The results show that the 1,3,5-trimethylbenzene (TMB) and triblock copolymer both have influences on the morphology of the aerogels, resulting in a change in the mechanical properties. The researchers further revealed the intrinsic relationship between the morphology and mechanical properties.

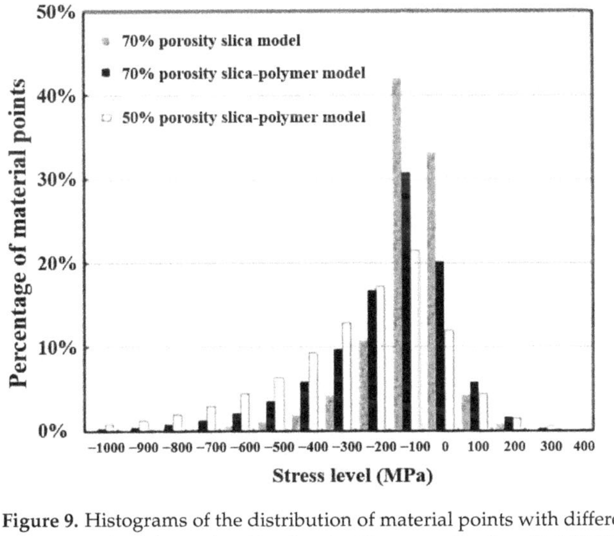

Figure 9. Histograms of the distribution of material points with different stress levels at 19% compressive strain deformation. Reprinted with permission from Ref. [152]: Copyright 2011, Elsevier.

Capadona et al. [137] reinforced the silica skeleton via the polymerization of the di-isocyanate using the amine-modified surface of a sol–gel-derived mesoporous silica network. Through the reaction with amines and urea linkages, the polymer was coated to the surface of the aerogel skeleton, which is shown in Figure 10. The results reveal that the highest density crosslinked aerogel had the highest stress at failure, exhibiting the highest modulus and crosslink.

As Figure 11 shows, Yang et al. [138] prepared the modified silica gels using the method of copolymerizing tetraethylorthosilicate with 3-aminopropyltriethoxy-silane. During the ambient pressure, the researchers successfully controlled the shrinkage of the silica aerogels. The experimental data demonstrate that the elastic modulus of the silica aerogel skeleton increases because of the incorporated polymers.

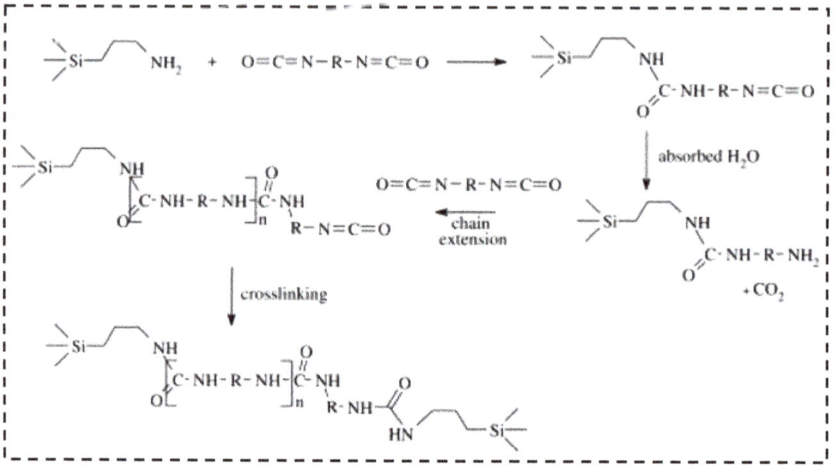

Figure 10. Synthetic route for amine-modified aerogel crosslinked with diisocyanates. Reprinted with permission from Ref. [137]: Copyright 2006, Elsevier.

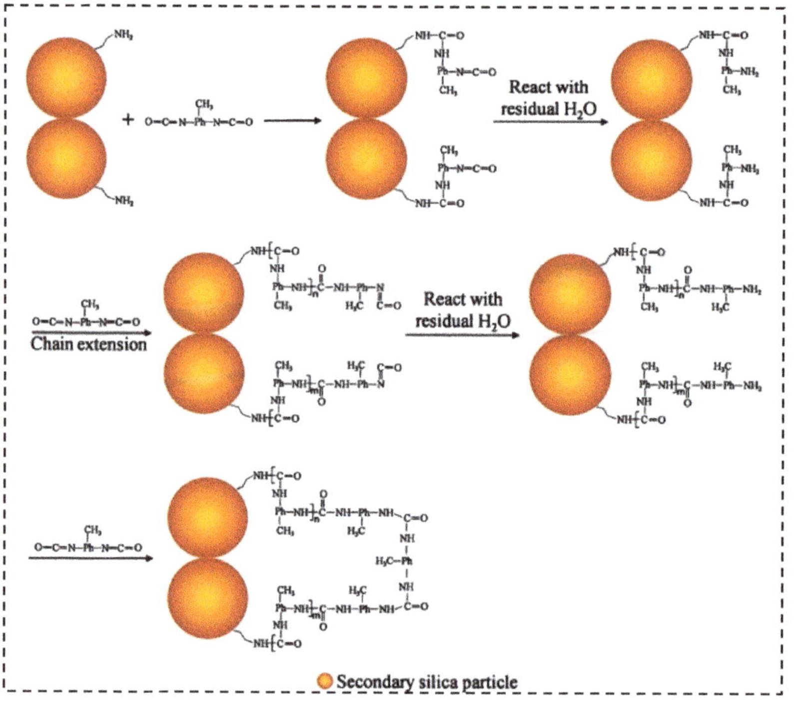

Figure 11. Reaction–modification of amine-modified aerogel using toluene diisocyanate. Reprinted with permission from Ref. [138]: Copyright 2011, Elsevier.

3.3. Polyurethane

Polyurethane (PU) is a typical foaming material, and its high thermal conductivity is an important issue that restricts sustainable development [153]. Based on the previous reports, the polyurethane-based hybrids prepared using the sol–gel approach showed excellent thermal insulating effectiveness and mechanical properties due to the inorganic and organic co-networks [154]. In general, there are a certain amount of hydroxyl groups remaining on the surface of the solid skeleton of the silica wet gel. Therefore, polyurea can form covalent bonds with silica wet gel and enhance the adhesion on the surface of the solid skeleton.

Cho et al. [140] prepared fabricated foldable silica aerogel/polyurethane composites (APCs), and the properties of the composites were theoretically verified via the proposed model. Figure 12 summarizes the morphological analysis of the PU1000 series, the schematic of the PU synthesis, and the fabrication process of the APCs. In Figure 12, the isocyanate-terminated prepolymers were synthesized by reacting with poly(tetramethylene ether glycol) (PTMG) and 2,4-diphenylmethane diisocyanate (MDI). Then, the prepared isocyanate-terminated PTMGs were chain-extended using 1–4 butanediol (BD). The results show that with an aerogel addition of 30%, the thermal conductivity of the APCs decreased by 72%, and the PU with a longer soft segment length demonstrated no breakage after bending.

Verdolotti et al. [139] synthesized organic–inorganic polyurethane-based hybrids, leading to an enhancement of the mechanical properties and thermal insulation via the isocyanate functional groups of IPTS reacted with OH of polyol under urethane bonds, as shown in Figure 13a. The researchers investigated the influences of mechanical behaviors on the aerogel-like siloxane domains. Figure 13b demonstrates the stress–strain curves of the foams ($HPUR_{ca1}$ and $HPUR_{ca2}$) compared to the pristine PUR. The Young's modulus of the foams ($HPUR_{ca1}$ and $HPUR_{ca2}$) achieved 30.17 MPa and 50 MPa. Moreover, the yield strength of these materials achieved 0.093 MPa and 2.15 MPa, respectively.

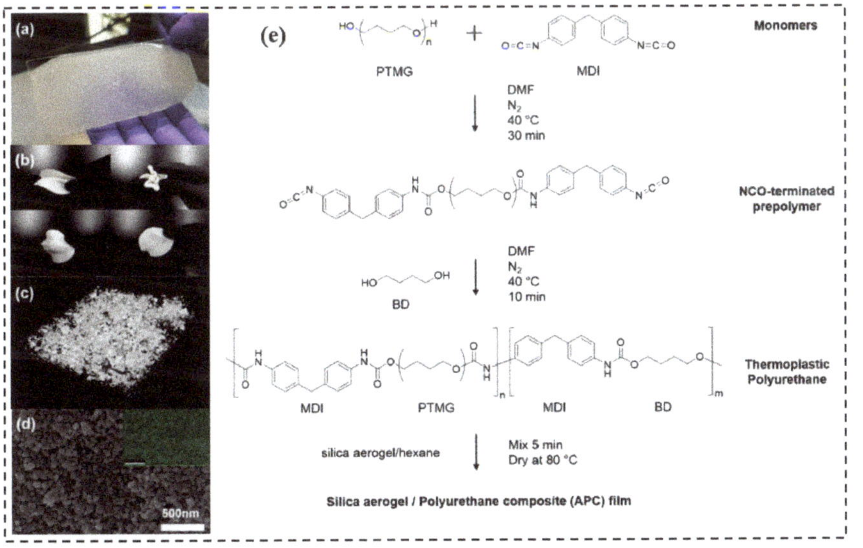

Figure 12. The photographs of (**a**) APC; (**b**) mechanical properties test; (**c**) m-CT image; (**d**) SEM and EDS image; (**e**) schematic of PU synthesis and fabrication process of APC. Reprinted with permission from Ref. [140]: Copyright 2019, Elsevier.

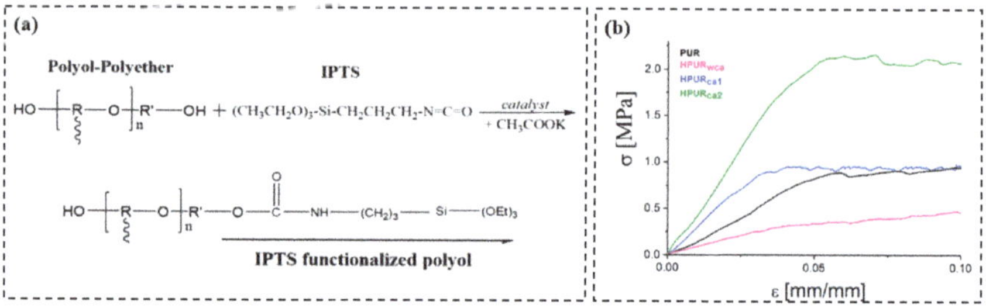

Figure 13. (**a**) Functionalization reaction between polyol-polyether and IPTS; (**b**) stress–strain curves of the foams. Reprinted with permission from Ref. [139]: Copyright 2021, Elsevier.

The researchers found that the flexibility of the hybrid aerogels are enhanced by the long-chain polymer molecules. Based on this theory, Duan et al. [141] prepared the mechanically reinforced hybrid silica aerogels using silane-end-capped urethane prepolymer and chain-extended polyurethane. The synthesis of prepolymer I, II, and polymer I, II are shown in Figures 14 and 15, respectively. The figures mainly explain the silane end groups participating in the silica network formation and the method that controls the amounts of polyurethane added to the aerogel network. The results show that the mechanical properties were enhanced via chain-extended polyurethane and the aerogels can suffer a 70% compressive strain with the addition of polymers.

The PU foams not only have lightness properties, but also demonstrate a continuous solid network, which can be used as reinforcements. Merillas et al. [142] reinforced the silica aerogel composites using reticulated polyurethane (PU) foams via ambient pressure drying and supercritical drying. Hexamethyldisilazane (HMDZ) was used to modify the surface of silica aerogels and the continuous network hybrid aerogel was formed using polyurethane. The results show that the elastic modulus increased from 130 to 450 kPa and the thermal conductivity was as low as 0.014 W/m·K.

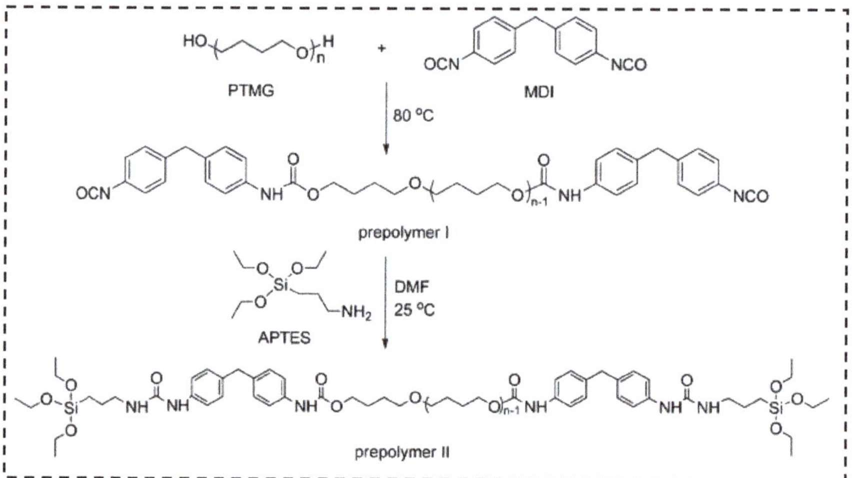

Figure 14. Synthesis of urethane prepolymer (prepolymer I) and APTES-end-capped prepolymer (prepolymer II). Reprinted with permission from Ref. [141]: Copyright 2013, American Chemical Society.

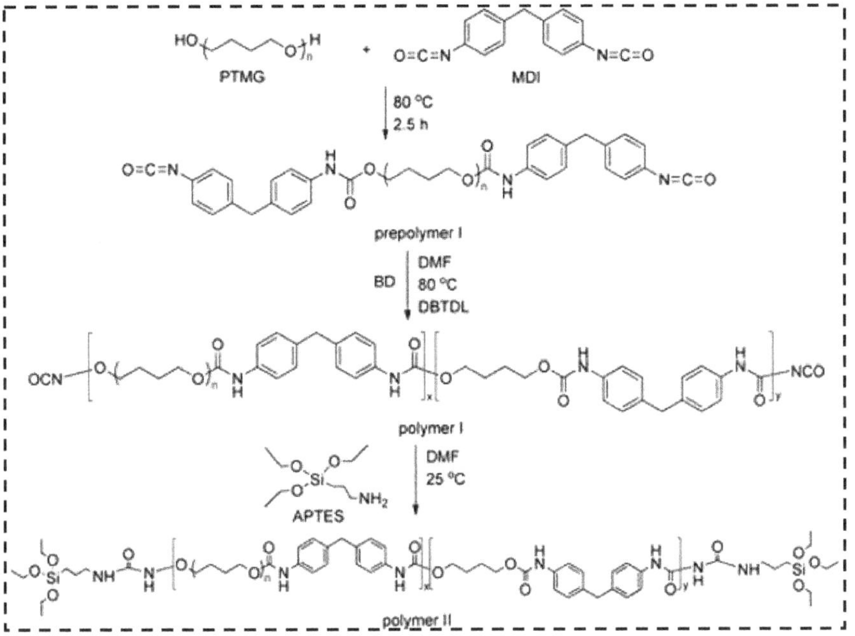

Figure 15. Synthesis of MDI-end-capped polyurethane (polymer I) and APTES-end-capped chain-extended polyurethane (polymer II). Reprinted with permission from Ref. [141]: Copyright 2013, American Chemical Society.

3.4. Polyimide

Compared with other organic constituents, the imide ring demonstrates a higher initial decomposition temperature [155]. Moreover, the chain structure constructed by the imide rings also create a polyimide (PI) with superior mechanical properties and a favorable chemical resistance [156]. Therefore, it is feasible to use polyimide to enhance the mechanical properties of silica aerogels.

Tian et al. [143] designed polyimide/silica composite aerogels using an integrated binary network via the in-situ synthesis. As shown in Figure 16a, the polyimide nanofiber aerogel (PINA) is used for the growth of the polymethylsilsesquioxane (PMSQ) network to synthesize the polyimide/silica (PSi) composite aerogel. The composite aerogels demonstrate excellent compressibility and flexibility, recovering from large compression (ε = 60%) and showing no collapsing. Meanwhile, the stress–strain curves (Figure 16b) demonstrate excellent compression recovery properties from 15%, 30%, 45% and 60% strain, respectively. In addition, the thermal conductivity of the polyimide/silica aerogel is as low as 0.0212 W/m·K and shows excellent resistance under 1200 °C.

Kantor et al. [144] synthesized heterogeneous polyimide–silica aerogels with low shrinkage by adding silica aerogel particles into a polyimide sol. The polyimide–silica aerogels exhibited heterogeneous structures and have properties of a high surface area over 609 m^2/g and a low thermal conductivity of 0.0175 W/m·K. Compared with general polyimide materials, the composite aerogels demonstrated potential commercial value. As shown in Figure 17, the BTC solution and silica aerogel powders occurred gelation in the polypropylene container, and the samples were obtained through aging and supercritical drying.

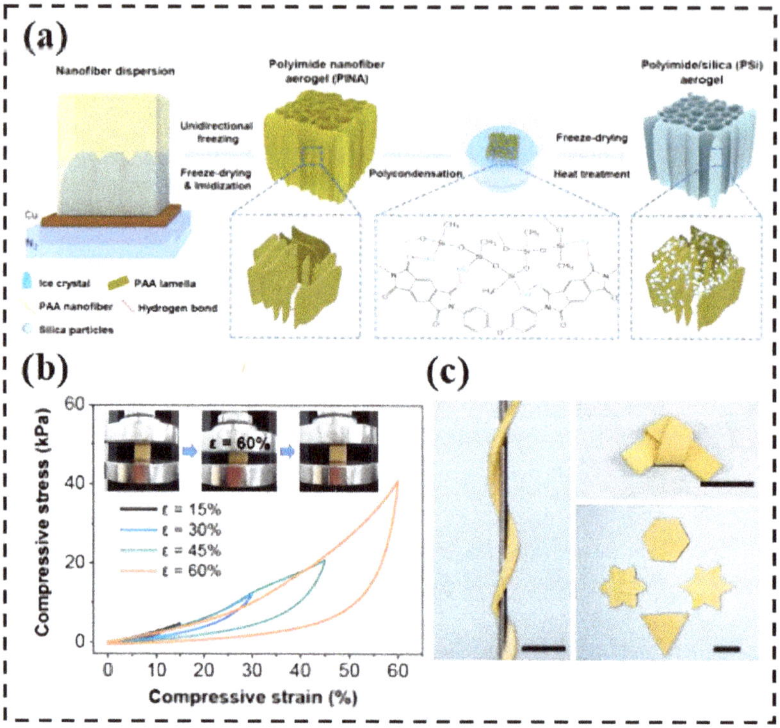

Figure 16. Preparation and composition of PSi aerogels: (**a**) schematic illustration of the preparation of PSi aerogels; (**b**) compressive stress–strain curves of PSi aerogel during loading–unloading cycles in the radial direction; (**c**) images of folded PSi-6 aerogel. Reprinted with permission from Ref. [143]: Copyright 2022, Elsevier.

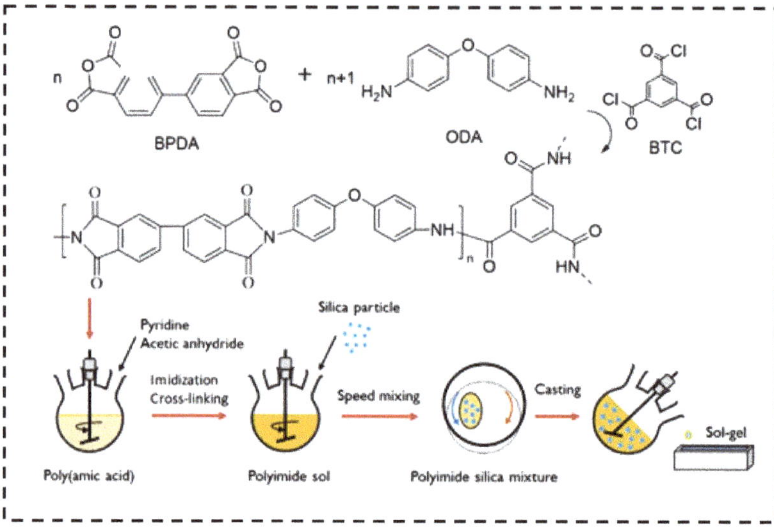

Figure 17. The preparation of PI-silica composites. Reprinted with permission from Ref. [144]: Copyright 2022, Elsevier.

Fei et al. [145] prepared the polyimide-crosslinked silica aerogels using different weight percentages of polyimide via the condensation reaction. The thermal conductivity of the aerogel is as low as 0.0306~0.0347 W/m·K, and the relatively high compressive strength can achieve a 1.03~3.82 MPa. Besides using polyimide to crosslink with silica aerogels to enhance their mechanical properties, using polyimide as a reinforcing phase is an efficient technical approach to improve the mechanical properties of silica aerogels. Fei et al. [146] used glass fiber and polyimide (PI) to reinforce the silica aerogel.

Zhang et al. [147] applied the "co-gel" strategy to fabricate the novel silica/polyimide (SiO_2/PI) nanocomposites. The SiO_2/PI nanocomposite aerogels exhibited excellent mechanical properties due to the hierarchically porous structure, such as a compressive modulus (1.96 MPa) and specific modulus (52.7 m^2/s^2). Moreover, the materials exhibited excellent flame resistance and low thermal conductivities between 25 °C and 300 °C.

3.5. Polystyrene

Polystyrene (PS) is a non-polar material that can improve the hydrophobic properties of silica aerogels. The silica precursors modified via amine, vinyl, and AIBN were also used to crosslink with PS to prepare the PS-reinforced silica aerogels.

Ilhan et al. [148] designed a new three-dimensional core–shell structure in which the PS was applied as the shell via the method of the free-radical polymerization process. Compared to the composite material prepared via polyurea and epoxy, the PS-crosslinked silica aerogels demonstrate a better hydrophobicity. Moreover, the silica aerogels showed excellent mechanical properties and maintained their integrity, while the thermal conductivity was as low as 0.041 W/m·K.

Maleki et al. [149] applied the growth of grafted polymers from the surface of silica gel to prepare the mechanically reinforced polymer–silica aerogels. The method of surface-initiated reversible addition–fragmentation chain transfer polymerization can significantly improve the compression strength of silica aerogels. Matias et al. [87] used polybutylacrylate (PBA) and polystyrene (PS) to prepare crosslinked flexible, monolithic, and superhydrophobic silica aerogels. Compared with the non-reinforced aerogel, the PBA-reinforced aerogel, MTMS-derived aerogel, and PS-reinforced aerogel demonstrated excellent Young's modulus values and compression strength, which can reach 91 kPa and 68 kPa, respectively. DeFriend et al. [150] used polystyrene beads to prepare mesoporous silica aerogel to investigate the influences of the surface area and pore volume on mechanical compression. These results demonstrate that the templating agents had a great effect on the compressive strength of the aerogels and that the concentrations were a great factor.

4. Conclusions

In this review, the synthesis chemistry and three main stages were introduced. Then, the process of sol–gel chemistry and the role of aging were also described in detail. The advantages and disadvantages of the three drying methods listed were also carefully analyzed and compared.

The five common polymers used to enhance the mechanical properties of silica aerogels were summarized. In the process of modification, the crosslinking agent has an important function. More importantly, the improvement in the mechanical properties is significantly influenced by process parameters such as time, temperature, and ratio. The linear density and shrinkage of the material increases significantly when the process parameters are not well controlled. This inevitably makes the aerogel lose the original advantages and causes a reduction in the mechanical properties. The silica aerogels' crosslinked polymers demonstrated good mechanical properties. In particular, polystyrene demonstrated a better performance of hydrophobicity due to its characteristics.

5. Outlook and Prospects

Compared with physical strengthening, the bonding tightness of silica aerogel particles strengthened via polymer crosslinking has the advantages of firmness and reliability. However, some problems still need to be further studied in the field of polymer reinforcement. Researchers should not only focus on the influence of polymers on the thermo-mechanical properties of aerogels during the research process, but should also carefully consider the effects of crosslinked polymers on other aspects of silica aerogels, such as the increase in density due to the addition of polymers, the poor flame retardancy of some polymers, and the difficult aging resistance of hybrid aerogels. The technology of eliminating the negative effects of polymers should be carefully studied. The functional role of silica in various fields should also be given more attention. Polymer crosslinked silica aerogels fully demonstrate the lightweight porous properties of aerogels and the related properties of polymers. Thus, this work will lay a solid and reliable foundation for the future development of multifunctional hybrid aerogels.

Author Contributions: Investigation, W.Z., Q.K., J.J., W.L., F.S. and Z.X.; resources, L.C., Q.K. and L.L.; writing—original draft, W.Z., L.C., Q.K., J.J. and F.S.; writing—review and editing, W.Z., L.C., Q.K., L.L., M.C., W.L., F.S. and Z.X.; supervision, W.Z., L.C., J.J. and M.C.; validation, W.Z., L.C., Q.K., L.L., M.C., J.J., W.L., F.S. and Z.X.; funding acquisition, L.L. and M.C.; project administration, L.L., M.C., J.J., W.L. and Z.X. All authors have read and agreed to the published version of the manuscript.

Funding: This work was supported by the National Natural Science Foundation of China (Grant No. 52004131), the National Natural Science Foundation of China (Grant No. 52204213), the Students' Scientific Research Training Program of the College of Emergency Management of Jiangsu University (JG-03-07), and the Education Reform Research and Talent Training Project of the College of Emergency Management of Jiangsu University (JG-01-18).

Institutional Review Board Statement: Not applicable.

Informed Consent Statement: Not applicable.

Data Availability Statement: Not applicable.

Conflicts of Interest: The authors declare no conflict of interest.

References

1. Li, Z.; Cheng, X.; He, S.; Huang, D.; Bi, H.; Yang, H. Preparation of ambient pressure dried MTMS/TEOS co-precursor silica aerogel by adjusting NH_4OH concentration. *Mater. Lett.* **2014**, *129*, 12–15. [CrossRef]
2. Du, A.; Zhou, B.; Zhang, Z.; Shen, J. A special material or a new state of matter: A review and reconsideration of the aerogel. *Materials.* **2013**, *6*, 941–968. [CrossRef] [PubMed]
3. Kim, C.H.J.; Zhao, D.; Lee, G.; Liu, J. Strong, machinable carbon aerogels for high performance supercapacitors. *Adv. Funct. Mater.* **2016**, *26*, 4976–4983. [CrossRef]
4. Hrubesh, L.W.; Coronado, P.R.; Satcher, J.H. Solvent removal from water with hydrophobic aerogels. *J. Non-Cryst. Solids* **2001**, *285*, 328–332. [CrossRef]
5. Su, L.; Niu, M.; Lu, D.; Cai, Z.; Li, M.; Wang, H. A review on the emerging resilient and multifunctional ceramic aerogels. *J. Mater. Sci. Technol.* **2021**, *75*, 1–13. [CrossRef]
6. Shen, J.; Zhang, P.; Song, L.; Li, J.; Ji, B.; Li, J.; Chen, L. Polyethylene glycol supported by phosphorylated polyvinyl alcohol/graphene aerogel as a high thermal stability phase change material. *Compos. Part B Eng.* **2019**, *179*, 107545. [CrossRef]
7. Yan, M.; Pan, Y.; Cheng, X.; Zhang, Z.; Deng, Y.; Lun, Z.; Gong, L.; Gao, M.; Zhang, H. "Robust–Soft" anisotropic nanofibrillated cellulose aerogels with superior mechanical, flame-retardant, and thermal insulating properties. *ACS Appl. Mater. Interfaces* **2021**, *13*, 27458–27470. [CrossRef]
8. Cuce, E.; Cuce, P.M.; Wood, C.J.; Riffat, S.B. Toward aerogel based thermal superinsulation in buildings: A comprehensive review. *Renew. Sustain. Energy Rev.* **2014**, *34*, 273–299. [CrossRef]
9. Wa, L.; Fengyun, L.; Fanlu, Z.; Mengjing, C.; Qiang, C.; Jue, H.; Weijun, Z.; Mingwei, M. Preparation of silica aerogels using CTAB/SDS as template and their efficient adsorption. *Appl. Surf. Sci.* **2015**, *353*, 1031–1036. [CrossRef]
10. Sun, H.; Xu, Z.; Gao, C. Multifunctional, ultra-flyweight, synergistically assembled carbon aerogels. *Adv. Mater.* **2013**, *25*, 2554–2560. [CrossRef]
11. Long, D.; Chen, Q.; Qiao, W.; Zhan, L.; Liang, X.; Ling, L. Three-dimensional mesoporous carbon aerogels: Ideal catalyst supports for enhanced H_2S oxidation. *Chem. Commun.* **2009**, 3898–3900. [CrossRef]
12. Lee, J.K.; Gould, G.L.; Rhine, W. Polyurea based aerogel for a high performance thermal insulation material. *J. Sol-Gel Sci. Technol.* **2009**, *49*, 209–220. [CrossRef]
13. Hou, X.; Mao, Y.; Zhang, R.; Fang, D. Super-flexible polyimide nanofiber cross-linked polyimide aerogel membranes for high efficient flexible thermal protection. *Chem. Eng. J.* **2021**, *417*, 129341. [CrossRef]

14. Merillas, B.; Villafañe, F.; Rodríguez-Pérez, M.Á. Improving the insulating capacity of polyurethane foams through polyurethane aerogel inclusion: From insulation to superinsulation. *Nanomaterials* **2022**, *12*, 2232. [CrossRef]
15. Kulkarni, A.; Jana, S.C. Surfactant-free syndiotactic polystyrene aerogel foams via pickering emulsion. *Polymer* **2021**, *212*, 123125. [CrossRef]
16. Wu, Y.; Wang, X.; Shen, J. Metal oxide aerogels for high-temperature applications. *J. Sol-Gel Sci. Technol.* **2023**, *106*, 360–380. [CrossRef]
17. Liao, J.; Liu, P.; Xie, Y.; Zhang, Y. Metal oxide aerogels: Preparation and application for the uranium removal from aqueous solution. *Sci. Total Environ.* **2021**, *768*, 144212. [CrossRef] [PubMed]
18. Gao, W.; Wen, D. Recent advances of noble metal aerogels in biosensing. *View* **2021**, *2*, 20200124. [CrossRef]
19. Nita, L.E.; Ghilan, A.; Rusu, A.G.; Neamtu, I.; Chiriac, A.P. New trends in bio-based aerogels. *Pharmaceutics* **2020**, *12*, 449. [CrossRef]
20. Paninho, A.B.; Mustapa, A.N.; Mahmudov, K.T.; Pombeiro, A.J.L.; Guedes Da Silva, M.F.C.; Bermejo, M.D.; Martín, Á.; Cocero, M.J.; Nunes, A.V.M. A bio-based alginate aerogel as an ionic liquid support for the efficient synthesis of cyclic carbonates from CO_2 and epoxides. *Catalysts* **2021**, *11*, 872. [CrossRef]
21. Wicklein, B.; Kocjan, A.; Salazar-Alvarez, G.; Carosio, F.; Camino, G.; Antonietti, M.; Bergström, L. Thermally insulating and fire-retardant lightweight anisotropic foams based on nanocellulose and graphene oxide. *Nat. Nanotechnol.* **2015**, *10*, 277–283. [CrossRef] [PubMed]
22. Kettunen, M.; Silvennoinen, R.J.; Houbenov, N.; Nykänen, A.; Ruokolainen, J.; Sainio, J.; Pore, V.; Kemell, M.; Ankerfors, M.; Lindström, T.; et al. Photoswitchable superabsorbency based on nanocellulose aerogels. *Adv. Funct. Mater.* **2011**, *21*, 510–517. [CrossRef]
23. Sajab, M.S.; Chia, C.H.; Chan, C.H.; Zakaria, S.; Kaco, H.; Chook, S.W.; Chin, S.X.; Noor, A.A.M. Bifunctional graphene oxide–cellulose nanofibril aerogel loaded with Fe(iii) for the removal of cationic dye via simultaneous adsorption and fenton oxidation. *Rsc Adv.* **2016**, *6*, 19819–19825. [CrossRef]
24. Maleki, H.; Durães, L.; García-González, C.A.; Del Gaudio, P.; Portugal, A.; Mahmoudi, M. Synthesis and biomedical applications of aerogels: Possibilities and challenges. *Adv. Colloid Interface Sci.* **2016**, *236*, 1–27. [CrossRef]
25. Maleki, H. Recent advances in aerogels for environmental remediation applications: A review. *Chem. Eng. J.* **2016**, *300*, 98–118. [CrossRef]
26. Ren, J.; Feng, J.; Wang, L.; Chen, G.; Zhou, Z.; Li, Q. High specific surface area hybrid silica aerogel containing POSS. *Microporous Mesoporous Mater.* **2021**, *310*, 110456. [CrossRef]
27. Schultz, J.M.; Jensen, K.I.; Kristiansen, F.H. Super insulating aerogel glazing. *Sol. Energy Mater. Sol. Cells* **2005**, *89*, 275–285. [CrossRef]
28. Gurav, J.L.; Jung, I.-K.; Park, H.-H.; Kang, E.S.; Nadargi, D.Y. Silica aerogel: Synthesis and applications. *J. Nanomater.* **2010**, *2010*, 409310. [CrossRef]
29. Long, H.; Harley-Trochimczyk, A.; Pham, T.; Tang, Z.; Shi, T.; Zettl, A.; Carraro, C.; Worsley, M.A.; Maboudian, R. High surface area MoS_2/graphene hybrid aerogel for ultrasensitive NO_2 detection. *Adv. Funct. Mater.* **2016**, *26*, 5158–5165. [CrossRef]
30. Kistler, S.S. Coherent expanded aerogels and jellies. *Nature* **1931**, *127*, 741. [CrossRef]
31. Kistler, S.S. Coherent Expanded-Aerogels. *J. Phys. Chem.* **1932**, *36*, 52–64. [CrossRef]
32. Zhao, C.; Yan, Y.; Hu, Z.; Li, L.; Fan, X. Preparation and characterization of granular silica aerogel/polyisocyanurate rigid foam composites. *Constr. Build. Mater.* **2015**, *93*, 309–316. [CrossRef]
33. Baetens, R.; Jelle, B.P.; Gustavsen, A. Aerogel insulation for building applications: A state-of-the-art review. *Energy Build.* **2011**, *43*, 761–769. [CrossRef]
34. Bheekhun, N.; Abu Talib, A.R.; Hassan, M.R. Aerogels in aerospace: An overview. *Adv. Mater. Sci. Eng.* **2013**, *2013*, 406065. [CrossRef]
35. Jones, S.M. Aerogel: Space exploration applications. *J. Sol-Gel Sci. Technol.* **2006**, *40*, 351–357. [CrossRef]
36. Ochoa, M.; Durães, L.; Beja, A.M.; Portugal, A. Study of the suitability of silica based xerogels synthesized using ethyltrimethoxysilane and/or methyltrimethoxysilane precursors for aerospace applications. *J. Sol-Gel Sci. Technol.* **2012**, *61*, 151–160. [CrossRef]
37. Mohammadi, A.; Moghaddas, J. Synthesis, adsorption and regeneration of nanoporous silica aerogel and silica aerogel-activated carbon composites. *Chem. Eng. Res. Des.* **2015**, *94*, 475–484. [CrossRef]
38. Perdigoto, M.L.N.; Martins, R.C.; Rocha, N.; Quina, M.J.; Gando-Ferreira, L.; Patrício, R.; Durães, L. Application of hydrophobic silica based aerogels and xerogels for removal of toxic organic compounds from aqueous solutions. *J. Colloid Interface Sci.* **2012**, *380*, 134–140. [CrossRef]
39. He, S.; Du, C.; Sheng, H.; He, C.; Liu, X.; Jin, X.; Chen, Q.; Tian, F. Ultrasensitive and self-powered multiparameter pressure–temperature–humidity sensor based on ultra-flexible conductive silica aerogel. *Gels* **2023**, *9*, 162. [CrossRef]
40. Feng, X.; Wang, L.; Yao, X.; Dong, H.; Wang, X.; Wang, Y. Trace water/amino-modified silica aerogel catalytic system in the one-pot sequential reaction of benzaldehyde dimethyl acetal and nitromethane. *Catal. Commun.* **2017**, *90*, 106–110. [CrossRef]
41. Ward, D.A.; Ko, E.I. Cheminform abstract: Preparing catalytic materials by the sol-gel method. *Cheminform* **1995**, *26*, 421–433. [CrossRef]
42. Ulker, Z.; Erkey, C. An emerging platform for drug delivery: Aerogel based systems. *J. Control. Release* **2014**, *177*, 51–63. [CrossRef]

43. Shang, L.; Lyu, Y.; Han, W. Microstructure and thermal insulation property of silica composite aerogel. *Materials* **2019**, *12*, 993. [CrossRef] [PubMed]
44. Madyan, O.A.; Fan, M.; Feo, L.; Hui, D. Physical properties of clay aerogel composites: An overview. *Compos. Part B Eng.* **2016**, *102*, 29–37. [CrossRef]
45. Leventis, N.; Sotiriou-Leventis, C.; Zhang, G.; Rawashdeh, A.-M.M. Nanoengineering strong silica aerogels. *Nano Lett.* **2002**, *2*, 957–960. [CrossRef]
46. Bertino, M.F.; Hund, J.F.; Zhang, G.; Sotiriou-Leventis, C.; Tokuhiro, A.T.; Leventis, N. Room temperature synthesis of noble metal clusters in the mesopores of mechanically strong silica-polymer aerogel composites. *J. Sol-Gel Sci. Technol.* **2004**, *30*, 43–48. [CrossRef]
47. Ma, H.-S.; Roberts, A.P.; Prévost, J.-H.; Jullien, R.; Scherer, G.W. Mechanicalstructure-propertyrelationshipofaerogels. *J. Non-Cryst. Solids* **2000**, *277*, 127–141. [CrossRef]
48. Meti, P.; Mahadik, D.B.; Lee, K.-Y.; Wang, Q.; Kanamori, K.; Gong, Y.-D.; Park, H.-H. Overview of organic–inorganic hybrid silica aerogels: Progress and perspectives. *Mater. Des.* **2022**, *222*, 111091. [CrossRef]
49. Parmar, K.R.; Dora, D.T.K.; Pant, K.K.; Roy, S. An ultra-light flexible aerogel-based on methane derived CNTs as a reinforcing agent in silica-CMC matrix for efficient oil adsorption. *J. Hazard. Mater.* **2019**, *375*, 206–215. [CrossRef]
50. Yi, Z.; Zhang, X.; Yan, L.; Huyan, X.; Zhang, T.; Liu, S.; Guo, A.; Liu, J.; Hou, F. Super-insulated, flexible, and high resilient mullite fiber reinforced silica aerogel composites by interfacial modification with nanoscale mullite whisker. *Compos. Part B Eng.* **2022**, *230*, 109549. [CrossRef]
51. Guo, H.; Meador, M.A.B.; Mccorkle, L.; Quade, D.J.; Guo, J.; Hamilton, B.; Cakmak, M.; Sprowl, G. Polyimide aerogels cross-linked through amine functionalized polyoligomeric silsesquioxane. *ACS Appl. Mater. Interfaces* **2011**, *3*, 546–552. [CrossRef] [PubMed]
52. Fan, Q.; Ou, R.; Hao, X.; Deng, Q.; Liu, Z.; Sun, L.; Zhang, C.; Guo, C.; Bai, X.; Wang, Q. Water-induced self-assembly and in situ mineralization within plant phenolic glycol-gel toward ultrastrong and multifunctional thermal insulating aerogels. *Acs Nano* **2022**, *16*, 9062–9076. [CrossRef] [PubMed]
53. Brinker, C.J.; Keefer, K.D.; Schaefer, D.W.; Ashley, C.S. Sol-gel transition in simple silicates. *J. Non-Cryst. Solids* **1982**, *48*, 47–64. [CrossRef]
54. Maleki, H.; Durães, L.; Portugal, A. An overview on silica aerogels synthesis and different mechanical reinforcing strategies. *J. Non-Cryst. Solids* **2014**, *385*, 55–74. [CrossRef]
55. Yang, M.; Li, J.; Man, Y.; Peng, Z.; Zhang, X.; Luo, X. A three-dimensional mullite-whisker network ceramic with ultra-light weight and high-strength prepared by the foam-gelcasting method. *J. Asian Ceram. Soc.* **2020**, *8*, 387–395. [CrossRef]
56. Borzęcka, N.H.; Nowak, B.; Gac, J.M.; Głaz, T.; Bojarska, M. Kinetics of mtms-based aerogel formation by the sol-gel method—experimental results and theoretical description. *J. Non-Cryst. Solids* **2020**, *547*, 120310. [CrossRef]
57. Rao, A.P.; Rao, A.V.; Gurav, J.L. Effect of protic solvents on the physical properties of the ambient pressure dried hydrophobic silica aerogels using sodium silicate precursor. *J. Porous Mater.* **2008**, *15*, 507–512. [CrossRef]
58. Xia, B.; Wang, Z.; Gou, L.; Zhang, M.; Guo, M. Porous mullite ceramics with enhanced compressive strength from fly ash-based ceramic microspheres: Facile synthesis, structure, and performance. *Ceram. Int.* **2022**, *48*, 10472–10479. [CrossRef]
59. Rao, A.V.; Kulkarni, M.M.; Amalnerkar, D.P.; Seth, T. Superhydrophobic silica aerogels based on methyltrimethoxysilane precursor. *J. Non-Cryst. Solids* **2003**, *330*, 187–195. [CrossRef]
60. García-González, C.A.; Alnaief, M.; Smirnova, I. Polysaccharide-based aerogels-promising biodegradable carriers for drug delivery systems. *Carbohydr. Polym.* **2011**, *86*, 1425–1438. [CrossRef]
61. Pierre, A.C.; Pajonk, G.M. Chemistry of aerogels and their applications. *Chem. Rev.* **2002**, *102*, 4243–4266. [CrossRef]
62. Zarzycki, J. Past and present of sol-gel science and technology. *J. Sol-Gel Sci. Technol.* **1997**, *8*, 17–22. [CrossRef]
63. Alié, C.; Ferauche, F.; Pirard, R.; Lecloux, A.J.; Pirard, J.-P. Preparation of low-density xerogels by incorporation of additives during synthesis. *J. Non-Cryst. Solids* **2001**, *289*, 88–96. [CrossRef]
64. Wu, G.; Wang, J.; Shen, J.; Yang, T.; Zhang, Q.; Zhou, B.; Deng, Z.; Bin, F.; Zhou, D.; Zhang, F. Properties of sol-gel derived scratch-resistant nano-porous silica films by a mixed atmosphere treatment. *J. Non-Cryst. Solids* **2000**, *275*, 169–174. [CrossRef]
65. Venkateswara Rao, A.; Bhagat, S.D. Synthesis and physical properties of TEOS-based silica aerogels prepared by two step (acid–base) sol–gel process. *Solid State Sci.* **2004**, *6*, 945–952. [CrossRef]
66. Kirkbir, F.; Murata, H.; Meyers, D.; Chaudhuri, S.R.; Sarkar, A. Drying and sintering of sol-gel derived large SiO_2 monoliths. *J. Sol-Gel Sci. Technol.* **1996**, *6*, 203–217. [CrossRef]
67. Harlick, P.J.E.; Sayari, A. Applications of pore-expanded mesoporous silicas. 3. Triamine silane grafting for enhanced co^2 adsorption. *Ind. Eng. Chem. Res.* **2006**, *45*, 3248–3255. [CrossRef]
68. Araki, S.; Doi, H.; Sano, Y.; Tanaka, S.; Miyake, Y. Preparation and CO_2 adsorption properties of aminopropyl-functionalized mesoporous silica microspheres. *J. Colloid Interface Sci.* **2009**, *339*, 382–389. [CrossRef]
69. Lee, C.J.; Kim, G.S.; Hyun, S.H. Synthesis of silica aerogels from waterglass via new modified ambient drying. *J. Mater. Sci.* **2002**, *37*, 2237–2241. [CrossRef]
70. Siouffi, A.-M. Silica gel-based monoliths prepared by the sol-gel method: Facts and figures. *J. Chromatogr. A* **2003**, *1000*, 801–818. [CrossRef]
71. Li, Q.; Afeworki, M.; Callen, N.M.; Colby, R.J.; Gopinadhan, M.; Nines Kochersperger, M.L.; Peterson, B.K.; Sansone, M.; Weston, S.C.; Calabro, D.C. Template-free self-assembly of mesoporous organosilicas. *Chem. Mater.* **2018**, *30*, 2218–2228. [CrossRef]

72. Nakanishi, K.; Minakuchi, H.; Soga, N.; Tanaka, N. Double pore silica gel monolith applied to liquid chromatography. *J. Sol-Gel Sci. Technol.* **1997**, *8*, 547–552. [CrossRef]
73. Tamon, H.; Kitamura, T.; Okazaki, M. Preparation of silica aerogel from TEOS. *J. Colloid Interface Sci.* **1998**, *197*, 353–359. [CrossRef]
74. Wong, J.C.H.; Kaymak, H.; Brunner, S.; Koebel, M.M. Mechanical properties of monolithic silica aerogels made from polyethoxydisiloxanes. *Microporous Mesoporous Mater.* **2014**, *183*, 23–29. [CrossRef]
75. Nguyen, B.N.; Meador, M.A.B.; Tousley, M.E.; Shonkwiler, B.; Mccorkle, L.; Scheiman, D.A.; Palczer, A. Tailoring elastic properties of silica aerogels cross-linked with polystyrene. *ACS Appl. Mater. Interfaces* **2009**, *1*, 621–630. [CrossRef]
76. Nadargi, D.Y.; Latthe, S.S.; Hirashima, H.; Rao, A.V. Studies on rheological properties of methyltriethoxysilane (MTES) based flexible superhydrophobic silica aerogels. *Microporous Mesoporous Mater.* **2009**, *117*, 617–626. [CrossRef]
77. Bhagat, S.D.; Rao, A.V. Surface chemical modification of TEOS based silica aerogels synthesized by two step (acid–base) sol–gel process. *Appl. Surf. Sci.* **2006**, *252*, 4289–4297. [CrossRef]
78. Zu, G.; Kanamori, K.; Shimizu, T.; Zhu, Y.; Maeno, A.; Kaji, H.; Nakanishi, K.; Shen, J. Versatile double-cross-linking approach to transparent, machinable, supercompressible, highly bendable aerogel thermal superinsulators. *Chem. Mater.* **2018**, *30*, 2759–2770. [CrossRef]
79. Hæreid, S.; Dahle, M.; Lima, S.; Einarsrud, M.-A. Preparation and properties of monolithic silica xerogels from TEOS–based alcogels aged in silane solutions. *J. Non-Cryst. Solids* **1995**, *186*, 96–103. [CrossRef]
80. Tajiri, K.; Igarashi, K.; Nishio, T. Effects of supercritical drying media on structure and properties of silica aerogel. *J. Non-Cryst. Solids* **1995**, *186*, 83–87. [CrossRef]
81. Nguyen, T.H.; Mai, N.T.; Reddy, V.R.M.; Jung, J.H.; Truong, N.T.N. Synthesis of silica aerogel particles and its application to thermal insulation paint. *Korean J. Chem. Eng.* **2020**, *37*, 1803–1809. [CrossRef]
82. Yuan, X.; Xu, L.; Pan, H.; Shen, Y.; Wang, L.; Xie, M. Eco-friendly approach for preparation of water-based superhydrophobic silica aerogels via ambient pressure drying. *Mater. Res. Express* **2021**, *8*, 015021. [CrossRef]
83. Shao, Z.; He, X.; Cheng, X.; Zhang, Y. A simple facile preparation of methyltriethoxysilane based flexible silica aerogel monoliths. *Mater. Lett.* **2017**, *204*, 93–96. [CrossRef]
84. Lamy-Mendes, A.; Malfait, W.J.; Sadeghpour, A.; Girão, A.V.; Silva, R.F.; Durães, L. Influence of 1D and 2D carbon nanostructures in silica-based aerogels. *Carbon* **2021**, *180*, 146–162. [CrossRef]
85. Wu, X.; Man, J.; Liu, S.; Huang, S.; Lu, J.; Tai, J.; Zhong, Y.; Shen, X.; Cui, S.; Chen, X. Isocyanate-crosslinked silica aerogel monolith with low thermal conductivity and much enhanced mechanical properties: Fabrication and analysis of forming mechanisms. *Ceram. Int.* **2021**, *47*, 26668–26677. [CrossRef]
86. Li, S.; Zhang, L.; Li, J.; Wu, Z.; Yang, C. Silica nanowires reinforced self-hydrophobic silica aerogel derived from crosslinking of propyltriethoxysilane and tetraethoxysilane. *J. Sol-Gel Sci. Technol.* **2017**, *83*, 545–554. [CrossRef]
87. Matias, T.; Varino, C.; de Sousa, H.C.; Braga, M.E.M.; Portugal, A.; Coelho, J.F.J.; Durães, L. Novel flexible, hybrid aerogels with vinyl- and methyltrimethoxysilane in the underlying silica structure. *J. Mater. Sci.* **2016**, *51*, 6781–6792. [CrossRef]
88. Yang, Z.; Yu, H.; Li, X.; Ding, H.; Ji, H. Hyperelastic and hydrophobic silica aerogels with enhanced compressive strength by using VTES/MTMS as precursors. *J. Non-Cryst. Solids* **2019**, *525*, 119677. [CrossRef]
89. He, S.; Sun, G.; Cheng, X.; Dai, H.; Chen, X. Nanoporous SiO_2 grafted aramid fibers with low thermal conductivity. *Compos. Sci. Technol.* **2017**, *146*, 91–98. [CrossRef]
90. Guo, H.; Nguyen, B.N.; Mccorkle, L.S.; Shonkwiler, B.; Meador, M.A.B. Elastic low density aerogels derived from bis[3-(triethoxysilyl)propyl]disulfide, tetramethylorthosilicate and vinyltrimethoxysilane via a two-step process. *J. Mater. Chem.* **2009**, *19*, 9054–9062. [CrossRef]
91. Meador, M.A.B.; Weber, A.S.; Hindi, A.; Naumenko, M.; Mccorkle, L.; Quade, D.; Vivod, S.L.; Gould, G.L.; White, S.; Deshpande, K. Structure-property relationships in porous 3d nanostructures: Epoxy-cross-linked silica aerogels produced using ethanol as the solvent. *ACS Appl. Mater. Interfaces* **2009**, *1*, 894–906. [CrossRef] [PubMed]
92. Randall, J.P.; Meador, M.A.B.; Jana, S.C. Polymer reinforced silica aerogels: Effects of dimethyldiethoxysilane and bis(trimethoxysilylpropyl) amine as silane precursors. *J. Mater. Chem. A* **2013**, *1*, 6642–6652. [CrossRef]
93. Zhang, Y.; Shen, Q.; Li, X.; Wang, L.; Nie, C. Facile preparation of a phenyl-reinforced flexible silica aerogel with excellent thermal stability and fire resistance. *Mater. Chem. Front.* **2021**, *5*, 4214–4224. [CrossRef]
94. Estella, J.; Echeverría, J.C.; Laguna, M.; Garrido, J.J. Effects of aging and drying conditions on the structural and textural properties of silica gels. *Microporous Mesoporous Mater.* **2007**, *102*, 274–282. [CrossRef]
95. Hong, J.-Y.; Yun, S.; Wie, J.J.; Zhang, X.; Dresselhaus, M.S.; Kong, J.; Park, H.S. Cartilage-inspired superelastic ultradurable graphene aerogels prepared by the selective gluing of intersheet joints. *Nanoscale* **2016**, *8*, 12900–12909. [CrossRef]
96. Einarsrud, M.-A.; Haereid, S. Preparation of transparent, monolithic silica xerogels with low density. *J. Sol-Gel Sci. Technol.* **1994**, *2*, 903–906. [CrossRef]
97. Omranpour, H.; Motahari, S. Effects of processing conditions on silica aerogel during aging: Role of solvent, time and temperature. *J. Non-Cryst. Solids* **2013**, *379*, 7–11. [CrossRef]
98. He, F.; Zhao, H.; Qu, X.; Zhang, C.; Qiu, W. Modified aging process for silica aerogel. *J. Mater. Process. Technol.* **2009**, *209*, 1621–1626. [CrossRef]
99. Rashid, A.B.; Shishir, S.I.; Mahfuz, M.A.; Hossain, M.T.; Hoque, M.E. Silica aerogel: Synthesis, characterization, applications, and recent advancements. *Part. Part. Syst. Charact.* **2023**, *40*, 2200186. [CrossRef]

100. Wu, G.-P.; Yang, J.; Wang, D.; Xu, R.; Amine, K.; Lu, C.-X. A novel route for preparing mesoporous carbon aerogels using inorganic templates under ambient drying. *Mater. Lett.* **2014**, *115*, 1–4. [CrossRef]
101. Leventis, N.; Palczer, A.; Mccorkle, L.; Zhang, G.; Sotiriou-Leventis, C. Nanoengineered silica-polymer composite aerogels with no need for supercritical fluid drying. *J. Sol-Gel Sci. Technol.* **2005**, *35*, 99–105. [CrossRef]
102. Chen, H.-B.; Hollinger, E.; Wang, Y.-Z.; Schiraldi, D.A. Facile fabrication of poly(vinyl alcohol) gels and derivative aerogels. *Polymer* **2014**, *55*, 380–384. [CrossRef]
103. Chen, H.-B.; Liu, B.; Huang, W.; Wang, J.-S.; Zeng, G.; Wu, W.-H.; Schiraldi, D.A. Fabrication and properties of irradiation-cross-linked poly(vinyl alcohol)/clay aerogel composites. *ACS Appl. Mater. Interfaces* **2014**, *6*, 16227–16236. [CrossRef] [PubMed]
104. Tang, Q.; Wang, T. Preparation of silica aerogel from rice hull ash by supercritical carbon dioxide drying. *J. Supercrit. Fluids* **2005**, *35*, 91–94. [CrossRef]
105. Maleki, H.; Durães, L.; Portugal, A. Development of mechanically strong ambient pressure dried silica aerogels with optimized properties. *J. Phys. Chem. C* **2015**, *119*, 7689–7703. [CrossRef]
106. Zuo, L.; Zhang, Y.; Zhang, L.; Miao, Y.-E.; Fan, W.; Liu, T. Polymer/carbon-based hybrid aerogels: Preparation, properties and applications. *Materials* **2015**, *8*, 6806–6848. [CrossRef]
107. Zhu, J.; Xie, J.; Lü, X.; Jiang, D. Synthesis and characterization of superhydrophobic silica and silica/titania aerogels by sol–gel method at ambient pressure. *Colloids Surf. A Physicochem. Eng. Asp.* **2009**, *342*, 97–101. [CrossRef]
108. Ślosarczyk, A.; Barełkowski, M.; Niemier, S.; Jakubowska, P. Synthesis and characterisation of silica aerogel/carbon microfibers nanocomposites dried in supercritical and ambient pressure conditions. *J. Sol-Gel Sci. Technol.* **2015**, *76*, 227–232. [CrossRef]
109. Błaszczyński, T.; Ślosarczyk, A.; Morawski, M. Synthesis of silica aerogel by supercritical drying method. *Procedia Eng.* **2013**, *57*, 200–206. [CrossRef]
110. Meador, M.A.B.; Alemán, C.R.; Hanson, K.; Ramirez, N.; Vivod, S.L.; Wilmoth, N.; Mccorkle, L. Polyimide aerogels with amide cross-links: A low cost alternative for mechanically strong polymer aerogels. *ACS Appl. Mater. Interfaces* **2015**, *7*, 1240–1249. [CrossRef]
111. Ku, H.; Wang, H.; Pattarachaiyakoop, N.; Trada, M. A review on the tensile properties of natural fiber reinforced polymer composites. *Compos. Part B Eng.* **2011**, *42*, 856–873. [CrossRef]
112. Boday, D.J.; Stover, R.J.; Muriithi, B.; Keller, M.W.; Wertz, J.T.; Defriend Obrey, K.A.; Loy, D.A. Strong, low-density nanocomposites by chemical vapor deposition and polymerization of cyanoacrylates on aminated silica aerogels. *ACS Appl. Mater. Interfaces* **2009**, *1*, 1364–1369. [CrossRef]
113. Li, X.; Wang, J.; Zhao, Y.; Zhang, X. Superhydrophobic polyimide aerogels via conformal coating strategy with excellent underwater performances. *J. Appl. Polym. Sci.* **2020**, *137*, 48849. [CrossRef]
114. Kim, H.S.; Jang, J.-U.; Lee, H.; Kim, S.Y.; Kim, S.H.; Kim, J.; Jung, Y.C.; Yang, B.J. Thermal management in polymer composites: A review of physical and structural parameters. *Adv. Eng. Mater.* **2018**, *20*, 1800204. [CrossRef]
115. Mahadik, D.B.; Jung, H.-N.; Han, W.; Cho, H.H.; Park, H.-H. Flexible, elastic, and superhydrophobic silica-polymer composite aerogels by high internal phase emulsion process. *Compos. Sci. Technol.* **2017**, *147*, 45–51. [CrossRef]
116. Maleki, H.; Durães, L.; Portugal, A. A new trend for development of mechanically robust hybrid silica aerogels. *Mater. Lett.* **2016**, *179*, 206–209. [CrossRef]
117. Wei, T.-Y.; Lu, S.-Y.; Chang, Y.-C. Transparent, hydrophobic composite aerogels with high mechanical strength and low high-temperature thermal conductivities. *J. Phys. Chem. B* **2008**, *112*, 11881–11886. [CrossRef]
118. Meador, M.A.B.; Vivod, S.L.; Mccorkle, L.; Quade, D.; Sullivan, R.M.; Ghosn, L.J.; Clark, N.; Capadona, L.A. Reinforcing polymer cross-linked aerogels with carbon nanofibers. *J. Mater. Chem.* **2008**, *18*, 1843–1852. [CrossRef]
119. Kim, M.; Eo, K.; Lim, H.J.; Kwon, Y.K. Low shrinkage, mechanically strong polyimide hybrid aerogels containing hollow mesoporous silica nanospheres. *Compos. Sci. Technol.* **2018**, *165*, 355–361. [CrossRef]
120. Novak, B.M.; Auerbach, D.; Verrier, C. Low-density, mutually interpenetrating organic-inorganic composite materials via supercritical drying techniques. *Chem. Mater.* **1994**, *6*, 282–286. [CrossRef]
121. Leventis, N.; Sadekar, A.; Chandrasekaran, N.; Sotiriou-Leventis, C. Click synthesis of monolithic silicon carbide aerogels from polyacrylonitrile-coated 3d silica networks. *Chem. Mater.* **2010**, *22*, 2790–2803. [CrossRef]
122. Guise, M.T.; Hosticka, B.; Earp, B.C.; Norris, P.M. An experimental investigation of aerosol collection utilizing packed beds of silica aerogel microspheres. *J. Non-Cryst. Solids* **2001**, *285*, 317–322. [CrossRef]
123. Boday, D.J.; Keng, P.Y.; Muriithi, B.; Pyun, J.; Loy, D.A. Mechanically reinforced silica aerogel nanocomposites via surface initiated atom transfer radical polymerizations. *J. Mater. Chem.* **2010**, *20*, 6863–6865. [CrossRef]
124. Meador, M.A.B.; Capadona, L.A.; Mccorkle, L.; Papadopoulos, D.S.; Leventis, N. Structure-property relationships in porous 3d nanostructures as a function of preparation conditions: isocyanate cross-linked silica aerogels. *Chem. Mater.* **2007**, *19*, 2247–2260. [CrossRef]
125. Randall, J.P.; Meador, M.A.B.; Jana, S.C. Tailoring mechanical properties of aerogels for aerospace applications. *ACS Appl. Mater. Interfaces* **2011**, *3*, 613–626. [CrossRef]
126. Rezaei, S.; Jalali, A.; Zolali, A.M.; Alshrah, M.; Karamikamkar, S.; Park, C.B. Robust, ultra-insulative and transparent polyethylene-based hybrid silica aerogel with a novel non-particulate structure. *J. Colloid Interface Sci.* **2019**, *548*, 206–216. [CrossRef]

127. Lenarda, M.; Chessa, G.; Moretti, E.; Polizzi, S.; Storaro, L.; Talon, A. Toward the preparation of a nanocomposite material through surface initiated controlled/"living" radical polymerization of styrene inside the channels of mcm-41 silica. *J. Mater. Sci.* **2006**, *41*, 6305–6312. [CrossRef]
128. Rezaei, S.; Zolali, A.M.; Jalali, A.; Park, C.B. Novel and simple design of nanostructured, super-insulative and flexible hybrid silica aerogel with a new macromolecular polyether-based precursor. *J. Colloid Interface Sci.* **2020**, *561*, 890–901. [CrossRef]
129. Salimian, S.; Malfait, W.J.; Zadhoush, A.; Talebi, Z.; Naeimirad, M. Fabrication and evaluation of silica aerogel-epoxy nanocomposites: Fracture and toughening mechanisms. *Theor. Appl. Fract. Mech.* **2018**, *97*, 156–164. [CrossRef]
130. Salimian, S.; Zadhoush, A.; Talebi, Z.; Fischer, B.; Winiger, P.; Winnefeld, F.; Zhao, S.; Barbezat, M.; Koebel, M.M.; Malfait, W.J. Silica aerogel-epoxy nanocomposites: Understanding epoxy reinforcement in terms of aerogel surface chemistry and epoxy–silica interface compatibility. *ACS Appl. Nano Mater.* **2018**, *1*, 4179–4189. [CrossRef]
131. Albooyeh, A.; Bayat, M.; Rafieian, P.; Dadrasi, A.; Khatibi, M.M. Silica aerogel/epoxy nanocomposites: Mechanical, vibrational, and morphological properties. *J. Appl. Polym. Sci.* **2020**, *137*, 49338. [CrossRef]
132. Domènech, B.; Mata, I.; Molins, E. Tuning the structure and the mechanical properties of epoxy-silica sol-gel hybrid materials. *RSC Adv.* **2016**, *6*, 10736–10742. [CrossRef]
133. Çok, S.S.; Ünal, H.Y.; Koç, F.; Pekbey, Y.; Gizli, N. Ionic liquid functionalized silica aerogels as reinforcing agents for epoxy nanocomposites. *J. Inorg. Organomet. Polym. Mater.* **2021**, *31*, 2445–2458. [CrossRef]
134. He, J.; Li, X.; Su, D.; Ji, H.; Qiao, Y. High-strength mullite fibers reinforced ZrO_2–SiO_2 aerogels fabricated by rapid gel method. *J. Mater. Sci.* **2015**, *50*, 7488–7494. [CrossRef]
135. Selver, E.; Öztaş, B.; Uçar, M.; Uçar, N.; Baydoğan, M.; Altay, P.; Geygel, B. Mechanical and thermal properties of glass/epoxy composites filled with silica aerogels. *Plast. Rubber Compos.* **2021**, *50*, 371–383. [CrossRef]
136. Churu, G.; Zupančič, B.; Mohite, D.; Wisner, C.; Luo, H.; Emri, I.; Sotiriou-Leventis, C.; Leventis, N.; Lu, H. Synthesis and mechanical characterization of mechanically strong, polyurea-crosslinked, ordered mesoporous silica aerogels. *J. Sol-Gel Sci. Technol.* **2015**, *75*, 98–123. [CrossRef]
137. Capadona, L.A.; Meador, M.A.B.; Alunni, A.; Fabrizio, E.F.; Vassilaras, P.; Leventis, N. Flexible, low-density polymer crosslinked silica aerogels. *Polymer* **2006**, *47*, 5754–5761. [CrossRef]
138. Yang, H.; Kong, X.; Zhang, Y.; Wu, C.; Cao, E. Mechanical properties of polymer-modified silica aerogels dried under ambient pressure. *J. Non-Cryst. Solids* **2011**, *357*, 3447–3453. [CrossRef]
139. Verdolotti, L.; Oliviero, M.; Lavorgna, M.; Santillo, C.; Tallia, F.; Iannace, S.; Chen, S.; Jones, J.R. "Aerogel-like" polysiloxane-polyurethane hybrid foams with enhanced mechanical and thermal-insulating properties. *Compos. Sci. Technol.* **2021**, *213*, 108917. [CrossRef]
140. Cho, J.; Jang, H.G.; Kim, S.Y.; Yang, B. Flexible and coatable insulating silica aerogel/polyurethane composites via soft segment control. *Compos. Sci. Technol.* **2019**, *171*, 244–251. [CrossRef]
141. Duan, Y.; Jana, S.C.; Lama, B.; Espe, M.P. Reinforcement of silica aerogels using silane-end-capped polyurethanes. *Langmuir* **2013**, *29*, 6156–6165. [CrossRef] [PubMed]
142. Merillas, B.; Lamy-Mendes, A.; Villafañe, F.; Durães, L.; Rodríguez-Pérez, M.Á. Silica-based aerogel composites reinforced with reticulated polyurethane foams: Thermal and mechanical properties. *Gels* **2022**, *8*, 392. [CrossRef] [PubMed]
143. Tian, J.; Yang, Y.; Xue, T.; Chao, G.; Fan, W.; Liu, T. Highly flexible and compressible polyimide/silica aerogels with integrated double network for thermal insulation and fire-retardancy. *J. Mater. Sci. Technol.* **2022**, *105*, 194–202. [CrossRef]
144. Kantor, Z.; Wu, T.; Zeng, Z.; Gaan, S.; Lehner, S.; Jovic, M.; Bonnin, A.; Pan, Z.; Mazrouei-Sebdani, Z.; Opris, D.M.; et al. Heterogeneous silica-polyimide aerogel-in-aerogel nanocomposites. *Chem. Eng. J.* **2022**, *443*, 136401. [CrossRef]
145. Fei, Z.; Yang, Z.; Chen, G.; Li, K. Preparation of tetraethoxysilane-based silica aerogels with polyimide cross-linking from 3, 3′, 4, 4′-biphenyltetracarboxylic dianhydride and 4, 4′-oxydianiline. *J. Sol-Gel Sci. Technol.* **2018**, *85*, 506–513. [CrossRef]
146. Fei, Z.; Yang, Z.; Chen, G.; Li, K.; Zhao, S.; Su, G. Preparation and characterization of glass fiber/polyimide/SiO_2 composite aerogels with high specific surface area. *J. Mater. Sci.* **2018**, *53*, 12885–12893. [CrossRef]
147. Zhang, X.; Ni, X.; Li, C.; You, B.; Sun, G. Co-gel strategy for preparing hierarchically porous silica/polyimide nanocomposite aerogel with thermal insulation and flame retardancy. *J. Mater. Chem. A* **2020**, *8*, 9701–9712. [CrossRef]
148. Faysal Ilhan, U.; Fabrizio, E.F.; Mccorkle, L.; Scheiman, D.A.; Dass, A.; Palczer, A.; Meador, M.B.; Johnston, J.C.; Leventis, N. Hydrophobic monolithic aerogels by nanocasting polystyrene on amine-modified silica. *J. Mater. Chem.* **2006**, *16*, 3046–3054. [CrossRef]
149. Maleki, H.; Durães, L.; Portugal, A. Synthesis of mechanically reinforced silica aerogels via surface-initiated reversible addition-fragmentation chain transfer (raft) polymerization. *J. Mater. Chem. A* **2015**, *3*, 1594–1600. [CrossRef]
150. Defriend, K.A.; Espinoza, B.; Patterson, B. Templating silica aerogel with polystyrene to improve their mechanical properties. *Fusion Sci. Technol.* **2007**, *51*, 693–700. [CrossRef]
151. Meador, M.A.B.; Fabrizio, E.F.; Ilhan, F.; Dass, A.; Zhang, G.; Vassilaras, P.; Johnston, J.C.; Leventis, N. Cross-linking amine-modified silica aerogels with epoxies: Mechanically strong lightweight porous materials. *Chem. Mater.* **2005**, *17*, 1085–1098. [CrossRef]
152. Fu, B.; Luo, H.; Wang, F.; Churu, G.; Chu, K.T.; Hanan, J.C.; Sotiriou-Leventis, C.; Leventis, N.; Lu, H. Simulation of the microstructural evolution of a polymer crosslinked templated silica aerogel under high-strain-rate compression. *J. Non-Cryst. Solids* **2011**, *357*, 2063–2074. [CrossRef]

153. Li, M.-E.; Wang, S.-X.; Han, L.-X.; Yuan, W.-J.; Cheng, J.-B.; Zhang, A.-N.; Zhao, H.-B.; Wang, Y.-Z. Hierarchically porous sio2/polyurethane foam composites towards excellent thermal insulating, flame-retardant and smoke-suppressant performances. *J. Hazard. Mater.* **2019**, *375*, 61–69. [CrossRef] [PubMed]
154. Cimavilla-Román, P.; Pérez-Tamarit, S.; Santiago-Calvo, M.; Rodríguez-Pérez, M.Á. Influence of silica aerogel particles on the foaming process and cellular structure of rigid polyurethane foams. *Eur. Polym. J.* **2020**, *135*, 109884. [CrossRef]
155. Liaw, D.-J.; Wang, K.-L.; Huang, Y.-C.; Lee, K.-R.; Lai, J.-Y.; Ha, C.-S. Advanced polyimide materials: Syntheses, physical properties and applications. *Prog. Polym. Sci.* **2012**, *37*, 907–974. [CrossRef]
156. Li, X.; Wang, J.; Zhao, Y.; Zhang, X. Template-free self-assembly of fluorine-free hydrophobic polyimide aerogels with lotus or petal effect. *ACS Appl. Mater. Interfaces* **2018**, *10*, 16901–16910. [CrossRef]

MDPI AG
Grosspeteranlage 5
4052 Basel
Switzerland
Tel.: +41 61 683 77 34

Molecules Editorial Office
E-mail: molecules@mdpi.com
www.mdpi.com/journal/molecules

www.ingramcontent.com/pod-product-compliance
Lightning Source LLC
Chambersburg PA
CBHW061823210926
51850CB00006B/786